FUNGI : DIVERSITY AND BIOTECHNOLOGY

Editors

M.K. Rai

Department of Biotechnology,
Amravati University,
Amravati - 444 602 (Maharashtra)

and

S.K. Deshmukh

Department of Natural Products,
Nicholas Piramal Research Centre,
1A, 1B & 1C, Nirlon Complex,
Off Western Express Highway,
Next to Food Bazar, Goregaon (East)
Mumbai (Maharashtra)

Published by:

SCIENTIFIC PUBLISHERS (INDIA)
5A, New Pali Road, P.O. Box 91
Jodhpur 342 001, India
E-mail: info@scientificpub.com
Website: www.scientificpub.com

Print : 2021

ISBN : 978-81-72334-03-1
eISBN: 978-93-87913-61-5

Printed in India

FOREWORD

The science of mycology had beginning in the eighteenth century. By now, fungi are the concern of the taxonomists, morphologists, geneticists, ecologists, phytopathologists, physicians, biochemists, molecular biologists, human doctors and commercial microbiologists.

"*Fungi : Diversity and Biotechnology*" provides a broad overview of mycology: it explores the diversity of keratinophilic, marine and those involved in bio-deterioration causing losses in paper industry and role of fungal biotechnology in pulp and paper industry. Considering the sustainability of environment, ectomycorrhiza and arbuscular mycorrhizal (AMF) fungi can be applied as biofertilizers in nurseries and plantations. Moreover, answers for questions of bioremediation with special regards to handling toxic metals in waste-water, recycling of agro-wastes, etc. has been incorporated.

The contributions to industrial applications include different fields of fungal biotechnology, such as, biotransformation, industrial enzymology and vaccination for fungal infections. The reader can gather information about biology of edible mushrooms and *Phansomba,* a folk medicinal mushroom. The increasing importance of biological control against plant pathogens account for reviewing the role of *Trichoderma* species as potential biocontrol agent against soil-borne and other fungus parasites; fungi can be applied as insecticides and for the control of malaria - vector - mosquitoes. The opportunistic mycotic infections caused by species of *Candida, Cryptococcus* and *Aspergillus* species are increasing in human hosts rendered susceptible by some predisposing factors, like AIDS infection. The readers should find aeromycology and prospects of keratinophilic fungi as useful topics.

It is hoped that this book will enrich the background of those who have already basic mycological knowledge and will provide contributions to current trends in mycology.

All the contributors confess :

"*Inter silvas fungorum querere verum!*" (free version of Horatius Epist. 2.2.45)

György J. Kövics
Professor of Plant Pathology,
Debrecen University, Hungary

PREFACE

Fungi have been very important to human beings from the time immemorial. On one hand, these fungi are used traditionally in fermentation, as food, for preparation of antibiotics, as bioremediators, biocontrol agents and for enriching the soil as biofertilizers. Some fungi, for example, *Curvularia lunata* can transform organic compounds and still others can produce enzymes like amylases and xylanases. In the field of paper and pulp technology also, there are many opportunities. The biotechnological potential of keratinophilic fungi for the degradation of the keratin can be utilized. The marine fungi are now being used for the discovery of antimicrobials. On the other hand, fungi cause diseases of plants, as well as animals and human beings. The opportunistic mycotic infections in cancer and AIDs patients are increasing with fast rate due to their immunocompromising capacity. Furthermore, some fungi occur in air and forms a major part of allergens. Taking into consideration these facts, there is need to discover new antifungal molecules.

Due to their immense utility, research on diversity of fungi is gaining ground and many mycologists are now concentrating on their applied aspects to utilize the fullest potential. There are several pockets/ places in India which are still unexplored! Hence, there is a pressing need to make a thorough survey of diverse group of fungi.

Efforts have been made in this book, to cover both basic and applied aspects particularly applications of fungi. In 21st century, fungi will play a pivotal role in developing sustainable mycotechnology as chemical technology has already shown its ill-effect to the environment and jeopardized the existence of mankind. The only alternative is vaccination against fungal infections, which would give a new insight into strategic management of mycotic infections.

The present book would be very useful to mycologists, plant pathologists, medical mycologists, environmentalists, pharmacologists, food technologists and non-governmental organizations and those who believe in microbe-based eco-friendly technologies.

We are indebted to Professor R.C. Rajak, Director of Basic Sciences, Bundelkhand University Jhansi (U.P.); and to Professor, S.C. Agrawal, Head, Department of Applied Microbiology and Biotechnology, Dr. H.S. Gaur University, Sagar, M.P., who graciously offered suggestions for this book. We thank to all the contributors, for the submission of valuable manuscripts. Thanks are also due to my colleagues — Drs N.J. Chikhale, P.A. Wadegaonkar, P.V. Thakare and Anita Patil. MKR. thanks Sarika Shende, Preeti Wagh, Sunita Bansod, Prajakta Deshmukh, Sonali Wankhede, Girish Tidke and Ravindra Ade, who checked manuscripts and helped in type-setting.

We are grateful to Dr. Sudhir Patil, Vice-Chancellor, Amravati University, Amravati and to Dr. Somesh Sharma, Chief Scientific officer, Quest Institute of Life Sciences, Nicholas Piramal India Limited, Mumbai for inspiration.

Finally, MKR thanks his children — Shivangi, Shivani and Aditya for their support and patience during editing work.

M.K. Rai

S.K. Deshmukh

CONTRIBUTORS

Agrawal, S.C., Department of Applied Microbiology and Biotechnology, Dr. H.S. Gour University, Sagar (M P).

Ahlawat, O.P., National Research Center for Mushroom, (Indian Council of Agricultural Research), Solan-173 213.

Ahmad, Absar, Division of Biochemical Sciences, National Chemical Laboratory, Pune, 411 008.

Arora, J.K., Department of Plant Pathology, Punjab Agricultural University, Ludhiana-141 004.

Barve, Yoganand Y., Post Graduate Department of Botany, Nagpur University campus, Amravati road, Nagpur-440 010; *E-mail*: yog28in@yahoo.com

Bhosle, Shekhar, Department of Botany, University of Pune, Pune – 411 007, M.S.

Borse, B.D., Department of Botany, Arts, commerce and Science College, Shindkheda, 425 406, Maharashtra

Borse, K.N., Department of Botany, Arts, commerce and Science College, Shindkheda, 425 406, Maharashtra

Chaturvedi, Pratiksha, Forest Pathology Division, Tropical Forest Research Institute, P.O.-R.F.R.C, Jabalpur - 482 021, M.P.

Chincholkar, S.B., Department of Microbiology, School of Life Sciences, North Maharashtra University, PB. 80, Jalgaon 425 001; *E-mail:*chincholkar_sb@ indiatimes.com

Dave, S.R., Department of Microbiology, School of Sciences, Gujarat University, Ahmedabad 380 009; *E-mail:* shaileshrdave@hotmail.com

Deshmukh, S.K., Department of Natural Products, Nicholas Piramal Research Centre, 1A, 1B & 1C, Nirlon Complex, Off Western Express Highway, Next to Food Bazar, Goregaon (East), Mumbai - 400 063, India, *E-mail:* sdeshmukh@nicholaspiramal.co.in

Deshpande, M.V., Biochemical Sciences Division, National Chemical Laboratory, Pune-411 008, *E-mail:* mvdesh@dalton.ncl.res.in

Dhawan, Shashi, Biodeterioration Division, National Research Laboratory for Conservation of Cultural Property, E/3, Aliganj, Lucknow - 226 024

Gugnani, H.C., Department of Medical Mycology, Vallabhbhai Patel Chest Institute, University of Delhi, Delhi-110 007; *E-mail:* harishgugnani@ hotmail. com, harishgugnani@yahoo.com

Dhaliwal, H.S., Department of Plant Pathology, Punjab Agricultural University, Ludhiana-141 004

Jain, Sudhir, Department of Applied Microbiology, College of Life Sciences, Cancer Hospital and Research Institute, Gwalior 474 009, M.P. *E-mail:* sudhirkjain1@rediffmail.com

Jain, P.C., Department of Applied Microbiology and Biotechnology, Dr. Hari Singh Gour Vishwavidyalaya, Sagar- 470 003.

Jamaluddin, Forest Pathology Division, Tropical Forest Research Institute, P.O.-R.F.R.C, Jabalpur-482 021, M.P.

Jobanputra, A.H., Department of Microbiology, School of Life Sciences, North Maharashtra University, PB. 80, Jalgaon 425 001

Kango, Naveen, Department of Applied Microbiology and Biotechnology, Dr. H.S. Gour Vishwavidyalaya, Sagar-470 003; *E-mail:* navinkango13@rediffmail.com

Kumar, Manoj, Department of Applied Microbiology and Biotechnology, Dr. H.S. Gour University, Sagar (M P).

Kumar, Sanjeev, Department of Microbiology, University of Delhi South Campus, Benito Juarez Road, New Delhi-110 021

Kushwaha, R.K.S., Department of Botany, Christ Church College, Kanpur 208 001, *E-mail* : rks_kushwaha@yahoo.co.in

Lachke, Anil, Division of Biochemical Sciences, National Chemical Laboratory, Pune, 411 008; *E-mail:* lachke@dalton.ncl.res.in

Lamrood, Prasad, Department of Botany, University of Pune, Pune – 411 007, M.S.

Mohan, Chander, Department of Plant Pathology, Punjab Agricultural University, Ludhiana-141004.

Mohan, V., Division of Forest Protection, Institute of Forest Genetics and Tree Breeding, Coimbatore - 641 002; *E-mail:* ifgtb@res.in

Nigam, Neeta, Biodeterioration Division, National Research Laboratory for Conservation of Cultural Property, E/3, Aliganj, Lucknow – 226 024

Pande, Department of Environmental science, Dr. B.A. Marathawada University, Aurangabad, Aurangabad- 431 004 (M.S). *E-mail:* sssbp3_agd@sancharnet.in

Pawar, N.S., Department of Botany, Arts, commerce and Science College, Shindkheda, 425 406. Maharashtra

Prakash, G., Department of Biochemical Engineering and Biotechnology Indian Institute of Technology, New Delhi

Qureshi, Shaista, Microbiology Research Lab., Danielson College, Chhindwara-480 001, M.P.

Rai, M.K., Department of Biotechnology, Amravati University, Amravati-444 602, Maharashtra

Rai, R.D., National Research Centre for Mushroom, (Indian Council of Agricultural Research), Solan-173 213, India

Sandhu, S., Fungal Biotechnology and Invertebrate Pathology Lab., Department of Biological Sciences, R.D. University, Jabalpur, Madhya Pradesh; *E-mail :* ssandhu@rediffmail.com and sardulsinghs@yahoo.co.uk

Satyanarayana, T., Department of Microbiology, University of Delhi South Campus, Benito Juarez Road, New Delhi-110 021; Phone: 091-11-4102008; Fax: 091-11-6885270, *E-mail:* tsnarayana@vsnl.net

Sayyed, R.Z., Department of Microbiology, School of Life Sciences, North Maharashtra University, PB. 80, Jalgaon 425 001

Thakre, R.P., Post Graduate Department of Botany, Nagpur University campus, Amravati road, Nagpur-440 010, *E-mail:* rpthakre@yahoo.com

Thakur, M.P., Department of Plant Pathology, Indira Gandhi Agricultural University, Raipur - 492 006, Chhatisgarh, *E mail* : mp_thakur@yahoo.com

Thind, T., Department of Plant Pathology, Punjab Agricultural University, Ludhiana-141 004. *E-mail* : tsthind@pau.edu

Tilak, S.T., 18, Vidyasagar Society, Near Mahesh Society, Bibvewadi, Pune – 411 037, *E-mail:* tilak_aerobiol@yahoo.co.in

Tripathi, Neetu, Department of Botany, Christ Church College, Kanpur 208 001

Vaidya, Jitendra, Department of Botany, University of Pune, Pune – 411 007. M.S., *E-mail* – Jitendra@unipune.ernet.in

Vikrant, P., Fungal Biotechnology and Invertebrate Pathology Lab., Department. of Biological Sciences, R.D. University, Jabalpur, Madhya Pradesh.

Vyas, Santosh, Division of Biochemical Sciences, National Chemical Laboratory, Pune, 411 008.

CONTENTS

CHAPTER - 1

FUNGAL BIOTECHNOLOGY : OPPORTUNITIES IN PULP AND PAPER MANUFACTURING

Santosh Vyas, Absar Ahmad and Anil Lachke*

Division of Biochemical Sciences, National Chemical Laboratory, Pune, 411008

***E-mail**: *lachke@dalton.ncl.res.in*

INTRODUCTION

Paper manufacture is one of the largest industries of the world. The demand for paper is rising at a fast rate. The global market for paper was about 80 million metric tons per annum at the beginning of 1980s, that is reached to 350 million metric tons by the year 2003 (Prasad, 2002; Changing tastes in paper products, 2001) The paper industry that is manufacturing above $ 200 billion worth paper annually is unfortunately considered as one of the most polluting industries of the world. Most of the paper manufacturing steps utilizes hazardous chemicals like alkali, chlorinated bleaching agents, reducing agents, surfactants, defoamers, deinking agents, sizing agents, binders, colourants and different additives (Changing tastes in paper products, 2001). These chemicals have harmful impact on the environment. They increase chemical oxygen demand in wastewater generated during the paper manufacturing process. As a result the technologists are looking forward to develop the novel, ecofriendly ways of manufacturing the pulp and paper. Fortunately the fungal biotechnology is offering various ecofriendly ways to minimize or avoid the use of "dirty" and harmful chemicals in paper and pulp industries (Table 1).

Fungi are excellent producers of extracellular enzymes of industrial importance. Recently, several fungal enzymes are found to be promising for developing ecofriendly technologies for paper and pulp manufacturing. These enzymes includes xylanases (Bajpai and Bajpai, 1992; Viikari *et al.*, 1994, 1996; Bissoon, *et al.*, 2002), mannanases (Gübitz, *et al.*, 1996; Tenkanen, *et al.*,1997; Oksanen, *et al.*, 2000) for biobleaching of paper and pulp, cellulases for enzymatic deinking (Jeffries *et al.*, 1994; Morkbak and Zimmermann 1998; Gubitz *et al.*, 1998a, 1998b; Vyas and Lachke 2003), and enhanced drainage rates (Stork, *et al.*, 1995), laccases and ligninases for delignification of pulp (Call, 1993; Call and Mücke, 1997; Jiménez *et al.*, 1997; Sealey and Ragauskas 1998; Ire *et al.*, 1989; Balakshin *et al.*, 2001), lipases, esterases for pith removal in the pulp and in deinking of waste paper (Ire *et al.*, 1989; Fischer and Messner 1992a; Morbak *et al.*, 1999) and pectinases for various fibre handling operatiions and enzymatic retting of flax fibers (Reid and Ricard 2000; Akin *et al.*, 2001; Evans *et al.*, 2002). Fungi are also useful in biomechanical pulping (Johnsrud *et al.*, 1987; Kirk *et al.*, 1990; Blanchette *et al.*, 1992; Akhtar *et al.*, 1992; Richard *et al.*, 1997; Messner and Srebotnik, 1994; Akhtar *et al.*, 1999). Several of these processes are commercialized and some of them are emerging rapidly. Some of the fundamental and applied aspects of these enzyme-based processes in pulp and paper manufacturing are presented herewith.

Table 1. Role of Biotechnology in Paper and Pulp Industry

Process	Mechanism	Microorganism Used
Fibre handling	Pectin removal using pectinanses	*Aspergillus* sp.
Biopulping	Using fungi for lignin breakdown (ligninase)	*Ceriporiopsis subvermispora; Lentinus edodes*
Enzymatic pitch control	Sticky lipid removal by lipases	*Candida cylindracea*
Biobleaching	Xylan breakdown by xylanases	*Trichoderma* sp., *Trametous versicolor*
Sizing	Starch modification with amylases	—
Refining	Removal fine particles with cellulases	*Trichoderma* sp.
Slime removal	Levan hydrolysis	—
Enzymatic deinking	Removal of ink particles from recycled fibers with alkaline active cellulases	*Humicola insolens*
Effluent treatment	Removal of lignosulfonates using microbial cultures	*Trametous versicolor, Aspergillus* sp.

THE PULP

Pulp is the product of mechanical or chemical breakdown of fibrous cellulosic materials, more or less in to component fibres. Pulps can be classified as chemical pulps and mechanical pulps.

Chemical Pulps: This contributes nearly 70-75% of the pulp manufactured for paper production all over the world. In chemical pulping process, the chemicals are used for removal of lignin and other wood components from cellulosic fibres. The wood chips are mixed with appropriate amounts of chemicals and cooked around 150°C at about 1.0 M PA pressure. The chemical pulps are preferred all over the world because of their high strength properties. However, the process is very energy intensive and requires large capital investments and with lower pulp yields. The process also produces polluting effluents that are difficult to process.

Mechanical Pulps: Mechanical pulp is made by grinding wood logs and chips against revolving abrasive stone. The process uses large amount of electrical energy and mainly used for low-density softwood species. When mechanical pulp is used to make paper they show fines entrapped between them that affects strength of the paper. The mechanical pulps contain lot of lignin and other wood components that tend to turn yellow on exposures to sunlight. All around the world attempts are made to improvise mechanical pulping process as they give higher pulp yields with much lesser pollutants. The major objectives are reduction on electricity costs, removing lignin from the mechanical pulp and improvisation in paper strength.

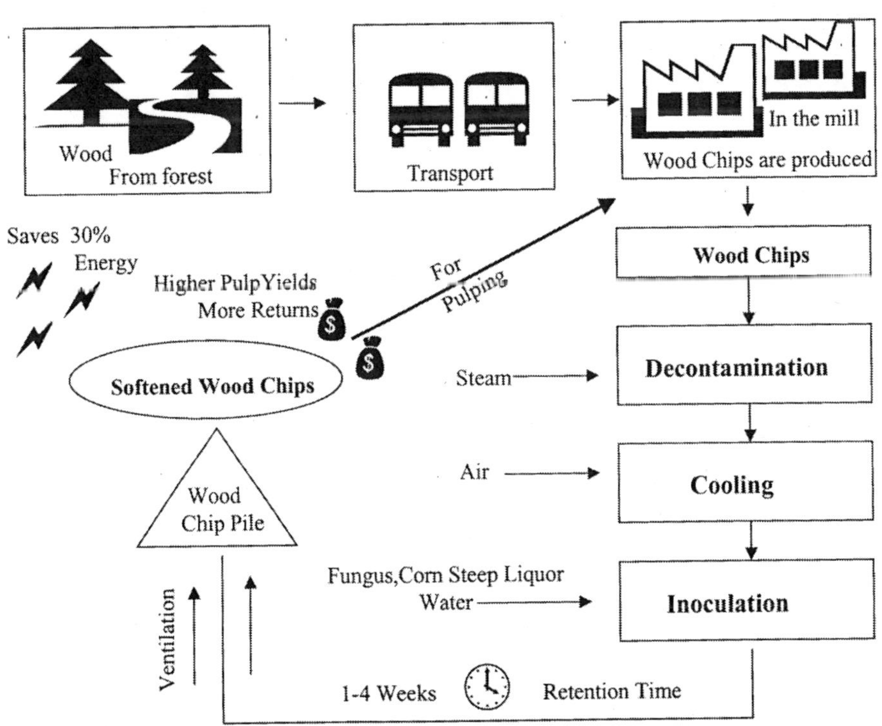

Fig. 1. Ecofriendly Biopulping with Fungi

BIOPULPING WITH FUNGI

Biopulping can be defined as the treatment of lignocellulosic materials with lignin degrading fungi prior to pulping. It utilizes natural tendency of fungi to colonize on wood and degrade lignin. It is already established that the pulping with fungi reduces paper-processing time and increases pulp yields by 30%. Biopulping has also shown to reduce the requirement of necessary chemicals. The process also lowers the cost of electricity by 33% (Kirk *et al.,* 1990; Messner and Srebotnik, 1994; Akhtar *et al.,* 1999). The ecofriendly biopulping process can be divided in to several steps as outlined in Fig 1.

Selection of an suitable fungus is the key in the overall process. Fungi generally attack cellulose component and affect pulp yields as well as paper quality. It is possible to eliminate this problem by choosing cellulase less mutants of ligninolytic fungi like *Ceriporiopsis subvermispora, Sporotrichum pulverulentum* and other white rot fungi like *Lentinus edodes* (Blanchette *et al.,* 1992; Akhtar *et al.,*1992). These fungi can naturally penetrate hardwood as well as softwood species. Sporulation of fungal strain is an another important criterion for the selection of fungi.

The wood chips are always contaminated with various bacteria and fungal spores at their surfaces. This affects the biopulping process. In order to get rid of these contaminants surface decontamination of wood chips is done by passing atmospheric steam on the chips for 15-30 seconds. Then the cooling of wood chips to ambient temperature is necessary before inoculating them with biopulping fungus. The fungal inoculum can be developed by supplementing growing fungus with cost-effective nutrients such as corn steep liquor. This inoculum is diluted with water and sprayed at pile of wood chip in such a way that fungus gets uniformly distributed at each chip. The ventilation, humidity and temperature of the wood chip pile are adjusted in a way that will promote fungal growth. The biopulping with fungus requires about 1-4 week's period before the chips are processed by usual mechanical pulping process (Akhtar *et al.,* 1999). The paper manufactured in this way showed higher strength properties over that of mechanically produced pulp (Akhtar *et al.,* 1992). Higher pulp yields, lower amounts of polluting effluents, savings in time and energy requirements and higher paper quality are the ultimate objectives of any pulp and paper industry. "Biopulping" with fungi can prove excellent way to achieve this. The technology is already in the market and no doubt, it is becoming a trend in the industry. The attempts are also going on to apply the biopulping process to non-wood pulping stock like bagasse (Johnsrud *et al.,* 1987). The development is highly significant for Indian paper industry and other developing countries, as sugarcane bagasse is the most commonly used raw material for paper manufacturing.

FUNGAL ENZYMES FOR PULP AND PAPER INDUSTRY

Pulp is composed of cellulose, hemicellulose and lignin. The enzymes are specific, safe and easy to control. They can accelerate the reaction by replacing the harsh chemicals and are biodegradable. Thus, the selected enzymes are more advantageous than chemicals for the desired modification of the pulp. The enzymes used in paper industry are lipases, hemicellulases, amylases, esterases, cellulases, ligninases and many more from different fungal sources (Bajpai, 1999). Here the choice of enzymes to be used depends on the type of the process and the type of pulp.

The pulp and paper industry is a large and growing part of the world's economy. Its consumption is continuously increasing globally. In general, the industry is very capital-intensive with a narrow profit margin. To keep up with the increasing demand for pulp and paper and to meet increasingly stringent environmental regulations, the industry has been constantly looking towards technological improvements. The current market size is probably only 1-2 per cent of total enzyme sales (Godfrey and West, 1996). But, driven largely by market and environmental demands for less chlorinated products and by-products, the pulp and paper industry is cited as the fastest growing market for industrial enzymes. In the United States it is projected to grow by 15 per cent a year for the next ten years. Cautious estimates indicate an enzyme market size of $14 million in the United States and $11 million in Europe in 1994.

FUNGAL CELLULASES

Cellulases are produced by wide variety of bacteria, fungi, actinomycetes, aerobes and anaerobes, mesophiles and thermophiles. However, fungi are the most studied organisms with respect to cellulose degradation. Generally, fungi hydrolyse cellulose to simple soluble sugars and utilizes them as a carbon source in their metabolism. It has been observed that, fungal cellulases differ in their molecular characteristics such as molecular weight, amino acid composition and sequence, isoelectric point and carbohydrate content etc. They also differ in their capacities to adsorb on the cellulose, substrate specificity and even in their catalytic activities. Despite of such variability the general mechanism for cellulose hydrolysis is shown Fig. 2.

Generally, a typical cellulolytic complex includes a variety of hydrolytic and oxidative enzymes. Hydrolytic enzymes such as endoglucanase [1,4-β-D glucan glucanohydrolase, (EC 3.2.1.4)]; cellobiohydrolase (CBH), [1,4-β-D glucan cellobiohydrolase; (EC 3.2.1.91), 1,4-β-D-glucan glucohydrolase, (EC 3.2.1.74) and β-glucosidase [1,4-β-D glucoside glucohydrolase, (EC 3.2.1.21)] are involved in degradation of crystalline cellulose to glucose. The enzyme action is generally initiated by random acting endoglucanases at amorphous regions within cellulose chain to produce cellooligosaccharides. The cellobiohydrolases then act sequentially by removing cellobiose units from non-reducing ends of the cellulose chains. Finally, β-glucosidase completes the hydrolysis by degrading cellobiose and cellodextrins to glucose (Eveleigh, 1987). Oxidative enzymes, such as cellobiose dehydrogenase (EC 1.1.5.1) (Westermark and Errickson, 1974), Cellobiose oxidase (EC 1.1.99.18) (Morpeth, 1985) have been characterized from cellulolytic fungus *Phanerochaete chrysosporium*. Oxidative enzymes have been found to induce hydrolytic components of cellulase complex. Lactonases [(D-glucano-1,5, lactonohydrolase (EC 3.1.1.17)] are responsible for hydrolysis of inhibitors of cellulase system (Bruchmann *et al.*, 1987).

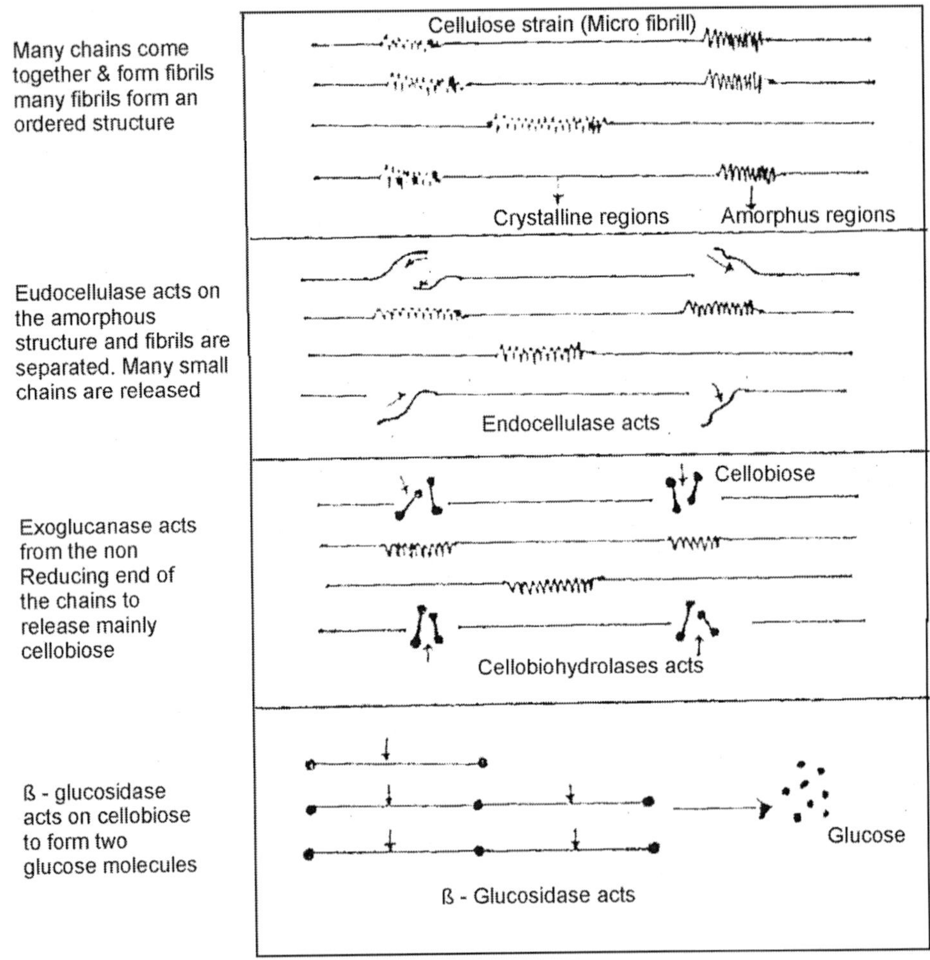

Fig. 2. Cellulase and their mode of action.

Our knowledge about fungal cellulase complex is mainly due to studies on hypersecretory fungi namely *Trichoderma reesei* (Kubicek, 1992; Kubicek *et al.*, 1993), *Penicillium pinophilum* (Bhat *et al.*, 1989),

Penicillium funiculosum (Wood and McCrae, 1982), *Fusarium oxysporum* (Christakopoulos *et al.,* 1995a, 1995b). *Aspergillus niger* (Okada,1985), *Sclerotium rolfsii* (Lachke and Deshpande, 1988) and *Humicola* sp. (Schulein, 1997). These fungi generally produce all three major components of cellulolytic systems exracellularly for complete hydrolysis of crystalline cellulose to glucose; but the proportion of different cellulolytic components varies considerably.

Cellulases in Pulp and Paper Industries

The mechanical process to improve paper strength by enhancing fibrillation and inter fibre bonding is called as beating or refining of pulp. Cellulases can be used for 'Biobeating' or Pulp refining (Kent and Jeffries 1996). The first process of refining paper fibrils with cellulases from white rot fungus *Trametes suaveolens* was patented in 1968 by Yerkes, W.D. (Yerkes, 1968). The process was successful in reducing beating time. Presently, various commercial formulations such as Pergalase, Maxazyme, Celluclast 1.5L are available in market for this application.

Enhancement of drainage rates with cellulase enzymes can be another success story of cellulases in paper industry. Drainage rate determines the loss of water during paper production. Low drainage rate severely affects machine speed that ultimately results in high-energy consumption with low-productivity. This problem is more critical when secondary fibres are used for paper production. Secondary fibre pulp with the commercial cellulase preparation can dramatically enhance drainage rates of recycled fibers (Pommier *et al.,* 1990). Commercial cellulase preparations such as Maxazyme 2000, Cellulase 250 P and SP 249 are available in the market for this application.

'Biodeinking' with Cellulases

The recovery and utilization of waste papers is increasing all over the world. The major difficulty in recycling of paper is removal of inks. Varieties of ink are used for different printing. These are considered as contact or non-contact ink types. Each of them requires different kind of statergies for deinking. Any modern office produces waste paper that includes laser printed papers and photocopier papers. Toners and laser printing inks are synthetic nylon or polyester based polymers with embedded carbon black that do not disperse readily. Toners are difficult to disperse than conventional oil based inks. Toners remain as large, flat rigid particles that separate poorly, from fibres during flotation and washing stages. Costly dewatering and dispersion steps followed by additional flotation and washing are required to improve the deinking of laser printed paper. Here is the real challenge for paper technologists because none of the conventional deinking methods can remove these inks from paper. Moreover, it is one of the most valuable grade papers available in market. In the enzymatic deinking of mixed office waste paper cellulases play a key role in releasing thermally fused toners from fibre surfaces (Jeffries *et al.,* 1994; Morkbak and Zimmermann, 1998;

Gubitz *et al.,* 1998a, 1998b; Vyas and Lachke, 2003). The process of removing ink and other contaminants from recycled fibre is termed as deinking.

The most favoured fungal sources for research and production of cellulases are *Aspergillus niger, Trichoderma reesei, Penicillium* sp., *Fusarium* sp. *and Humicola* sp. Each of these microorganisms can produce different kinds of cellulase components that differs in their mode of action as well as properties like activity towards crystalline cellulose, activity and stability in acid or alkaline pH. Different kinds of cellulases work best under different operational conditions. Acid cellulases that hydrolyze cellulose in acidic pH range of 3-6 can be used in acid sized paper. Alkaline cellulases that are effective between pH 8-10 are preferable since alkaline sized papers are more acceptable owing to their high fibre flexibility. Calcium carbonate is often used in the paper as a filler and brightner. It dissociates from fibres in alkaline environments. This also leaves fibres with alkaline pH hence alkaline active and stable cellulases are necessary to process them for deinking. This automatically saves additional acid treatment required for acid cellulases to work. However, microbial alkaline cellulases are very rare (Durand *et al.,* 1984). A few *Bacillus* sp. produce cellulases that are active in alkaline pH. Unfortunately bacterial cellulases lack complete cellulase complex. Thermal stability of enzyme is an additional important prerequisite for enzyme complex to be used for development of a commercial process. These all requirements led to shift research focus on screening of alkalophilic fungi that are capable of producing thermostable cellulases and are operative in a pH range of 8-10. Novozyme 342 from Novo Nordisk industries, Denmark is one of the commercially available enzyme preparations for deinking process. It is obtained on a large scale from micromycetes fungus *Humicola insolense.* This enzyme complex is active in a broad pH range of 5-10 (Schulein, 1997).

How Enzymes work during Deinking

Understanding behavior of the enzymes towards the fibre is essential to minimize the related negative effects on the strength of the paper as well as quality. The role of cellulases in the deinking is still not clear. Jeffries *et al.* (1994) suggested that the enzymes having higher filter paper degrading activities are effective in deinking. However, Qin *et al.,* (1998, 2000) have suggested that the filter paper degrading activity of the cellulase complex had detrimental effect on paper strength. According to their studies when cellulases with different cellobiohydrolases and endo-glucanases activities from *Aspergillus* L22 and *Trichoderma pseudo-koningii* S28 were used they have different effects on the fibre properties. During deinking process individual endoglucanase treatment does not result in the decrease of average fibre length or coarseness, which is important to modify the pulp fiber. The synergetic effect of the endoglucanases and cellobiohydrolases could lead to the cross breakage, and this is probably the major cause for the decrease in the paper strength. Gubitz *et al.* (1998 a,b) treated laser-printed wastepaper individually and with combinations of purified endoglucanases from *Gloeophyllum sepiarium* and *Gloeophyllum trabeum* and found that pure endoglucanases were responsible for most of the success in deinking operations. Although all enzyme treatments increased freeness and slightly decreased intrinsic fibre strength, the strength of the handsheet's was maintained or marginally improved. The nature of the individual endoglucanase plays an important role in this process. According to their observations different deinking efficiencies and strength properties were measured with both the endoglucanases, those are closely related enzymes in terms of molecular properties and substrate specificities. We have found that the alkaline active cellulases from alkalotolerant *Fusarium* sp. are useful for deinking of MOW paper (Vyas and Lachke, 2003). We have presented a schematic diagram that indicates the probable mechanism of the enzymatic deinking process by cellulases (Fig. 3). During biodeinking process, cellulases act in different ways. It is well understood that endoglucanase initiate action on cellulose. The role of endoglucanase in overall process of cellulose degradation is to split the cellulose fibers at several amorphous sites and generates innumerous non-reducing ends of the chain. This action supports loosening of fibres which in turn helps in releasing ink particles from MOW paper during flotation deinking process in the presence of surfactants. Secondly, cellulases act at the frazzled surfaces of cellulose fibres and release short fibers from it. This capability of releasing short fibres can be assigned to mechano-chemical property of endoglucanase components. It can be interpreted that this short fiber forming activity also removes residual fibres from the toner surfaces. Finally endoglucanases release fibres from surface of hairy toners, that in turn enhances the flotation efficiencies by increasing hydrophobicity of ink particles. From the data obtained during our work it seems that the randomly acting endoglucanases are useful for overall deinking process. The viscometric analysis revealed that Endo B from alkalotolerant *Fusarium* sp. is more random in its action and has more access to the surfaces of cellulose fibers (Vyas and Lachke, 2003; Vyas *et al.,* 2003). In the flotation deinking process the pulp is diluted to 1% consistency. The air bubbles are injected through the pulp from the bottom. The Ink particles that are more hydrophobic get attached to bubbles that rise to the top creating froth layer. This layer is swept off by skimming action. Here the stability of air bubble is the most important factor (Vyas *et al.,* 2001). Generally, ink particles are 40-50 μ in size. Additional chemicals termed as 'collectors' are needed if the ink chemicals are smaller in size. All the ink particles can't be removed in a single run. So the process is repeated couple of times. The mechanism of enzymatic deinking suggested that the enzyme molecules reduce the hydrodynamic drag that in turn increases filtration and flotation rates. In flotation process, the air bubble rising to the top carries ink particles along with cellulose fibres. These fibres exert negative effects on flotation efficiencies. Cellulases simply 'shaves' fibres from the ink particles where fibres will settle down and ink particles will adhere to the air bubble in much better way experiencing much less drag from interaction with water. Thus, the ink particles can be carried to the top of the tank much easily. In short, the cellulase enzyme complex reduces the hydrodynamic drag produced due to the attachment of fibres to the floating air bubbles (Vyas *et al.*, 2001). The enzyme complex can also reduce specific surface area of fibres, thereby lowering the interaction with contaminants. The enzyme activity enhances the draining resulting in enhancing filtration or flotation process. Once the flotation cycles are over, the pulp passes through different micro screens (0.15 mm) under pressure and the accepted fibres undergo subsequent washing steps. This results in removal of stickies and other contaminants.

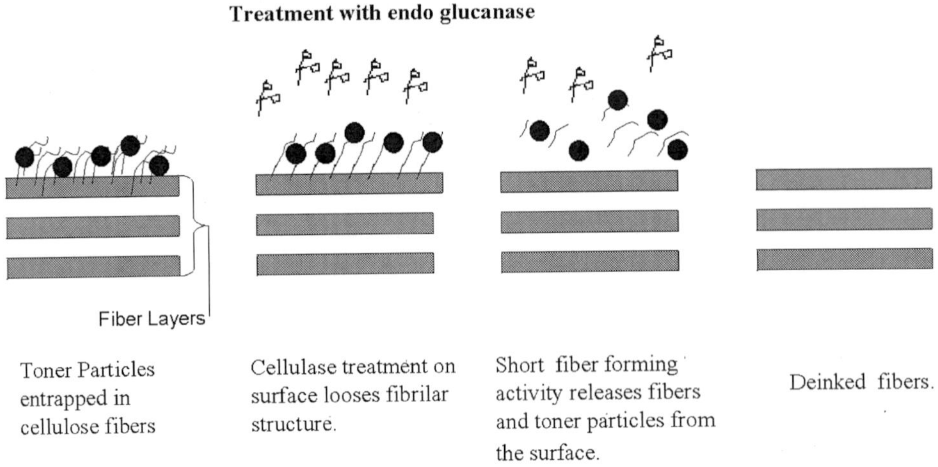

Treatment with endo glucanase

Fiber Layers

Toner Particles entrapped in cellulose fibers

Cellulase treatment on surface looses fibrilar structure.

Short fiber forming activity releases fibers and toner particles from the surface.

Deinked fibers.

Fig. 3. Mechanism of Biodeinking by Cellulases

Advantages of Enzymatic Deinking

The deinking with enzymes works best with chemically pulped fibre as compared to mechanically pulped fibre. This is because lot of lignin is associated with mechanically pulped fibres. As compared to water deinked pulps the enzyme releases upto 95% of toner particles from recycled fibres. The properties of deinked pulp such as tensile index, burst index, tear index are comparable to pulps from conventional deinking procedures. The enzymatic process preserves fibre integrity and also shows improved pulp brightness. Effluent samples from the deinking plants indicated lower oxygen demands as well as less toxicity than that of traditional processes.

The enzymatic "biodeinking" process requires fewer amounts of chemicals. This is a double benefit as it saves cost of waste treatment along with saving cost of chemicals required for the process. It also avoids landfill disposal costs. Enzyme deinked pulp shows high drainage rates resulting in faster machine speed and less energy cost. The technology for paper and pulp manufacturing is changing rapidly. This industry is becoming a major prospective market for the enzyme industry. However, enzymatic deinking is relatively a new process. Many of the paper and pulp industries all over the world are aware of this technology. They have even developed some unique pilot plants to explore the ecofriendly enzymatic technology of the future. The technologists from Forest Products Laboratory (Madison, WI, USA) have prepared some cost analysis of enzymatic deinking. It is approximately US$ 1.60 to 2.40 per 1000 Kg of the recovered paper. In some cases, depending upon the quality of the waste paper, the expenses can be as high as US$ 18 per 1000 Kg of the raw materials. In comparison, conventional deinking, chemical cost is between US$ 3.30 to 8.00 when sodium hydroxide (1%) is used alone or in combination with hydrogen peroxide. Considering many factors the cost of conventional deinking process and proposed enzymatic process will not differ much. Thus, the biodeinking with cellulases can fulfill the demands for better quality papers in ecofriendly way. Indeed it is the novel ecofriendly way of recycling the wastepaper.

HEMICELLULASES

Hemicellulose is one of the largest components of wood pulp after cellulose. Unlike cellulose it is more complex in its structure and composition. It is a heteropolymer that is made up of chains of xylan, mannan, galactan and arabinan. (Prade, 1995). In nature various microorganisms are capable of degrading hemicellulose to simpler sugars. Hemicellulose degrading enzymes are known as glycanases and they include, xylanases, mannanases, galactanases and arabinanases (Bastawade, 1992). Xylan is the major component of hemicellulose. It is a homopolymer of xylose units. Xylan is degraded by group of hydrolytic enzymes that are involved in hydrolysis of glycosidic bonds of xylan and esterases that ruptures ester linkages for efficient breakdown of xylan (Christov and Prior, 1993; Kulkarni *et al*, 1999). The hydrolases includes endoxylanase (EC 3.2.1.8) that randomly hydrolyse xylan backbone in to shorter xyloligosaccharides and β-xylosidases (EC 3.2.1.37) which cleaves xylooliglsaccharides to xylose. α-L-Arabinofuranosidase (EC 3.2.1.55) and α glucuronidase (EC 3.2.1.1) removes arabinose and 4-*o*-Methylglucuronic acid substituents respectively from xylan backbone are important side chain cleaving xylan hydrolases. The generalized scheme of xylan breakdown is demonstrated in Fig 4.

Although xylanses are widely distributed in nature, fungal xylanases have attracted researcher over the years due to simpler nutritional requirements of fungi and their hypersecretory nature. Xylanases are often secreted along with cellulases and are studied from various fungi such as *Aspergillus niger, A. foetidus, A. Oryzae* (Christov *et al.*, 1999), *Themomyces lanuginosus* and *T. aurantiacus* (Alam *et al.*, 1994), *Trichoderma koningii* (Wood and McCrae, 1986), *T. reesei* (Dekkar, 1985), *Humicola lanuginosa* (Anand and Vithayathil, 1990).) *Fusarium oxysporum* (Singh *et al.*, 1995), *Schizophyllum commune* (Haltrich and

Walter Steiner, 1994). A few yeast strains like *Cryptococcus albidus* (Biely and Petrakova, 1984), *Aureobasidium pullulans* (Leathers, 1989) have also been reported to produce extracellular xylanases.

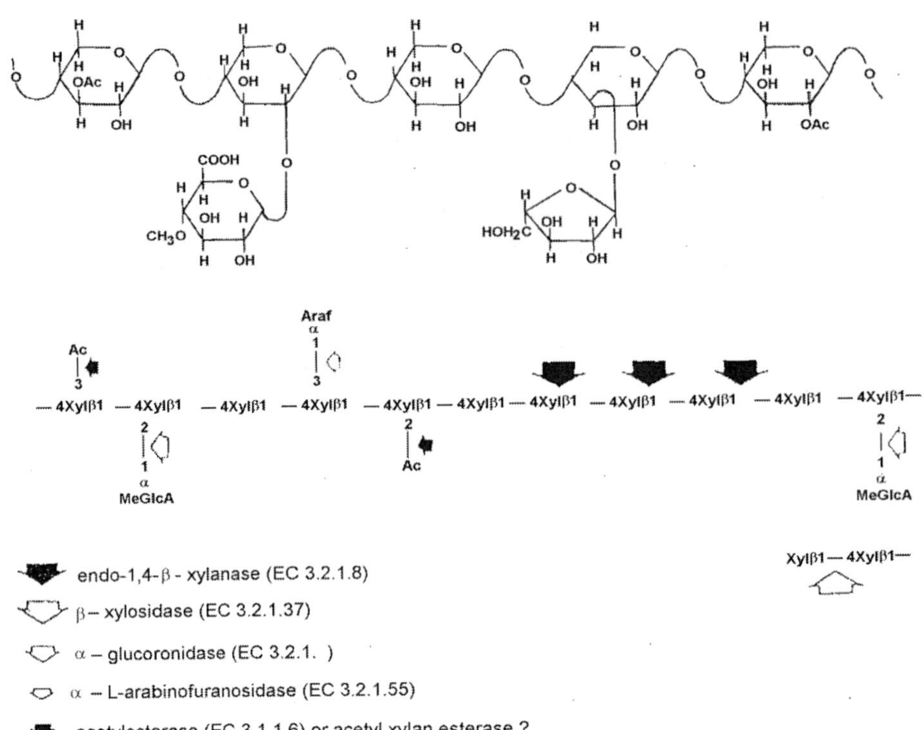

Fig. 4. Enzymatic Hydrolysis of Hemicellulose

Xylanases in pulp bleaching

It is been observed by various workers that the native structure of hemicellulose gets largely modified during alkaline pulping process (Viikari *et al.*, 1994, 1996). During the pulping process xylan component of hemicellulose gets readsobed at cellulose surface. This surface adsorbed xylan binds soluble lignin to form insoluble complexes. When these insoluble complexes reprecipitate on cellulose fibres the pulp becomes black. Also the insoluble chromophore complexes formed between residual lignin and other carbohydrates in the pulp are very difficult to extract as they are covalently attached to pulp matrix.

The process of removing lignin and chromophores from the pulp is called as bleaching. The paper maker uses chlorine and chlorine dioxide to bleach the pulp. However, this results in formation of hazardous chlorinated hydrocarbons as well as toxic dioxins. Hence, the focus of research and development work in the field has shifted more towards the improving the chemical process that will be ecofriendly as well as cost effective. Xylanase can be used as alternative biobleaching agent to hazardous chlorinated chemicals. With the recent environmental concerns on chlorine bleaching and lack of inexpensive, effective process, xylanases are gaining significant commercial importance. Different mechanisms have been proposed in order to explain bleach-boosting effects of xylanases (Ragauskars *et al.,* 1994; Buchert *et al.,* 1996; Suurnakki *et al.,* 1996; Wong *et al.,* 1996; Jeffries and Davis, 1998). However, according to an accepted mechanism, the xylanases act simultaneously at various xylan-containing linkages in the pulp that ultimately results in improved brightness (Viikari *et al.,* 1994, 1996). The probable mechanism of the process is presented in Fig. 5.

Lignin,Hemicellulose and Cellulose in close association in wood pulp.

Xylanases act on various lignin containing linkages during biobleaching process

Cellulose fibres

L:Lignin
H:Hemicellulose
C:Cellulose

Fig. 5. Mechanism of biobleaching with Xylanases

Enzyme characteristics

For successful application of xylanases in biobleaching process, they should lack the activity towards cellulosic substrates and should be functional at alkaline pH (Viikari *et al.,* 1994, 1996). Several welknown xylanase producing fungi often co-secrets cellulolytic enzymes that may adversely affect paper quality. Hence, all over the world the attempts are going on to search new microbial systems that can produce alkaline active cellulase free of xylanolytic activity. Xylanases are produced by large variety of microorganisms but the fungi that are capable of growing at alkaline pH and secreting alkali-stable extracellular xylanases have been very little investigated (Biely, 1985). Fungal cellulase free xylanases have been reported from *Cephalosporium* Sp. (Bansod and Rele, 1998). In our work we have obtained an alkaline active, alkali stable cellulase free xylanase from an alkalotolerant fungus *Chloridium* sp. The studies revealed that the enzyme can be produced on cheap agricultural / forest waste residues. The fungus also utilized dairy waste whey and produces commercially important cellulase free xylanase (Ahmad *et al.,* 1998). The preliminary studies indicated that the enzyme could be useful in pulp bleaching.

The mechanism of bleaching with xylanases

Different mechanisms have been proposed in order to explain bleach-boosting effects of xylanases during enzymatic treatments on the pulp (Ragauskars *et al.,* 1994; Buchert *et al.,* 1996; Suurnakki *et al.,* 1996; Wong *et al.,* 1996; Jeffries and Davis, 1998). However, most generally accepted mechanism suggests that the xylanase act simultaneously at various xylan containing linkage (Viikari *et al.,* 1996). Lignin-carbohydrate complexes are generated in the pulping process that acts as physical barriers to the entry of bleaching chemicals. Xylanases facilitate the removal of these complexes that in turn results in improved brightness. The xylanases also act on relocated xylan component. That action helps in increasing porosity for the entry of bleaching agents. The kappa number reduction during xylanase action is found independent of xylooligosacharides removal from xylan. However, good correlation was observed with release of hexenuronic acid residues from the pulp samples (Jeffries and Davis, 1998).

For the successful application of xylanases in the biobleaching process understanding of structural functional relationship of enzyme with its substrate becomes very important. It has been found that xylanases having similar properties such as molecular weight, pH optima and substrate specificities behave differently in its action during prebleaching operations. This is been attributed to their binding abilities towards pulp fibrils (Tenkanen *et al.,* 1995).

As the carboxylic groups are present at the surface of fibres they carry net negative charge at neutral pH. The interaction of enzymes with fibres is affected by the charge of the protein. The enzymes are totally bound to fibres when pH is below their pI, but can not bound at pH values above pI (Viikari *et al.,* 1994, 1996). This observation was also reported by other researchers during their studies with *Trichoderma* xylanases (Tenkanen *et al.,* 1992; Torronen and Rouvinen, 1995). From these studies it can be concluded that, because of their higher binding capacities the enzyme having basic pI are more effective in the prebleaching operations. The *Chloridium* xylanase obtained in our work possesses pI 8.31 and have found potentially useful in bleaching of bagasse pulp at alkaline conditions.

MANNANASES

As mentioned earlier, xylan and glucomannan form the basic backbone polymer of wood hemicellulose. Galactoglucomannan are the major groups of mannans in the cell wall of higher plants. The backbone is composed of β 1-4 linked D-glucose and D-mannose units that are distributed randomly within the molecules. D-galactose side chains are attached to mannose or glucose via α 1-6 bonds (Timmel, 1967). *O*-Acetyl galactoglucomanann are the principle hemicelluloses in the soft wood (about 20%). Hardwood contains low glucomannan 2-5% no galactose or acidic side chains are present (Timmel, 1967).

Glucomannan is a major hemicellulosic constituent in the softwood. During kraft pulping high amounts of glucomannans are dissolved and degraded (Casebier and Hamilton, 1965). Reprecipitation of mannan occurs in lesser amount than that of xylan (Sjostrom, 1977). Compared to xylanases, mananananase are more heterogeneous group of enzymes produced by different fungi like *Sclerotium rolfsi* (Gübitz *et al.,* 1996), *T. reesei* (Tenkanen *et al.,* 1997; Oksanen *et al.,* 2000) *Aspergillus niger* (Ademark *et al.,* 1998). The main hydrolysis products from galacto-mannan and glucomannan are mannobiose, mannotriose and various mixed oligosaccharides. The hydrolysis yield depends on the degree of substitution as well as distribution of substituents. The hydrolysis of glucomanan is also affected by glucose to mannose ratio (McClery, 1991). Studies on *T. reesei* mannanases revealed that they have multidomain structure similar to cellulolytic enzymes (Tenkanen *et al.,* 1995). The protein contains the catalytic core domain, which is connected by linker to cellullose binding domain. (Stalbrand *et al.,* 1995). Mannanases have been found to be more specific with respect to their substrates and only a few mannanases have been shown to hydrolyse glucomannan in softwood pulp.

T. reesei mannanases have been used successfully for bleach boosting effects in kraft pulp processing (Paice *et al.,* 1992). The studies carried out by Finnish researchers have suggested that mannanases treatment was effective in increasing bleachability of conventional pine kraft fibres only when the outer surface material of fibres was mechanically removed prior to treatments (Suurnakki *et al.,* 1996). The exact mechanism for mannanases in bleaching of kraft pulp is not fully understood. The probable mechanism suggests that underneath the outermost surface layer of kraft fibres and the surface of modified pulp the glucomannan is more closely located to lignin and that the enzymatic removal of glucomannan increases the bleachability of lignin.

PECTINASES

Pectic substances occurring in nature are made up of galactouronic residues linked by α 1-4 linkages. Depending on the form of backbone chain they are broadly classified as protopectin, pectic acid and pectin (Be-Miller, 1986). Pectin in native form is located in the plant cell wall and it is linked with other structural polysaccharides. Pectin is the key component in conferring rigidity to plant cell wall. During pulping process pectin substances remain attached to cellulose microfibrills by means of hemicellulose and facilitates water retention (Pifferi *et al.,* 1989).

Pectinases is the group of enzymes that catalyze degradation of pectin. Based on substrate specificity and mode of action, pectinases can be broadly classified as pectin esterases, pectin depolymerases and protopectinases (Kashyap *et al.,* 2001). Although these enzymes are characterized for different applications depolymerising enzymes have attracted more attention for their ecofriendly application in paper and pulp industry (Reid and Ricard, 2000).

Pectinases are reported from various microbial sources (Kashyap *et al.,* 2001). Fungal pectinases have been studied from *Aspergillus niger* (Acuna-Arguelles *et al.,* 1995), *Penicillium frequentans* (Borin *et al.,* 1996) *Sclerotium rolfsii* (Channe and Shewal, 1995), *Rhizoctonia solani* (Marcus *et al.,* 1986). However, these enzymes are optimally active in acidic conditions and their application at several paper making process becomes less feasible. Fortunately, alkaline active pectinases have been obtained from *Penicilium italicum* (Alana *et al.,* 1990) and *Aspergillus* sp. (Leda *et al.,* 2000).

Pectinases for pulp and paper processing

Enzymatic retting

All over the world, the forests are depleting due to over exploitation for various purposes. Under these conditions cellulose from non-wood sources such as agricultural residues, bast fiber plants like flax, hemp, can prove significant for paper manufacturing. In these sources cellulose is glued by middle lamella pectin and woody core of lignified fibres (Morvan *et al.,* 1990). Thus for isolation of these cellulose fibres intensive process are required. Traditionally this has been achieved by using mixed culture microbial process. These processes suffer from weather dependence. Also the bad odors are a released in effluents during retting by traditional methods (Van Sumere, 1992). Therefore, there is a need for effective retting process. Microbial pectinase can release cellulosic fibres from fibre bundles (Akin *et al.,* 1998; Zhang *et al.,* 2000). The exact mechanism for how enzyme act during enzymatic process of retting is not exactly known. It was previously thought that it requires combination of xylanases, mannanases and pectinases for release of cellulose fibres from flax. However, latter studies using pectinase preparations from *Rhizopus oryzae* overruled the involvement of hemicellulases in the process (Henricksson *et al.,* 1999). Similarly, studies carried out using purified pectinases from *Aspergillus* sp. demonstrated release of polysacharides from flax (Henricksson *et al.,* 1997).

Shives are small bundles of fibres that are not separated in to individual fibres during pulping. They appear as splinters that are darker than the pulp. Pectinase when used with combination of hemicellulase have proven successful in removal of shives (Tolan *et al.,* 1994). The process is found more effective when enzyme treatment carried out after alkaline soaking of pulp. Overall the process resulted in increased fibrillation, decreased shive content and better sheet forming properties of pulp (Tolan *et al.,* 1994). Removal of bark is important step in processing of wood because even small amounts of bark residues cause darkening of high quality pulp. The layer between wood and bark is called cambium. It is made up mainly pectin and little amount of lignin. Application of pectinolytic enzymes before pulping requires up to 80%

less energy consumption for debarking of pulp during mechanical pulping process (Viikari *et al.,* 1989).

In modern papermaking process retention aids are added to the pulp to keep fine and fillers in the paper sheet and to speed up the drainage of the water. These retention aids generally includes cationic polymers. These polymers form the bridges between the negatively charged surfaces of fines and filler particles. Negatively charged dissolving anionic materials such as pectin or polygalactouronic acids combines with cationic retention aids resulting in increasing their demands (Sundberg *et al.,* 1998). Hexamers and longer chain of polygalactouronic acid shows more binding to retention aids as compared to monomers, dimers of polyglactouronic acids (Thornton, 1994). The pectinases can depolymerise long chain polymers of galactouronoic acid and lowers cationic demands in the paper making process (Viikari *et al.,* 1996). Novozyme 863, SP 249 are the commercial enzyme preparation available for this applications in paper mills.

LIGNINASES

Lignin is widely distributed renewable aromatic polymer in nature. Making up to 15-30% of woody cell wall of gymnosperm (softwood), and angiosperm (hardwood) plants. Lignin is complex and variable biopolymer of *p* hydroxyphenyl propane units, linked through many C-O-C and C-C bonding. It is basically formed by random coupling of radical sp. arising from the peroxides mediated dehydrogenation of three cinnamyl alcohol derivatives; p-coumaryl, coniferyl and sinapyl alcohols. Due to their phenolic nature, they are extremely resistant to microbial attack and bind together cellulose fibres to provide strength. The degradation of lignin in the natural environment is a slow process and results only in its partial conversion to soil humus. Variety of bacteria, fungi and yeast has been studied for their lignin degradation. However fungi are excellent lignin degraders. Lignin's produced by fungi have been classified as Lignin peroxidase (LiP, EC 1.11.1.7), Manganese peroxidase (MnP, EC 1.11.1.13) and Laccase (benzenediol : oxygen oxidoreductase EC 1.10.3.2) and other H_2O_2 generating enzymes (Kirk and Farrell, 1987; Fiechter, 1993; Garg and Modi, 1999). Lignin peroxidases (Kelly and Reddy, 1986) catalyses H_2O_2 dependent oxidation of variety of phenolic acid and non phenolic model compounds. These reactions include oxidative cleavage of most abundant β*o*-4, Cα-Cβ linkages and other bonds in lignin and its model components.

Manganese peroxidase requires free magnous ions for their activity. They are H_2O_2 dependent and first catalyses oxidation of Mn II to Mn III, which subsequently oxidize the various phenolic compounds. Laccase oxidizes phenolic units in lignin in the presence of oxygen. It can also oxidize nonphenolic units in the presence of low molecular wt. mediators. Laccase can catalyze the alkyl phenyl and Cα-Cβ cleavage of phenolic lignin dimers and dimethoxylation of several lignin model compounds (Kirk and Farrell, 1987; Fiechter, 1993; Garg and Modi, 1999).

Other H_2O_2 producing enzyme like glucose oxidases (Kelly and Reddy 1986), glyoxyl oxidases (Kersten and Kirk, 1987), methanol oxidases are important as they generate H_2O_2 for utilization by extracellular peroxidases during the degradation of lignin.

Lignin in woody as well as herbaceous plants occurs as a closely linked to other carbohydrate moieties mainly the hemicellulose. For making of bright writing, printing grade paper residual lignin in kraft pulp must be removed by bleaching. Despite of extensive research on ligninolytic fungi and their lignin degrading enzymes these enzymes were not applied successfully in various applications. This is mainly due to complexity of lignin containing compounds. *In vivo* lignin degradation occurs with the help of synergistic action of various ligninases along with different mediators but *in vitro* single enzymes are not able to mimic the complete biological system. Recently laccase and manganese peroxidases have shown promising results (Call, 1993; Call and Mücke, 1997; Suurnakki *et al.,* 2001). Commercial enzyme preparation, Lignozyme (GmbH, Germany) has improved the performance of mediator system of laccase by *Coriolus versicolor* (Call, 1993). The mediator used is low molecular weight ecofriendly chemical hydroxybenzotriaozole. Treatment with laccase mediator system causes significant *k* number reduction and reasonable reduction in treatment times. The laccase enzyme requires oxygen as co- substrates for oxidation reaction. It is thought that laccase while oxidizing chemical mediator generates strong oxidizing co-mediator that is the real bleaching agent (Crestini *et al.,* 2001). The operating conditions for enzyme action are temperature 40-65°C, pH 4-7, pulp consistency 1-20%, pressure 1-14 bar and duration 1-4 hr. The enzyme can be used for bleaching of softwood, hardwood as well as pulp obtained from agricultural plant fibers (Bajpai, 1999). The laccase mediator enzyme system can be useful for bleaching of softwood pulp. The results obtained showed that reduction in *k* number as well as increase in brightness of pulp.

Ligninases in effluent treatment

White rot fungi, such as *P. chrysosporium, T.versicolor, Lentitus edodes* have been successfully applied in the degradation and decolorization of bleach plant effluents (Garg and Modi, 1999; Bajpai, 2001). The basic mechanism involved in the process may be adsorption on the mycelium or metabolism by lignin degrading enzymes. Factors affecting the fungal treatment of pulp mill effluents includes concentration of nutrients, pH, temperature and dissolved oxygen of system (Bajpai, 2001). Fungi have been immobilized in calcium alginate gel, fluidized bed and in bioreactor, however fungal systems used for decolorization of effluent waste water suffers with few drawbacks such as additional carbon source is needed for fungal growth in these effluents. Secondly, the process is not sustainable and needs fresh inoculum to be replenished every time.

Recent understanding of fungal ligninolytic systems can solve few of these problems. Ligninases have been well characterized from white rot fungi and has already shown potential to remove colour and adsorbable organic halogen (Aox) (Paice and Jurasek, 1984; Hammel, 1989; Bergbaur *et al.,* 1991). Call, 1991 have shown that laccases from *T. versicolor* can remove 70-90% of the lignin from the wastewater. Laccases immobilized on carbon support was used by Davis and Burns (1990). The results showed fast removal of polychlorinated compounds from effluents. Attempts are going on to understand the extact mechanism of lignin breakdoown in white rot fungi. Isolation of various lignin peroxidases (LiP) and manganese peroxidase (Mn P) systems and their characterization for application in waste water processing is going to be the key in order to get rid of polluting harmful effluents of pulp and paper industry.

OTHER FUNGAL ENZYMES

Lipases

Pitch is the major problem in softwood pulp. Pitch is a term used collectively for hydrophobic component of pulp that includes resin, resin acid, triglycerides, waxes etc. Pitch formation has been usually observed after chlorine bleaching process. The pitch formation results in sticky deposits on papermaking machines. It also affects water-holding capacities of pulp. And results in holes or tearing of paper.

Lipases have been used for reduction of pitch. Lipases obtained from *Candida cylindrica* act on triglycerides that are responsible for formation of these adhesives in pulp. Fischner and Messner (1992b) have suggested that enzymatic hydrolysis of triglycerides prior to bleaching, and then liberated glycerides can be extracted by sodium hydroxide solution during washing stage. Enzymatic process can remove up to 90% of triglycerides in 3 hr at 37°C (Ire *et al.,* 1989; Davis and Burns, 1990; Fischer and Messner, 1992a). The process has already commercialized. The enzyme formulation named as Resinase is available in the market for the same reason.

Lipase preparations have been found useful in deinking of magazine waste papers. Generally inks used in magazine printing are based on soy oil or other oil base. During enzymatic treatment these inks are degraded by lipases and can be easily separated by washing and flotation steps (Fischer and Messner, 1992b; Morbak *et al.,* 1999).

Amylases are the starch hydrolyzing enzymes. Starch is generally used as filler during paper manufacture. Amylases have been found useful for deinking and improving brightness of the paper during recycling of waste paper (Zollner *et al.,* 1998; Elegir *et al.,* 2000). Similar to pitch formation, accumulation of slime or stickies in the pulp because of microbial contamination of pulp causes problem in paper making process. The major problem in the slime formation is their characterization. If the chemical composition of slime is identified it is possible to develop enzyme based biocides for their removal from pulp.Microbial levan hydrolysing enzymes hydrolyse slime produced by different bacterial species and improvises the machine speed.

Several fungal enzymes such as esterases, resinases, lipases, xylanases mannanases, pectinases and laccases are gaining significant importance in deinking processes of recycling of waste paper (Bajpai, 1997). Moreover, these enzymes are commercially produced by fungal systems. The fermentation technology for enzyme production is changing rapidly. Novel types of fermenters have been designed for large-scale enzyme production. These developments in biochemical engineering along with advancement of fungal biotechnology has surprisingly resulted in bringing down the cost of the enzymes for pulp and paper manufacturing.

RECENT ADVANCES IN FUNGAL BIOTECHNOLOGY

The technology in paper processing is changing rapidly. Expectations for better quality paper and development of ecofriendly processes for manufacture of paper and pulp are going to be important drives for growth of paper industry in future. Biotechnology can play important role in fulfilling both the objectives of the industry. Fungal enzymes can serve as important tools in developing ecofriendly process for industry. There are about 72,000 named species of fungi and about 1500 more species are listed every year (Bennett and Faison, 1997). Although fungi are recognized as "factories" for production of healthy, robust industrial enzymes, surprisingly most of the commercial enzymes are produced from a very limited fungal genera. Most of the fungal kingdom is still unexplored territory and opportunities lies with biotechnologists to explore and exploit fungal biodiversity that can serve as tremendous resource for novel enzymes. The isolation and identification of extremophilic fungi is already underway and several commercial enzymes for pulp and paper industry have been produced and characterized from them. Thermophillic fungi such as *Chaetomium thermophilae* (Erikssen and Goksoyr, 1977), *Humicola griseus* (Hayashida and Mo, 1986), *Myceliopthora thermophila* (Roy *et al.*, 1990), *Sporotrichum thermophilae* (Coutts and Smith, 1976) have been found as excellent producers of cellulases. Thermally stable endoxylanases are studied from *Thermomyces sp* (Purkarthofer *et al.*, 1993), *Thermoascus aurantiacus* (Khandke *et al.*, 1989), *Humicola lanuginosa* (Anand *et al.*, 1990). A commercially important laccase enzyme has been purified from thermophillic fungus *Chaetomium thermophilium* (Chefetz *et al.*, 1998). Unlike thermophiles, alkalophillic fungi are rare. Few of the alkalo-tolerent fungi like *Fusarium* sp. (Vyas and Lachke, 2003), *Chloridium* sp. (Ahmad *et al.*, 1998), *Cepahalosporium* sp (Bansod and Rele, 1998) have been found to produce alkaline active cellulases, cellulase free alkaline xylanases that can be used in various operations of pulp and paper manufacturing.

Progress in molecular techniques, especially expression cloning, molecular screening, protein engineering has revolutionized screening and production of enzymes that can be applied for pulp and paper manufacturing. Approximately, 95% of microbial population is considered as non-culturable on artificial media (DeLong, 1996). With recent advancement in molecular biology it is possible to screen genomic DNA directly from the given source without prior isolation of micro-organism. This has opened up possibilities of screening for novel microbial enzymes. With the new approach based on "database mining" using bioinformatics tools it is possible to explore large fungal biodiversity for novel biocatalysts. Various attempts are going on all over the world and the results obtained are encouraging. Example based on screening for industrially important family 45 cellulase enzyme is self-explanatory. This enzyme is commercially obtained from fungus *Humicola sp*. The routine activity based screening showed that enzyme can only be obtained from few other fungi like *Fusarium oxysporum*, *Myceliopthora thermophila* and *Acremonium* sp. These all fungi belong to pyrenomycetus class of Ascomycota. When the molecular studies carried out by using degenerate primers it was found that family 45 cellulase genes occurred in 56 different fungal species belonging to 41 genera (Dalboge and Lene Lange, 1998).

Biotechnological applications of enzyme in paper and pulp industry require thermostable enzyme preparation with wide pH, temperature range. Since the availability of ideal enzyme preparation is limited the application of protein engineering to these enzymes have gained significant importance. Protein engineering has already shown success in generating cellulases, xylanases with desired properties. Recently cellulases with altered substrate specificity, altered pH and temperature profiles have been engineered with techniques in protein engineering. With the recent knowledge of protein 3D structures, amino acid sequence databases and the domain structure studies on different hydrolytic enzymes the protein engineering of these enzymes is practically possible.

CONCLUSION

The pulp and paper industry is a large and growing part of the world's economy. To keep up with the increasing demand for pulp and paper and to meet increasingly stringent environmental regulations, the industry has been constantly looking towards technological improvements.

The fungal biotechnology is offering various ecofriendly ways to avoid use of harmful chemicals, and saving on energy as well as water in paper and pulp industries. Attempts are going on to understand and apply the fundamental and applied aspects of enzyme-based processes in pulp and paper manufacturing. The same industry is cited as the fastest growing market for industrial enzymes. The developments in biochemical engineering along with advancement of fungal biotechnology has surprisingly resulted in bringing down the cost of the enzymes for pulp and paper manufacturing. Although fungi are recognized as "factories" for production of healthy, robust industrial enzymes, most of the fungal kingdom is still unexplored territory and opportunities lies with biotechnologists to explore and exploit fungal biodiversity that can serve as tremendous resource for novel enzymes. Progress in molecular techniques, especially expression cloning, molecular screening, protein engineering has revolutionized screening and production of enzymes that can be applied for pulp and paper manufacturing. However, the greatest challenge lies ahead to successfully adopt and apply these enzymes based process for pulp and paper manufacturing.

ACKNOWLEDGEMENT

S.V. is thankful to the Council of Scientific and Industrial Research (CSIR), Government of India, for financial assistance.

REFERENCES

Acuna-Arguelles, M.E., Guttierrez-Rajas, M., Viniegra-Gonzalez, G., and Favela Torres E. (1995). Production and properties of three pectinolytic activities produced by *A. niger* in submerged and solid state fermentation, *Appl. Microbiol. Biotechnol.* 43: 808-814.

Ademark, P., Varga, A., Medve, J., Harjunpää, V., Drakenberg, T., Tjerneld, F. and Stålbrand, H. (1998). Softwood hemicellulose-degrading enzymes from *Aspergillus niger*: Purification and properties of a –mannanase. *J. Biotechnol.* 63 (3): 199-210.

Ahmad, A., Lachke, A. H., Sathivel, C. and Kinhikar, A.G. (1998). A Process for the preparation of Alkalistable xylanase. Indian Patent Filed (NF358/98)

Akin, D.E, Foulk, J.A, Dodd, R.B. and McAlister, III D. D. (2001). Enzyme- retting of flax and characterization of processed fibers. *J.Biotechnol.* 89 (2-3) : 193-203.

Akhtar, M. Attridge, M.C., Myer, G.C., Kirk, T.K. and Blanchette, R.A. (1992). Biomechanical pulping of loblolly pine with different strains of white rot Fungus *Cerioporiopsis subvermispora. Tappi J.* 75: 105-109.

Akin, D.E., Henriksson, G., Morrison, W.H., and Ericksson, K-E.L. (1998). Enzymatic retting of flax In: Ericksson K-E.L.,Cavaco-Paulo A. (Eds) Enzyme applications in fibre processing,Acs Symposium ser 687, American Chemical Society, Washington D.C.

Alana, A., Alkorta, I., Dominguez, J. B., Llama, M.J. and Sera, J.L. (1990). Pectin Lyase activity in an *Penicillium Italicaum* strain. *Appl. Environ.Microbiol.*56 (12):3755- 3759.

Alam M, Gomes I., Mohiuddin G., Hoq M.M. (1994).Production,characterization of thermostable xylanases by *Thermomyces lanuginosus* and *Thermoascus aurantiacus. Enz. Microb. Technol.* 16: 298- 302.

Anand, L., Krishnamurthy, S. and Vithayathil, P.J. (1990). Purification and properties of xyalanase from thermophillic fungus *Humicola lanuqinosa. Arch. Biochem. Biophys.* 276: 546 - 553.

Anand, L. and Vithayathil, P.J. (1990). Production of extracellular xylanase by thermophillic fungus *Humicola lanuginosa I.J.Exptl.Biol.* 4:434-437.

Akhtar, M., Horn, E.G., Lentz, M. J., Scott, G.M, Sykes, M.S., Meyers, G.C., Kirk, T.K. and Swaney, R.E, (1999). Toward Commercialization of Biopulping Paper Age (Jan): 1-4.

Bajpai, P. (1997). Enzymatic Deinking, *Adv.Appl. Microbiol.*45,241-269.

Bajpai P. (1999). Application of enzymes in the pulp and paper industry. *Biotechnol Prog.*15, (2):147-157.

Bajpai, P. (2001). Microbial degradation of pollutants in pulp mill effluents. *Adv. Appl. Microbiol.* 48: 79-134.

Bajpai, P. and Bajpai, P.K. (1992). Biobleaching of kraft pulp. *Process Biochem* 27(6): 319-325.

Balakshin, M., Chen, C.L., Gratzl, J.S., Kirkman, A.G. and Jakob, H. (2001). Biobleaching of pulp with dioxygen in laccase mediator system effect of variables on the reaction kinetics. *J. Mol. Catal. B: Enzymatic* 16 (34):205-215

Bansod S.M. and Rele M.V. (1998). Alkaline xylanase from an *Alkalotolerant Cephalosporium* sp. *J. Biochem Mol. Biol.& Biophys.*2:121-128.

Bastawade, K. B. (1992). Xylan structure, microbial xylanases and their mode of action *W. J. Microb. Technol.* 8 : 353-68.

Be-Miller, J.N. (1986). An Introduction to pectins: Structure and properties. In: Fishman, M.L, Jem J.J. (Eds.) Chemistry and Functions of Pectins, Acs Symposium series 310, American Chemical Society, Washington DC.

Bennett, J.W. and Faison, B.D. (1997). Use of Fungi in biodegradation. In: Environmental Microbiology, ASM press, Washington.

Bergbaur, M., Eggert, C. and Kraepelin, G. (1991). Degradation of chlorinated lignin compounds in bleach plant effluents by white rot fungus *Trametes versicolor. Appl. Microbiol. Biotechnol.* 35: 105-107.

Bhat, K.M., McCrae, S.I. and Wood, T.M. (1989). The endo-(1-4)-β-D-glucanase system of *Penicillium pinophilum* cellulase. isolation, purification and characterization of five major endoglucanase components. *Carbohydr.Res.* 190: 279-229.

Biely, P. (1985). Microbial Xylanolytic Systems. *Trends Biotechnol.*3(11):286-289.

Biely P. and Petrakova E. (1984). Novel inducers of the xylan degrading enzyme system of *Cryptococcus albidus, J.Bact.* 160 :408-412.

Bissoon, S., Christov, L. and Singh, S. (2002). Bleach boosting effects of purified xylanase from *Thermomyces lanuginosus* SSBP on bagasse pulp. *Process Biochem* 37 (6): 567-572.

Blanchette, R.A., Burnes, T.E., Eerdman M.M. and Akhtar, M. (1992). Evaluating Isolates of *Phanerochaete Chrysosporium* and *Cereporiopsis subvermispora* for the use of biological pulping process, *Holzforchung* 46:109-115.

Borin, M.D.F, Said, S. and Fonseca M.J.V. (1996). Purification and characterization of an extracellular endopolygalactouronase from *Penicillium frequentas, J. Agric. Food Chem.* 44:1616-1620.

Bruchmann, E.-E., Schach, H.S. and Graf, H. (1987). Role and properties of lactonase in cellulase system. *Biotech. Appl. Biochem.* 9 : 146-159.

Buchert J., Carlsson, G., Viikari L. and Strom, G. (1996). Surface Characterization of unbleached kraft pulps by enzymatic peeling and ESCA. *Holzforchung* 50 (1): 69-74.

Call, H. P. (1993) Process for producing cellulose from lignin containing raw material using enzyme or micro organism while monitoring and maintaining the redox potential. US Pat No. 5,203,964.

Call, H. P. and Mücke, I. (1997). History, overview and applications of mediated lignolytic systems, especially laccase-mediator-systems (Lignozym®-process). *J. Biotechnol.* 53 (2-3): 163-202.

Casebier, R.L. and Hamilton, J. L. (1965). Alkaline degradation of glucomannans and galactomannans. *Tappi J.* (48) : 664-669.

Changing tastes in paper products (2001). In: Performance Chemicals Europe. May- June 2001. 42-46.

Channe P.S., Shewal J.G. (1995). Pectinase production by *Sclerotium rolfsii*: Effect of culture conditions. *Folia Microbiol.* (40): 111-117.

Chefetz, B., Chen, Y. and Hadar, Y. (1998). Purification and characterization of laccase from *Chaetomium thermophilum* and its role in humification, *Appl. Env. Microbiol.* 64 :3175-3179.

Christakopoulos, P., Kekos, D., Macris, B.J., Claeyssens, M. and Bhat M.K. (1995a). Purification and Characterization of less randomly acting endo -1,4-β-D-glucanase from culture filtrates of *Fusarium* oxysporum. *Arch. Biochem. Biophys.* 316 : 428-433.

Christakopoulos, P., Kekos, D., Macris B. J., Claeyssens, M., Bhat, M. K. (1995b). Purification and mode of action of low molecular mass endo -1,4-β-D- glucanase from *Fusarium* oxysporum, *J. Biotechnol.*, 39: 85-93.

Christov, L.P. and Prior, B.A. (1993). Esterases of Xylan-Degrading microorganisms: Production, Properties and Significance. *Enz. Microb. Technol.* 15 : 460-475.

Christov, L.P., Szakacs, G., and Balakrishnan, H. (1999). Production, Partial Characterization and use of cellulase free xylanases in pulp bleaching. *Process Biochem.* 34: 511 - 517.

Coutts, A.D. and Smith, R.E. (1976). Factors influencing production of cellulases by *Sporotrichum thermophile. Appl. Env. Microbiol.* 31:819-825.

Crestini, C., Agyropoulous, D.S. and Vikkari, L. (2001). Oxidation of lignin by the laccase mediator system : The role of 1- Hyroxybenzotriazole In 8th International conference on biotechnology in pulp and paper industry. Vahala, P, and Lantto, R. Eds. Helsinki, pp 119-121.

Davis, S. and Burns, R.G. (1990). Decolorization of phenolic effluents by soluble and immobilized phenoloxidases. *Appl. Microb. Biotechnol.* 32: 721 - 726.

Dalboge, H. and Lene Lange (1998).Using molecular techniques to identify new microbial catalysts, *TIBTECH* 16: 265-72.

Dekkar, R.H.F. (1985). Biodegradation of hemicellulose. In Biosyntheis and Biodegradation of wood components.eds. Higuchi T., Academic Press Tokyo : 505-33.

DeLong, E.F. (1996). In: Micribial Diversity in Time and Space, Colwell R.R., Simidu U. and Ohwada K. Eds :125-34.

Durand, H, Soucaille, P. and Tiraby, G. (1984). Comparative study of cellulases and hemicellulases from four fungi: mesophiles *Trichoderma reesei* and *Penicillium* sp.and thermophiles *Thielavia terrestris* and *Sporotrichum cellulophilum. Enz. Microb Technol.* 6 : 175-80.

Elegir, G., Panizza E. and Canetti, M. (2000). Neutral enzyme-assisted deinking of xerographic office waste with a cellulase /amylase mixture. *Tappi J.* 83 (11): 71.

Erikssen, J. and Goksoyr, J. (1977). Cellulases from *Chaetomium thermophile* Var. *dissiutum. Eur. J. Biochem.* 77: 445-475.

Evans, J.D., Akin, D.E. and Foulk, J.A. (2002). Flax-retting by polygalacturonase-containing enzyme mixtures and effects on fiber properties. *J. Biotechnol.* 97 (3): 223-31.

Eveleigh, D.E. (1987). Cellulase: a perspective. *Phil. Trans. R. Soc. Lond.,* 321:435-7.

Fiechter, A. (1993). Function and synthesis of enzymes involved in lignin degradation. *J. Biotechnol.* 30 (1): 49-55.

Fischer, K. and Messner, K. (1992a). Reducing troublesome pitch in pulp mills by lipolytic enzymes. *Tappi J.* 75: 130-134.

Fischer, K. and Messner, K. (1992b). Adsorption of lipase on pulp fibers during biological pitch control in paper industry. *Enz. Microb. Technol.* 14: 470-473.

Garg, S.K. and Modi, D.R. (1999). Decolorization of Pulp-Paper Mill Effluents by White - Rot Fungi. *Crit. Rev. Biotechnol.* 19 (2): 85-112.

Godfrey, T. and S. West (1996). In: Industrial Enzymology 2nd edition, Macmillan Nature, United Kingdom.

Gübitz, G.M., Hayn, M., Sommerauer, M. and Steiner, W. (1996) Mannan degrading enzymes from *Sclerotium rolfsii*: characterisation and synergism of two endo - mannanases and a -mannosidase. *Biores.Technol.* 58 (2): 127-35.

Gubitz, G.M., Mansfield, S.D., Bohm, D. and Saddler, J.N. (1998a). Effect of endoglucanases and hemicellulases in magnetic and flotation deinking of xerographic and laser-printed papers. *J. Biotechnol.* 65 : 209-215.

Gubitz, G.M., Mansfield, S.D. and Saddler, J.N. (1998b). Effectiveness of two endoglucanases from *Gloeophyllum* species in deinking mixed office waste paper. Int. Conf Biotechnol Pulp Ind, 47th, 1998; C : 135-138.

Gübitz, G.M., Schnitzhofer, W., Balakrishnan, H. and Steiner, W. (1996). Two mannanases from *Sclerotium rolfsii* in total chlorine free bleaching of softwood kraft pulp. *J. Biotechnol* 50 (2-3) : 181-188.

Haltrich, D. and Walter Steiner (1994). Formation of xylanase by *Schizophyllum commune:* Effect of medium components. *Enz. Microb.Technol* 16 : 229-235.

Hammel, K.E. (1989).Organopollutant degradation by ligninolytic fungi, *Enz. Microb.Technol.* 11: 776-777.

Hayashida, S. and Mo, K. (1986). Production and characteristics of avicel disintegrating endoglucanase from a protease-negative *Humicola grisea* var. *thermoidea* mutant. *Appl. Environ. Microbiol.* 51: 1041-1046.

Henricksson, G., Akin, D.E., Rigsby, L.L., Patel, N. and Eriksson, K-E. L. (1997). Influence of chelating agents on enzymatic retting of flax. *Tex. Res. J.* 67:829-836

Henricksson, G., Akin, D.E., Slomczynski, D. and Eriksson, K-E.L. (1999). Production of highly efficient enzymes for flax retting by *Rhizomucour pusillus, J.Biotechnol.*68, 115-123.

Ire, Y., Matsukura, T., and Hata, K. (1989). Process for preparing mechanical pulps and paper containing said pulp, European Patent E.P.374700 A2.

Jeffries, T.W., Klungness, J.H., Marguerite, S. and Cropsey, K.R. (1994). Comparison of enzyme enhanced with conventional deinking of xerographic and laser-printed paper. *Tappi J.* 77: 173-179.

Jeffries T.W. and Davis, M. (1998) Mechanism of kappa reduction and colour removal. In Proceedings of the 7th International conference on Biotechnology In pulp and Paper industry, Vancouver, BC, Canada, June 16-19 (C) :41-43.

Jiménez, L., Martínez, C., Pérez, I. and López, F. (1997). Biobleaching procedures for pulp from agricultural residues using *Phanerochaete chrysosporium* and enzymes, *Process Biochem* 32 (4) : 297-304.

Johnsrud, S.C., Fernandez, N., Lopez, P., Guitierrz, I, Saez, A. and Ericksson, K.E. (1987). Properties of fungal pretreated high yield Bagasse pulps. *Nordic pulp and paper res.* J. 2 : 47-52.

Kashyap, D.R., Vohra, P.K., Chopra, S. and Tewari, R. (2001). Applications of pectinases in the commercial sector:a review. *Biores.Technol.*77(3):215-27.

Kelly, R.L. and Reddy, C.A. (1986). Purification and Chracterization of glucose oxidase from ligninolytic cultures of *Phenerochaete Chrysosporium. J. Bacteriol.*166:169-74.

Kent Kirk, T. and Jeffries, T.W. (1996), Role of microbial enzymes in pulp and Paper Processing, In: Enzymes for Pulp and Paper Processing, ACS symposium Series. 655, Ch. 1:1-13.

Kersten, P.J. and Kirk, T.K. (1987). Involvement of a new enzyme Glyoxal oxidase, in extracellular H_2O_2 production by *Phanerochaete Chrysosporium J. Bacteriol.* 169: 2195-2201.

Khandke, K.M, Vithayathil, P.J. and Murthy, S.K. (1989). Purification of xylanase, β-glucosidase, endocellulase and exocelluase from thermophillic fungus *Thermoascus aurianticus. Arch. Biochem. Biophys.* 274: 491-500.

Kirk, T.K., Burgess, R.R. and Koning, J.W. (1990). Use of fungi in pulping wood : An overview of biopulping research In: frontiers in Industrial Mycology. Proceedings of Industrial Mycology Symposium 1990, June 25-26 Madison, W. I. Leatham, G.F. Ed. Routledge Chapman and Hall, New York, N.Y. 99 - 111.

Kirk, T.K. and Farrell, R.L. (1987). Enzyme combustion the microbial degradation of lignin. *Ann. Rev. Microbiol.* 41: 465-505.

Kubicek, C.P. (1992). The cellulase proteins of *T. reesei* : the structure, multiplicity mode of action and regulation formation. *Adv. Biochem. Eng.* 45: 1-27.

Kubicek, C.P., Messner, R., Gruber, F., Mach, R.L. and Kubicek-Pranz, E.M. (1993). The *Trichoderma reesei* cellulase regulatory puzzle from the interior of a secretory fungus. *Enz. Microb. Technol.* 15: 90-99.

Kulkarni, N., Shende, A. and Rao, M. (1999). Molecular and biotechnological aspects of xylanases. *FEMS Microbiol. Rev.* 23: 411-456.

Lachke, A. H. and Deshpande, M.V. (1988). *Sclerotium rofsii.* Status in cellulase research. *FEMS Microbiol. Rev.* 54 :177-184.

Leathers, T.D. (1989). Purification and properties of xylanase from *Aureobasidium J. Ind. Microbiol.* 4: 341-348.

Leda, R., Castilho, Ricardo, A., Medronho and Tito, L.M. Alves (2000). Production and extraction of pectinases obtained by solid state fermentation of agro industrial residues with *Aspergillus niger. Biores.Technol.* 71(1): 45-50.

Marcus, L., Barash, I., Sneh, B., Koltin, Y. and Finker A., (1986). Purification and characterization of pectolytic enzymes produced by virulent and hypovirulent isolates of *Rhizoctonia solani* Kuhn. *Physiol. Mol. Plant Pathol.* 29: 325-336.

McClery, B.V. (1991). In: Enzymes in Biomass conversion: Leatham G.F., Himmel M. Eds: ACS Symp. Ser. 460: American Chemical Society : Washington, 437.

Messner, K. and Srebotnik, E. (1994). Biopulping : An Overview of developments in an environmentally safe paper-making technology. *FEMS Microbiol. Rev.* 13: 351-364.

Morkbak, A. L. and Zimmermann, W. (1998). Deinking of recovered mixed office paper by treatment with a single endo-glucanase I from *Humicola insolens* and a neutral surfactant. *Prog. Pap. Recycl.* 73 :33-37.

Morbak, A.L., Degn, P. and Zimmermann, W. (1999). Deinking of soy bean oil based ink printed paper with lipases and a neutral surfactant, *J. Biotechnol.* 67 (2-3) : 229-236.

Morpeth, F.F. (1985). Some properties of cellobiose oxidase from white rot fungus *Sporotrichum pulverulentum. Biochem.J.* 228: 557-564.

Morvan, C., Jauneau, A., Flaman, A., Millete, J. and Demarty, M. (1990). Degradation of flax polysaccharides with purified endo-polygalactouronase. *Carbohydr. Polymers,*13 (2) : 149-163.

Okada, G., (1985). Purification and properties of cellulase from *Aspergillus niger. Agric. Biol. Chem.* 49: 1257.

Oksanen T., Pere J., Paavilainen L, Buchert J. and Viikari L. (2000). Treatment of recycled kraft pulps with *Trichoderma reesei* hemicellulases and cellulases, *J. Biotechnol* 78 (1) : 39-48.

Paice, M.G, Gurnagul, N., Page, D. H. and Jurasek, L. (1992) Mechanism of Hemicellulose-directed prebleaching of kraft pulps. *Enz. Microb. Technol.*, 4 : 272-276.

Paice, M.G. and Jurasek, L. (1984). Peroxidase catalysed colour removal from bleach plant effluent, *Biotechnol Bioeng.* 26:477-480.

Pifferi, P.G., Buska, G., Manenti, I., Lo., Presti, A. and Spagna, G. (1989). Immobilization of pectinesterase on γ-alumina for the treatment of juices. *Belgian J. Food Chem. Biotechnol.* 44: 173-82.

Pommier J.C., Goma, G., Fuentes, J.L. and Rousset, C. (1990). Using enzymes to improve the process and the product quality in reycycled paper industry. *Tappi* 73 (12): 197-202.

Prade, R.A. (1995). Xylanases: from Biology to Biotechnology. *Biotechnology and Genetic Engineering Reviews* 13 :101-131.

Prasad, N.S.K. (2002). An overview of the Paper Chemicals Industry, Part 2 The Paper Industry, *Chemical Weekly*, Dec. 2 :169-176.

Purkarthofer, H., Sinner, M. and Steiner, W. (1993). Cellulase free Xylanse from *Thermomyces lanuginosus* : Optimisation of production in submerged and solid state fermentation. *Enz. Microb.Technol.*,15:677-682.

Qin, M., Gao, P., Qu, Y., Fu, Y., Shao, Z. and Wang, Q. (1998). Physical Characteristics of enzymically modified fibers from old newsprint. Emerging Technol. Pulping Papermaking Fast-Grow. Wood, Proc. Int. Symp. 1998, 462- 473.

Qin, M., Gao, P., Shao, Z., Fu, Y. and Wang, Q. (2000). Enzymatic deinking of old newsprint. *Zhongguo Zaozhi (Chinese)* 19 (4): 7-13.

Ragauskars, A.J., Poll, K.N. and Cesternino, A. (1994). Effects of xylanase pretreatment procedures on non chlorine bleaching.*Enz.Microb.Technol.*16:492-495.

Reid, I. and Ricard, M. (2000). Pectinase in papermaking: solving retention problems in mechanical pulps bleached with hydrogen peroxide, *Enz. Microb. Technol.* 26 (2-4):115-123.

Richard, J.E., Chandra, P. and Saddler, J.N. (1997). Effects of a fungal treatment on the brightness and strength properties of a mechanical pulp from Douglas-fir. *Biores Technol.* 61 (1): 61-68.

Roy, S.K., Dey, S.K., Raha, S.K. and Chakrabarty, S.L. (1990). Purification and properties of an extracellular endoglucanase from *Myceliopthora thermophila* D-14. (ATCC 48104) *J. Gen. Microbiol.* 136: 1967-1971.

Schulein, M. (1997). Enzymatic properties of cellulases from *Humicola insolens*, *J. Biotechnol* 57 :71-81.

Sealey, J. and Ragauskas, A.J. (1998). Residual lignin studies of laccase-delignified kraft pulps. *Enz. Microb. Technol.* 23 (7-8) :422-426.

Singh, A., Kuhad, R.C. and Kumar, M. (1995). Xylanase production by a hyper-xylanolytic mutant of *Fusarium Oxysporum. Enz. Microb. Technol.* 17:551-53.

Sjostrom, E. (1977). The behavior of wood polysaccharides during alkaline pulping process. *Tappi J.* (60) :151-54.

Stalbrand, H., Saloheimo, A., Vehmaanpera, J., Henrissat, B. and Penttilla, M. (1995). Cloning and expression in *Saccharomyces cerevisiae* of *Trichoderma reesei* β mannanase gene containing cellulose binding domain. *Appl. Environ. Microbiol.* 61(3):1090-1097.

Stork, G., Pereira, H., Wood, T. M., Dusterhoft, E., Toft, A. and Puls J. (1995). Upgrading recycled pulps using enzymatic treatment. *Tappi J* 78: 79-88.

Sundberg, K.E., Sundberg, A.C., Thornton, J.W. and Holmbalm, B.R. (1998). Pectic acids in production of wood containing paper. *Tappi J.* 81(7):131-136.

Suurnakki, A., Heijnesson, A., Buchert, J., Westermark U. and Viikari, L. (1996). Effect of pulp surfaces on enzyme aided bleaching of kraft pulp. *J. pulp paper sci* 22 (3): J 91-96.

Suurnakki, A., Rittstieg, K., Viikari, L. and Buchert, J. (2001). Understanding of reactions in laccase and laccase mediator treatments of lignin model compounds and PGW pulp, 8th International conference on biotechnology in pulp and paper industry. Vahala, P, and Lantto. R. Eds. Helsinki 121-123.

Suurnakki, A., Heijnesson, A., Buchert, J., Viikari, L. and Westermark, U. (1996) Chemical characterization of the surface layers of unbleached pine and birch kraft pulp fibers. *J. Pulp Paper Sci.* 22: 43-47.

Tenkanen, M., Buchert, J. and Viikari L. (1995). Binding of hemicellulases on isolated polysaccharide substrates. *Enz. Microb.Technol,* 17: 499-505.

Tenkanen, M., Makkonen, M., Perttula, M., Viikari, L. and Teleman, A. (1997). Action of *Trichoderma reesei* mannanase on galactoglucomannan in pinc kraft pulp, *J. Biotechnol.,* 57 (1-3):191- 204.

Tenkanen, M., Puls, J. and Poutanen, K. (1992).Two major xylanases of *Trichoderma reesei Enz. Microb. Technol* 14: 566- 574.

Thornton, J.W. (1994). Enzymatic degradation of polygalactouronoic acids released from mechanical pulp during peroxide bleaching. *Tappi J.* 77(3): 161 - 167.

Timmel, E. (1967). Recent progress in the chemistry of wood hemicelluloses. *Wood. Sci. Technol.* 1.45-87.

Tolan, J.S., Guenette, M., Thibault, L. and Winstanley, C. (1994). The use of novel enzyme treatment to improve the efficiency of shive removal by bleaching. *Pulp and Paper Can.,* 95(12) : 96-101.

Torronen, A. and Rouvinen, J. (1995). Structural comparison of two major endo - 1,4 xylanase from *T. reesei. Biochemistry.*34: 847-856.

Van Sumere, C.F. (1992). Retting of flax with special reference to enzymatic retting In: Sharma H.S.S., Van Sumere, C.F. Eds., The Biology and Processing of Flax. M. Publications Relfast, Northern Ireland : 157-98.

Viikari, L., Kantelinen, A., Sundquist, J. and Linko, M. (1994). Xylanases in bleaching: From an Idea to the Industry. *FEMS Microbiol Rev.* 13: 335-350.

Viikari, L., Ratto, M., Kantelinen, A., (1989). Method for the debarking of logs, Finnish patent application. 896, 291.

Viikari, L., Suurnakki, A. and Buchert, J. (1996). Enzyme-aided bleaching of kraft pulp : Fundamental mechanisms and practical applications, In: Enzymes for Pulp and Paper Processing. *ACS symposium Series* 655, 2 : 15-24.

Vyas, S., Absar, A. and Lachke, A. (2003). The mechanism of deinking of mixed office waste paper by alkaline active fungal cellulases. National Symposium on "Prospecting of Fungal Diversity and Emerging Technologies" & 29Th Annual Meeting of Mycological Society of India. Agharkar Reasearch Institute, Pune (6-7 Feb. 2003).

Vyas, S. and Lachke, A. (2003). Biodeinking of mixed office waste paper by alkaline active cellulases from alkalotolerant *Fusarium* sp. *Enz. Microb. Technol.* 2 : 236-45.

Vyas, S., Pandhare, A., Yogi, V., Palankar, V., Mantri, M. and Lachke, A. (2001). A Laboratory Scale Flotation Cell Prototype for Efficient Enzymatic Deinking of Waste Papers. All India Biotech Association 2001: Conference on Biotechnology - The Science and the Business, IIT, Delhi, New Delhi (28-30 Sept. 2001).

Westermark, U. and Errickson, K.E. (1974). Celllobiose: Quinone oxido reductase - A new wood-degrading enzymes from white rot fungi. *Acta chem Scand.* 28:209-214.

Wong, K.K.Y., Yokota, S., Saddler, J.N. and DeJong, E. (1996). Enzymatic hydrolysis of lignin-carbohydrate complexes isolated from kraft pulp. *J. Wood Chem Technol.*16 : 121-138.

Wood, T.M. and McCrae, S. I. (1982). Purification and some properties of a (1,4)-β-D-glucan glucohydrolase associated with the fungus *P. funiculosum. Carbohydr.res.* 110, 291-303.

Wood, T.M. and McCrae, S.T. (1986). Studies of two low molecular weight endo-1,4 β Xylanase, constitutively synthesized by cellulolytic fungus *Trichoderma koningii. Carb. Res.* 148: 321-30.

Yerkes, W.D. (1968). The process for the digestion of cellulosic material by enzymatic action of *Trametes suaveolens*, United states patent No, 3, 406, 089.

Zhang, J., Henriksson, G. and Johansson, G. (2000). Polygalactouronase is the component in Enzymatic retting of flax, *J. Biotechnol.* (81): 85-89.

Zollner, H, Schroeder, K., and Leland, R. (1998). Enzymic deinking of nonimpact printed white office paper with α-amylase, *Tappi J.* 81 (3) : 166-170.

CHAPTER - 2

INDIAN KERATINOPHILIC FUNGAL FLORA : A REVIEW

Neetu Tripathi and R.K.S. Kushwaha

Department of Botany, Christ Church College, Kanpur 208 001

E-mail- rks_kushwaha@yahoo.co.in

INTRODUCTION

The fungi grouped under dermatophytes are essentially related to diseases of the cutinized and keratinized tissue of man and animals. Distinctions can be made between the wide range of these fungi which are potentially keratinolytic because they possess proteolytic enzymes capable of degrading keratinized substrate in pure culture and keratinophilic organisms that show a high competitive ability for keratin substrate in natural habitat. These fungi appeared to be restricted to keratinous substrate in soil and many workers studied the colonization of keratinous baits in soil. The isolation of keratinophilic fungi from diverse habitats/substrates like soil, birds, feathers, nests, animal horns and hooves have given valuable information for their distribution in India.

During the last two-three decades there have been many reports on the isolation of dermatophytes and related keratinophilic fungi from Indian soils (Puri, 1961; Padhye, 1961; Padhye and Thirumulachar, 1962; Sarkar, 1962 a, b; Randhawa and Sandhu, 1963, 1964; Pawar *et al.,* 1963 and Mahapatra and Gugnani, 1964).

Dey and Kakoti (1955) were the first Indian workers who isolated *Microsporum gypseum* from soil of Dibrugarh, Assam. The group of Thirumalachar and Padhye isolated keratinophilic fungi mainly from Maharashtra. After the comprehensive publications of Randhawa and Sandhu (1965) and Garg (1966b) from Delhi and Rajasthan, other groups have started to work on isolation of these fungi. Later the extensive isolation work on these keratinophilic fungi started at the school of Sagar, Bhubneshwar, Jabalpur and followed at Lucknow, Jammu, and Agra at several scattered places.

The scattered reports on Indian keratinophilic flora are not compiled so far. It is, therefore, this is for the first time we made comprehensive listing on the isolation of keratinophilic and related fungi including their substrate of isolation. This review contains as possible all the reported fungi, which were isolated by hair baiting method or from keratinous substrate irrespective of their keratinolytic ability. However, two taxonomic reviews were published. One of them (Kushwaha, 2000) deals with species of *Chrysosporium* and another (Kushwaha and Tripathi, 2004) redescribed pathogenic fungi.

1. *Absidia corymbifera*

Human hair, nail. Nigam (1987), Nigam and Kushwaha (1989 b).

2. *Absidia cylindrospora*

Unknown. Rajak *et al.* (1992).

3. *Absidia glauca*

Chicken feathers. Singh *et al.* (1994).

4. *Absidia spinosa*

Human hair. Ramesh and Hilda (1999).

5. *Acitheca purpurea*

Bird feather. Dixit and Kushwaha (1991 b).

6. *Acremonium hennebertii*

Human hair. Parihar (1998).

7. *Acremonium implicatum*

Human hair. Katiyar (1998), Parihar (1998), Katiyar and Kushwaha (2000), Katiyar *et al.* (2000).

8. *Acremonium obclavatum*

Human hair. Nigam and Kushwaha (1989 a), Nigam and Kushwaha (1993).

9. *Acremonium* sp.

Human hair, horse hair, cow hair, animal hair, chicken feathers, wool, nail, horn. Nigam (1987), Nigam and Kushwaha (1989 b), Saxena (1993), Gupta (1994), Nigam *et al.* (1994), Ghawana *et al.* (1997), Katiyar (1998), Parihar (1998), Katiyar and Kushwaha (2000), Katiyar *et al.* (2000), Ghosh and Bhatt (2000), Srivastava *et al.* (2001), Govil *et al.* (2001), Bhadauria and Kushwaha (2002).

10. *Acrodontium album*

Human hair. Kushwaha (1976).

11. *Ajellomyces dermatitidis*

Bird feather. Dixit and Kushwaha (1991 b).

12. *Allescheria boydii*

Unknown. Padhye and Thirumulachar (1968).

13. *Alternaria alternata*

Human hair, bird feather, nail. Dixit (1991), Ramesh and Hilda (1999), Ghosh and Bhatt (2000).

14. *Alternaria lutea*

Buffalo hair. Singh *et al.* (1994).

15. *Alternaria tenuis*

Human nail, animal skin. Singh *et al.* (1994).

16. *Alternaria* sp.

Skin. Mitra *et al.* (1998).

17. *Aphanoascus fulvescens*

Human hair, chicken feathers. Nigam and Kushwaha (1985), Nigam (1987), Saxena (1993), Parihar (1998), Ghosh and Bhatt (2000).

18. *Aphanoascus gertleri*

Human hair. Nigam and Kushwaha (1985), Nigam (1987).

19. *Aphanoascus keratinophilus*

Human hair, chicken feathers, cow hair, nail. Katiyar (1998), Parihar (1998), Katiyar and Kushwaha (2000), Katiyar *et al.* (2000).

20. *Aphanoascus terreus*

Human hair, bird feather, nails. Sur and Ghosh (1980), Jain (1983), Jain *et al.* (1985), Dixit (1991), Saxena (1993), Kushwaha (1995), Katiyar (1998), Parihar (1998), Katiyar and Kushwaha (2000), Katiyar *et al.* (2000). As *Keratinophyton terreum* Human hair, horn, hooves, sheep hair, feather. Randhawa and Sandhu (1964), Randhawa and Sandhu (1965), Kushwaha (1976), Jain and Agrawal (1977), Singh and Agrawal (1983), Deshmukh and Agrawal (1983 b), Kushwaha (1984), Srivastava (1985), Srivastava *et al.* (2001), Govil *et al.* (2001).

21. *Aphanoascus* sp.

Human hair. Nigam and Kushwaha (1989a), Saxena (1993), Ghosh and Bhatt (2000).

22. *Apinisia queenslandica*

Human hair, bird feather. Dixit (1991), Saxena (1993).

23. *Arachniotus flavoluteus*

Unknown. Roy *et al.* (1972).

24. *Arthroderma benhamiae*

Human hair. Kushwaha and Agrawal (1978), Agnihotri and Agrawal (1989), Saxena (1993).

25. *Arthroderma ciferrii*

Human hair. Agnihotri and Agrawal (1989), Saxena (1993), Parihar (1998).

26. *Arthroderma cuniculi*

Cow hair, nail. Agnihotri and Agrawal (1989), Parihar (1998).

27. *Arthroderma curreyi*

Human hair. Dixit and Kushwaha (1990).

28. *Arthroderma flavescens* conidial state *Trichophyton flavescens*

Human hair, horn, nail, bird feather. Nigam (1987), Nigam and Kushwaha (1987), Nigam and Kushwaha (1989a).

29. *Arthroderma gertleri* conidial state *Trichophyton vanbreuseghemii*

Human hair, horn. Nigam (1987), Nigam and Kushwaha (1987), Nigam and Kushwaha (1989a).

30. *Arthroderma quadrifidum*

Human hair. Randhawa and Sandhu (1965), Saxena (1993).

31. *Arthroderma simii*

Human hair. Padhye and Thirumalachar (1967), Deshmukh (2002a,c).

32. *Arthroderma tuberculatum* conidial state *Chrysosporium tuberculatum*

Human hair. Padhye *et al.* (1966), Sur and Ghosh (1980), Nigam (1987).

33. *Arthroderma uncinatum*

Human hair. Saxena (1993).

34. *Arthroderma* sp.

Human hair. Saxena (1993).

35. *Arthrographis kalrai*

Human hair. Nigam and Kushwaha (1990).

36. *Aspergillus candidus*

Feather. Kaul and Sumbali (1999 b), Kaul and Sumbali (2000 a).

37. *Aspergillus deflectus*

Feather. Kaul and Sumbali (1999 b), Kaul and Sumbali (2000 a).

38. *Aspergillus flavus*

Human hair, buffalo hair, bird feather, nail, animal skin. Nigam (1987), Singh *et al.* (1994), Nigam *et al.* (1994), Kaul and Sumbali (1999 a, b, 2000 a, b).

39. *Aspergillus fumigatus*

Human hair, chicken feathers, horn, nail, wool. Kushwaha (1976), Nigam (1987), Singh *et al.*(1990), Dixit (1991), Singh *et al.* (1994), Ghawana *et al* . (1997).

40. *Aspergillus luchuensis*

Leather. Nigam *et al.* (1994).

41. *Aspergillus nidulans*

Human hair, bird feather, horn, nail. Kushwaha (1976), Dixit (1991), Kaul and Sumbali (1999 a), Kaul and Sumbali (2000 b).

42. *Aspergillus niger*

Human hair, bird feather, chicken feathers, horn, wool, nail, animal skin, leather. Kushwaha (1976), Nigam (1987), Dixit (1991), Singh *et al.* (1994), Nigam *et al.* (1994), Ghawana *et al.* (1997).

43. *Aspergillus nigricans*

Buffalo hair. Singh *et al.* (1994).

44. *Aspergillus ochraceous*

Chicken feathers, nail. Singh *et al.* (1994), Kaul and Sumbali (1999 b, 2000 a).

45. *Aspergillus quercinus*

Nail, wool. Kushwaha (1976), Kushwaha (1984).

46. *Aspergillus sachhari*

Horn, nail. Kushwaha (1976), Dixit (1991).

47. *Aspergillus sulfureus*

Horn, nail, bird feather, leather. Kushwaha (1976), Nigam (1987), Dixit (1991), Nigam *et al.* (1994).

48. *Aspergillus sydowii*

Feather. Kaul and Sumbali (1999 b, 2000 a).

49. *Aspergillus terreus*

Human hair. Ramesh and Hilda (1999).

50. *Aspergillus terricola*

Buffalo hair. Singh *et al.* (1994).

51. *Aspergillus ustus*

Human hair, buffalo hair, bird feather horn, wool. Kushwaha (1984), Dixit (1991), Singh *et al.* (1994).

52. *Aspergillus versicolor*

Feather. Kaul and Sumbali (1999 b), Kaul and Sumbali (2000 a).

53. *Aspergillus* sp.

Human hair, chicken feathers, buffalo hair, nail, wool, skin. Nigam (1987), Nigam and Kushwaha (1989 b), Singh *et al.* (1994), Ghawana *et al.* (1997), Mitra *et al.* (1998).

54. *Auxarthron brueneum*

Not known. Jain and Agrawal (1977).

55. *Auxarthron conjugatum*

Human hair, bird feather. Jain (1983), Dixit (1991), Dixit and Kushwaha (1991 b), Deshmukh and Agrawal (2003).

56. *Auxarthron reticulatum*

Bird feather. Dixit and Kushwaha (1991 b).

57. *Auxarthron umbrinum*

Chicken feathers, buffalo hair. Singh *et al.* (1994), Singh *et al.* (1996).

58. *Beauveria bassiana*

Skin. Mitra *et al.* (1998).

59. *Blastomyces dermatidis*

Unknown. Randhawa *et al.* (1983).

60. *Botryotrichum keratinophilum*

Substrate: Human hair, bird feather. Kushwaha (1976), Kushwaha and Agrawal (1976), Singh and Agrawal (1983), Dixit (1991).

61. *Botryotrichum piluliferum*

Human hair, horn, nails. Kushwaha and Agrawal (1975), Kushwaha (1976), Dixit (1991), Parihar (1998).

62. *Botrytis* sp.

Animal skin. Singh *et al.* (1994).

63. *Cephaliophora* sp.

Human hair, nail. Nigam (1987), Nigam and Kushwaha (1989 b).

64. *Cephalosporium acremonium*

Bird feather, horn, nail. Kushwaha (1976), Dixit (1991).

65. *Cephalosporium curtipes*

Human hair. Ramesh and Hilda (1999).

66. *Chaetomium caprinum*

Bird feather. Dixit (1991).

67. *Chaetomium cochlioides*

Bird feather. Dixit (1991).

68. *Chaetomium crispatum*

Bird feather. Dixit (1991).

69. *Chaetomium funicolum*

Bird feather. Dixit (1991).

70. *Chaetomium globosum*

Human hair, bird feather, horn, nail. Kushwaha (1976), Kushwaha (1984), Nigam (1987), Nigam and Kushwaha (1989 b), Dixit (1991), Kaul and Sumbali (1999 a), Kaul and Sumbali (2000 b).

71. *Chaetomium indicum*

Human hair, bird feather. Kushwaha (1976), Dixit (1991).

72. *Chaetomium subterraneum*

Human hair. Ramesh and Hilda (1999).

73. *Chaetomium* sp.

Chicken feathers, bird feather, nail. Dixit (1991), Singh *et al.* (1994).

74. *Choenephora* sp.

Chicken feathers, animal skin. Singh *et al.* (1990), Singh *et al.* (1994).

75. *Chrysosporium* anamorph of *Arthroderma cuniculi*

Human hair, sheep hair, nail, feather. Singh *et al.* (1994), Ramesh and Hilda (1999), Srivastava *et al.* (2001), Govil *et al.* (2001), Deshmukh (2002 b).

76. *Chrysosporium* anamorph of *Arthroderma tuberculatum*

Bird feather, chiken feather. Gupta (1994), Deshmukh and Shukla (2000), Deshmukh (2002c) and as *Myceliophthora* anamorph of *Arthroderma tuberculatum.*

Horn. Nigam and Kushwaha (1985), Nigam (1987), Nigam and Kushwaha (1987), Nigam and Kushwaha (1989 a).

77. *Chrysosporium* anamorph of *Arthroderma curreyi*

Human hair, bird feather, sheep hair. Dixit (1991), Srivastava *et al.* (2001), Govil *et al.* (2001).

78. *Chrysosporium* anamorph of *Neogymnomyces demonbreunii*

Human hair, chicken feathers, buffalo hair, human nail. Singh *et al.* (1990), Saxena (1993), Gupta (1994).

79. *Chrysosporium* anamorph of *Nannizziopsis vriesii*

Human hair, sheep hair, feather. Saxena (1993), Srivastava *et al.* (2001), Govil *et al.* (2001).

80. *Chrysosporium* anamorph of *Pectinotrichum llanense*

Buffalo hair, nail. Singh *et al.* (1990), Singh *et al.* (1994).

81. *Chrysosporium articulatum*

Human hair. Katiyar (1998).

82. *Chrysosporium asperatum*

Unknown. Jain *et al.* (1985).

83. *Chrysosporium carmichaelii*

Human hair, bird feather. Nigam and Kushwaha (1985), Nigam (1987), Nigam and Kushwaha (1987), Nigam and Kushwaha (1989 a), Dixit and Kushwaha (1990), Dixit (1991), Dixit and Kushwaha (1991 a), Bahuguna (1993), Katiyar (1998), Katiyar and Kushwaha (2000), Katiyar *et al.* (2000).

84. *Chrysosporium cuniculi*

Human nails. Parihar (1998).

85. *Chrysosporium farinicola*

Human hair. Nigam (1987), Nigam and Kushwaha (1989 a), Dixit (1991), Bahuguna (1993), Ghosh and Bhatt (2000).

86. *Chrysosporium geophilum*

Human hair. Srivastava (1985), Kushwaha and Srivastava (1989).

87. *Chrysosporium georgii*

Human hair, sheep hair. Saxena (1993), Katiyar (1998), Katiyar and Kushwaha (2000), Katiyar *et al.* (2000), Srivastava *et al.* (2001), Govil *et al.* (2001). As *Trichophyton georgiae* Sheep hair, feather. Srivastava *et al.* (2001), Govil *et al.* (2001).

88. *Chrysosporium indicum*

Human hair, animal hair, cow hair, buffalo hair, bird feather, chicken feathers, nail, horn, hooves, wool, animal skin. Randhawa and Sandhu (1965), Garg (1966 b), Padhye *et al.* (1966, 1967), Sur and Ghosh (1980), Verma *et al.* (1982), Singh and Agrawal (1983), Jain (1983), Deshmukh and Agrawal (1983 a, b), Srivastava (1985), Jain *et al.* (1985), Nigam (1987), Nigam and Kushwaha (1989a), Dixit and Kushwaha (1990), Singh *et al.* (1990), Nigam and Kushwaha (1991), Dixit (1991), Dixit and Kushwaha (1991 a), Bahuguna (1993), Bhadauria

(1993), Saxena (1993), Singh *et al.* (1994), Nigam *et al.* (1994), Ghawana *et al.* (1997), Katiyar (1998), Parihar (1998), Deshmukh (1999), Kaul and Sumbali (1999 a), Parihar and Kushwaha (2000), Katiyar and Kushwaha (2000), Kaul and Sumbali (2000 b), Katiyar *et al.* (2000), Deshmukh and Shukla (2000), Deshmukh *et al.* (2000), Srivastava *et al.* (2001), Govil *et al.* (2001), Deshmukh (2002 b,c), Bhadauria and Kushwaha (2002), Deshmukh and Agrawal (2003).

89. *Chrysosporium inops*

Human hair. Saxena (1993).

90. *Chrysosporium keratinophilum*

Human hair, chicken feather, bird feather, cow hair, horse hair, nail, buffalo hair, animal skin, sheep hair. Randhawa and Sandhu (1965), Garg (1966b), Roy *et al.* (1972), Jain (1983), Deshmukh and Agrawal (1985), Jain *et al.* (1985), Nigam and Kushwaha (1986), Nigam (1987), Nigam and Kushwaha (1989a), Singh *et al.* (1990), Bahuguna (1993), Bhadauria (1993), Saxena (1993), Gupta (1994), Singh *et al.* (1994), Nigam *et al.* (1994), Kushwaha (1995), Kaul and Sumbali (1997), Katiyar (1998), Parihar (1998), Mitra *et al.* (1998), Ramesh and Hilda (1999), Kaul and Sumbali (1999a), Kaul and Sumbali (1999b), Kaul and Sumbali (2000 a), Kaul and Sumbali (2000b), Katiyar and Kushwaha (2000), Ghosh and Bhatt (2000), Katiyar *et al.* (2000), Deshmukh *et al.* (2000), Srivastava *et al.* (2001), Govil *et al.* (2001), Deshmukh (2002a), Deshmukh (2002c), Bhadauria and Kushwaha (2002), Deshmukh and Agrawal (2003).

91. *Chrysosporium lobatum*

Human hair. Bhadauria (1993), Srivastava (1995), Deshmukh (1999), Deshmukh (2002 b, c).

92. *Chrysosporium lucknowense*

Human hair, bird feather. Garg (1966 a), Deshmukh and Agrawal (1983 b), Jain *et al.* (1985), Nigam (1987), Nigam and Kushwaha (1989 a), Dixit (1991), Bahuguna (1993), Bhadauria (1993), Deshmukh and Shukla (2000).

93. *Chrysosporium merdarium*

Human hair, chicken feathers, buffalo hair, nail. Deshmukh *et al.*, (1985), Nigam (1987), Nigam and Kushwaha (1989 a), Singh *et al.* (1990), Parihar (1998), Singh *et al.* (1994).

94. *Chrysosporium mephiticum*

Human hair. Sigler *et al.* (1986), Deshmukh (2002 c).

95. *Chrysosporium pannicola*

Human hair, bird feather, chicken feathers, cow hair, nail. Nigam (1987), Nigam and Kushwaha (1989 a), Nigam and Kushwaha (1991), Bahuguna (1993), Gupta (1994), Katiyar (1998), Parihar (1998), Katiyar and Kushwaha (2000), Katiyar *et al.* (2000), Deshmukh and Shukla (2000), Deshmukh (2002 b,c), Deshmukh and Agrawal (2003). *Chrysosporium pannicola* was reported as *Chrysosporium evolceanui.* Human hair, bird feather, nail. Randhawa and Sandhu (1965), Garg (1966 b), Garg (1966c), Padhye *et al.* (1966), Sur and Ghosh (1980), Deshmukh and Agrawal (1983a), Srivastava (1985), Nigam (1987), Nigam and Kushwaha (1989a), Dixit and Kushwaha (1990), Dixit (1991), Dixit and Kushwaha (1991a), Bahuguna (1993), Bhadauria (1993), Gupta (1994), Nigam *et al.* (1994). It was also reported as *Trichophyton evolceanui* Unknown. Randhawa and Sandhu (1965).

96. *Chrysosporium pannorum*

Human hair, bird feather. Dixit and Kushwaha (1991a), Bhadauria (1993), Kaul and Sumbali (1997), Katiyar (1998), Kaul and Sumbali (1999b), Ramesh and Hilda (1999), Kaul and Sumbali (2000a), Deshmukh *et al.* (2000).

97. *Chrysosporium pseudomerdarium*

Cow hair. Parihar (1998).

98. *Chrysosporium queenslandicum*

Human hair, bird feather, chicken feathers, buffalo hair, animal hair, cow hair, horse hair. Nigam and Kushwaha (1985), Nigam (1987), Nigam and Kushwaha (1987), Nigam and Kushwaha (1989a), Singh *et al.* (1990), Dixit (1991), Dixit and Kushwaha (1991a), Rajak *et al.* (1991), Malviya *et al.* (1992), Bahuguna (1993), Bhadauria (1993), Saxena (1993), Gupta (1994), Kaul and Sumbali (1997), Katiyar (1998), Parihar (1998), Kaul and Sumbali (1999 a), Kaul and Sumbali (1999 b), Kaul and Sumbali (2000a), Kaul and Sumbali (2000 b), Katiyar and Kushwaha (2000), Katiyar *et al.* (2000), Deshmukh and Shukla (2000), Deshmukh (2002c), Bhadauria and Kushwaha (2002), Deshmukh and Agrawal (2003). It was also reported as *Chrysosporium crassitunicatum,* Human hair, horn, Chicken feathers, and animal skin. Kushwaha and Agrawal (1976), Kushwaha and Agrawal (1977 b), Deshmukh and Agrawal (1983 b), Nigam (1987), Nigam and Kushwaha (1989a), Singh *et al.* (1994).

99. *Chrysosporium* state of *Ctenomyces serratus*

Human hair, chicken feather. Garg (1966c), Dixit and Kushwaha (1990), Dixit (1991), Deshmukh *et al.* (2000), Deshmukh (2002a), Deshmukh (2002 b), Deshmukh (2002c).

100. *Chrysosporium sulfureum*

Human hair, chicken feathers, cow hair, horn, nail. Nigam and Kushwaha (1985), Nigam (1987), Nigam and Kushwaha (1987), Nigam and Kushwaha (1989a), Parihar (1998).

101. *Chrysosporium synchronum*

Unknown. Srivastava (1995).

102. *Chrysosporium tropicum*

Human hair, bird feather, buffalo hair, chicken feathers, animal hair, nail, horn, hooves, wool, animal skin. Randhawa and Sandhu (1965), Garg (1966b), Garg (1966c), Padhye *et al.* (1966), Roy *et al.* (1972), Kushwaha and Agrawal (1976), Jain and Agrawal (1977), Sur and Ghosh (1980), Jain (1983), Singh and Agrawal (1983), Deshmukh and Agrawal (1983 a,b), Kushwaha (1982, 1984), Srivastava (1985), Jain *et al.* (1985), Nigam and Kushwaha (1986), Iyer *et al.* (1989), Nigam and Kushwaha (1989 a), Awasthi and Gotewal (1990), Dixit and Kushwaha (1990), Singh *et al.* (1990), Dixit (1991), Dixit and Kushwaha (1991 a), Bahuguna (1993), Bhadauria (1993), Saxena (1993), Gupta (1994), Singh *et al.* (1994), Nigam *et al.* (1994), Saidi *et al.* (1994), Kushwaha (1995), Kaul and Sumbali (1997), Jahan (1997), Ghawana *et al.* (1997), Katiyar (1998), Parihar (1998), Deshmukh (1999), Kaul and Sumbali (1999 a), Kaul and Sumbali (1999 b), Ramesh and Hilda (1999), Kaul and Sumbali (2000 a), Kaul and Sumbali (2000 b), Katiyar and Kushwaha (2000), Katiyar *et al.* (2000), Deshmukh and Shukla (2000), Ghosh and Bhatt (2000), Deshmukh *et al.* (2000), Srivastava *et al.* (2001), Govil *et al.* (2001), Deshmukh (2002 a,b,c), Bhadauria and Kushwaha (2002), Deshmukh and Agrawal (2003).

103. *Chrysosporium tropicum* I

Human hair, horn, nail, chicken feathers, wool. Kushwaha (1976), Kushwaha (1984), Kushwaha (1985), Nigam (1987).

104. *Chrysosporium tropicum* II

Horn. Kushwaha (1976), Kushwaha (1985), Nigam (1987).

105. *Chrysosporium tropicum* III

Horn. Kushwaha (1976), Kushwaha (1995), Nigam (1987).

106. *Chrysosporium tropicum* IV

Human hair. Kushwaha (1976), Kushwaha (1995), Nigam (1987).

107. *Chrysosporium xerophilum*

Human hair, nail. Singh *et al.* (1990), Katiyar (1998), Katiyar and Kushwaha (2000), Katiyar *et al.* (2000), Ghosh and Bhatt (2000), Bhadauria and Kushwaha (2002).

108. *Chrysosporium zonatum*

Human hair, chicken feathers, buffalo hair, horse hair, human nails. Katiyar (1998), Parihar (1998), Deshmukh (1999), Katiyar and Kushwaha (2000), Katiyar *et al.* (2000), Deshmukh and Shukla (2000), Ghosh and Bhatt (2000), Deshmukh (2002 c), Deshmukh and Agrawal (2003). It was also reopted as *Chrysosporium gourii* Human hair. Jain *et al.* (1993).

109. *Chrysosporium* sp. I

Human hair, bird feather, chicken feathers, buffalo hair, animal hair, wool, nail, animal skin. Randhawa and Sandhu (1965), Garg (1966 b), Padhye *et al.* (1966), Roy *et al.* (1972), Verma *et al.* (1982), Jain (1983), Singh and Agrawal (1983), Deshmukh and Agrawal (1983 a), Deshmukh and Agrawal (1983 b), Srivastava (1985), Deshmukh (1985), Jain *et al.* (1985), Nigam (1987), Nigam and Kushwaha (1989 a), Dixit and Kushwaha (1990), Singh *et al.* (1990), Dixit and Kushwaha (1991 a), Bahuguna (1993), Bhadauria (1993), Saxena (1993), Gupta (1994), Singh *et al.* (1994), Nigam *et al.* (1994), Ghawana *et al.* (1997), Deshmukh (1999), Ramesh and Hilda (1999), Deshmukh and Shukla (2000), Deshmukh *et al.* (2000), Deshmukh (2002 c).

110. *Chrysosporium* sp. II

Wool. Roy *et al.* (1972), Deshmukh and Agrawal (1983 a), Deshmukh and Agrawal (1983 b), Ghawana *et al.* (1997), Deshmukh (1999), Deshmukh and Shukla (2000), Deshmukh *et al.* (2000).

111. *Chrysosporium* sp. III

Wool. Deshmukh and Agrawal (1983 b), Ghawana *et al.* (1997), Deshmukh *et al.* (2000).

112. *Circinella muscae*

Feather, buffalo hair. Kaul and Sumbali (1999 b), Kaul and Sumbali (2000 a).

113. *Cladosporium chlorocephalum*

Human hair. Kushwaha (1976).

114. *Cladosporium cladosporioides*

Human hair. Ramesh and Hilda (1999), Ghosh and Bhatt (2000).

115. *Cladosporium herbarum*

Human hair, bird feather. Dixit (1991), Ramesh and Hilda (1999).

116. *Cladosporium trichoides*

Human nail. Singh *et al.* (1990).

117. *Cladosporium werneckii*

Unknown. Pawar *et al.* (1965).

118. *Coccidiodis immitis*

Chicken feathers. Singh *et al.* (1994), Singh *et al.* (1996).

119. *Cryptococcus neoformans*

Unknown. Padhye and Thirumalachar (1967).

120. *Ctenomyces serratus*

Chicken feathers. Randhawa and Sandhu (1965), Garg (1966 b), Padhye *et al.* (1966), Padhye *et al.* (1967), Deshmukh and Agrawal (1983 a), Jain *et al.* (1985), Nigam (1987), Nigam and Kushwaha (1989 a), Kushwaha (1995), Parihar (1998), Deshmukh (1999), Deshmukh and Agrawal (2003). Its anamorphic state was also reported as *Myceliophthora vellerea* Human hair. Saxena (1993), Gupta (1994), Katiyar (1998), Katiyar and Kushwaha (2000), Katiyar *et al.* (2000).

121. *Cunninghamella echinulata*

Human nail, animal skin. Singh *et al.* (1994).

122. *Cunninghamella elegans*

Human hairs, chicken feathers, bird feather, nail. Kushwaha (1976), Singh and Agrawal (1983), Nigam (1987), Nigam and Kushwaha (1989 b), Dixit (1991).

123. *Cunninghamella verticillata*

Human hair. Ramesh and Hilda (1999).

124. *Curvularia geniculata*

Chicken feathers, buffalo hair, horn. Kushwaha (1976), Singh *et al.* (1994).

125. *Curvularia indica*

Human hair, horn. Kushwaha (1976), Kushwaha (1984), Dixit (1991).

126. *Curvularia lunata*

Human hair, bird feather, nail. Nigam (1987), Nigam and Kushwaha (1989 b), Dixit (1991), Kaul and Sumbali (1999 a), Kaul and Sumbali (2000 b).

127. *Curvularia subulata*

Horn, nail. Kushwaha (1976).

128. *Curvularia* sp.

Animal skin. Singh *et al.* (1994), Mitra *et al.* (1998).

129. *Cylindrocarpon radicicola*

Buffalo hair, chicken feathers, and animal skin. Singh *et al.* (1994).

130. *Cylindrocarpon* sp.

Human nail. Singh *et al.* (1990).

131. *Dimargaris* sp.

Human hair, chicken feathers, nail. Nigam (1987), Nigam and Kushwaha (1989 b), Singh *et al.* (1990).

132. *Dreschlera indica*

Human nail. Singh *et al.* (1994).

133. *Dreschlera sativum*

Human hair. Ramesh and Hilda (1999).

134. *Emmonsia parva*

Sheep hair, feather. Srivastava *et al.* (2001), Govil *et al.* (2001).

135. *Epidermophyton floccosum*

Sheep hair, feather. Gokhle *et al.* (1959 b), Garg (1966 b), Klokke and Durairaj (1967), Sundaram and Subramaniyam (1981), Srivastava *et al.* (2001), Govil *et al.* (2001).

136. *Fusariella concinna*

Horn. Kushwaha (1976), Kushwaha (1984).

137. *Fusarium brachygibbosum*

Horn. Kushwaha (1976).

138. *Fusarium chlamydosporum*

Bird feather, horn, nail. Kushwaha (1976), Dixit (1991).

139. *Fusarium lateritium*

Buffalo hair. Singh *et al.* (1994).

140. *Fusarium moniliforme*

Human hair, bird feather, nail. Dixit (1991), Kaul and Sumbali (1999 a), Kaul and Sumbali (1999 b), Ramesh and Hilda (1999), Kaul and Sumbali (2000 a), Kaul and Sumbali (2000 b).

141. *Fusarium nivale*

Chicken feathers, horn. Kushwaha (1976).

142. *Fusarium oxysporum*

Human hair, buffalo hair, bird feather, horn, nail, animal skin. Kushwaha (1976), Dixit (1991), Singh *et al.* (1994), Ramesh and Hilda (1999).

143. *Fusarium poe*

Horn. Kushwaha (1976).

144. *Fusarium roseum*

Human hair, bird feather, nail. Kushwaha (1976), Dixit (1991).

145. *Fusarium semitectum*

Buffalo hair, chicken feathers, nail, animal skin. Singh *et al.* (1994).

146. *Fusarium solani*

Human hair, feather. Ramesh and Hilda (1999), Kaul and Sumbali (1999 b), Kaul and Sumbali (2000 a).

147. *Fusarium sporotrichoides*

Substrate: Horn. Reference: Kushwaha (1976).

148. *Fusarium* sp.

Substrate: Human hair, buffalo hair, chicken feathers, horn, nail, animal skin. Reference: Kushwaha (1976), Nigam (1987), Nigam and Kushwaha (1989 b), Singh *et al.* (1990), Singh *et al.* (1994), Nigam *et al.* (1994).

149. *Geotrichum candidum*

Substrate: Chicken feathers. Reference: Singh *et al.* (1994), Singh *et al.* (1996).

150. *Geotrichum* sp.

Substrate: Human hair, bird feather. Reference: Dixit (1991), Katiyar (1998), Katiyar and Kushwaha (2000), Katiyar *et al.* (2000).

151. *Geomyces pannorum*

Substrate: Human hair. Reference: Dixit (1991), Katiyar and Kushwaha (2000), Katiyar *et al.* (2000), Deshmukh and Shukla (2000), Deshmukh (2002 a).

152. *Gliocladium agrawalli*

Substrate: Human hair. Reference: Kushwaha and Agrawal (1976).

153. *Gliocladium catanulatum*

Substrate: Human hair. Reference: Ramesh and Hilda (1999).

154. *Gliocladium deliquescence*

Substrate: Horn, wool. Reference: Kushwaha (1976).

155. *Gliocladium penicilloides*

Substrate: Wool. Reference: Kushwaha (1976).

156. *Gliocladium roseum*

Substrate: Horn. Reference: Kushwaha (1976), Kushwaha (1984).

157. *Gliocladium virens*

Substrate: Human hair, feather. Reference: Kaul and Sumbali (1997), Kaul and Sumbali (1999 b), Kaul and Sumbali (2000 a).

158. *Graphium penicilloideus*

Substrate: Unknown. Reference: Rajak *et al.* (1991), Malviya *et al.* (1992).

159. *Gymnascella citrina*

Substrate: Unknown. Reference: Roy *et al.* (1972).

160. *Gymnascella dankaliensis*

Substrate: Human hair, animal skin. Reference: Roy *et al.* (1972), Mitra *et al.* (1998), Deshmukh (2002 b).

161. *Gymnascella hylinospora*

Substrate: Human hair. Reference: Katiyar (1998), Katiyar and Kushwaha (2000), Katiyar *et al.* (2000), Deshmukh (2002 b) and reported as *Narasimhella hyalinospora* Substrate: Human hair. Reference: Parihar (1998). As *Pseudoarachniotus hyalinosporus* Substrate: Unknown. Reference: Roy *et al.* (1972) and as *Rollandina hyalinospora* Substrate: Human hair. Reference: Ghosh and Bhatt (2000).

162. *Gymnascella marginospora*

Substrate: Bird feather. Reference: Dixit and Kushwaha (1991 b).

163. *Gymnascella nodulosa as Pseudoarachniotus nodulses*

Substrate: Unknown. Reference: Roy *et al.* (1972).

164. *Gymnoascus demonbreunii*

Substrate: Human hair. Reference: Saxena (1993).

165. *Gymnascoideus petalosporus*

Substrate: Human hair. Reference: Ghosh and Bhatt (2000) As *Gymnoascus petalosporus*

Substrate: Human hair. Reference: Agnihotri and Agrawal (1989), Saxena (1993).

166. *Helminthosporium oryzae*

Substrate: Nail. Reference: Dixit (1991).

167. *Helminthosporium sativum*

Substrate: Nail. Reference: Kushwaha (1976).

168. *Helminthosporium* sp.

Substrate: Animal skin. Reference: Singh *et al.* (1994).

169. *Humicola fusco-atra*

Substrate: Human hair. Reference: Kushwaha (1976).

170. *Keratinomyces ajelloi*

Substrate: Unknown. Reference: Padhye and Thirumulachar (1962), Randhawa and Sandhu (1965), Garg (1966b), Verma *et al.* (1982), Deshmukh and Agrawal (1983, 2003). And as *Trichophyton ajelloi* Substrate: Human hair. Reference: Saxena (1993), Deshmukh (2002 a).

171. *Malbranchea aurantiaca*

Substrate: Human hair. Reference: Jain (1983), Deshmukh and Agrawal (1983 b), Deshmukh (1999), Ghosh and Bhatt (2000), Deshmukh (2002 b).

172. *Malbranchea chrysosporoidea*

Substrate: Chicken feathers, bird feather. Reference: Singh *et al.* (1994), Singh *et al.* (1996), Kaul and Sumbali (1999 a), Kaul and Sumbali (2000 b), Deshmukh (2002 c).

173. *Malbranchea flava*

Substrate: Human hair. Reference: Kaul and Sumbali (1997), Kaul and Sumbali (1999 b), Kaul and Sumbali (2000 a).

174. *Malbranchea flavorosea*

Substrate: Cow skin, animal skin. Reference: Singh *et al. (*1994), Singh *et al.* (1996).

175. *Malbranchea fulva*

Substrate: Human hair, buffalo hair, nail. Reference: Singh *et al.* (1994), Singh *et al.(*1996), Deshmukh (2002 b), Deshmukh (2002 c).

176. *Malbranchea gypsea*

Substrate: Human nails. Reference: Singh *et al. (*1994), Singh *et al.* (1996), Singh (1999), Bhadauria and Kushwaha (2002).

177. *Malbranchea pulchella*

Substrate: Human hair, bird feather, horn, nail. Reference: Kushwaha (1976), Kushwaha and Agrawal (1977 a), Dixit (1991), Dixit and Kushwaha (1991 a), Katiyar (1998), Katiyar and Kushwaha (2000), Katiyar *et al.* (2000), Deshmukh and Shukla (2000).

178. *Malbranchea sulfurea*

Substrate: Reference: Deshmukh and Agrawal (1982).

179. *Malbranchea* sp.

Substrate: Human hair, chicken feathers horse hair, cow hair, animal skin, and nail. Reference: Singh *et al.* (1990), Bhadauria (1993), Saxena (1993), Gupta (1994), Singh *et al.* (1994), Katiyar (1998), Parihar (1998), Katiyar and Kushwaha (2000), Katiyar *et al.* (2000), Deshmukh (2002 a).

180. *Memnoniella echinulata*

Substrate: Human hair. Reference: Dixit (1991).

181. *Microascus manganii*

Substrate: Human hair, bird feather. Reference: Kaul and Sumbali (1997), Kaul and Sumbali (1999 a), Kaul and Sumbali (1999 b), Kaul and Sumbali (2000 a), Kaul and Sumbali (2000 b).

182. *Microsporum audouinii*

Substrate: Sheep hair, feather. Reference: Gokhle *et al.* (1959 a), Gokhle *et al.* (1959 b), Srivastava *et al.* (2001), Govil *et al.* (2001).

183. *Microsporum baullardii*

Substrate: Sheep hair, feather. Reference: Srivastava *et al.* (2001), Govil *et al.* (2001).

184. *Microsporum canis*

Reference: Randhawa and Sandhu (1965), Sundaram and Subramaniyam (1981), Deshmukh *et al.* (2000), Deshmukh and Agrawal (2003).

185. *Microsporum cookei*

Substrate: Unknown. Reference: Randhawa and Sandhu (1965), Iyer *et al.* (1989), Deshmukh *et al.* (2000).

186. *Microsporum equinum*

Substrate: Human hair. Reference: Dixit (1991), Dixit and Kushwaha (1992).

187. *Microsporum fulvum*

Substrate: Human hair, bird feather, horn, hooves, wool, sheep hair. Reference: Kushwaha (1976), Jain (1983), Kushwaha (1984), Srivastava (1985), Nigam (1987), Nigam and Kushwaha (1989 a), Dixit (1991), Dixit and Kushwaha (1991 a, 1992), Bhadauria (1993), Saxena (1993), Ghawana (1997), Ghawana *et al.* (1997), Katiyar (1998), Katiyar and Kushwaha (2000), Katiyar *et al.* (2000), Srivastava *et al.* (2001), Govil *et al.* (2001).

188. *Microsporum gallinae*

Substrate: Chicken feathers. Reference: Deshmukh and Shukla (2000).

189. *Microsporum gypseum*

Substrate: Human hair, buffalo hair, chicken feathers, bird feather, horn, nail, wool, animal skin. Reference: Dey and Kakoti (1955), Randhawa *et al.*(1959), Desai *et al.* (1959), Puri (1961), Padhye (1961), Randhawa and Sandhu (1965), Garg (1966 b), Padhye *et al.* (1966), Padhye *et al.* (1967), Roy *et al.* (1972), Kushwaha (1976), Sur and Ghosh (1980), Jain (1983), Jain *et al.* (1985), Sundaram and Subramaniyam (1981), Deshmukh and Agrawal (1983 a,b), Kushwaha (1984), Nigam (1987), Nigam and Kushwaha (1989 a), Iyer *et al.* (1989), Awasthi and Gotewal (1990), Singh *et al.* (1990), Dixit and Kushwaha (1990), Dixit (1991), Dixit and Kushwaha (1991a), Dixit and Kushwaha (1992), Bahuguna (1993), Bhadauria (1993), Saxena (1993), Gupta (1994), Singh *et al.* (1994), Singh *et al.* (1996), Ghawana (1997), Ghawana *et al.* (1997), Katiyar (1998), Parihar (1998), Mitra *et al.* (1998), Ramesh and Hilda (1999), Katiyar and Kushwaha (2000), Katiyar *et al.* (2000), Ghosh and Bhatt (2000).

190. *Microsporum gypseum complex*

Substrate: Human hair, bird feather, animal skin. Reference: Dixit and Kushwaha (1992), Ranganathan *et al.* (1998), Deshmukh (1999), Deshmukh *et al.* (2000), Deshmukh (2002 a, b, c), Deshmukh and Agrawal (2003).

191. *Microsporum nanum*

Substrate: Human hair. Reference: Gokhle *et al.* (1959 a), Jain *et al.* (1985), Ghawana (1997), Ramesh and Hilda (1999), Deshmukh (2002 a).

192. *Microsporum racemosum*

Substrate: Human hair, chicken feathers. Reference: Singh *et al.* (1990), Dixit (1991), Dixit and Kushwaha (1992).

193. *Microsporum vanbreuseghemii*

Substrate: Human hair, buffalo hair. Reference: Sethi *et al.* (1967), Singh *et al.* (1990), Dixit (1991), Parihar (1998), Deshmukh (2002 a).

194. *Microsporum* sp. I

Substrate: Human hair, chicken feathers, nail. Reference: Kushwaha and Agrawal (1976), Nigam (1987), Nigam and Kushwaha (1989 a), Bhadauria (1993), Saxena (1993), Dixit and Kushwaha (1991 a), Dixit and Kushwaha (1992).

195. *Microsporum* sp. II

Substrate: Human hair. Reference: Dixit and Kushwaha (1992).

196. *Monosporium apiospermum*

Substrate: Unknown. Reference: Pawar *et al.* (1963).

197. *Mortierella* sp.

Substrate: Human hair, nail. Reference: Nigam (1987), Nigam and Kushwaha (1989 b).

198. *Mucor globosus*

Substrate: Human nail. Reference: Singh *et al.* (1994).

199. *Mucor mucedo*

Substrate: Human nail, animal skin. Reference: Singh *et al.* (1994).

200. *Mucor pusillus*

Substrate: Human hair, nail. Reference: Nigam (1987), Nigam and Kushwaha (1989 b).

201. *Mucor* sp.

Substrate: Human hair, bird feather, nail, animal skin. Reference: Dixit (1991), Singh *et al.* (1994).

202. *Mycelia sterilia*

Substrate: Human hair, chicken feathers, horn, nail, wool. Reference: Kushwaha (1976), Singh *et al.* (1990), Dixit (1991), Dixit and Kushwaha (1991 a), Singh *et al.* (1994).

203. *Myceliophthora* anamorph of *Corynascus novoguineensis*

Substrate: Human hair, horn. Reference: Nigam and Kushwaha (1985), Nigam (1987), Nigam and Kushwaha (1987).

204. *Myceliophthora lutea*

Substrate: Human nail, chicken feathers. Reference: Singh *et al.* (1994).

205. *Myceliophthora* sp.

Substrate: Human hair, chicken feathers ,bird feather, human nail. Reference: Dixit and Kushwaha (1990), Singh *et al.* (1990), Dixit (1991), Bhadauria (1993).

206. *Myriodontium keratinophilum*

Substrate: Bird feather. Reference: Deshmukh (2002c, 2004).

207. *Myxotrichum cancellatum*

Substrate: Bird feather. Reference: Dixit and Kushwaha (1991 b).

208. *Myxotrichum* sp.

Substrate: Horn. Reference: Nigam (1987), Nigam and Kushwaha (1989 a).

209. *Nannizzia fulva*

Substrate: Human hair. Reference: Saxena (1993),Dixit (1991).

210. *Nannizzia gypsea*

Substrate: Human hair. Reference: Padhye *et al.* (1967), Srivastava (1985), Dixit (1991), Saxena (1993).

211. *Nannizzia incurvata*

Substrate: Unknown. Reference: Padhye *et al.* (1967).

212. *Nannizzia* sp.

Substrate: Human hair. Reference: Jain *et al.* (1985), Saxena (1993).

213. *Neogymnomyces demonbreunii*

Substrate: Bird feather. Reference: Dixit and Kushwaha (1991 b).

214. *Onygena* sp.

Substrate: Horn. Reference: Nigam (1987), Nigam and Kushwaha (1989 a).

215. *Paecilomyces fusisporus*

Substrate: Human hair, horn. Reference: Kushwaha (1976), Bhadauria and Kushwaha (2002).

216. *Paecilomyces lilacinum*

Substrate: Human hair, bird feather. Reference: Dixit (1991).

217. *Paecilomyces varioti*

Substrate: Human hair, horn. Reference: Kushwaha (1976), Ghosh and Bhatt (2000).

218. *Penicillium aurantiobruneum*

Substrate: Wool. Reference: Kushwaha (1976).

219. *Penicillium aurantiogriseum*

Substrate: Feather. Reference: Kaul and Sumbali (1999 b), Kaul and Sumbali (2000 a).

220. *Penicillium chrysogenum*

Substrate: Human hair, bird feather, nail. Reference: Dixit (1991), Nigam *et al.* (1994), Ramesh and Hilda (1999), Kaul and Sumbali (1999 a), Kaul and Sumbali (2000 b).

221. *Penicillium citrinum*

Substrate: Human hair, bird feather ,nail. Reference: Dixit (1991), Nigam *et al.* (1994), Kaul and Sumbali (1999 b), Ramesh and Hilda (1999), Kaul and Sumbali (2000 a), Ghosh and Bhatt (2000).

222. *Penicillium corylophilum*

Substrate: Human hair. Reference: Ghosh and Bhatt (2000).

223. *Penicillium duclauxi*

Substrate: Horn, nail. Reference: Kushwaha (1976).

224. *Penicillium funiculosum*

Substrate: Human hair. Reference: Ramesh and Hilda (1999)

225. *Penicillium griseofulvum*

Substrate: . Reference: Kaul and Sumbali (1999 b, 2000 a).

226. *Penicillium janthinellum*

Substrate: . Reference: Ghosh and Bhatt (2000).

227. *Penicillium lilacinum*

Substrate: Horn. Reference: Kushwaha (1976, 1984), Ghosh and Bhatt (2000).

228. *Penicillium rubrum*

Substrate: Wool. Reference: Kushwaha (1976).

229. *Penicillium thomii*

Substrate: Human hair. Reference: Ramesh and Hilda (1999).

230. *Penicillium* sp.

Substrate: Human hair, buffalo hair, chicken feathers, animal skin, wool. Reference: Nigam (1987), Nigam and Kushwaha (1989 b), Singh *et al.* (1994), Ghawana *et al.* (1997), Mitra *et al.* (1998).

231. *Pilobolus* sp.

Substrate: Human hair. Reference: Nigam (1987), Nigam and Kushwaha (1989 b).

232. *Rhizopus globosus*

Substrate: Buffalo hair, human nail, animal skin. Singh *et al.* (1994).

233. *Rhizopus nigricans*

Substrate: Animal skin. Reference: Singh *et al.* (1994).

234. *Rhizopus oryzae*

Substrate: Human hair, bird feather, nail. Reference: Dixit (1991).

235. *Rhizopus stolonifer*

Substrate: Human hair, nail. Reference: Nigam (1987), Nigam and Kushwaha (1989 b), Ramesh and Hilda (1999).

236. *Rhizopus* sp.

Substrate: Human nail, animal skin, wool. Reference: Singh *et al.* (1994), Ghawana *et al.* (1997).

237. *Rhizomucor pusillus*

Substrate: Unknown. Reference: Rajak *et al.* (1992)

238. *Rollandina vriesii*

Substrate: Human hair. Reference: Saxena (1993).

239. *Scopulariopsis brevicaulis*

Substrate: Human hair, bird feather, buffalo hair, animal skin, human nail. Reference: Singh *et al.* (1990), Malviya *et al.* (1991), Rajak *et al.* (1991), Malviya *et al.* (1992), Singh *et al.* (1994), Kaul and Sumbali (1997), Kaul and Sumbali (1999 a), Kaul and Sumbali (1999 b), Kaul and Sumbali (2000 a), Kaul and Sumbali (2000 b).

240. *Sepedonium maheshwarianum*

Substrate: Bird feather. Reference: Kaul and Sumbali (1999 a, b).

241. *Shanorella spirotricha*

Substrate: Bird feather. Reference: Dixit and Kushwaha (1991 b).

242. *Syncephalastrum racemosum*

Substrate: Human hair, bird feather, chicken feathers, nail. Reference: Kushwaha (1976), Nigam (1987), Nigam and Kushwaha (1989 b), Dixit (1991), Ramesh and Hilda (1999).

243. *Taralomyces trachyspermus*

Substrate: Unknown. Reference: Hasija *et al.* (1990).

244. *Tieghemiomyces* sp.

Substrate: Human hair. Reference: Nigam (1987), Nigam and Kushwaha (1989 b).

245. *Torula sp.*

Substrate: Human hair. Reference: Nigam (1987), Nigam and Kushwaha (1989 b).

246. *Trichoderma harzianum*

Substrate: Horn. Reference: Rao *et al.* (1975), Kushwaha (1976).

247. *Trichoderma hematum*

Substrate: Human hair, horn. Reference: Kushwaha (1976).

248. *Trichoderma viride*

Substrate: Horn. Reference: Kushwaha (1976).

249. *Trichophyton equinum*

Substrate: Human nail, sheep hair, feather. Reference: Jain (1983), Singh *et al.* (1994), Srivastava *et al.* (2001), Govil *et al.* (2001).

250. *Trichophyton flavescens*

Substrate: Human hair. Reference: Nigam and Kushwaha (1989 a), Dixit (1991), Gupta (1994).

251. *Trichophyton mentagrophytes*

Substrate: Human hair, chicken feathers, bird feather, horn, nail, wool, animal skin. Reference: Behl and Sharma (1957), Gokhle *et al.* (1959a), Gokhle *et al.* (1959b), Desai *et al.* (1959), Puri (1961), Randhawa and Sandhu (1965), Garg (1966b), Klokke and Durairaj (1967), Kushwaha (1976), Kushwaha and Agrawal (1976), Sundaram and Subramaniyam (1981), Deshmukh and Agrawal (1983 a,b), Jain (1983), Kushwaha (1984), Jain *et al.* (1985), Agnihotri and Agrawal (1989), Dixit and Kushwaha (1990), Dixit (1991), Dixit and Kushwaha (1991a), Bhadauria (1993), Saxena (1993), Mitra *et al.* (1998), Ranganathan *et al.* (1998), Ramesh and Hilda (1999), Deshmukh *et al.* (2000), Deshmukh (2002 b), Deshmukh and Agrawal (2002).

252. *Trichophyton rubrum*

Substrate: Human hair, animal skin, sheep hair, feather. Reference: Behl and Sharma (1957), Srivastava and Gupta (1958), Desai *et al.* (1959), Gokhle *et al.* (1959 a), Garg (1966 b), Klokke and Durairaj (1967), Kushwaha (1976), Kushwaha and Agrawal (1976), Sundaram and Subramaniyam (1981), Jain (1983), Deshmukh and Agrawal (1983 b), Kushwaha (1984), Bhadauria (1993), Mitra *et al.* (1998), Ranganathan *et al.* (1998), Srivastava *et al.* (2001), Govil *et al.* (2001).

253. *Trichophyton shoelensis*

Substrate: Unknown. Reference: Behl and Sharma (1957), Desai *et al.* (1959), Sundaram and Subramaniyam (1981).

254. *Trichophyton simii*

Substrate: Human hair, human nails, wool, animal skin, feather. Reference: Klokke and Durairaj (1967), Padhye and Thirumalachar (1967), Srivastava (1985), Dixit (1991), Singh *et al.* (1996), Ghawana (1997), Ghawana *et al.* (1997), Mitra *et al.* (1998), Ranganathan *et al.* (1998), Deshmukh and Shukla (2000), Govil *et al.* (2001).

255. *Trichophyton terrestre*

Substrate: Human hair, cow skin, animal skin, sheep hair. Reference: Padhye (1968), Singh and Agrawal (1983), Dixit (1991),

Saxena (1993), Singh *et al.* (1994), Singh *et al.* (1996), Mitra *et al.* (1998), Ramesh and Hilda (1999), Deshmukh and Shukla (2000), Srivastava *et al.* (2001), Govil *et al.* (2001), Deshmukh (2002 a, b, c), Deshmukh and Agrawal (2003).

256. *Trichophyton tonsurans*

Substrate: Unknown. Reference: Desai *et al.* (1959), Gokhle *et al.* (1959 a), Randhawa and Sandhu (1965), Garg (1966b), Klokke and Durairaj (1967), Sundaram and Subramaniyam (1981), Verma *et al.* (1982).

257. *Trichophyton vanbreuseghemii*

Substrate: Human hair, Human nail. Reference: Nigam and Kushwaha (1989 a), Dixit (1991), Bahuguna (1993), Singh *et al.* (1994).

258. *Trichophyton verrucosum*

Substrate: Unknown. Reference: Desai *et al.* (1959).

259. *Trichophyton violaceum*

Substrate: Unknown . Reference: Behl and Sharma (1957), Desai *et al.* (1959), Klokke and Durairaj (1967), Sundaram and Subramaniyam (1981).

260. *Trichophyton viride*

Substrate: Human hair. Reference: Ramesh and Hilda (1999).

261. *Trichophyton* sp.

Substrate: Human hair, bird feather, nail. Reference: Singh and Agrawal (1983), Awasthi and Gotewal (1990), Dixit (1991), Bhadauria (1993).

262. *Trichothecium roseum*

Substrate: Buffalo hair. Reference: Singh *et al.* (1994), Ghosh and Bhatt (2000).

263. *Uncinocarpus orissi*

Substrate: Human hair. Reference: Ghosh and Bhatt (2000).

264. *Uncinocarpus reesii*

Substrate: Bird feather. Reference: Dixit and Kushwaha (1991b), Deshmukh (2002 a), Ghosh and Bhatt (2000).

265. *Verticillium lecanii*

Substrate: Wool. Reference: Kushwaha (1976), Kushwaha (1984).

266. *Verticillium luteoalbum*

Substrate: Human hair. Reference: Ramesh and Hilda (1999).

267. *Verticillium saksenii*

Substrate: Human hair, nail and feather. Reference: Kushwaha (1980).

268. *Veticillium tenuipes*

Substrate: Human hair, cow hair. Reference: Katiyar (1998), Parihar (1998), Katiyar and Kushwaha (2000), Katiyar *et al.* (2000).

269. *Verticillium* sp.

Substrate: Chicken feathers. Reference: Kushwaha (1976).

270. *Xynophila mephitalis*

Substrate: Bird feather. Reference: Dixit and Kushwaha (1991 b).

REFERENCES

Agnihotri, S.K. and Agrawal S.C. (1989). Keratinophilic fungi and related dermatphytes from the soils of Madhya Pradesh (India). *Proc. Nat. Acad. Sci India* 59(B): 193-196.

Awasthi, P.B. and Gotewal, S. (1990). Prevalence of keratinophilic fungi in certain soils of Rohilkhand division in relation to edaphic factors conditions. *Proceedings of the National Academy of Sciences*. 76:37.

Bahuguna, S. (1993). Hair perforation leading to its degradation by keratinophilic fungi. Ph. D. Thesis, Kanpur University, India.

Behl, P.N. and Sharma, M.D. (1957). Incidence of mycotic infections in Delhi. *Indian Journal of Dermatology*. 3:5-7.

Bhadauria, S. S. (1993). Studies on antagonism between keratinophilic fungi and dermatophytes. Ph.D. Thesis, Kanpur University, India.

Bhadauria, D. and Kushwaha, R.K.S. (2002). Keratinophilic fungi from soils of hills and their keratinolytic ability. In: *Frontiers of fungal diversity in India* (Prof. Kamal Festscrift). (Rao, G.P., Manoharachary, C., Bhat, D.J., Rajak, R.C. and Lakhanpal, T.N. Eds.) International Book Distribution Company, Lucknow, India. pp 252-270.

Desai, S., Marquis, L. and Bhatt, M.L.A. (1959). Mycological study on common superficial mycoses. *Bulletin Calcutta School of Tropical Medicine*. 7:37.

Deshmukh, S. K. (1985). Isolation of dermatophytes and other keratinophilic fungi from soil of Mussoorie (India). *Mykosen*. 28: 98-101.

Deshmukh, S. K. (1999). Keratinophilic fungi isolated from soils of Mumbai, India. *Mycopathologia*. 146:115-116.

Deshmukh, S. K. (2002 a). Incidence of dermatophytes and other keratinophilic fungi in the glacier bank soils of the Kashmir valley (India). *Mycologist*. 16: 165-167.

Deshmukh, S. K. (2002 b). Incidence of keratinophilic fungi from selected soils of Kerala state (India). *Mycopathologia*. 156: 177-181.

Deshmukh, S. K. (2002 c). Isolation of dermatophytes and other keratinophilic fungi from Karnala bird sanctuary, Maharashtra (India). *Journal of Basic Applied Mycology*. 2: 194-196.

Deshmukh, S.K. (2004). Keratinophilic Fungi From feathers of pigeon (*Columbia livea*) in Thane district of Maharashtra India. *Mycoses*. (In Press)

Deshmukh, S.K. and Agrawal, S.C. (1982). *Malbranchea sulfurea* (Miehe) Sigler and Carmichael – A new record from India. *Current Science*. 51:785-786.

Deshmukh, S.K. and Agrawal, S.C. (1983 a). Isolation of keratinophilic fungi from coastal habitats of Goa (India). *Kavaka.* 11:53-54.

Deshmukh, S.K. and Agrawal, S.C. (1983 b). Prevalence of dermatophytes and other keratinophilic fungi in soil of M.P. (India). *Mykosen.* 26:574-577.

Deshmukh, S.K. and Agarwal S.C. (1985) Degradation of human hair by some dermatophytes and other keratinophilic fungi. *Mykosen.* 28:463-466.

Deshmukh, S.K. and Agrawal, S.C. (2003). Isolation of dermatophytes and other keratinophilic fungi from soils of Jammu. *Mycoses.* 46: 226-228.

Deshmukh, S.K., Agrawal, S.C. and Jain, P.C. (2000). Isolation of dermatophytes and other keratinophilic fungi from soil of Mysore (India). *Mycoses.* 43:55-57.

Deshmukh, S.K. Jain, P.C. and Agrawal, S.C. (1985). *Chrysosporium merdarium* (Link) Carmichael. A new record from India. *Acta Botanica Indica.* 13:107-108.

Deshmukh, S.K. and Shukla R. V. (2000). Isolation of keratinophilic fungi from poultry farm soils of Chhattisgarh (India). *Kavaka.* 28: 55-58.

Dey, N. C. and Kakoti, L. M. (1955). *Microsporum gypseum* in India. *Journal of Indian Medical Assocciation.* 25:160-164.

Dixit, A.K. and Kushwaha, R.K.S. (1990). Keratinophilic fungi of Andman Islands, India. *Indian Journal of Microbiology.* 30:349-350.

Dixit, A.K. (1991). Occurrence and antagonistic studies of keratinophilic fungi with special reference to *Microsporum* species. Ph.D. Thesis, Kanpur University, India.

Dixit, A.K. and Kushwaha, R.K.S. (1991a). Occurrence of keratinophilic fungi on Indian birds. *Folia Microbiologia.* 36: 383-386.

Dixit, A.K. and Kushwaha, R.K.S. (1991b). Some new ascomycetous keratinophilic fungi on Indian birds. *Geobios New Reports.* 10:58-60.

Dixit, A.K. and Kushwaha, R.K.S. (1992). Hyphal growth and sporulation of six species of *Microsporum* Gruby. *Geobios.* 19: 259-262.

Garg, A.K. (1966a). An addition to genus *Chrysosporium* Corda. *Mycopathologia.* 30: 221-224.

Garg, A.K. (1966 b). Isolation of Dermatophytes and other keratinophilic fungi from soils in India. *Sabouraudia.* 4:259-264.

Garg, A.K. (1966c). Occurrence of keratinophilic *Chrysosporium* Corda in Indian soils. *Mycopathologia Mycologia et Applicata.* 29: 189-192.

Ghawana, V.K. (1997). Studies on decomposition of wool: using keratin decomposing fungi and its possible role in soil fertility. Ph.D. Thesis, DEI, Dayalbagh, Agra, India.

Ghawana, V.K., Srivastava, J.N. and Kushwaha, R.K.S. (1997). Some observations on fungal succession during decomposition of wool in soil. *Mycoscience.* 38:79-81.

Ghosh, G.R. and Bhatt, S. (2000). Keratinophilic fungi from Chilka Lake-side soil Orissa (India). *Indian Journal of Microbiology.* 40: 247-254.

Gokhle, B.B., Padhye, A.A. and Thirumalachar, M.J. (1959a). Observation on mycoses in patients examined at the Sasoon Hospital, Poona. *Bulletin Calcutta School of Tropical Medicine.* 7:41.

Gokhle, B.B., Padhye, A.A. and Thirumalachar, M.J. (1959b). In: Fungal diseases in India (Proc. Symp. Held at school of Tropical Medicine, Calcutta) (Pub. 1962) : 100-107.

Govil, N. (1999). Survey on keratinophiles from the belt of hilly (Shimla, H.P) and plain (Agra, U.P.) areas. M.Sc. Dissertation, DEI, Dayalbagh, Agra, India.

Govil, N., Bhatnagar, V.P., Kumar, A., Mathur, M. and Srivastava, J.N. (2001). Biodiversity of keratinophyles in Indian hills (Shimla, H.P.) and plains (Agra, U.P.). *Journal of Indian Botanical Society.* 80: 183-186.

Gupta, M. (1994). Decomposition of hair and feathers by keratinophilic fungi. Ph. D. Thesis Kanpur University, India.

Hasija, S.K., Malviya, H. and Rajak, R.C. (1990). Keratinolytic ability of some fungi isolated from gelatin factory campus Jabalpur (M.P.) *Proceedings of the National Academy of Sciences.* 601: 305-309.

Iyer, R., Sharma, S., Williamson, M.D. and Joshi S. (1989). Studies on the occurrence of keratinophilic fungi in soils of Jaipur. *Indian Phytopathology.* 42:199-201.

Jahan, F. (1998). Keratin decomposition by *Chrysosporium tropicum.* Ph.D. Thesis, Kanpur University, India.

Jain, M., Shukla, P.K. and Srivastava, O.P. (1985). Keratinophilic fungi and dermatophytes in Lucknow soils and their global distribution. *Mykosen.* 28:98-101.

Jain, P.C. and Agrawal, S.C. (1977). Keratinophilic fungi from the soils of Mount Abu (India). *Geobios.* 4:136-138.

Jain, P.C. (1983). Prevalence of keratinophilic fungi in soils of Madhya Pradesh and Uttar Pradesh, India.*Geobios New Reports.* 2:10-13.

Jain, P.C, Deshmukh, S.K. and Agrawal, S.C.(1993). *Chrysosporium gourii,* Jain, Deshmukh and Agrawal, Sp. Nov. *Mycoses.* 36:77-79.

Katiyar, S. (1998). The keratinomycetes from bottom sediments and their hair invasion ability. Ph.D. Thesis; Kanpur University, India.

Katiyar, S. and Kushwaha, R.K.S. (2000). Human hair colonizing fungi in water sediments of India. *Mycopathologia.* 152: 81-84.

Katiyar, S., Parihar, P. and Kushwaha, R.K.S. (2000). Keratinomycetes from bottom sediments of high altitude. *Mycopathologia.* 166-167

Kaul, S. and Sumbali, G. (1997). Keratinolysis by poultry farm soil fungi. *Mycopathologia.* 139:137-140.

Kaul, S. and Sumbali, G. (1999 a). Production of extra cellular keratinases by keratinophilic fungal species inhabiting feathers of living poultry birds (*Gallus domesticus*) : A comparison. *Mycopathologia.* 146:19-24.

Kaul, S. and Sumbali, G. (1999b). Impact of some ecological factors on the occurrence of poultry soil-inhabiting keratinophiles. *Mycopathologia.* 143:155-159.

Kaul, S. and Sumbali, G. (2000a). Keratinophilic fungi from poultry farm soil of Jammu, India. *Mycologists*. 14:289-291.

Kaul, S. and G. Sumbali, (2000b). Keratinophilic fungi from feathers of Indian poultry birds. *Mycologists*. 14: 13-15.

Klokke, M.H. and Durairaj, P. (1967). The causal agent of superficial mycoses isolated in rural areas of South India. *Sabouraudia*. 5:153-158.

Kushwaha, R.K.S. (1976). Studies on keratinophilic fungi from soil. Ph.D. Thesis, University of Saugar, Sagar (India).

Kushwaha, R.K.S. (1980). A new species of *Verticillium*. *Current Science*. 49: 948-949.

Kushwaha, R.K.S. (1982). Four atypical strain of *Chrysosporium tropicum*. Mycological Society of America News Letter. 33:37.

Kushwaha, R.K.S. (1984). Morphological studies of some dermatophytes and keratinophilic fungi. *Bulletin Botanical Society of University of Saugar*.30 and 31: 48-51.

Kushwaha, R.K.S. (1985). Unusual *Chrysosporium tropicum* in Indian Soil. *Indian Journal of Plant Sciences*. 3: 41-42.

Kushwaha, R. K.S. (1995). Biodeterioration of feathers by keratinomycetes isolated from a museum of spain. In : *Biodeterioration of Cultural Property 3*. Eds: C. Aranyanak and C. Singhasiri. Bangkok, Thailand. 289-302.

Kushwaha, R.K.S. (2000). The genus *Chrysosporium*, its physiology and biotechnological potential. In : *Biology of dermatophytes and other keratinophilic fungi*. Eds. R.K.S Kushwaha and J. Guarro. *Revista Iberoamericana de Micologia, Spain*. 66-76.

Kushwaha, R.K.S. and Agrawal, S.C. (1975). A new keratinophilic species of *Botryotrichum*. *Transactions of Japan Mycological Society*.16: 454-458.

Kushwaha, R.K.S and Agrawal, S.C. (1976). Some keratinophilic fungi and related dermatophytes from soil. *Proceedings of the Indian National Science Academy*. 42 :102-110.

Kushwaha, R.KS. and Agrawal, S.C. (1977a). Isolation of *Malbranchea pulchella* from poultry farm soil in India. *Mykosen*. 20: 163-164.

Kushwaha, R.K.S. and Agrawal, S.C. (1977b). *Chrysosporium crassitunicatum* Sp. Nov., A new keratinophilic fungus. *Transactions of the British Mycological Society*. 68:464-467.

Kushwaha, R.K.S. and Agrawal, S.C. (1978). Occurrence of *Arthroderma benhamiae* Ajello and Cheng in Indian Soils. *Indian Journal of Microbiology*. 18:125.

Kushwaha, R.K.S and Srivastava, J.N. (1989). A new species of *Chrysosporium*. *Current Science*. 58: 970-971.

Kushwaha, R.K.S. and Tripathi, N. (2004). Human pathogenic fungi : A review. In : *Animal and Human Pathogenic Fungi*. Ed : R.K.S. Kushwaha. Scientific Publishers (India). 185-250.

Mahapatra, L.N. and Gugnani, H.C. (1964). Studies on strains of *Microsporum gypseum* isolated from soil. *Mycopathologia et Mycologia Applicata*. 22: 175-181.

Malviya, H.K., Rajak, R.C. and Hasija, H.K. (1991). Keratin degradation by *Scopulariopsis brevicaulis. Journal of Indian Botanical Society.* 70: 213-216.

Malviya, H.K., Rajak, R.C. and Hasija, H.K. (1992). Synthesis and regulation of extracellular keratinase in three fungi, isolated from the grounds of a gelatin factory, Jabalpur, India. *Mycopathologia.* 120:1-4.

Mitra, S.K., Sikdar, A., and Das, P. (1998). Dermatophytes isolated from selected ruminants in India. *Mycopathologia.* 142:13-16.

Nigam, N. and Kushwaha, R.K.S. (1985). Eight new keratinophilic fungal records from India. *Proceedings of National Academy of Sciences India.* 55th Ann. Sess. 26.

Nigam, N. and Kushwaha, R.K.S. (1986). Thermotolerance of keratinized spores of *Chrysosporium tropicum* and *Chrysosporium keratinophilum. Indian Journal of Plant Sciences.* 4:5-8.

Nigam, N. (1987). Studies on *Chrysosporium* and allied fungi from soil. Ph.D. Thesis, Kanpur University, Kanpur, India.

Nigam, N. and Kushwaha, R.K.S. (1987). Seven new keratinophilic fungal records from India. *Kavaka.* 15:29-31

Nigam, N. and Kushwaha, R.K.S. (1989a). Some new reports on keratinophilic fungi. *Current Science.* 58: 1374.

Nigam, N. and Kushwaha, R.K.S. (1989b). Occurrence of non-keratinophilic fungi on keratin. *Transactions of Japan Mycological Society.* 30: 1-8.

Nigam, N. and Kushwaha, R.K.S. (1990). Occurrence of *Arthrographis kalrai* on human hair. *National Academy of Sciences Letters.* 13: 63.

Nigam, N. and Kushwaha, R.K.S. (1991). Scanning electron microscopy of *Chrysosporium pannicola. Cryptogamic Botany.* 2/3: 153-154.

Nigam, N. and Kushwaha, R.K.S. (1993). Ecology of soil inhabiting keratinophilic fungi with special reference to *Chrysosporium.* In : *Fungal Ecology and Biotechnology,* Eds: B. Rai, D.K. Arora, N.K. Dubey and P.D. Sharma, Rastogi Publication. 173-182.

Nigam, N., Dhawan, S. and Nair, M.V. (1994). Deterioration of feather and leather objects of some Indian museum by keratinophilic and non keratinophilic fungi. *International Biodeterioration and Biodegradation.* 33:145-152.

Padhye, A.A. (1961). *Microsporum gypseum* from soil in Poona. *Prof. Agharkar Commemoration Volume.* 117-119.

Padhye, A.A. (1968). *Trichophyton terrestre* from poultry farm soil in India. *Mykosen.* 11: 725-726.

Padhye, A.A. and Thirumalachar, M.J. (1962). Isolation of *Keratinomyces ajelloi* from soil of Bombay in India. *Current Science.*31:100-101.

Padhye, A.A. and Thirumalachar, M.J. (1967). Isolation of *Trichophyton simii* and *Cryptococcus neoformans* from soil in India. *Hindustan Antibiotic Bulletin.* 10: 200-202.

Padhye, A.A. and Thirumalachar, M.J. (1968). Distribution of *Allescheria boydii* Shear in soil of Maharashtra state. *Hindustan Antibiotic Bulletin.* 10: 200-201.

Padhye, A.A., Mishra, S.P. and Thirumalachar, M.J. (1966). Occcurrence of soil inhabiting dermatophytes and other keratinophilic fungi from soil in Poona. *Hindustan Antibiotic Bulletin.* 9: 90-93.

Padhye, A.A, Pawar, V.H., Sukapure, R.S. and Thirumalachar, M.J. (1967). Keratinophilic fungi from marine soils of Bombay, India Part I. *Hindustan Antibiotic Bulletin.* 10:138-141.

Parihar, P. (1998). Exploitation of non pathogenic keratinophilic fungi for their potential to degrade keratin. Ph.D. Thesis, Kanpur University, Kanpur, India.

Parihar, P. and Kushwaha, R.K.S. (2000). Rapid degradation of human hair by *Chrysosporium indicum* isolated from soil of Thailand. In : *Microbes : Agriculture, Industry and Environment.* Eds: D.K.Maheshwari, R.C. Dubey, G. Prasad and Navneet. 177-181.

Pawar, V.H., Padhye, A.A. and Thirumalachar, M.J. (1963). Isolation of *Monosporium apiospermum* from marine soils in Bombay. *Hindustan Antibiotic Bulletin.* 6:50-53.

Pawar, V.H., Raholkar, P.W. and Thirumalachar, M.J. (1965). *Cladosporium werneckii* isolated from marine habitat. *Hindustan Antibiotic Bulletin.* 8:19-20.

Puri, H.R. (1961). Isolation of dermatophytes from India. *Bulletin of Calcutta School Tropical Medicine* 8:121-122.

Rajak, R.C., Parwekar, S., Malviya, H.K. and Hasija, S.K. (1991). Keratin degradation by fungi isolated from the grounds of a gelatin factory campus in Jabalpur, India. *Mycopathologia.* 114: 83-87.

Rajak, R.C., Malviya, H.K., Deshpande, H. and Hasija, S.K. (1992). Keratinolysis by *Absidia cylidrospora* and *Rhizomucor pusillus* : a biochemical proof. *Mycopathologia.* 118: 109-114.

Ramesh, V.M. and Hilda, A. (1999). Incidence of keratinophilic fungi in the soil of primary school and public parks of Madras city, India. *Mycopathologia.* 143:139-145.

Randhawa, H.S. and Sandhu, R.S. (1963). On the occurrence of two new species of *Trichophyton* in Indian soil. *Mycopathologia.* 20:225-228.

Randhawa, H.S. and Sandhu, R.S. (1964). *Keratinophyton terreum* gen. Sp. Nov. a keratinophilic fungus from soil in India. *Sabouraudia.* 3:251-256.

Randhawa, H.S. and Sandhu, R.S. (1965). A survey of soil inhabiting dermatophytes and related keratinophilic fungi of India. *Sabouraudia.* 4: 71-79.

Randhawa, H.S., Nath, A. and Vishwanathan, R. (1959). Isolation of *Microsporum gypseum* (Bodin) Guiart and Grigorakis from soil in Delhi. *Science and Culture.* 25:326-327.

Randhawa, H.S., Khan, Z.U. and Gaur, S.N. (1983). *Blastomyces dermatitidis* in India: First report of its isolation from clinical material. *Sabouraudia.* 21: 215-221.

Ranganthan, S., Balajee, S.A.M., and Raja, S.M. (1998). A survey of dermatophytosis in animals in Madras, India. *Mycopathologia.* 140:137-140.

Rao, D.P.C, Kushwaha, R.K.S. and Agarwal, S.C. (1975). *Trichoderma harzianum* (Rifai) a new record from India. *Science and Culture.* 41: 604.

Roy, K., Ghosh, G.R. and Dutta, S.K. (1972). Keratinophilic fungi and the prevalence of dermatomycosis in Orissa, India. *Sabouraudia.* 10:218-229.

Saidi, A., Bhatt, S., Richard, J.L., Sikdar, A. and Ghosh, G.R. (1994). *Chrysosporium tropicum* as a probable cause of mycoses of poultry in India. *Mycopathologia.* 125:143-147.

Saksena, V. (1993). Studies on anamorphs and teleomorphs of *Arthroderma* and related fungi. Ph. D. Thesis, Kanpur University, Kanpur, India.

Sarkar, A. (1962a). Notes on two species of *Trichophyton* from Indian soils. *Bulletin of Calcutta School Tropical Medicine Hygiene.* 10: 75-76.

Sarkar, A. (1962b) Perfect state of fructification of *Microsporum gypseum* in unsterile soil. *Bulletin of Calcutta School Tropical Medicine Hygiene.* 10: 76-77.

Sethi, K.K., Randhawa, H.S., Kurup, P.V. and Ajello, L. (1967). Isolation of *Microsoporum vanbreuseghemii* from soil in India. *Sabouraudia.* 6:81-82.

Sigler, L., Guarro, J. and Punsola, L. (1986). New keratinophilic species of *Chrysosporium. Canadian Journal of Botany.* 64: 1212-1215.

Singh, B.S., Singh, D.V., Singh, B.G. and Sinha, R.J. (1990). Incidence of keratinophilic fungi and related dermatophytes in Agra soils. *Advances in Plant Sciences.* 3:8-15.

Singh, C.J. (1999). Exocellular proteases of *Malbranchea gypsea* and their role in keratin deterioration. *Mycopathologia.* 143:147-150.

Singh, C. J., Singh, B.G. and Singh, B.S. (1994). Keratinophilic fungi of Ghana Birds Sanctury, Bharatpur (Rajasthan). *Advances in Plant Sciences.* 7: 280-291.

Singh, C. J., Singh B.G. and Singh B.S. (1996). Morphological, keratin colonizing and degrading characters of some keratinophilic fungi and related dermatophytes isolated from Keoladeo National Park, Bharatpur. *Mycotaxon.* 59: 467-473

Singh, K.V. and Agrawal, S.C. (1983). Keratinophilic fungi from some substrates in central India. *Geobios.* 10:165-171.

Srivastava, O.P. and Gupta, R.N. (1958). Survey of human skin fungi in Lucknow and effect of diaryl sulphones, sulphoxides and sulphides on *Trichophyton rubrum. Journal of Scientific and Indian Research.* 17:87-90.

Srivastava, J.N. (1985). Studies on soil inhabiting dermatophytes and keratinophilic fungi. Ph.D. Thesis, Kanpur University, Kanpur, India.

Srivastava, A. (1995). Ecophysiological studies on the soil borne dermatophytes and keratinophilic fungi. Ph. D. Thesis DEI, Dayalbagh, Agra, India.

Srivastava, J.N., Kumar, A., Kumar M. and Govil, N. (2001). Perforator behaviour of keratinophiles from soil of Agra and Shimla. *Journal of Mycology and Plant Pathology.* 31: 318-321.

Sundram B.M. and Subramaniam (1981). Studies on human pathogenic fungi and keratinophilic fungi from soils. *Indian Journal of Microbiology.* 21.

Sur, B. and Ghosh, G.R. (1980). Keratinophilic fungi from Orissa, I. Isolation from soil. *Sabouraudia.*18: 269-274.

Verma, T.N., Sinha, B.K. and Das, U.L. (1982). Isolation of keratinophilic fungi from soil in Bihar (India). *Mykosen.* 25: 449-452.

CHAPTER – 3

MARINE FUNGI FROM MAHARASHTRA : DIVERSITY AND TAXONOMY

B.D. Borse, K.N. Borse and N.S. Pawar

Department of Botany, Arts, commerce and Science College,
Shindkheda, 425 406. Maharashtra

INTRODUCTION

Marine mycology was established as a discrete science in the late 1930's following the work of mycologists like Sparrow (1937) and Barghoorn and Linder (1944). Although the first obligate marine fungus *Halotthia posidoniae* (Durieu and Montagne) Kohlmeyer, was recorded on the rhizomes of the sea grass *Posidonia oceanica* (Durieu and Montagne, 1869). Sparrow (1937) reported the occurrence of a novel group of zoosporic fungi, the thraustochytrids in marine algae and paved the way to studies of marine fungi. Barghoorn and Linder (1944) discovered variety of unique ascomycetes and deuteromycetes that inhabit wood submerged in the sea. These two articles gave a tremendous impetus to the studies of marine fungi all over the world.

Kohlmeyer (1974) defines those fungi as 'marine', which grow and reproduce in marine habitat; he further divided them into obligate and facultative marine fungi. The obligate marine fungi are those that grow and sporulate exclusively in marine habitats; the facultative marine fungi are those from fresh water or terrestrial habitats that are able to grow and possibly also to sporulate in the marine environment as well.

These fungi occur predominantly in intertidal zone of ocean and estuarine habitats, although a few of them are found in the deep sea, off shore and in brackish waters. They often appear on drift wood deposited on beaches, intertidal wood caught between natural rocks or stones functioning for coastal protection, stationary wood such as harbor timber, decaying parts of mangroves, salt marsh plants, sea grasses, algae, animals and calcareous remains of animals. Some are pathogens of algae and animals. Propagules of some species are found in the sea foam and beach sand samples. They appear to be cosmopolitan in their distribution with only a few species having been shown to be endemic to any particular region. Some tropical species however tend to be associated only with mangroves or other phanerogams which are restricted to tropical waters. Their distribution is mainly limited by dissolved oxygen and temperature, less so by salinities of the marine waters.

Several workers have attempted ecologically oriented studies of marine fungi. Their role in the degradation of wood has been extensively investigated. More than 150 species have been reported growing on wood and other cellulosic substrates submerged in the sea. The marine fungi colonizing mangrove substrates are somewhat different from those colonizing submerged parts of mangroves. Jones and Alias (1997) listed about 165 species of marine fungi on mangroves. The mangrove marine fungi now constitute the second largest group after wood inhabiting marine species.

It is very important to learn how the marine fungi are collected from their habitats. Samples of drift wood deposited on sandy beaches, intertidal wood caught between rocks, intertidal stationary harbor timber, submerged dead and decaying parts of mangroves and salt marsh plants (stems, exposed roots, proproots, pneumatophores, leaves, seedlings) are collected in polythene bags during low tide. One half of the samples are examined directly and the remaining samples are incubated at room temperature in the plastic boxes containing a layer of wet blotting papers or sterile sand moistened with sea water and examined at periodic intervals. Sterile seawater is periodically sprayed to prevent drying out of samples. Foam samples are collected from sandy beaches for arenicolous species. Accumulated foam or scum is scooped up in a clean plastic vial or wide mouth bottle and fixed in 5% FAA on the spot to avoid germination of propagules.

About 72,065 species of fungi are known from all over the world (Hawksworth *et al.* 1995), but in contrast to this less than 750 species of marine fungi have been described from marine habitats. These comprise about 100 Chytridiomycota, Labyrinthulomycota and Oomycota and, 360 Ascomycota, 10 Basidiomycota, 74 Mitosporic fungi, 180 yeasts, 10 Lichens (Kohlmeyer and Kohlmeyer, 1979, Hyde and Sharma 2000a). The relatively low number of marine fungi may be due to the fact that large regions of the oceans have not been thoroughly investigated for their fungal diversity.

Almost three-quarters of all recognized marine fungi have been described in the last six decades. There have been a number of taxonomic and ecological reappraisals by Johnson and Sparrow (1961), Jones (1976), Kohlmeyer and Kohlmeyer (1979), Moss (1986) and Hyde and Pointing (2000). Kohlmeyer and Volkmann-Kohlmeyer (1991) and

Hyde and Sarma (2000b) have given excellent keys to the identification of the higher marine fungi.

Lots of work has been done on the diversity of marine fungi of various regions of the world. Jones (1968) initiated the work on marine fungi from Maharashtra coast (the Arabian Sea). Further, these fungi were studied by Late Professor S.D. Patil and his coworkers from University of Poona (Patil and Borse 1982, 1983 a-b, 1985, 1986; Borse 1984, 1985, 1987 a-e, 1988, 2000 a-b, 2002), Srivastava (1989, 1994), Borse *et al.* (1988), Borse and Srivastava (1988, 1994), Ramesh and Borse (1989), Borse and Hyde (1989), Hyde and Borse (1986 a-b). Kohlmeyer and Vittal (1986) reported few fungi from the coast of Maharashtra.

Fungi from other coastal states and Union territories of India were studied by Becker and Kohlmeyer (1958), Kohlmeyer (1959), Kohlmeyer *et al.* (1967), Jones (1968), Nair (1970), Raghukumar (1973, 1987, 1988a-c 1996), Raghukumar *et al.* (1988,1990), Raghukumar and Ch. Raghukumar (1988), Raghukumar and Balasubramanian (1991), Sathe and Raghukumar (1991), Ch Raghukumar (1986ac, 1987 ab, 1988, 1996) Ch Raghukumar *et al.* (1992), Kohlmeyer and Vittal (1986) Kohlmeyer and Volkman Kohlmeyer (1987) Ravikumar and Vittal (1987, 1991a-b, 1996), Ravikumar and Purushothaman (1988a-b), Hyde *et al.* (1992), Chinnaraj (1992, 1993), Chinnaraj and Untawle (1992), Sridhar and Kaveriappa (1988, 1991), Sridhar and Prasannarai (1993, 2001), Prasannarai and Sridhar (1997, 2001), Prasannarai *et al.* (1999, 2000), Ananda *et al.* (1998) Nandan *et al.* (1993) Borse *et al.* (1999a-b, 2000a-b, 2001a-c), Sarma and Vittal (2000, 2001), Sharma and Hyde (2000) Borse and Borse (2001), Borse and Pawar (2001) Patil and Borse (2001) and Ananda and Sridhar (2001) and Borse *et al.* (2002).

Various marine fungi have been reported from different habitats and localities of Maharashtra coast. The work done on these fungi from this area reflects a great diversity in fungal flora. During the last two decades as much as eighty-two species (60 Ascomycota, 3 Basidiomycota, 19 Mitosporic fungi) have been studied. Thirty seven species are found as the new additions to the Indian fungi while two genera (*Acrocordiopsis* Borse & Hyde, *Biatriospora* Hyde & Borse) and eight species (*Acrocordiopsis patilii* Borse & Hyde, *Aigialus mangrovei* Borse, *A. rhizophorae* Borse, *Biatriospora marina* Hyde & Borse, *Didymella avicenniae* Patil & Borse, *Halosarpheia ratnagiriensis* Patil & Borse, *Julella avicenniae* (Borse) Hyde, *Massarina velatospora* Hyde & Borse) are new to science. However, there is a lack of information on the zoosporic fungi, parasites of marine algae and animals needs to be investigated.

The information on marine fungi from the Indian subcontinent is still in its infancy and therefore, studies from different coastal regions of India are needed to explore not only the fungal diversity but also to gather knowledge on their substrate relationship, cultural characteristics and physiology, quantitative estimation, biotechnology, decomposition biology etc.

The present chapter is divided into three major parts, namely the key, the list of species, and the plates. The dichotomous key is mostly based on the propagules and leads to specific taxa. The list of species includes concise descriptions of major characters used in identification. The plates represent camera lucida drawing of all species. Whenever feasible, species of the same genus, or morphologically similar species were grouped together.

KEY TO THE MARINE FUNGI FROM MAHARASHTRA

Key to Ascomycota

1. Ascomata opening with cleft- like ostiole (hysterothecial), or discoid and hymenium completely exposed at maturity (apothecial) 2
1. Ascomata without opening (clestothecial) or with round ostiole (perithecial) 7
2. Ascomata embedded or superficial, not discoid, ostiole cleft-like 3
2. Ascomata superficial, discoid, hymenium exposed.. *Dactylospora haliotreopha* Fig. 1 I
3. Ascomata superficial, boat-shaped, ascospores with transverse septa only .... *Lophiostoma mangrovei* Fig. 2T
3. Ascomata immersed, ascospores muriform 4
4. Ascospores with 13 or more transverse septa 6
4. Ascospores with less than 13 transverse septa 5
5. Ascospores 16-27μm diameter, more than 8 transverse septa .... *Aigialus parvus* Fig. 3I
5. Ascospores 10-16 μm diameter, less than 8 transverse septa.. *Aigialus mangrovei* Fig. 3J
6. Ascospores 18-28 μm diameter, 70-95 μm long..... *Aigialus grandis* Fig. 3G
6. Ascospores 18-25 μm diameter, 80-103 μm long.. *Aigialus rhizophorae* Fig. 3H
7. Ascomata peridium with tree-like appendages.. *Dryosphaeria tropicalis* Fig. 2H
7. Ascomata peridium without tree-like appendages....................................8
8. Ascospores with hyaline, ribbon -like appendages all round the middle and at each apex, most species with a long spine–like appendages at each end (*Corollospora* sp.)..9
8. Ascospores without or with different appendages 14
9. Ascospores one-septate 10
9. Ascospores multi-septate...11
10. Ascospores brown................................ *Corollospora cinnamomea* Fig. 2L
10. Ascospores hyaline.................................... *Corollospora maritima* Fig. 2K
11. Ascospore diameter 13-26 μm .... *Corollospora colossa* Fig. 2N
11. Ascospore diameter less than 13 μm 12
12. Ascospores with apical spines .... *Corollospora lacera* Fig. 2M
12. Ascospores without apical spines 13

13. Ascospores more than 12 septate- -............ *Corollospora filiformis* Fig. 2O
13. Ascopsores less than 12 septate- - .... *Corollospora pulchella* Fig. 2P
14. Ascospores filiform, tapering at both ends into mucilage-containing, cell-like processes (*Lulworthia sp.*) - 15
14. Ascospores different, if filiform, without apical chambers- 16
15. Ascopsores longer than 500 μm- .... *Lulworthia grandispora* Fig. 3L
15. Ascospores shorter than 500 μm-- .... *Lulworthia sp.* Fig. 3M
16. Ascospores non-septate at maturity- 17
16. Ascospores with one and more septa at maturity-- 19
17. Ascospores dark brown, with longitudinal germ slit -.. *Halorosellina oceanica* Fig. 1C
17. Ascospores yellow-brown or olive-brown, without slit-......................... 18
18. Ascospores olive-brown, allantoid, 7-10 x 2-3 μm- ....*Eutypa bathurstensis* Fig. 1B
18. Ascospores yellow-brown, non-allantoid, 20-28 x 7-10 μm - ....*Rhizophila marina* Fig. 1A
19. Ascospores with one-septum at maturity - 20
19. Ascospores with one and more septa at maturity- 48
20. Ascospores septate near the base, larger cell dark, small cell light coloured- - ....*Zopfiella latipes* Fig. 1p
20. Ascospores septate near the middle or if asymmetrical, concolorous-... 21
21. Ascospores hyaline, light grayish, yellow, or yellowish brown at maturity- 22
21. Ascospores dark-coloured (brown or blackish) at maturity- 46
22. Ascospores without appendages or gelatinous or mucilaginous sheaths-- 23
22. Ascospores with appendages or sheath- 28
23. Ascomata orange-yellowish, ostiolate- 24
23. Ascomata colour different, ostiolate or non-ostiolate- 25
24. Ascospores with longitudinal ridges- .... *Kallichroma tethys.* Fig. 1H
24. Ascospores without longitudinal ridges- .... *Kallichroma glabrum* Fig. 1G
25. Ascospores longer than 60 μm-- ....*Bathyascus tropicalis* Fig. 3O
25. Ascospores shorter than 60 μm- 26
26. Asci bitunicate, thick-walled-- ....*Acrocordiopsis patilii* Fig. 1D
26. Asci unitunicate, thin walled -- 27
27. Asci with apical pore- .... *Lignincola longirostris* Fig.1F
27. Asci without apical pore- .... *Lignincola laevis* Fig. 1E
28. Ascospores without appendages, but enclosed in uniform gelatinous sheath- 29
28. Ascospores with appendages, if covered by a sheath, also additional fibrous appendages present-- 30
29. Ascomata over 1 mm diameter, superficial .... *Ascocratera manglicola* Fig. 3A

29. Ascomata less than 1 mm diameter, immerse-- .... *Didmymella avicenniae* Fig. 1J
30. Ascospores thick-walled- .... *Aniptodera chesapeakensis* Fig. 1L
30. Ascospores thin-walled- 31
31. Ascospores with one cap like appendage *Halosphaeria culcullata* Fig. 1K
31. Ascospores with more than one appendage - 32
32. Ascospores completely enclosed in a sheath, also fibrillar or strand like appendages present - .... *Nimbospora octonae* Fig. 2J
32. Ascospores different - 33
33. Ascospores appendages only terminal or subterminal (an additional gelatinous cover around the spore may occur) - 34
33. Ascospore appendages terminally and laterally attached- .... *Ceriosporopsis caduca* Fig. 1V
34. Ascospores with 3-4 thin, more or less rigid appendages at each apex- - 35
34. Appendages different- 37
35. Ascospores diameter less than 6 μm -- .... *Arenariomyces parvulus* Fig. 1Y
35. Ascospores diameter more than 7 μm -..36
36. Ascospores consistently with 3 appendages at each end, ascomatal surface smooth - ....*Arenariomyces trifurcates* Fig. 1W
36. Ascospores with 3 or 4 appendages at each end, ascomatal surface with spines --.........................*Arenariomyces majusculus* Fig. 1X
37. Mature ascospores enclosed in an exosporic sheath that is pierced at each apex by an elongated outward growing, tapering appendage- ....*Ceriosporopsis halima* Fig. 1U
37. Mature ascospores without sheath .... - 38
38. Ascospores with 2-5 appendages at each end - 39
38. Ascospores appendages different, not more than 1 at each end-- ...... --40
39. Two subterminal appendages at each ascospore apex, these pairs arranged at right angles to one another - .... *Antennospora quadricornuta* Fig. 1S
39. 3-4 (rarely 5) subterminal ascospore appendages at each ascospore apex - .... *Antennospora salina* Fig. 1T
40. Appendages semiglobose or lenticular, apical- 41
40. Appendages hamate, lateral--... 43
41. Ascospore diameter 18 μm or less- -.............*Halosarpheia fibrosa* Fig. 2C
41. Ascospore diameter less than 13 μm- 42
42. Ascospores longer than 15 μm- .... *Halosarpheia marina* Fig. 2D
42. Ascospores shorter than 15 μm -............... -*Halosarpheia minuta* Fig. 2E
43. Ascospores diameter 11 μm or less- 44
43. Ascospores diameter over 11 μm- 45
44. Ascospores 20-34 x 7-11 μm, freshwater or estuarine species- .... *Halosarpheia retorquens* Fig. 2F

44. Ascospores 21-28 x 8 –11 μm, marine species- .... *Halosarpheia viscosa* Fig. 2G
45. Ascospores longer than 47 μm-- .... *Halosarpheia ratnagiriensis* Fig. 2A
45. Ascospores 47 μm or shorter- .... *Halosarpheia abonnis* Fig. 2B
46. Ascospores verrucose to verruculose-- .... *Verruculina enalia* Fig. 1M
46. Ascospores smooth or striate, not verrucose- - 47
47. Ascospores smooth- .... *Salsuginia ramicola* Fig. 10
47. Ascospores distinctly striate-- ....*Lineolata rhizophorae* Fig. 1N
48. Ascospores with transverse septa only- 49
48. Ascospores muriform-- 60
49. Ascospores without appendages or gelatinous sheaths-- .... *Leptoshaeria australiensis* Fig. 2R
49. Ascospores with appendages or gelatinous sheaths-- 50
50. Ascospores hyaline at maturity--- 51
50. Ascospores coloured at maturity-- 58
51. Mature ascospores completely enclosed in a uniform gelatinous sheath - 52
51. Mature ascospores without gelatinous sheath, but with 3-4 radiating acuminate appendages- .... *Torpedospora radiata* Fig. 2 I
52. Ascospores with 1-3 septa-.. 53
52. Ascospores with 3 or more septa- 56
53. Ascospore length 45 μm or more- 54
53. Ascospore shorter than 45 μm- 55
54. Ascomata conical, superficial, over 1mm diameter- .... *Ascocratera manglicola* Fig. 3A
54. Ascomata + or- immersed under a clypeus, smaller than 1 mm- .... *Massarina velatospora* Fig. 3B
55. Ascospores 34-47 x 9-13 μm - ....*Massarina thalassiae* Fig. 3C
55. Ascospores 25-30 x 7-10 μm - .... *Saccardosella mrinospora* Fig. 2Q
56. Ascospores 3 - septate with distinct longitudinal striation, central cells darker than end cells- ....*Asteroshariella striataspora* Fig. 3D
56. Ascospores without striations - 57
57. Ascospores 3-5 septate, third cell from apex largest- ....*Phaeospharia neomaritima* Fig. 3 E
57. Ascospores 5-septate- .... *Quintaria lignatilis* Fig. 2S
58. Ascospores non-septate in the central, 2-4 septa near each apex, with hyaline refractive apices- .... *Biatriospora marina* Fig. 3F
58. Ascospores septate in the center, apical cells hyaline or lighter coloured than the brown middle cells- 59
59. Asci 2-spored- .... *Savoryella paucispora* Fig. 1R
59. Asci 8-spored...- *Savoryella lignincola* Fig. 1Q
60. Ascospores with gelatinous sheath .... *Julella avicenniae* Fig. 3N
60 Ascospores without gelatinous sheath .... *Pleospora pelagica* Fig. 3K

II. Key to Basidiomycota

1. Basidiocarps puffball-like .... *Nia vibrissa* Fig. 3Q
1. Basidiocarps funnel-shaped...2
2. Basidiocarps 500-2500 μm high .... *Calathella mangrovei* Fig. 3R
2. Basidiocarps 350-500 μm high .... *Halocyphina villosa* Fig. 3P

III. Key to Mitosporic Fungi

1. Conidia borne on or in fruit bodies (Coelomycetes) 2
1. Conidia borne on hyphae (Hyphomycetes) 6
2. Conidia non-septate 3
2. Conidia multiseptate 4
3. Composite pycnidia with central ostioles, under the bark of *Rhizophora* .... *Cystospora rhizophorae* Fig 4J
3. Pycnidia different, on other substrates .... *Phoma sp.* Fig. 4I
4. Conidia muriform 5
4. Conidia with transverse septa only, in *Salicornia*, 20 x 7 μm ....*Ascochyta salicorniae* Fig. 4A
5. Conidia with 3 transverse septa, without gelatinous sheath .... *Camarosporium roumeguerii* Fig. 4C
5. Conidia with 5 or more transverse septa, with distinct gelatinous sheath .... *Camarosporium palliatum* Fig. 4B
6. Conidia one-celled .... *Periconia prolifica* Fig. 4M
6. Conidia several-celled 7
7. Conidia with 3 or more radiating arms .... *Clavatospora bulbosa* Fig. 4H
7. Conidia without radiating arms 8
8. Conidia 1-2 celled .... *Trichochadium alopallonellum* Fig. 4O
8. Conidia multicellular 9
9. Conidia with transverse septa only 10
9. Conidia muriform or irregular conglomerates of cells 18
10. Conidia more or less straight 11
10. Conidia helicoid 13
11. Conidia 2-5 septate 12
11. Conidia 1-2 septate .... *Bacteridesmium linderi* Fig. 4Q
12. Conidia 2-5 septate .... *Trichocladium achrasporium* Fig. 4N
12. Conidia 2-3 septate ....*Trichocladium opacum* Fig. 4P
13. Apical cell broader than basal cell (*Cirrenalia*) 14
13. Apical cell not broader than Basal cell.. 17
14. Conidia not constricted at the septa ....*Cirrenalia pygmea* Fig. 4F
14. Conidia constricted at the septa 15
15. Conidia 6-12 septate .... *Cirrenalia trropicalis* Fig. 4G
15. Conidia less than 6 septate 16

16. Conidia reddish-brown, coiled 12-24 μm in diameter .... *Cirrenalia macrocephala* Fig. 4E
16. Conidia without reddish tint, coiled 30 μm or more in diameter .... *Cirrenalia basiminuta* Fig. 4D
17. Conidial filament turns to produce a terminal regular spiral .... *Zalerion maritimum* Fig. 4R
17. Conidial filament turns to a lateral, variable spiral .... *Zalerion varium* Fig. 4S

LIST OF SPECIES ISOLATED FROM MAHARASHTRA COAST

Ascomycota

(1) *Acrocordiopsis patilii* Borse & Hyde

Ascocarps 1-3 x 1-2 mm; Asci 150-250 x 12-16 μm; Ascospores 16-25 x 10-16 μm, one septate, hyaline, ovoid or ellipsoidal and with one large guttule per cell.

Substrate: Stem of *Rhizophora mucronata*, stem of *Avicennia marina* and *Avicennia officinalis.*

Localities: Bankot, Malvan, Ratnagiri and Revadanda.

(2) *Aigialus grandis* Kohlmeyer & Schatz

Ascocarps 950-1200 μm; Asci 400-500 x 26-36 μm; Ascospores 70-95 x18-28 μm, ellipsoidal, muriform, 14-17 transverse-septa, 1-4 longitudinal-septa in all but the end cells, yellow-brown, apical cells light brown, gelatinous cap around the apical and subapical cells.

Substrate: Bark and pneumatophores of *Sonneratia alba* and *S. apetala*, prop roots and seedlings of *R. mucronata.*

Localities: Bankot, Dabhol, Jaigad, Malvan, Ratnagiri and Srivardhan.

(3) *A. mangrovei* Borse

Ascocarps 600-850 x 600-800 μm; Asci 300-425 x 20-30 μm; Ascospores 35-55 x 10-16 μm, muriform, 6-7 transverse-septa, 1-2 longitudinal-septa per segment, except in the end cells, yellow-brown.

Substrate: Stem, roots and pneumatophores of *A. marina.*

Localities: Bankot, Dahanu, Harne, Ratnagiri, Revadanda and Revas.

(4) *A. parvus* Schatz and Kohlmeyer

Ascospores 800-1250 x 950-1300 μm; Asci 300-400 x 30-35 μm; Ascospores 50-70 x 20-25 μm, 10-11 transverse-septa, 1-3 longitudinal-septa, gelatinous cap around apical and subapical cell.

Substrates: Intertidal wood, stem, and roots of *A. marina.*

Localities: Malvan, Revadanda, Revas and Vasai

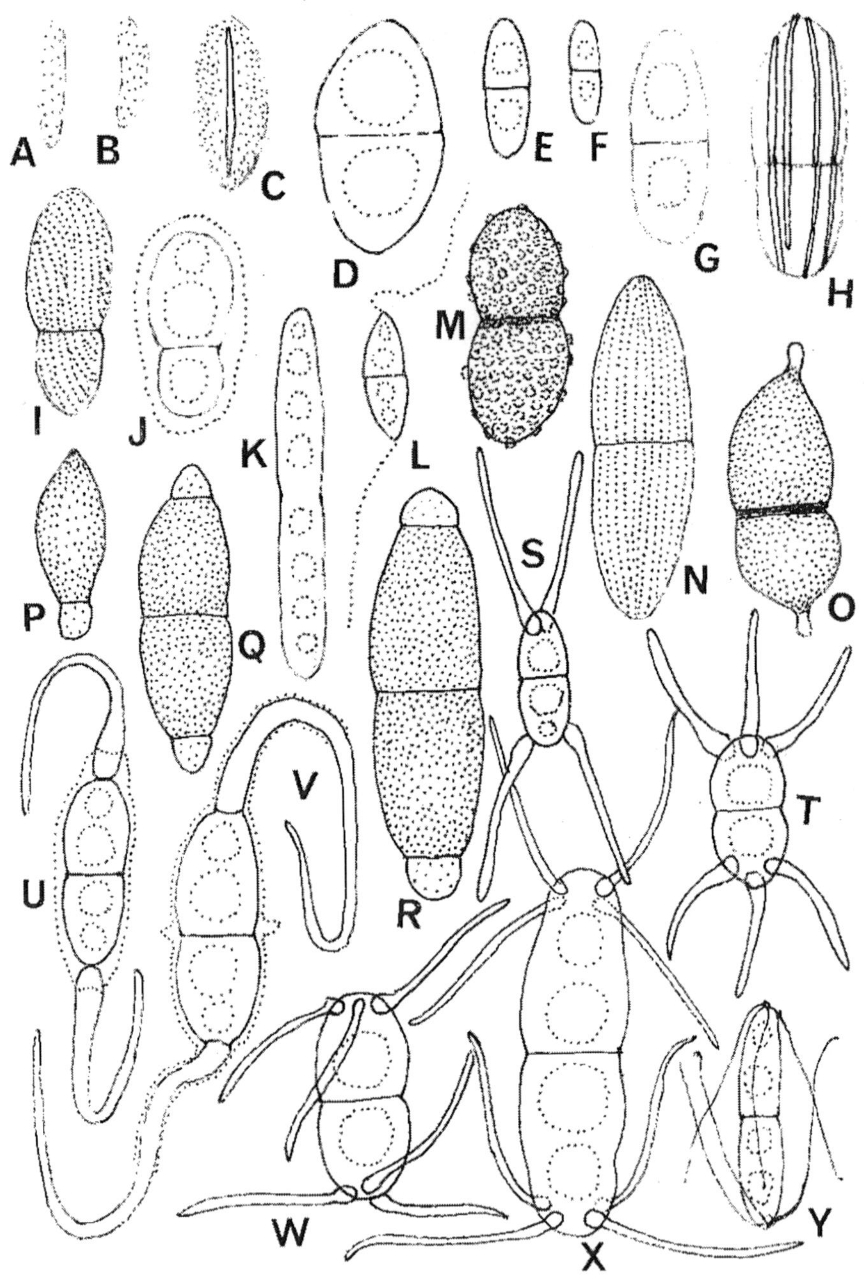

Fig. 1. *Ascosproes. A. Rhizophila marina, B. Eutypa bathurstensis, C. Halorosellina oceanica, D. Acrocordiopsis patilii, E. Lignincola laevis, F. L. longirostris, G. Kallichroma glabrum, H. K. tethys, I. Dactylospora haliotrepha, J. Didymella avicenniae, K. Halosphaeria cucullata, L. Aniptodera chesapeakensis, M. Verruculina enalia, N. Lineolata rhizophorae, O. Salsuginea ramicola, P. Zopfiella latipes, Q. Savoryella lignicola, R. S. paucispora, S. Antennospora quadricornuta, T. A. salina, U. Ceriosporopsis halima, V. C. caduca, W. Arenariomyces trifurcatus, X. A. majusculus, Y. A. parvulus.*

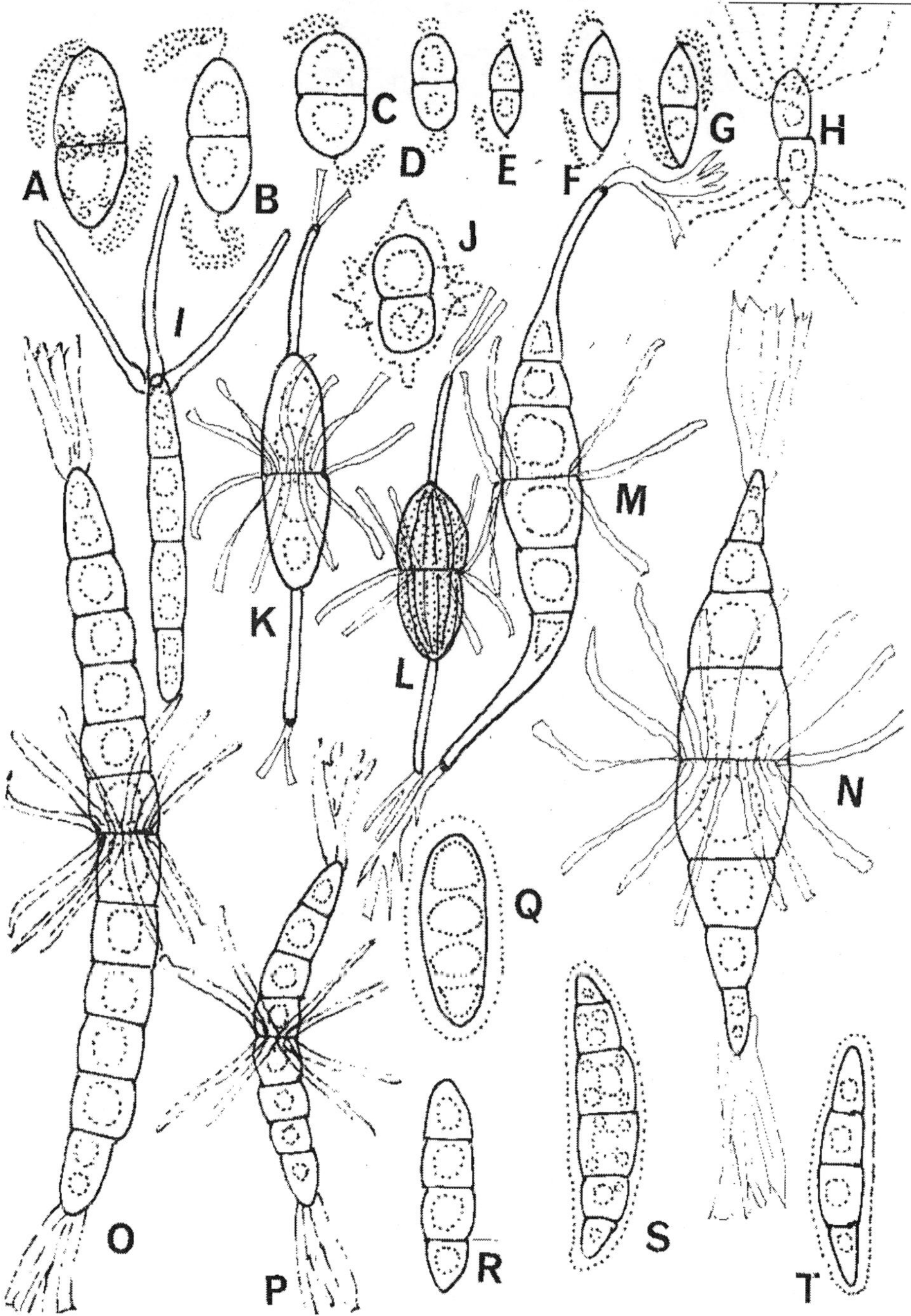

Fig. 2. *Ascospores . A. Halosarpheia ratnagiriensis, B. H. abonnis, C. H. fibrosa, D. H. marina, E. H. minuta, F. H. retorquens, G. H. viscosa, H. Dryosphaera tropicalis, I. Torpedospora radiata, J. Nimbospora octonae, K. Corollospora maritima, L. C. cinnamomea, M. C. lacera, N. C. collosa, O. C. filiformis, P. C. pulchella, Q. Saccardoella marinaospora, R. Leptosphaeria australiensis, S. Quintaria lignatilis, T. Lophiostoma mangrovei.*

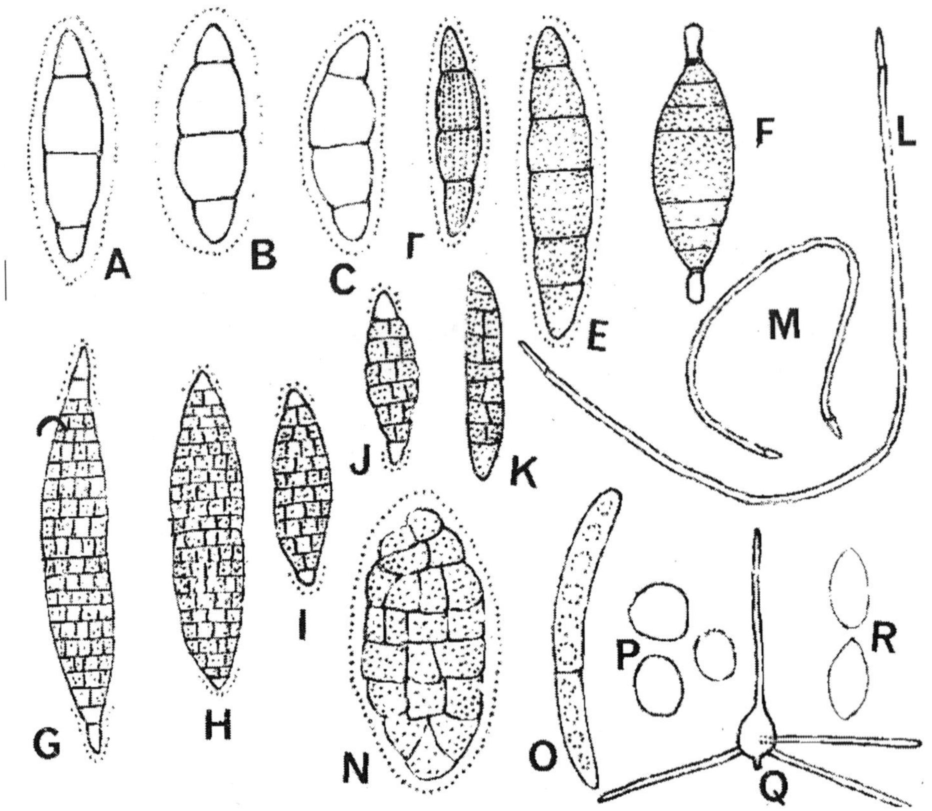

Fig. 3. *Ascospores. A. Ascocratera manglicola, B. Massarina velatospora, C. M. thalassiae, D. Astrosphaerella striatispora, E. Phaeosphaeria neomaritima, F. Biatriospora marina, G. Aigialus grandis, H. A. rhizophorae, I. A. parvus, J. A. mangrovei, K. Pleospora pelagica, L. Lulworthia grandispora, M. Lulworthia sp. N. Julella avicenniae, O. Bathyascus tropicalis, Basidiospores. P. Halocyphina violosa, Q. Nia vibrissa, R. Calathella mangrovei.*

(5) *A. rhizophorae* Borse

Ascocarps 900-1200 x 950-1100 μm; Asci 350-475 x 30-40 μm; Ascospores 70-98 x 18-25 μm, 13-15 trans-septa, 1-4 longisepta, except in the end cells; brown.

Substrate: Prop roots of *R. marina.*

Localities: Dabhol, Jaigad.

(6) *Aniptodera chesapeakensis* Shearer and Miller

Ascocarps 200-300 x 175-275 μm; Asci 70-120 x 15-20 μm; Ascospores 22-28 x 7-12 μm, ellipsoidal, one-septate, hyaline, thick-walled, appendages present or absent; appendages composed of a single coiled or folded filament

Substrates: Intertidal wood, prop roots of *R. mucronata.*

Localities: Bankot, Revadanda, and Vengurla

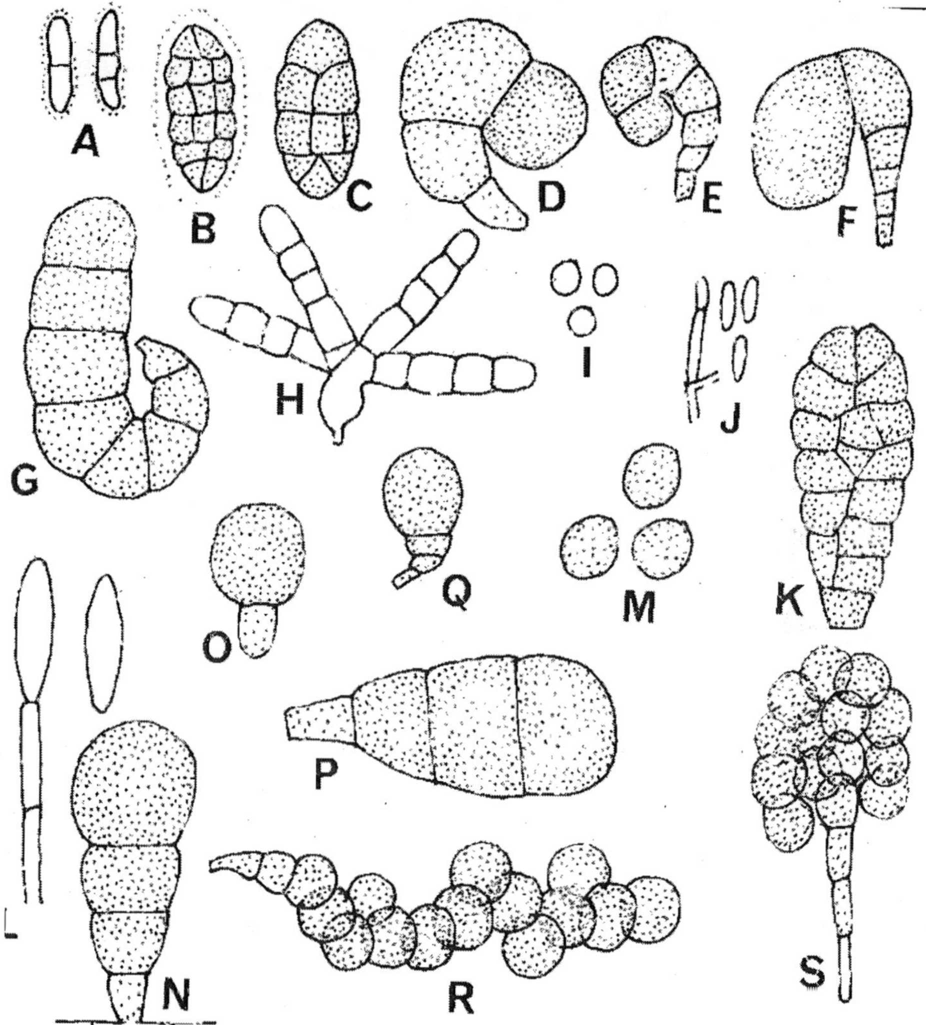

Fig. 4. *Conidia. A. Ascochyta salicorniae, B. Camerosporium pallatium, C. C. roumerguerii, D. Cirrenalia basiminuta, E. C. macrocephala, F. C. pygmea, G. C tropicalis, H. Clavatospora bulbosa, I. Phoma sp., J. Cytopsora rhizophorae, K. Dictyosporium pelagicum, L. Phomopsis mangrovei, M. Periconia prolifica, N. Trichocladium achrasporum, O. T. alopallonellum, P. T. opacum, Q. Bactrodesmium linderi, R. Zalerion maritimum, S. Z. varium.*

(7) *Antennospora quadricornata* (Cribb & Cribb) Johnson

Ascocarps 150-250 x 150-275 μm; Asci not observed as they are deliquescing; Ascospores 20-35 x 8-12 μm, ellipsoidal, one-septa, hyaline pairs of appendages at right angles to one another at each end.

Substrate: Intertidal wood, prop roots, and seedling of *R. mucronata*, stem of *A. marina*, pnumatophore of *S. alba*.

Localities: Bankot, Dabhol, Devgad, Mahim, Malvan, Ratnagiri, Reddi, Revadanda, Revas, and Vengurla.

(8) *A. salina* (Meyers) Yusoff, Jones and Moss

Ascocarps 150-300 x 150-250 μm; Asci 60-80 x 30-35 μm; Ascospores 20-28 x 9-12 μm, ellipsoidal, one-septate, hyaline, 3-4 radiating appendages at each end.

Substrate: Intertidal wood, stem of *A. marina*.

Localities: Malvan, Ratnagiri, Revadanda, Revas, Vengurla, and Vasai.

(9) *Arenariomyces majusculus* Kohlmeyer and Kohlmeyer

Ascocarps 100-150 μm; Asci 45-60 x 17-20 μm; Ascospores 30-40 x 10-14 μm, one-septate, hyaline, cylindrical, 3-4 subterminal appendages at both ends.

Substrate: Intertidal wood, drift-wood of *A. marina* with sand grains.

Localities: Ratnagiri

(10) *A. parvulus* Koch

Ascocarps 45-60 μm, globose, brown black; Asci not seen; Ascospores 16-18 x 3-4 μm, cylindrical, one-septate, hyaline with 3 terminal appendages at both ends.

Substrate: Intertidal wood, drift seedling of *R. mucronata* with sand grains.

Localities: Malvan.

(11) *A. trifurcatus* Hohnk

Ascocarps 100-300 μm, globose; Asci 60-80 x 20-23 μm, Ascospores 24-38 x 7-16 μm, ellipsoidal, one-septate, hyaline with three subterminal appendages at both ends.

Substrate: Intertidal wood, drift leaves of *R. mucronata* with sand grains.

Localities: Alibag, Ganapatipule, Malvan, and Revadanda.

(12) *Ascocratera manglicola* Kohlmeyer

Ascocarps 675-850 x 1000-1300 μm; Asci 250-260 x 20-28 μm; Ascospores 45-60 x 14-20 μm, 1-3 septate, hyaline with gelatinous sheath.

Substrate: Damaged stem of living tree of *A. marina*

Localities: Mumbai, Revdanda, and Vasai.

(13) *Asterosphaerella striatispora* (Hyde) Hyde

Ascocarps 100-300 x 200-450 μm, Asci 140-200 x 10-15 μm; Ascospores 31-38 x 6-9 μm, 3-septate, brown, central cells darker than end cells and with distinct longitudinal striations.

Substrate: Drift seedling of *R. mucronata*.

Localities: Malvan and Dabhol.

(14) *Bathyascus tropicalis* Kohlmeyer

Ascocarps 150-200 x 125-150 µm; Asci 90-110 x 20- 25 µm; Ascospores 75-110 x 6-9 µm, elongated, curved, one-septate below the center and hyaline.

Substrate: Intertidal wood, drift wood of *A. marina.*

Localities: Alibag, Malvan.

(15) *Biatriospora marina* Hyde and Borse

Ascocarps 650 - 850 x 350 - 500 µm; Asci 175-400 x 22-40 µm; Ascospores 55-80 x 16-25, µm fusiform, 2-4 septate towards each end, dark brown with chamber at the ends.

Substrate : Damaged stem of living trees of *A. marina* and *A. officinalis.*

Localities: Malvan, Ratnagiri, Revadanda, and Vasai.

(16) *Ceriosporopsis caduca* Jones and Zainal

Ascocarps 100-150 x 100-130 µm; Asci not seen; Ascospores 28-32 x 10-12 µm, ellipsoidal, one-septate, hyaline, single subcylindrical, tapering appendage (40-70 x 3-6 µm) at each end.

Substrate : Intertidal wood, drift wood of *A. marina.*

Localities: Malvan, Revadanda, and Srivardhan.

(17) *C. halima* Linder

Ascocarps 200-400 x 150-450 µm; Asci 60-80 x 15-20 µm; Ascospores 18-30 x 6-12 µm, ellipsoidal, one-septate, hyaline, with gelatinous sheath pierced at each apex by subcylindrical tapering appendage.

Substrate: Intertidal wood, drift wood of *A. marina* and prop roots of *R. mucronata.*

Localities: Alibag, Dabhol, Mahim, Malvan, and Ratnagiri.

(18) *Corollospora cinnamomea* Koch

Ascocarps 75-100 µm, globose ; Asci not seen; Ascospores 20-25 x 6-9 µm, one-septate, brown , spines 12-16 µm long.

Substrate: Drift wood, Ascocarps on sand grains attached to the drift prop roots of *R. mucronata.*

Localities: Ganapatipule, Malvan, and Ratnagiri.

(19) *C. colossa* Nakagiri and Tokura

Ascocarps 300-500 µm, globose ; Asci not seen; Ascospores 60-100 x 15-25 µm, 7-septate, without spines.

Substrate : Drift wood, on sand grains attached to the seedlings of *R. mucronata.*

Localities: Ratnagiri.

(20) *C. filiformis* Nakagiri

Ascocarps 150-225 μm, globose; Asci 100-125 x 20-25 μm; Ascocarps 90-110 x 6-8 μm, fusiform, 10-15 septate, hyaline, without spines.

Substrate : Drift wood.

Localities: Ratnagiri and Ganapatipule.

(21) *C. lacera* Linder

Ascocarps 100-150 μm, globose; Asci 80-110 μm x 15-20 μm; Ascospores 30-60 x 10-16 μm, 5-septate , hyaline, spines 10-14 μm long.

Substrate : Intertidal wood, Ascospores in foam samples.

Localities: Ganapatipule.

(22) *C. maritima* Werdermann

Ascocarps 100-300 μm, globose, Asci 80-140 x 20-50 μm; Ascospores 20-35 x 6-12 μm, one-septate, hyaline and spines 10-20 x 1-1.5 μm.

Substrates: Drift wood, drift pneumatophores of *A. marina*, drift prop roots and leaves of *R. mucronata.*

Localities: Alibag, Dabhol, Ganapatipule, Malvan, Revadanda, and Revas.

(23) *C. pulchella* Kohlmeyer, Schmidt and Nair

Ascocarps 100-250x μm, globose; Asci 100-140 x 20-30 μm; Ascospores 55-100 x 7-12 μm, fusiform, curved, 7-septate, hyaline, without spines.

Substrate: Drift wood, drift prop roots, and seedlings of *R. mucronata*

Localities: Malvan, Ratnagiri, Revas, and Vengurla.

(24) *Dactylospora haliotreopha* (Kohlm. and Kohlm.) Hafellner

Ascocarps 200-300 x 400-1000 μm; Asci 70-100 x 17-20 μm; Ascospores 18-30 x 8-12 μm, one-septate in the lower third, brown with forked longitudinal striations.

Substrate: Stems and roots of *A. marina* and *A. oficinnalis*, stem of *Ceriops tagal*, prop roots of *R. mucronata.*

Localities: Alibag, Bankot, Dabhol, Ganapatipule, Harne, Malvan, Ratnagiri, Revadanda, and Vengurla.

(25) *Didymella avicenniae* Patil and Borse

Ascocarps 250-275 x 275-350 μm; Asci 140-190 x 28-32 μm; Ascospores 25-32 x 10-15, hyaline, ellipsoidal, 1-septate with gelatinous sheaths.

Substrate: Roots of *A. marina.*

Localities: Revas.

(26) *Dryoshaera tropicalis* Kohlmeyer and Kohlmeyer

Ascocarps 150-225 μm, globose, appendaged; Asci 40-50 x 20-25 μm; Ascospores 12-14x 4-5 μm, ellipsoidal, 1-septate, hyaline with cap-like or many subterminal appendages at each ends.

Substrate: Driftwood, drift prop roots of *R. mucronata.*

Localities: Malvan.

(27) *Eutypa bathurstensis* Hyde and Rappaz

Ascocarps 300-600 x 500-800 μm; Asci 30-45 x 8-10 μm ; Ascospores 7-10 x 2-3 μm ; allantoid and olive-brown.

Substrate : Stem of *A. marina*

Localities: Ratnagiri

(28) *Halorosellinia oceanica* Whalley, Jones, Hyde and Lassioe.

Ascocarps 400-600 μm x subglobose; Asci 180-220 x 140-145 μm; Ascospores 18-20 x 8-10 μm, one-celled, reddish brown, elliptic, with germination slit.

Substrate: Stem of *A. marina, S. alba,* prop root, and seedling of *R. mucronata*

Localities: Bankot, Dabhol, Malvan, Ratnagiri, Revadanda, and Revas.

(29) *Halosarpheia abonnis* Kohlmeyer

Ascocarps 400-500 x 300-400 μm; Asci 200-250 x 28-40 μm ; Ascocarps 30-45 x 14-18 μm; ellipsoid me-septate, bipolar appendages hamate unwinding to form long filament.

Substrate:- Intertidal wood, on stem of *A. marina.*

Localities: Bankot, Revadanda.

(30) *H. fibrosa* Kohlmeyer & Kohlmeyer

Ascocarps 350-450 x 325-450 μm; Asci 150-200 x 30-45 μm; Ascocarps 25-40 x 15-20 μm; broad ellipsoidal, me-septate, hyaline appendages bipolar cap like first, uncoiling to form long fine filaments.

Substrate: Intertidal wood, stem of *A. marina,* prop root of *R. mucronata.*

Localities: Jaigad, Revas.

(31) *H. marina* (Cribb and Cribb) Kohlmeyer

Ascocarps 200-275 μm, globose; Asci 75-110 x 20-25 μm; Ascospores 20-24 x 10 -12 μm, ellipsoidal, hyaline, one- septate, with gelatinous cap-like appendages at poles.

Substrate: Prop roots of *R. mucronata.*

Localities: Bankot.

(32) *H. minuta* Leong

Ascocarps 225-425 x 175-300 μm; Asci 40- 60 x 10-14 μm; Ascospores 10-14 x 4-6 μm, ellipsoidal, hyaline one-septate, with apical appendages, unfurling in water into long filament.

Substrate: Prop roots of *R. mucronata*

Localities: Malvan.

(33) *H. ratnagiriensis* Patil and Borse

Ascocarps 450-600 x 250-300 μm; Asci 200-375 x 50- 75 μm; Ascospores 48-65 x 18-29 μm, ellipsoidal, one septate, hyaline, appendages bipolar, first hamate, then uncoiling to form long fine filament.

Substrate: Prop root and seedling of *R. mucronata*, pneumatophore of *S. alba.*

Localities: Bankot, Ratnagiri.

(34) *H. retorquens* Shearer and Crane

Ascocarps 350-450 x 400-480 μm; Asci 62-85 x 18-25 μm; Ascospores 18-30 x 7-10 μm; ellipsoidal, hyaline one-septate; appendages bipolar, hamate, finely unwinding to form long fine filament

Substrate: Prop roots and seedling of *R. mucronata.*

Localities: Bankot, Jaigad.

(35) *H. viscosa* (Schmidt) Shearer & Crane ex Kohlm. & Kohlm.

Ascocarps 240-250 x 180- 200 μm; Asci 50- 100 x 15-28 μm; Ascospores 21-28 x 8-11 μm, hyaline, ellipsoidal, appendaged, one-septate; Appendages bipolar, first hamate, then uncoling to form long fine filament.

Substrate: Prop root of *R. mucronata.*

Localities: Malvan.

(36) *Halosphaeria cucullata* (Kohlmeyer) Kohlmeyer

Ascocarps 200-275 x 150-250 μm; Asci not seen; Ascospores 25-50 x 6-9 μm, cylindrical, one-septate, hyaline; Appendages not observed.

Substrate: Stem of *A. marina.*

Localities: Revas.

(37) *Jullela avicenniae* (Borse) Hyde

Ascospores 225-375 x 250-400 μm; Asci 150-175-28-38 μm; Ascospores ellipsoidal, muriform, yellow- brown, 5-9 trans-septa, 2-3 longisepta per segment with gelatinous sheath.

Substrate: Roots and pneumatophores of *A. marina.*

Localities: Bankot, Jaigad, Malvan, Ratnagiri, Revadanda, and Revas.

(38) *Kallicroma glabrum* (Kohlm.) Kohlm. & Volkm.- Kohlmeyer

Ascocarps 300-350 x 325-450 μm; Asci 90-100 x 15- 23 μm; Ascospores 17-26 x 8-13 μm, hyaline, one septate, ellipsoidal, longitudinal ridges absent.

Substrate: Stem of *A. marina.*

Localities: Revadanda.

(39) *Kallicroma tethys* (Kohlm. & Kohlm.) Kohlm. & Volkm.-Kohlm.

Ascocarps 300-350 x 400 μm, Asci 90-100 x 15-25 μm; Ascospores 17-24 x 8-10 μm, hyaline, one-septate, ellipsoidal, longitudinal ridges present.

Substrate: Stem of *A. marina.*

Localities: Revadanda.

(40) *Leptoshaeria australiensis* (Cribb & Cribb) Hughes

Ascocarps 125-160 x 100-135 μm; Asci 75-100 x 12-16 μm; Ascospores 20-35 x 6- 10 μm, ellipsoidal, 3-septate and hyaline.

Substrate: Intertidal wood. Prop root and seedlings of *R. mucronata*

Localities: Bankot, Jaigad, Revadanda, and Srivardhan.

(41) *Lignincola laevis* Hohnk

Ascocarps 150-250 μm, ellipsoidal; Asci 50-70 x 15-20 μm; ascospores 16-24 x 6-8 μm, hyaline, ellipsoidal one –septate.

Substrates : Intertidal wood, roots and stems of *A. marina,* pneumatophores of *S. alba.*

Localities: Alibag, Bankot, Ganapatipule, Malvan, Ratnagiri, Reddi, Revadanda, Revas, and Vengurla.

(42) *L. longirostris* (Cribb & Cribb) Kohlmeyer

Ascocarps 100-200 μm, globose; Asci 50-80 x 17-21 μm; Ascospores 13-20 x 4-8 μm, hyaline, one-septate and ellipsoidal.

Substrates : Stem of *A. marina,* pneumatophores of *S. alba.*

Localities: Malvan, Ratnagiri.

(43) *Lineolata rhizophorae* (Kohlm. & Kohlm.) Kohlmeyer

Ascocarps 300-400 x 200-350 μm; Asci 150-175 x 14-18 μm; Ascospores 23-32 x 9-12 μm, one-septate, dark –brown and ellipsoidal.

Substrate : Stem of *Ceriops tagal.*

Localities: Ganapatipule.

(44) *Lophiostoma mangrovei* Kohlmeyer and Vittal

Ascocarps 775-950 x 1000 – 1400 μm, Asci 155 – 218 x 13-20 μm, Ascospores 37-55 x 6-11 μm, fusiform, one – septate, later 3-septate, hyaline with gelatinous sheath.

Substrate : Stem of *A. marina.*

Localities: Mumabai, Revadanda, and Vasai.

(45) *Lulworthia grandispora* Meyers

Ascocarps 500-525 x 225-300 µµm, Asci not seen; Ascospores 575-700 x 4-5 µm, filamentous, curved, hyaline with apical chambers.

Substrates : Roots and pneumatophores of *A. marina*; prop roots and seedlings of *R. mucronata*; pneumatophores of *S. alba*.

Localities: Bankot, Dabhol, Jaigad, Malvan, Ratnagiri, and Srivardhan.

(46) *Lulworthia* sp.

Ascocarps 150-300 x 125-250 µm; Asci not seen; Ascospores 340-380 x 4-5 µm, filamentous , curved , hyaline with apical chambers.

Substrate: Roots of *A. marina*; Seedlings of *R. mucronata*

Localities: Malvan, Ratnagiri, Revadanda, and Revas.

(47) *Massarina thalassiae* Kohlmeyer and Volkm. Kohlmeyer

Ascocarps 270-425 x 240-335 µm; Asci 100-160 x 28-34 µm; Ascospores 34-47 x 9-13 µm, ellipsoidal, 1-3 septate, hyaline with gelatinous sheath.

Substrate : Stem of *S. apetala*.

Localities: Srivardhan.

(48) *M. velatospora* Hyde and Borse

Ascospores 700-1200 µm, subglobose; Asci 220-320 x 23-34 µm; Ascospores 45-56 x 14-19 µm; fusiform, 1-3 septate, hyaline, surrounded by gelatinous sheath.

Substrate : Prop roots of *R. mucronata*, stem of *C. tagal*.

Localities: Bankot, Dabhol, Ganapatipule, Harne, Jaigad, Ratnagiri, Malvan, Srivardhan.

(49) *Nimbospora octonae* Kohlmeyer

Ascocarps 550-850 x 800-950 µm, Asci 130-150 x 20-25 µm; Ascospores 18-28 x 9-15 µm, ellipsoidal, one-septate, hyaline, gelatinous sheath enclosing 6-7 equatorial appendages and similar one at each pole.

Substrate : Prop roots of *R. mucronata*.

Localities: Malvan.

(50) *Phaeosphaeria neomaritima* (Gessner and Kohlm.) Shoemaker & Babcock

Ascocarps 120-180 x 100-160 µm; Asci 100-150 x 20-30 µm; Ascospores 35-40 x 10-12 µm, fusiform, yellowish to yellow brown, 5-6 septate with gelatinous sheaths.

Substrates : Stem of *S. apetala*; prop root of *R. mucronata*.

Localities: Srivardhan, Dabhol.

(51) *Pleospora pelagica* Johnson

Ascocarps 500-800 x 550-700 µm; Asci 275-325 x 25-30 µm; Ascospores 35-52 x 10-15 µm, 7-9 transsepta, 1 longiseptum in 1 to 8 segments, yellow brown.

Substrate : Drift mangrove wood.

Localities: Malvan.

(52) *Qurintaria ligniatilis* (Kohlm.) Kohlm. & Volkm.-Kohlmeyer

Ascocarps 200-450 x 250-400 µm; Asci 150-225 x 20-30 µm; Ascospores 35-60 x 10-15 µm, fusiform, five septate, young ascospores hyaline, mature ones yellow-brown, gelatinous sheath present.

Substrate : Prop roots of *R. mucronata.*

Localities: Bankot, Jaigad, and Malvan.

(53) *Rhizophila marina* Hyde

Ascocarps 500-950 x 425-700 µm; Asci 74-100 x 13-24 µm; Ascospores 20-28 x 7-10 µm, ellipsoidal, one-celled, yellow-brown at maturity.

Substrate : Prop roots of *R. mucronata.*

Localities: Dabhol.

(54) *Saccardoella marinospora* Hyde

Ascocarps 800-1000 x 650-950 µm; Asci 190-200 x 10-12 µm; Ascospores 25-30 x 7-10 µm, cylindrical, hyaline, three-septate, surrounded by a thin sheath.

Substrate : Prop roots of *R. mucronata.*

Localities: Bankot.

(55) *Salsuginea ramicola* Hyde

Ascocarps 1000-2500 x 500-1500 µm; Asci 440-500 x 30-35 µm; Ascospores 60-70 x 25-30 µm; 2-celled, obovoid, brown-black, hyaline apical germ pores.

Substrates : Stems of *A. marina, A. officinalis, S. alba, S. apetala.*

Localities: Bankot, Dabhol, Malvan, Ratnagiri, and Srivardhan.

(56) *Savoryella lignicola* Jones and Eaton

Ascocarps 170-350 x 120-250 µm; Asci 100-180 x 16-24 µm; Ascospores 24-36 x 8-12 µm, ellipsoidal, three-septate, central cells brown, apical cells smaller and hyaline.

Substrate : Drift prop roots of *R. mucronata.*

Localities: Sandy beaches of Malvan.

(57) *S. paucispora* (Cribb and Cribb) Koch

Ascocarps 90-135 x 220-250 µm; Asci 80-106 x 16-23 µm; Ascospores 44-50 x 12-17 µm, triseptate, ellipsoidal, central cells brown, apical cells smaller and hyaline.

Substrate : Drift branch of *A. marina*.

Localities: Sandy beach of Malvan.

(58) *Torpedospora radiata* Meyers

Ascocarps 100-340 x 100-350 µm; Asci 100-180 x 12-16 µm; Ascospores 30-52 x 4-9 µm, clavate, triseptate, hyaline, appendaged; Appendages 3-5 radiating on the lower end,

Substrate : Drift wood of *A. marina*.

Localities: Alibag, Malvan, Mumbai, and Ratnagiri.

(59) *Verruculina enalia* (Kohlm.) Kohlm. and Volkm.-Kohlm.

Ascocarps 100-300 x 100-140 µm; Asci 125-150 x 12-15 µm; Ascospores 18-22 x 8-12 µm, ellipsoidal, one-septate, dark brown with surface verrucose.

Substrates: Roots of *A. marina*, prop roots of *R. mucronata*, stem of *C. tagal.*

Localities: Dabhol, Dogged, Ganapatipule, Jaigad, Harne, Malvan, Ratnagiri, Reddi, Revadanda, Revas, Vasai, and Vengurla.

(60) *Zopfiella latipes* (Lundquist) Malloch and Cain

Ascocarps 200-600 µm, globose; Asci 80-120 x 12-18 µm; Ascospores 20-30 x 10-13 µm, one-septate in the lower third, Apical cell brown and lower cell hyaline.

Substrate : Roots of *A. marina*.

Localities: Bankot.

Basidiomycota

(1) *Calathella mangrovi* Jones and Agerer

Basidiocarps cyp helloid, 800-2000 x 400-1000 µm, club like, yellowish, old ochre-yellow, leathery and tomentose; Basidia 60-80 x 5-8 µm, Basidiospores 8-13 x 6-10 µm, one-celled, smooth, hyaline and elliptical.

Substrate : Prop roots and seedlings of *R. mucronata*.

Localities: Bankot, Dabhol, Malvan, and Ratnagiri.

(2) *Halocyphina villosa* Kohlmeyer and Kohlmeyer

Basidiocarps cyphelloid, 400-500 x 350-450 µm, Basidia 15-20 x 5-7 µm; Basidiospores 8-10 x 7-9 µm, globose, one-celled, hyaline.

Substrates : Roots of *A. marina*, prop roots of *R. mucronata*, pneumatophores of *S. alba*.

Localities: Bankot, Dabhol, Ganapatipule, Jaigad, Malvan, Ratnagiri, Revadanda, Revas and Srivardhan.

(3) *Nia vibrissa* Moore and Meyers

Basidiocarp 1-3 mm, subglobose; Basidia not seen; Basidiospores 9-15 x 6-11 μm, ovoid, one-celled, appendaged, hyaline; appendages one at the apex and 3-5 subterminal radiating around the base.

Substrates : Prop roots of *R. mucronata*, pneumatophores of *A. marina.*

Localities: Alibag, Malvan and Revadanda.

Mitosporic fungi

(1) *Ascochyta salicorniae* Magnus

Pycnidia 100-180 x 90-210 μm; Conidia 10-19 x 4-7 μm, cylindrical, 1-3 septate, hyaline to yellowish with mucilaginous sheath.

Substrate : Dead and decayed stems of *Salocornia brachiata.*

Localities: Alibag.

(2) *Bacterodesmium linderi* (Shearer)Palm & Stewart

Conidia 19-29 x 9-20 μm, 1-2 septate , base of the conidia curved, dark brown apical cell, basal cell light-brown.

Substrate : Prop root of *R. mucronata.*

Localities: Dabhol.

(3) *Camarosporium palliatum* Kohlmeyer and Kohlmeyer

Pycnidia 120-190 x 100-190 μm; Conidia 20-34 x 9-20 μm, muriform, ellipsoidal, 5-trans-septa, 3-4 longisepta, dark-brown with a gelatinous sheath.

Substrate : Stems of *S. brachiata.*

Localities: Alibag.

(4) *C. roumeguerii* Saccardo

Pycnidia 90-210 x 85-260 μm; Conidia 10-20 x 7-13 μm, ovoid, muriform, 1-3 trans-septa, 1-2 longisepta, olive-brown.

Substrate : Stems of *S. brachiata.*

Localities: Alibag.

(5) *Cirrenalia basiminuta* Raghukumar and Zainal

Conidia 3-5 septate, cells increasing in size and pigmentation from base to apex, Apical cell 9-16 x 9-14 μm, basal cell.

Substrates : Prop roots and seedlings of *R. mucronata.*

Localities: Deogad, Malvan.

(6) *Cirrenalia macrocephala* (Kohlm.) Meyers & Moore

Conidia 2-7 septate, ¼ - 1 times contorted, apical cell 6-14 μm diam., reddish fuscous.

Substrate : Intertidal wood, drift mangrove wood.

Localities: Dabhol.

(7) *C. pygmea* Kohlmeyer

Conidia 3-4 septate, one-half to one times contorted, apical cell 16-23 μm, black or fuscous.

Substrates : Prop roots and seedlings of *R. mucronata*, stem of *C. tagal.*

Localities: Bankot, Dabhol, Ganapatipule, Malvan, and Srivardhan.

(8) *C. tropicalis* Kohlmeyer

Conidia 6-12 septate, one to one and half times contorted, apical cell 10-20 μm, umber to reddish –brown.

Substrate : Prop roots of *R. mucronata.*

Localities: Malvan.

(9) *Clavatospora bulbosa* (Anast.) Nakagiri and Tubaki

Conidia tetraradiate, divergent upper arms 20-70 x 4-7 μm, light brown.

Substrates : Prop roots of *R. mucronata*, pneumatophores of *S. alba*.

Localities: Bankot, Ganapatipule, Mahim, Malvan, Reddi, Revadanda, Revas, and Srivardhan.

(10) *Cytospora rhizophorae* Kohlmeyer and Kohlmeyer

Pycnidia 0.5-0.8 x 1.5-1.7 mm; Locules 370-500 x 100-300 μm; conidia 3-6 x 1-1.5 μm, ellipsoidal, one-celled, hyaline or yellow, discharged as cirri or beads.

Substrates : Prop roots and seedlings of *R. mucronata.*

Localities: Dabhol, Malvan.

(11) *Dictyosporim pelagicum* (Linder) Hughes

Conidia 12-60 x 9-25 μm, ellipsoidal, multi septate, 3-8 irregular rows of cells with brown-black colour.

Substrate : Wood of *A. marina.*

Localities: Mumbai.

(12) *Periconia prolifica* Anastasiou

Conidia 6-13 μm, light to dark brown and one celled.

Substrates : Wood of *S. alba*, roots of *A. marina*, stem of *C. tagal.*

Localities: Malvan, Ratnagiri and Revas.

(13) *Phoma* sp.

Coelomycetes resembling *Phoma* are frequently collected in the marine and estuarine habitats on a variety of substrates . No effort has been made so far to separate and describe the different species.

Substrate : Intertidal, drift and mangrove wood.

Localities: Revadanda, Dabhol and Ratnagiri.

(14) *Phomopsis mangrovei* Hyde

Pycnidia 250-400 x 130-200 μm; conidia 11-18 x 3-4 μm, hyaline, ellipsoid, aseptate.

Substrate : Prop roots of *R. mucronata.*

Localities: Bankot, Dabhol, and Malvan.

(15) *Trichocladium achrasporum* (Meyers & Moore) Dixon

Conidia 20 - 34 x 20 - 24 μm; 2 - 5 septate, dark-brown.

Substrate : Drift wood.

Localities: Malvan and Dabhol.

(16) *Trichocladium alopalionellum* (Meyers & Moore) Kohlm. & Volkm.- Kohlm.

Conidia 11-20 x 5-9 μm, 1-2 celled, fuscous.

Substrates : Prop root and seedling of *R. mucronata*, pneumatophores of *S. alba.*

Localities: Bankot, Dabhol, Jaigad, Malvan, Ratnagiri, and Vengurla.

(17) *T. opacum* (Corda) Hughes

Conidia 19-25 x 13-17 μm, 2-3-septate, dark-brown, smooth walled and ovate.

Substrates : Prop root and seedling of *R. mucronata.*

Localities: Deogad, Malvan.

(18) *Zalerion maritimum* (Linder) Anastasiou

Conidia 4-27 septate, 1-6 times coiled, conidial filament 6-12 μm, fuscous or almost black,

Substrate : Prop roots of *R. mucronata.*

Localities: Jaigad, Malvan, and Vengurla.

(19) *Z. varium* Anastasiou

Conidial coil irregular, 15-65 x 14-56 μm, composed of 10-30 cells, brown to dark brown.

Substrates : Wood of *A. marina, A. officinalis.*

Localities: Bankot, Dabhol, Jaigad, Malvan, Ratnagiri.

CONCLUSION

Becker and Kohlmeyer (1958) for the first time studied marine fungi of India. During the last five decades as much as 166 species have been recorded from Indian marine waters (Borse, 2002). However, there is lack of information on the lower fungi, parasite of algae, sea grasses and animals and it needs to be studied. Data on marine fungi from India is still in its infancy, such studies from different coastal regions of India are needed to explore not only fungal diversity but also to gather data on their substrate relationship, cultural studies, physiology, quantitative estimation, biotechnology, decompositive biology, etc. These fungi are not only unique for their habitats and magnificent propagules types but also are of great significance in the organic matter decomposition, as pathogens and in the energy budget of the marine ecosystems.

REFERENCES

Ananda, K., Prasannarai, K. and Sridhar, K. (1998). Occurrence of higher marine fungi on marine animal substrates of some beaches along with western coast of India. *Ind J. Mar. Sci.* 27: 233-236.

Ananda, K. and Sridhar, K. (2001). *Aniptodera indica* a new species of mangroove inhabiting ascomycetes from western coast of India. *J. Environ. Biol.* 22: 283-286.

Barghoorn, E.S. and Linder, D.H. (1944). Marine fungi: their Taxonomy and Biology. *Farlowia* 1 : 395 – 433.

Becker, G. and Kohlmeyer, J. (1958). Deterioration of wood by marine fungi and its special significance for fishing crafts. *J. Timber Dry. Preserv. Asso. India.* 4: 1-10.

Borse, B.D. (1984). Marine fungi from India – I. *Indian Bot. Reptr.* 3: 156-157.

Borse, B.D. (1985). Marine fungi from India – II. *Curr. Sci.* 54: 881-882.

Borse, B.D. (1987a). Marine fungi from India-IV. *Trans. Mycol. Soc. Japan* 28 : 55-61.

Borse, B.D. (1987b). Marine fungi from India –V. *Curr. Sci.* 56: 1109-1111.

Borse, B.D. (1987c). Marine fungi from India- VI. *Indian Bot. Reptr.* 6: 108 -110.

Borse, B.D. (1987d). Marine fungi from India- VII. *Indian Bot. Reptr.* 6: 62-64.

Borse, B.D. (1987e). New species of *Aigialus* from India. *Trans. Br. Mycol. Soc.* 88: 424-426.

Borse, B.D. (1988). Frequency of occurrence of marine fungi from Maharashtra coast, India. *Ind. J. Mar. Sci.* 17: 165-167.

Borse, B. D. (2000a). Marine fungi from India – X. *Savoryella* Jones & Eaton (Ascomycotina). In: *Ecology of fungi* (Eds. Bhat, D.J. & S. Raghukumar), Pp. 163-165, Goa University Press, Goa.

Borse, B. D. (2000b). Marine fungi from Maharashtra (India) – V. A check-list. *J. Phytol. Soc.* 13: 123- 128.

Borse, B.D. (2002). Marine fungi from India – XI. A check- list. *J. Ind. Bot. Soc.* 81 : 149-158.

Borse, B.D. and Borse, K.N. (2001). New reports of marine ascomycetes from Orrisa India *Geobios* 28: 64-67.

Borse, B.D. Pawar, N.S. and Borse, K.N. (2002). Marine fungi from Orrisa (India) – 1. Genus *Corollospora* Werderman. *Geobios* 29 : 258-263.

Borse, B.D. and Hyde, K. D. (1989). Marine fungi from India – III. *Acrocordiopsis patilii* gen. et sp. nov. from mangrove wood. *Mycotaxon* 34: 535-540.

Borse, B.D., Kelkar, D.J. and Patil, A.C. (2000a). Frequency of occurrence of marine fungi from Pirotan Island (Gujarat) India. *Geobios* 27 : 145-148.

Borse, B.D., Nandan, S.N. and Shinde, D.N. (1999a). Higher marine fungi from Goa coast (India). *BRI's J. A. S. T.* 2: 52-55.

Borse, B.D., Patil, K.B. Patil, R.V. and Kelkar, D.J. (2000b). Marine fungi in foam, intertidal wood and dead *Avicennia marina* wood from Daman coast, India. *Geobios* 27: 42-44.

Borse, B.D., Patil, R.V. and Kelkar, D.J. (1996b). Marine fungi from Diu Island (India). BRI's J.A.S.T. 2:1-8.

Borse, B.D., and Pawar, N.S. (2001). *Carbosphaerella* and *Dryosphaera*. Two new generic record of marine ascomycetes from West Bengal (India). *Geobios*. 28: 117-120.

Borse, B.D., Pawar, N.S. and Borse, K.N. (2001b). Marine fungi from Sunderbans (India) 1. Sea foam spora. *J. Ind. Bot Soc.* 80: 237-240.

Borse, B.D., Ch. Ramesh and Shrivastava, A.D. (1988). Marine fungi from Maharashtra (India). *Ind. Bot. Reptr.* 7: 18-25.

Borse, B.D. and Shrivastava, A.D. (1988). Marine fungi from India – VIII. *Ind. Bot. Reptr.* 7: 26-30.

Borse, B.D. and Shrivastava, A.D. (1994). Marine fungi from India IX-Loculoascomycetes. *Ind. Bot. Reptr.* 13 : 1-3.

Borse, K.N., Pawar, N.S. and Borse, B.D. (2001c). Marine fungi from Orissa (India) II. Arenicolous group *BRI's J.A.S.T.* 4:8-12.

Chinnarai, S. (1992). Higher marine fungi from Lakshadweep Islands and a note on *Quintaria lignatilis. Crypto Mycol* 13: 312-319.

Chinnarai, S. (1993). Higher marine fungi from mangroves of Andaman and Nicobar Island *Sydowia.* 45: 109-115.

Chinnarai, S. and Untawale, A.G. (1992). Manglicolous fungi from India. *Mahasagar.* 25: 25-29.

Durieu, G. and Montagne, J.F.C. (1869). Pyrenomycetes Fr. In : *Exploration Scientifique del Algeric Botanique,* pp. 443-608. Paris.

Hawksworth, D.L., Kirk, P.M. Sutton, B.C. and Pegler, D.M. (1995). *Ainsworth and Bisby's Dictionary of the fungi* 8th ed. IMI, Egham, UK.

Hyde, K.D. and Borse, B.D. (1986a). Marine fungi from Seychelles-V. *Biatriospora marina* gen. et sp. nov. from mangrove wood. *Mycotaxon* 26 : 263-270.

Hyde, K.D. and Borse, B.D. (1986b). Marine fungi from Seychelles-VI. *Massarina velatospora,* a new Ascomycotina from mangrove wood. *Mycotaxon* 27 :161-169.

Hyde, K.D. and Pointing, S.B. eds. (2000). Marine Mycology – A Practical Approach. Fungal Diversity Press, Hong Kong.

Hyde, K.D. and Sarma, V.V. (2000a). Morphology and Taxonomy of higher marine fungi. In: *Marine Mycology – A Practical Approach.* (Eds. Hyde, K.D. and S. B. Ponting), Fungal Diversity Press, Hong Kong, Pp. 172-204.

Hyde, K.D. and Sarma, V.V. (2000b). A pictoral key to the identification of higher marine fungi In: *Marine Mycology : A practical Approach* (Eds. Hyde & Ponting), Fungal Diversity Press, Hong Kong.

Hyde, K.D., Vrijmoed, L.L.P., Chinnaraj, S., and Jones, E.B.G. (1992). *Massarina armatispora* sp. nov., a new intertidal ascomycetes from mangroves. *Bot. Mar.* 35: 325-328.

Johnson, T.W. Jr. and Sparrow, F.K. Jr. (1961). *Fungi in Oceans and Estuaries* . Cramer, New York , pp. 668.

Jones, E.B.G. ed. (1968). Marine fungi. *Curr. Sci.* 37: 378 - 379.

Jones, E.B.G. ed. (1976). *Recent Advances In Aquatic Mycology.* Wiley, New York.

Jones, E.B.G. and Alias, S.A. (1997). Biodiversity of mangrove fungi. In : *Biodiversity of Tropical Microfungi* (ed. K.D. Hyde), Hong Kong University Press, Hong Kong, pp. 71-92.

Kohlmeyer, J. (1959). Newfunde holzbesiedelndner Meeresplilze. *Nova Hedwigia* 1 : 77-99.

Kohlmeyer, J. (1974). On the definition and taxonomy of higher marine fungi. *Veroff. Inst. Meeresforsch. Bremerh.* 5 : 263-286.

Kohlmeyer, J. and Kohlmeyer, E. (1979). *Marine Mycology. The Higher Fungi* Academic Press, New York, pp. 690.

Kohlmeyer, J., Schmidt, I. and Nair, N.B. (1967). Eine neue *Corollospora* (Ascomycete) aus dem Indien ozean und der Ostee. *Ber. Dt. Bot . Ges.* 80 : 98-102.

Kohlmeyer, J. and Vittal, B.P.R. (1986). *Lophiostoma mangrovis*, a new marine ascomycetes from the tropics. *Mycologia* 78 : 485-489.

Kohlmeyer, J. and Volkmann-Kohlmeyer, B. (1987). Marine fungi from Aldabra, the Galopagos and other tropical Islands. *Can. J. Bot.* 65 : 571-582.

Kohlmeyer, J. and Volkmann-Kohlmeyer, B. (1991). Illustrated key to the filamentous higher marine fungi. *Bot. Mar.* 34 : 1-61.

Moss, S.T. ed. (1986). *The Biology of Marine Fungi.* Cambridge University Press, Cambridge, pp. 382.

Nair, N.B. (1970). The problem of timber destroying organisms along the Indian coast In, *Proc. Int. Cong.* Mar. Coross. Fanling. IInd. 1968 pp 1-7.

Nandan, S.N. Shinde, D.N. and Borse, B.D. (1993). Marine fungi from Goa Coast (India). *Biol. Ind.* 4: 29-34.

Patil, K. B. and Borse, B. D. (2001). Studies on higher marine fungi from Gujarat coast (India). *Geobios* 28 : 41-44.

Patil, S. D. and Borse, B.D. (1982). Marine fungi from Maharashtra (India)-1. The genus *Halosarpheia* J. & E. Kohlmeyer . *Ind. Bot. Reptr.* 1: 102-106.

Patil, S. D. and Borse, B. D. (1983a). Marine fungi from Maharashtra (India)-II. Some arenicolous fungi. *Ind. Bot. Reptr.* 2: 86-87.

Patil, S. D. and Borse, B.D. (1983b). Marine fungi from Maharashtra (India)-III. Some fungi from mangroves. *Ind. Bot. Reptr.* 2: 56-58.

Patil, S.D. and Borse B.D. (1985). Marine fungi from Maharashtra (India)-IV. Some Loculoascomycetes. *Trans. Mycol. Soc. Japan* 26 : 271-276.

Patil, S.D. and Borse, B.D. (1986). Marine fungi from Indian mangroves. In : *The Mangroves* (ed. L.J. Bhosale), Shivaji University Press, Kolhapur, pp. 151-152.

Prasannarai K. and Sridhar K.R. (1997). Effect of incubation period of drift wood on the occurrence of marine fungi. *Ind. J. Mar. Sci.* 26: 380-382.

Prasannarai, K. and Sridhar, K.R. (2001). Diversity and abundance of higher marine fungi on woody substrates along the west coast of India. *Curr. Sci.* 81: 304-316.

Prasannarai, K., Ananda, K. and Sridhar, K.R. (1999). Intertidal fungi in Mangalore harbour, South India. *Bot. Mar.* 42: 117-122.

Prasannarai, K. Ananda, K. and Sridhar, K.R. (2000). *Corollospora indica* , a new marine fungal species from west coast of India. *J. Environ. Biol.* 21: 235-239.

Raghukumar, C. (1986a). *Chytridium polysiphoniae*, a fungal pathogen on the red alga *Centroceras clavatum* (Agarth) Montagne from Goa. *Ind J. Mar. Sci.* 15: 42-44.

Raghukumar, C. (1986b). Thraustochytrid fungi associated with marine algae. *Ind J. Mar. Sci.* 15: 121-122.

Raghukumar, C. (1986c). Fungal parasite of marine green algae. *Cladophora* and *Rhizoclonium Bot. Mar.* 29: 289-297.

Raghukumar, C. (1987a). Fungal parasite of marine green algae from Mandapan (South India) *Dis. aquat Org.* 3: 137-145.

Raghukumar, C. (1987b). Fungal parasite of green alga *Chaetomorpha media* . *Dis. aquat Org.*3: 147-150.

Raghukumar, C. (1988). Changes in the marine algae *Chaetomorpha media* on infection by a fungal pathogen. *Bot. Mar.* 31: 311-315.

Raghukumar, C. (1996). Zoosporic fungal parasites of marine biota In Advances in zoosporic fungi (ED. R. Dayal) M.D. Publications Pvt Ltd. New Delhi, pp. 61-80

Raghukumar, C., Nagarkar, S. and Raghukumar, S. (1992). Association of thraustochytrid and fungi with living marine algae *Mycol. Res.* 96: 542-546.

Raghukumar, S. (1973). Marine Lignicolus fungi from India *Kavaka* 1: 73-85.

Raghukumar, S. (1987). Occurrence of the thraustochytrid, *Corallochytrium limacisporium gen et sp.* nov. in the coral reef lagoons of the Lakshadweep Island in the Arabian Sea. *Bot. Mar.* 30: 83-89.

Raghukumar, S. (1988a). Some interesting fungi from the west coast of India. *Kavaka* 14: 77-80.

Raghukumar, S.(1988b). *Schizochytrium mangrovei* Sp. nov. thranstochytrid from mangroves from India. *Trans. Br. Mycol. Soc.* 90: 627-631.

Raghukumar, S. (1988c). Detection of the thranstochytrid protist *Ulkenia visurgensis* in a hybroid using Immunofluorescence . *Mar. Biol.* 97: 53-58.

Raghukumar, S. (1996). Fungi in the marine realm: status challenges and prospects. *Kavaka.* 24: 25-34.

Raghukumar, S. and Balasubramanian, R. (1991). Occurrence of thranstochytrid fungi in coral mucus. *Ind. J. Mar. Sci.* 20: 176-181.

Raghukumar, S. and Raghukumar, C. (1988). A report on the occurrence of thranstochytrid species in Indian water. *Kavaka.* 16:40-44.

Raghukumar, S., Raghukumar, C. and Rajendran, A. (1990). Abundance of thranstochytrid fungi in Arabian Sea. Esth. *Coast. Shelf. Sci.* 31: 351-358.

Raghukumar, S., Zainal, A. and Jones, B.G. (1988). *Cirrenalia basiminuta*, a new lignicolous marine Deuteromycetes from the tropics. *Mycotaxon* 31: 163-170.

Ramesh, Ch. and Borse, B. D. (1989). Marine fungi from Maharashtra coast (India) . *Acta. Bot. Indica* 17 : 143-146.

Ravikumar, D. R. and Purushothaman, A. (1988a). *Cirrenalia tropicales*, an addition to Indian marine fungi. *Curr. Sci.* 57: 674.

Ravikumar, D.R. and Purushothaman, A. (1988b). *Corollospora intermeida*, a lignicolous marine fungus from India. *Curr. Sci.* 57: 898-899.

Ravikumar, D.R. and Vittal, B.P.R. (1987). Studies on mangrove fungi of India I Notes on some Ascomycetes. *Kavaka.* 15: 99-103.

Ravikumar, D.R. and Vittal, B.P.R. (1991a). Two *Corollospora* species new to India. *Kavaka*19: 17-18.

Ravikumar, D.R. and Vittal, B.P.R. (1991b). A new species of *Bathyascus* from Indian mangroves. *Mycol. Resh.* 95: 370-371.

Ravikumar, D.R. and Vittal, B.P.R. (1996). Fungal diversity on decomposing biomass of mangroves plant *Rhizophora* in Pichavaram estuary, East coast of India. *Indian J. Mar.Sci.* 25:142-144.

Sarma, V.V. and Hyde, K.D, (2000). *Trispora mandoviana* sp. nov. from Chorao mangrooves, Goa, the west coast of India. *Austra Mycol* 19: 52-56.

Sarma, V.V. and Vittal, B.P.R. (2000). Bio-diversity of mangrove fungi on different substrata of *Rhizophora apiculata* and *Avicennia* sp. from Godavari and Krishna delta, east coast of India. *Fungal Diversity.* 5: 23-41.

Sharma, V.V. and Vittal, B.P.R. (2001). Bio-diversity of manglicolous fungi on selected plants in the Godavari and Krishna delta, east coast of India. *Fungal Diversity.* 6: 115-130.

Sathe, V. and Raghukumar, S. (1991). Fungi and their biomass in detritus of the sea grass *Thalassia hemprichii* (Ehrengerg) Ascherson. *Bot Mar.* 34: 271-277.

Sparrow, F.K. Jr. (1937). The occurrence of saprophytic fungi in marine mud. *Biol. Bull.* 73 : 242-248.

Shridhar, K.R. and Kaveriappa, K.M. (1988). Occurence and survival of aquatic hyphomycetes in blackish and sea water. *Archiv fur Hydrobiologie* 113: 153-160.

Shridhar, K.R. and Kaveriappa, K.M. (1991). A note on marine fungi from Mangalore coast *Mahasagar,* 24: 63-66.

Shridhar, K.R. and Prasannarai, K. (1993). Marine fungi on decaying wood from Mangalore coast *Mahasagar,* 26: 33-38.

Sridhar, K.R. and Prasannarai, K. (2001). Biogeography and biodiversity of higher marine fungi in tropics - a review. Ecology, Environment and Conservation 7 : 219-234.

Srivastava, A.D. (1989). Marine fungi from Bombay-III. Some Loculoascomycetes. *Ind. Bot. Reptr.* 8 : 16-19.

Srivastava, A.D. (1994). Marine fungi from Bombay IV. *Ind. Bot. Reptr.* 13: 92-93.

CHAPTER – 4

FUNGAL DIVERSITY IN PAPER MATERIALS AND ITS ROLE IN BIODETERIORATION

Shashi Dhawan and Neeta Nigam

Biodeterioration Division, National Research Laboratory for Conservation of Cultural Property, E/3, Aliganj, Lucknow - 226024

INTRODUCTION

Biotic factors, such as fungi, a specialized group of microorganism are well known dangerous agents responsible for paper deterioration (Dhawan and Agrawal, 1986) and paper ageing phenomenon (Havermans, 1994), but they are usually underestimated in comparison with abiotic factors, both environmental (light, temperature, oxygen, humidity changes, air pollutants) and related to manufacturing processes, such as modern high yield chemo-thermo-mechanical pulp (Lanting, 1990; Priest and Stanley, 1994). Since paper of high resistance is generally characterized by minimal alteration of its chemical components with time, most paper making processes are commonly focused on improving the resistance of paper to abiotic factors.

On the contrary, fungi are well known and dangerous microbes responsible for paper deterioration, because of their hydrolytic enzyme activity (Zabel and Morvell,1992; Coughlan *et al.,* 1993; Wood and Gorcia Compayo, 1994 and Teeri and Koivula, 1995). In particular cellulose activities cause the most severe damages on paper and recently it has been reported that the amorphous fraction of cellulose (the main component of paper) is the most affected by endoglucanase enzymatic attack of fungi (Capitani *et al.,*1998). However, different kinds of paper are characterized by qualitative/quantitative differences that can affect the biotic and abiotic paper deterioration process. It is also well known that the mechanism of paper deterioration due to micro fungi can change according to the characteristics of paper.

There are several effects of fungal activity on paper. Primarily the major part of the cellulolytic fungi needs glucose and nitrogenous compounds to synthesize their own amino acids. Glucose is obtained by an enzymatic process, in which the cellulose polymers are broken up into smaller glucose molecules by the aid of cellulose, an intricate exo-enzyme complex produced by fungi (Caneva *et al.*, 1991).

As enzymatic break down contains, the organism causing oxidative process excretes acidic metabolites. The oxidation of the cellulose molecule causes fluffy and porous structure of weakened mechanical strength, discoloration (Szeepanowska and Lovett, 1992) and increased sensitivity of light. Not all the papers are equally susceptible to fungal attack. Paper with a high percentage of cellulose fibers is more resistant than low quality, heavily milled products.

The phenomenon known as "foxing" and which has been subject to innumerable investigations is according to the most recent findings, believed to be the deed of micro fungi.

Not all of the paper attacking fungi is cellulolytic. Since paper in the production phase, and/or by handling and use acquire various amounts of sizing coatings, dirt and debris, the nutritional requisite of other fungi are often rapidly fulfilled (Kowalik, 1980, 1984). It is even reported that certain conservation methods and materials introduce and facilitate fungal development in paper eg., resizing with gelatin and methyl cellulose (Florian, 1996).

In addition to cellulose paper also certain lignin, hemicelluloses, pectin, waxes, tannin, proteins and minerals as suitable substrates of microbial growth. Further more paper may contain resinous materials, fillers, dyestuffs and other various impurities during the time of production of paper, which forms the base for the nutrition of microbes. Microbes growing on paper are fungi, bacteria and actinomycetes. These organisms multiplying on paper utilize intentional additives of adventitious impurities and prepare substrate of typical cellulolytic microorganisms.

The microbial deterioration of paper material like manuscripts both illustrated and written, books archival materials, decorative wall paper etc is a serious problem throughout the world in museum and libraries, archives etc. (Baynes-Cope and Allsopp, 1984; Dhawan, 1986; Bronislaw Zyska, 1997; Ricelli *et al.*, 1999; Nitterus, 2000 and Rebrikova and Manturovskaya, 2000). Scientists after prolonged observations have maintained that fungi are exceptionally abundant on the earth. They have variety of fermentive complexes and high ecological adaptability, which give them a leading place in damaging paper material.

Biodeterioration of paper materials is a worldwide problem due to the vast number of libraries, archives, museums spread all over world and large no. of that type of material are stored. Ewald (1966) has prepared a summary of investigations into microbial colonization of library materials in Western Europe.

Other published investigation of fungi colonizing library materials include by Strzelczyk (1968) Orlita (1977), Strzelcyk and Leznicka, (1981), Niuksha (1993). However, Wessel (1970), Zyska (1993, 1996) have summarized the current state of knowledge of microbiology of library materials. The storage of books is frequent source of problems both real and potential which requires the attention of the microbiologist. (Baynes Cope, 1981).

In museums we have very rich collection of paper paintings, which includes watercolors, gouaches and pastels. They have very good source of nutrition for fungi. Pastels are made with crayons consisting of pigments and a binder. Therefore, pastels are extremely difficult to store and preserve. In gouache paintings like the Indian miniatures there is a ground of gum and glue, is mixed with an inert material. Microorganisms attack is found on all these paintings (Fig. 1).

Fig. 1.

Fungi growing on paper produce black, brown and yellow etc. colored pigmentation which penetrates the paper (Nyuksha, 1956; Kathpalia, 1960; Kowalik, 1980; Chiraporn 1995) (Fig. 2). These stains are difficult to remove by the usual restoration methods.

The detailed list of different genera and their symptomology has been elaborated in Table 1.

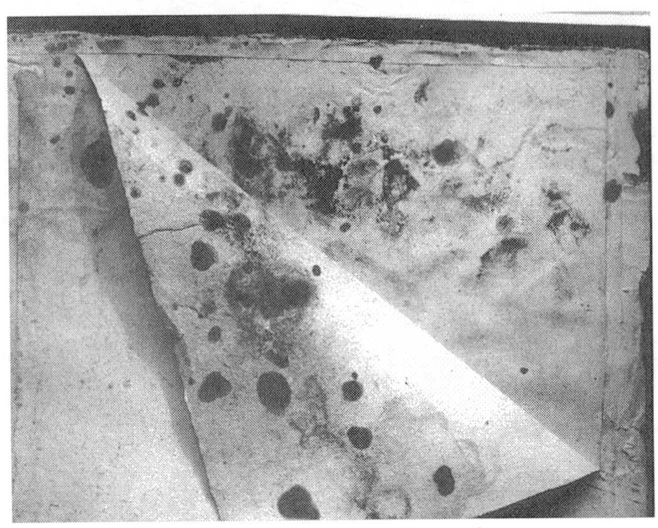

Fig. 2.

Table 1. Fungi isolated from paper materials and their symptoms

Fungi	Symptom	References
Acrostalgmus cinnabarinus	—	Kowalik *et al.* (1962, 1963)
Acrothecium sp	Olive green stain	Kowalik *et al.* (1962, 1963)
Alternaria sp.	Brown grayish spots	Kowalik *et al.* (1962, 1963)
Alternaria alternata	—	Bonaventura and Composano (1946); Niuksha (1956); Gallo (1961); Kowa-lik *et al.* (1962); Savulescu and Lazer (1971); Czerwinska & Kowalik (1979); Marcheva (1985); Chiraporn (1995)
Ascotricha guamensis	Black spots	Chiraporn (1995)
Aspergillus candidus	White or creamy spots	Nyuksha (1956, 1985); Orlita (1977); Czerwiniska and Kowalik (1979); Chiraporn (1995)
A. glaucus	Olive green spots	Chiraporn (1995)
A. flavus	Yellow green to brown spot	Nyuksha (1956, 1974); Smirnova (1962); Gallo (1961); Savulescu and Lazer (1971); Petrova and Erastov (1959); Orlita (1977); Czerwiniska & Kowalik (1979); Chiraporn, (1995)
A. fumigatus	In the beginning gray afterwards green and at last dull brown	Nyuksha (1956, 1974); Kowalik *et al.*, (1962); Savulescu and Lazer (1971); Petrova & Erastor (1959); Orlita (1977); Czerwinska & Kowalik (1979); Kowalik *et al.* (1962, 1963); Chiraporn (1995).

Continue....

Fungi	Symptom	References
A. niger	—	Bonaventure & Camposano (1946); Niuksha (1956, 1974); Kowalik and Sadurksha (1956), Kowalik *et al.* (1962, 1963); Gallo (1961); Savulescu and Lazer (1971); Orlita (1977); Czerwinska and Kowalik (1979)
A. ochraceus	—	Savulescu and Lazer (1971); Orlita (1977)
A. terreus	Light brown spot	Niuksha (1956, 1974); Czerwinska and Kowalik (1979), Chiraporn, (1995)
A. versicolour	—	Niuksha (1956, 1974); Beljakova (1960); Gallo (1961); Gallo and Strzelezyk (1971); Savulescu and Lazer (1971); Orlita (1977)
A. wentii	Light rosc	Kowalik ct *al.*, (1962,1963)
Aurobasidium pullulans	Black spot	Bavendamn (1991); Kowalik *et al.* (1962); Savulescu and Lazer (1971); Chiraporn (1995)
Botrytis cinerea	—	Niuksha (1956, 1974); Kowalik *et al.* (1962, 1963)
Botryotricum atrogriseum	Steel gray shade spots	Niuksha (1956); Savulescu and Lazer (1971); Kowalik *et al.* (1962, 1963)
Cephalosporium acremonium	Orange to rose tint	Kowalik *et al.* (1962, 1963)
Chaetomium bostrychodes	—	Florenzano (1948, 1949)
C.elatum	—	Niuksha (1956,1974); Gallo (1961); Kowalik *et al.* (1962); Savulescu and Lazer (1971)
C. globosum	Green black spot	Bonavetura and Composano (1946); Florenzano (1948); Niuksha (1956); Petrova and Erastov (1959); Beljakova (1960); Gallo (1961); Kowalik *et al.* (1962); Savulescu and Lazer (1971); Czerwinska and Kowalik (1979); Chiraporn (1995)
C. indicum	—	Florenzano (1949); Gallo (1961); Kowalik *et al.* (1962); Savulescu and Lazer (1971); Niuksha (1974)
C. murorum	—	Florenzano (1948); Savulescu and Lazer (1971)
Cladosporium cladosporioides	—	Chiraporn (1995)

Continue....

Fungi	Symptom	References
C. herbarum	Blackish green spot an paper	Bona ventura and Composano (1946); Niuksha (1956); Kowalik and Sadurksha (1956); Gallo (1961); Kowalik *et al.* (1962, 1963); Savulescu and Lazer (1971); Marcheva (1985)
Chaetomella horrida	Gray brown spot	Kowalik *et al.* (1962, 1963)
C. oxysporum	—	Chiraporn (1995)
Curvularia eragrostidis	Purple red spot	Kowalik *et al.* (1962, 1963); Chiraporn (1995).
C. geniculata	—	Kowalik *et al.* (1962, 1963); Chiraporn (1995)
C. lunata	Black spots	Kowalik *et al.* (1962, 1963); Chiraporn (1995)
Drechslera sp.	—	Chiraporn (1995)
Eidamella spinosa	Developed amaranthine spots	Kowalik *et al.* (1962, 1963)
Emericella nidulans	Purple red spot	Chiraporn (1995)
E. variecolor	Deep purple colour spot	Chiraporn (1995)
Epicoccum nigrum	—	Kowalik & Sadurksha (1956); Kowalik *et al.* (1962, 1963); Niuksha (1956)
Fusarium sp	Violet tints, blue spots, Pink violet spots and Yellowish pink spots	Kowalik *et al.* (1962, 1963); Chiraporn (1995)
Gliocladium roseum	Stained paper rose	Kowalik *et al.* (1962, 1963)
Gliomastix sp.	—	Chiraporn (1995)
Gliomastix convoluta	Stained black paper	Kowalik *et al.* (1962, 1963)
Haplographium fuligineum	—	Niuksha (1956); Beljakova (1960); Gallo (1961)
Helminthosporium sp.	Gray black spot	Kowalik *et al.* (1962, 1963)
Humicola sp.	—	Chiraporn (1995)
Memnoniella echinate	—	Chiraporn (1995)
Monilia sp	—	Chiraporn (1995)
Mucor racemosus	—	Niuksha (1956); Kowalik and Sadurska (1956); Beljakova (1960); Gallo (1961)
Mucor sp.	—	Chiraporn (1995)
Mycogane nigra	Green spot	Kowalik *et al.* (1962, 1963)
Neurospora sitophyla	—	Niuksha (1956); Petrova and Erastov (1959); Beljakova (1960)
Nigrospora sp	—	Chiraporn (1995)

Continue....

Fungi	Symptom	References
Paecilomyces variotii	Light yellow brown shade	Niuksha (1956, 1974); Beljakova (1960) Gallo (1961); Kowalik (1962, 1963); Czersinska & Kowalik (1979)
Paecilomyces sp.	—	Chiraporn (1995)
Penicillium aurantigriaeum	—	Niuksha (1956); Kowalik and Sadurska (1956); Kowalik *et al.* (1962); Savulaescu and Lazer (1971).
P. brevi compactu	—	Kowalik & Sadurska (1956); Savulescu and Lazer (1971); Czerwinska and Kowalik (1979, 1956)
P. chrysogenusm	—	Czerwinska and Kowalik (1980); Niuksha (1956); Savuleseu and Lazer (1971); Strzelczyk (1968); Orlita (1977); Voronina & Nazarova (1980)
P. decumbens	—	Niuksha (1956, 1974); Orlita (1977)
P.expansum	—	Bonaventura & Composano (1946); Niuksha (1956); Savulescu and Lazer (1971)
P. funiculosum	—	Niuksha (1956, 1974); Gallo (1961); Kowalik *et al.* (1962); Orlita (1977)
P. glabrum	—	Niuksha (1956); Beljakova (1960); Kowalik *et al.* (1962); Strzelczyk (1968); Savulescu and Lazer (1971)
P. oxalicum	—	Beljakova (1960); Savulescu and Lazer (1971); Niuksha (1974)
P. purpurogennum	Red colour spot	Chiraporn (1995)
P. roquefortii	—	Beljakova (1960); Niuksha (1956, 1974)
Phoma humicola	Gray brown to black spot	Kowalik *et al.* (1962, 1963)
Pleospora herbarum	—	Beljakova (1960); Niuksha (1956, 1974); Orlita (1977)
Rhizopus stolonifer	—	Niuksha (1956, 1974); Kowalik and Sadurska (1956); Savulescu and Lazer (1971)
Scopulariopsis brevicaulis		Niuksha (1956, 1974); Kowalik and Sadurska (1956); Kowalik *et al.* (1962)
Sepedonium sp.	Gray spot	Kowalik *et al.* (1962, 1963)
Spicaria sp.	Lilac spot	Kowalik *et al.* (1962, 1963)
Sporotricum bombycinum	—	Niuksha (1956); Kowalik & Sadurska (1956); Gallo (1961); Petrova and Erastov (1959); Czerwinska and Kowalik (1979)
S. polysporum	—	Niuksha (1956, 1974); Beljakova, (1960); Gallo (1997)

Continue....

Fungi	Symptom	References
Stachybotrys atra	Black spot	Gallo (1997); Beljakova (1960); Niuksha (1956); Chiraporn (1995); Kowalik *et al.* (1962, 1963).
Stemphylium botryosum	—	Marcheva (1985)
Trichocladium sp.	Stained greenish yellow or greenish rusty	Kowalik *et al.* (1962,1963)
Trichothecium roseum.	Initially white then pale rose stains	Niuksha (1956)
Trichoderma harzianum	—	Chiraporn (1995)
T. koningii	—	Gallo (1961); Niuksha (1956); Beljakova (1960)
T. viride	Light green stain	Kowalik and Sadurusha (1956); Kawalik *et al.* (1962, 1963)
Torula herbarum	Dark brown spots	Kowalik *et al.* (1962, 1963); Chiraporn (1995)
Verticillium sp.	—	Chiraporn (1995)

Effect of Fungi and its Mechanism in Deterioration of Paper

Surface damage

The microscopic film of condensed material when placed on the surface of the paper to create an image is sufficient to support fungal growth. A thin layer of pastel or watercolour is therefore an encouraging environment, which give rise to fungal development. Fungi grow by apical extension of hyphe in a radiant fashion. The rate of growth depends on the availability of nutrients provided by the paper and on environment factors. In addition to the nutritional factors fungal growth is limited by the degree of access of fungi to the substrate (Moo-Young *et al.,* 1983).

Fungi often adhere firmly to the surface and strong attachment is observed especially in the region of cracks and holes. The difficulty in removing fungi by rinsing has suggested that they are protected not only by irregularities of the surface to which they adhere, but their attachment also facilitates the production of adhesive like substances and/or anchor like hold fast structures. In fact many filamentous fungi secrete extra cellular polysaccharides with adhesive properties.

Various types of fungal colonies can be observed on paper objects. Some fungi tend to create brittle surface pads in contrast to the luxuriant growth of others. Some fungi are characterized by extremely abundant conidiophore formation and under proper conditions sporulate profusely, other grow sterile mycelia. Leathery colonies may survive for years while ever succulent types are less durable (Cochrane, 1958).

Some of the fungi enter into the structure of paper and medium by the formation of specific fruiting bodies such as pycnidia, acervuli, perithecia. The position of fruiting body can be superficial, erumpent or immersed (Anisworth, 1971). *Chaetomium (*ascomycete fungus) is easily recognized by conspicuous hair (setae) growing on perithecium. The perithecia are produced superficially without a stroma and are clearly attached to the paper by the hair. The coelomycete fungus *Phoma* produces pycnidia, which are buried in the paper fibers. *Pestalotia* produces discoid or cushion shaped fruiting structure called acervuli.

All the mentioned forms pycnidia, peritheciea, acervuli are dark brown in color or black and create surface accretion which obscure the image in case of paper paintings.

Structural damage

Fungi deteriorating paper materials adversely affected their longevity by the enzymatic digestion of cellulose fibers of paper component. The process of secretion occurs in advancement of hyphal tips and in the first stage of infestation, creates boreholes in the surface of paper. These facilitate the penetration of hold fast into the paper surface.

In order to visualize clearly the mechanism of fungal infestation on the cellulose chain, a short review of its molecular architecture is needed. Cellulose is a linear homopolymer of anhydroglucose units linked together by B-D-1-4 glucosidic bonds. Long glucan chains present in the molecule are associated in parallel arrangements with adjacent chains forming insoluble thread like micro fibrils (Brown *et al.,* 1981). The basic repeated unit in the molecular chain is anhydrous cellobiose (Atalla, 1981). The units' possess two different conformations described as crystalline and amorphous. The crystalline is highly organized and more resistant to fungal attack than amorphous.

The group of enzyme involved in cellulose hydrolysis is termed the cellulose system. The number of enzymes, however, is unknown. So far, three types of enzymes have been recognized (i) Five different endo, 1, 4-β glucanases which attack at random the 1,4-β linkage along the cellulose chain (ii) one enzyme exo – 1, 4-β glucanases, which separates cellobiose or glucose units from the non reducing end of the cellobiose (iii) two 1, 4-β glucosidases which hydrolyse cellobiose are water soluble cello-dextrin to glucose and gluconolactone (Eriksson, 1981; Wood, 1981; Kirk Kent, 1983).

Amorphous cellulose may be degraded by endo and exo glucanases separately. Crystalline cellulose is hydrolyzed by the synergistic action of Endo and exo glucanases (Brown *et al.,* 1981; Eriksson, 1981). There are various theories on the sequence of the action of each enzyme.

Thc visible effects of the enzymatic activity of fungi are thin weakened area which gradually become thinner and softer leading eventually to the complete loss of the paper. The structural damage in paper can be treated in the manner as mechanical abrasions and paper losses.

Marcheva (1985) studied the cellulose activity of the dark micromycctes which damage paper products (*Cladosporium herbarum,*

Stemphylium botryosum and Aternaria alternata) isolated from damaged documents of the Bulgarian state archives. For different species and enzymes the peak of activity ranged from 10-16 days of incubation. The addition of starch caused an increase of enzyme activity up to 9 folds in *A. alternata*.

As a result of microbial action on paper there is a considerable reduction in the alpha cellulose (Pietrykowaska, 1979) and an increase in copper number (Nyuksha, 1954; Strzelczyk and Leznicka, 1981). The weight of the cellulose usually shows only an insignificant decrease but its mechanical strength decreases considerably.

Pigmentation

Fungi deteriorating the paper, affect the surface characteristics of the image by various forms of discoloration caused by pigments produced by them. This observation may be applied to the majority of fungi imperfecti and in particular the genera *Aspergillus* and *Penicillium*. Pigments may be incrusted in spores present in mycelium and secreted to the substrate (Alexopoulos and Mims, 1979). Nyuksha (1956) studied the pigmentation of fungi growing on paper and reported the color of the pigment may change depending on the condition of the growth and properties of the paper. It was explained that *Gymnoascus sitosus* forms no pigment at all on the paper and colophony. On barytra paper it produces a fine crimson red color and on rag paper with starch sizing, lemon yellow. In the same way, *Penicillium pinophilium* forms a red pigment on rag paper with gelative sizing and on filter paper a pink shade. Later on, Kowalik *et al.* (1962, 1963) observed different kinds of pigmentation on paper that was attributed to growth of following fungi:

Acrostalgmus cinnabarinus, Acrothecium sp., *Alternaria* sp., *Aspergillus fumigatus, A. niger, A. wentii, Botryotrichum atrogriseum, Cephalosporium acremonium, Chaetomium horrida, Cladosporium herbarum, Curvularia lunata, Eidamella spinosa, Epicoccum nigrum, Fusarium* sp., *Gliomastix convouta, Gliochladium roseum, Helminthosporium* sp. *Mycogane nigra, Sepedonium* sp., *Spicaria* sp., *Stachybotrys atra, Torula* sp., *Trichoderma viride, Trithocladium* sp. *Trichothecium roseum, Phoma humicola.*

Nyuksha and Kossior (1976) demonstrated that during the cohabitation with other fungi A. *flavus*. Stimulates pigmentation of other fungi like *Myxotricum deflexum* or paper while suppresses the growth of V*erticillium tenerum, Stachybotrys charatarum,* etc.

Foster (1949) explained that the red colouration of kojic acid with iron is an outstanding example. However, Butcher (1968) explained that the colourless metabolic products of fungi discharged to the substrate react chemically with it and the paper yields a coloured compound. Chiraporn (1995) also studied the pigmentation and discolouration of paper by several fungi.

Foxing the paper

A characteristic brown rusty spotted discoloration found commonly on old books, lithographs, manuscripts, etc. is referred to as "foxing". The Japanese term for this is "Hoshi". A number of scientists have worked on this aspect (Chalk, 1960; Carter, 1972; Cain and Miller, 1982; Cain and Miller, 1983; Kenzo *et al.,* 1987; Arai, 1984, 1987a, b). but so far no lasting restoration treatment has been reported. In fact the reappearance of foxing is of much concern as this clearly indicates that the treatments being applied are not adequate.

Two possible reasons have been suggested; (i) the stain arise from the breakdown of metal impurities in the sheet with iron, the most frequently cited cause, (ii) The sufficient high humidity and available nutrients in the paper allow a specific fungus or fungi to thrive. Press, (1974) analyzed total iron concentration by x-ray fluorescence spectroscopic method.

It was concluded that the foxing is associated with low rather that higher iron latent. Tang (1978) suggested that in the formation of foxing spots, copper and iron seem to play a prominent role. It was higher in foxed spots. However, copper does not show clear effect. Since both iron and copper catalyze the oxidation of cellulose these agent may be primarily or completely responsible for the damage.

The other possibility may be involvement of fungal infection in the formation of foxed spots (Meynell and Newsam, 1978; Nol and Kenneth, 1983). When paper become sufficiently damp, fungal spores germinate and mycelial development begins particularly at sites relatively rich in nutrients. However, lesions evidently progress extremely slowly either because fungal growth is intermitted or because nutrient are scarce and exhausted before much growth is occurred. As result hyphae are generally scarce in the lesions or may be dead or absent from many parts of them and paper, therefore, never looks moldy. Meynell and Newsam (1978) and Nol and Kenneth (1983) isolated species of *Aspergillus, Penicillium* and *Gliocladium* from foxed stamps.

To confirm that there is a biological agency to cause foxing, various biochemical analyses were carried out (Arai *et al.,* 1968). It was presumed that the browning reaction takes place in paper at accumulation of moisture, caused by local condensation process in the book. It is expected that many factors influencing condensation and evaporation will play or role in this respect, additional research to be planned to verify the condensation theory experimentally (Ligterink *et al.,* 1989).

Arai *et al.* (1990) reported in detail that combination of components like malic acid, glucose, cello-oligo saccharides and amino acid, kept in particular environment can induce foxing and it was found that L-malic acid, glucose and y-amino-butyric acid are the main causes in the formation of foxing the optimum environment for the induction of foxing is relative humidity 75% and temperature 35^{0}C

FACTORS AFFECTING FUNGAL GROWTH

The microbial spores, which are always present in air, come in contact with paper materials and after getting proper humidity (more than 75%) and temperature (25-35^{0}C) starts to germinate. Anchabagze (1949) published a report of an investigation of manuscripts at the Georgian State Museum and enlisted 17 species of fungi from air and considered them to be the principal cause of the abundant contamination of the material. Nyuksha (1954a, 1954b) isolated fungi from books and from air of the library. Finally, the relationship between the mycoflora of paper was studied. Nyuksha (1956) again explained that many species do not take active part in the destruction of paper fiber. The particular species referred to were from *Mucorales, Monilia, sitophila* and many representatives of *Penicillium* variety *asymetrica*. These fungi are widely distributed in nature, and it is difficult in practice to protect the samples from them. Baynes Cope and Allsopp (1984) observed mold growth in libraries. Tripathi (1987) indicated the presence of 19 types of fungal spores in a library. Nitterus (2000) also reported the occurrence of fungi in archives and libraries. Finally, the books lose their leaf structure and become converted into a deformed solid mass.

Humidity

The single most important factor governing fungal growth is the presence of water. It has been common practice to recommend various humidity levels to avoid fungal bloom. However, these relative humidity levels eq. below 75% (65-70%), Thomson (1986), 70%, Stolow (1985), 50-60% Bevaringshand and Bogen (1994), Rebrikova and Manturovskaya (1993) cannot be considered completely safe, since fungal growth has been reported although R_H levels were maintained at, or below, supposedly non-favourable levels (Nyuksha, 1983, Gallo, 1985). Other authors Florian (1993) Scott (1994) stress the importance of considering the amount available water held within such hygroscopic materials such as paper wood and textiles. This availability or activity, of water (aw), naturally depends the amount of humidity in the surrounding air (R_H). The relationship between aw and R_H at equilibrium can be expressed as.

$$R_H\ (\%) = aw \times \frac{1}{100}$$

that is an aw of 0.65 correspond to an R_H of 65%. aw is defined as ratio of the vapor pressure of water over the material to the vapor pressure over water at the same temperature (Allsopp and Seal, 1986). A rise/fall in R_H of the microclimate surrounding the object consequently induces it to gain/lose moisture to restore equilibrium. The availability of water is a substrate expressed as aw, governs the growth of microorganism, generally having a tolerance in aw between 1.0 and / 0.55 (Florian, 1993), the aw being dependent an the hygroscopic capacity of the material which in turn is influenced by several factors, *e.g.* pH, presence of solutes, oxygen permeability, deterioration level etc. aw values being the most favourable, or optimum, for growth, seems mostly to be above 0.85, very few below 70 (Scott, 1994). Fungi as a part of their survival strategy are also believed to withstand low aw by producing polyols *e.g.* Glycerol, working as osmoregulators, water observers, thus enabling retention of rather large quantities of water.

It is of major importance for the survival of fungi that the regulation of osmotic pressure by the plasma membrane is maintained as otherwise important intracellular component would start leaking out, causing lysis. Some osmotolerant fungi are especially adapted to environments with high osmotic pressure/low aw, and maintain their growth even at such harsh conditions. *Aspergillus flavus* is reported to survive in very severe conditions, seemingly impossible for maintained growth and reproduction Nyuksha (1983).

pH

A majority of the micro fungi prefer a slightly acidic environment with a pH of 4-6, and by excreted metabolic waste products also able to change the pH of the habitat of not at an optimum, *e.g.* by production of oxalic acid (Caneva *et al.*, 1991). The formation of stains due to coloured products in spores, hyphae and excreted metabolites is believed to be pH dependent, with a maximum stain formation at pH 5 (Szeepanowska and Lovett, 1992).

Temperature

The effect of temperature an growth and survival of the fungi is of considerable importance since it dictates the rate of reactions catalyzed by enzymes. Three cardinal temperatures are usually determined, the minimum, below which no growth occurs, the optimum when growth is more rapid, the maximum, when the fungi are starting to get lethally injured through denaturation of enzymes, disruption of membranes, etc. The temperature below, which no growth occurs, is not to be understood as a lethal temperature.

Many fungi will survive for the period of at least several months of continuous sub zero temperature. Fluctuating freeze thaw cycle reduce viability, and finally exhaust even dormant spores, though initially activating (Florian, 1993). High temperature, such as moist heat (autoclaves) is quickly lethal, but is not relevant to conservation practice since the heat needed to kill the fungi would also be harmful to the treated objects.

Spores and hyphae in mature colonies showing active growth contain considerable amount of water and the majority are killed at temperature below $0^{0}C$, and above or close to $40^{0}C$. These are, however, exceptions and depending on the conditions at which growth of the microorganism is optimum, they are divided into different groups.

Thermophilic organisms colonize habitats of high temperatures >45^{0}C, optimum 55-65^{0}C having more heat stable enzymes, more saturated membrane lipids than mesophilic organisms.

Mesophilic organisms colonize habitats of moderate temperature, 20-45^{0}C, ordinary room temperature, many of these mesophilic organisms are pathogenic, showing optimum growth at 37^{0}C.

Psychrophilic organisms grow at low temperatures, below 15^{0}C (0^{0}C).

Oxygen and Light

For the growth of fungi oxygen is required therefore restricted aeration may alter the type of fungal growth and normal pigmentation.

The light stimulates the the formation of fruiting bodies as well as enhancing pigment production. Cochrane (1958) suspected that strong pigmentation in the presence of light has a protective function for the fungus.

Nutrients

Fungi can utilize almost any naturally occurring compound as a source of carbon and energy as long as the availability of water is sustained. Other elements required are oxygen, nitrogen, sulfur, potassium, magnesium and various trace elements and to some extent also vitamins. Since the fungi, as earlier mentioned, as a part of their survival strategy can alter their metabolism in order to compensate the lack of some nutritional element, it should be remembered that this often leads to alterations in structure, hue and persistence of produced pigments, reproduction strategies, etc.

CONCLUSIONS

Fungi are well known organisms for the deterioration of all types of paper materials. However, qualitative/quantitative differences that can affect the biotic and abiotic deterioration process characterize different kinds of paper. There are about 78 fungal species belong to different genera are so far reported from various types of paper materials. These fungi causes surface damage, structural damage (deterioration of the longevity of the paper fiber by the involvements of enzymes) and pigmentation/staining (affects the surface characteristics of the image). A characteristic brown rusty spots can be commonly found on old books, lithographs, etc., due to the sufficient high humidity and available nutrients in the paper. The combinations of L-malic acid and glucose, y-aminobutyric acid are the main causes in the formation of foxing. In the development of diverse group of fungi humidity, temperature, pH, oxygen, light and nutrients are responsible one or other way and complimenting each other.

REFERENCES

Alexopoulos, J.C. and Mims, W. Ch. (1979). *Introductory Mycology*, 3rd ed. John Wiley & Sons Inc, New York.

Allsopp, D. and Seal, K.J. (1986). *Introduction to Biodeterioration*. London. Arnold.

Anisworth, G.C. (1971). *Anisworth & Bisby's Dictionary of Fungi*, 6th ed, CMI Kew, England.

Arai, H. (2000). Foxing caused by fungi: Twenty five years of study. International Biodeterioration and Biodegradation. 46 : 181-188.

Atalla, R.H. (1981). *Chemistry and Biochemistry of Wood Based Processes and Products*, 1, Stockholm. 223-241 Elsevier, Amsterdam.

Baynes Cope, A.D. (1981). *Caring for Books and Documents*. British Museum Publications. Ltd.32 pp ISBN 017141 2006 5

Caneva, G., Nugari, M.P. and Salvadori, O. (1991). *Biology in the conservation of works of art Rome*: I CC ROM 26-43.

Capitani, D., Emanuele, M.C., Segre, A.L., Fanelli, C., Fabbri, A.A., Attanasio, D., Focher, C. and Capretti, G. (1998). Early detection of enzymatic attack on paper by NMR relaxometry, EPR Spectroscopy and X-ray power spectra. Nordic Pulp and Paper Res. J. 13 (2) : 95-100.

Carter, J. (1972). ABC for Book Collectors, 5th Ed. Hart Davies, London.

Chiraporn, A. (1995). Microscopical study of fungal growth on paper and textiles. Biodeterioration of cultural property 3. Proceeding of the 3rd international conference Bangkok, Thailand. 83-95.

Cochrane, V.W. (1958). *Physiology of Fungi*. John Wiley & Sons Inc., Now York.

Coughlan, M.P., Touhy, M.G., Filho, X, Puls, J., Claeyssens, M., Vrsanska, M. and M.M. Hughes (1993). Enzymological aspects of microbial emphasis on fungal system. Hemicellulose and Hemicellulase. Coughlan, M.P. and Hazle Wood, G. P. Eds. Portland Press London 53-85.

Czerwiniska, E. and Kowalik, R. (1979). Microbial deterioration of audiovisual collections. *Restaurator* 3: 63-80.

Dhawan, S. (1986). *Microbial deterioration of paper material. A literature review*. Govt. of India, Department of Culture, National Research Laboratory for Conservation of Cultural Property, Lucknow.

Dhawan, S. and Agrawal, O.P. (1986). Fungal flora of miniature paper paintings and lithographs. *International Biodeterioration* 22 (2): 95-99.

Ewald, G. (1966). Mikro organismen als Schadlinge in bibliotheken und archiven. *Biblio thek und Wissenschaft*. 3 : 13 –112.

Florian, M.L. (1993). Conidial fungi (mould) activity on artifact materials - A New look at prevention, control and eradication. ICOM committee for conservation of 10th Triennial meeting, Washington. 868-874.

Florian, M.L. (1996). The role of the conidia of fungi in fox spots. *Studies in Conservation*, 41 : 65-75.

Foster, J.W. (1949). *Chemical Activities of fungi*. Academic Press, Inc. New York.

Gallo, F. (1985). *Biological factors in the deterioration of books*. Technical notes Rome ICCROM.

Havermans, J.B. (1994). Environment et conservation de l' Ecrit, del' Image et du son. Acterdes douxiemes Journees Internationals d' Etudes de PARSAG, Paris. 16/20-5-94: 39-47.

Kathpalia, Y.P. (1960). *Deterioration and Conservation of paper.* Biological deterioration. *Indian Pulp. Paper* 15(1): 117-125.

Kirk Kent, T. (1983). Fungal Technology. 4, Edward Arnold, London.

Kowalik, R.I., Sadurska and E. Czerwinska. (1962). *Microbiological Deterioration of old Book Remedies. Bollentino,* dell Instituto di Pathlogia del O. 'Libro "Alfonso Gallo". 21:116-151.

Kowalik, R.I., Sadurska and E. Czerwinska (1963). Zniszczenia Papieru Prezez Mikroflore. Blok-Notes Muzeum Mickicwicza. 2:259-312.

Kowalik, R. (1980a). Microbiodeterioration of library materials. Part II. Microbiodecomposition of basic organic material. Chapter IV. *Restaurator* 4.: 135-219.

Kowalik, R. (1980b). Microbiodeterioration of library materials. *Restaurator* 4 : 99-114.

Kowalik, R. (1984). Microbiodeteriation of Library materials. *Restaurator* 6 : 61-115.

Ligterink, Frank, J., Povek, H.J. and Smith, W.H. (1989). *Restaurator* 3: 225-233

Moo-young, M., Moreiva, A.R. and Tengerdy (1983). Fungal Technol-4., Edward, Arnold, London.

Nitterus, M. (2000). Fungi in archives and Libraries. *Restaurator* 25-40.

Nol, Lea, Henis, Y. and Kenneth, R.G. (1983) Biological Factors of Foxing in Postage Stamp Paper. *Int. Bio. Bull.* 19: 19-25.

Nyuksha Yu. P. (1956). Microflora of books and paper. Bot. Lh. 41 (6): 797-809.

Nyuksha, Yu. P. (1983). Some Special cases of Biological Deterioration. *Restaurator* 5 : 177-182.

Orlita, A. (1977). The occurrence of fungi on book, leather, buildings from the Baroque period. *International Biodeterioration Bulletin.* 13:45-47.

Priest, D. J. and Stanley, J. (1994). The ageing of paper containing Chemithermo mechanical pulp. Environment et conservation de 1'Ec rit de 1' Image et du son. Actes des douxiemes Journees internationals de' Etudes de 1' ARSAG, Paris. 16/20-5-94: 154-160.

Rebrikova, N.L. and Manturovskaya, N.V. (1993). Study of factors facilitating the loss of viability of microscopic fungi in Library and museum collections. In ICOM 10th meeting Preprints, Washington. 887-890.

Rebrikova, N.L. and Manturovskaya, N.V. (2000). Foxing a new approach to an old problem *Restaurator.*85-100.

Ricelli, A., Fabbri, A.A., Fanelli, C., Menicagli, R., Samaritani, S., Pini, D., Rapaccini, S.M. and Salvadori, P. (1999). Fungal growth on samples of paper : Inhibition by new anti fungal. Restaurator 97-10

Scott, G. (1994). Moisture, Ventilation and mould growth. Preventive Conservation Practice, Theory and research. Ottawa: International Institute for Conservation of Historic & Artistic Works. 149-153.

Stolow, N. (1985). Conservation and Exhibition. Packing, Transport & Storage Consideration. London: Butterworths.

Strzelczyk, A.B. (1968). Influence of antifungal vapours on spore germination of fungi isolated from deteriorated old book. *Canadian Journal of Microbiology.* 14 : 901-906.

Strzelczyk, A.B. and Leznicka, S. (1981). The role of fungi and bacteria in the consolidation of books. *International Biodeterioration Bulletin.* 17: 57-67.

Szeepanowska, H. and Lovett, Ch. M. (1992). A study on the removal and prevention of fungal stains on paper. *J. Am. Inst. Cons.* 31 : 147-160.

Teeri, T.T. and Koivula, A. (1995). Cellulose degradation by native and engineered fungal cellulase carbohydrates in Europe: 28-33

Thomson, G. (1986). The museum environment London: Butterworth.

Voronina, L.I., Nazarova, O.N. and Petushkova Yu. P. (1980) Desinfection and straigh-tening of parchment damaged by micro-organisms. In: Restaurator. International journal for the preservation of library and archival material (Kopenhagen) 4 (1980) S. 91-97.

Wessel, C.J. (1970). Environmental factors affecting the permanance of library materials. In Deterioration Band Preservation of Library Materials. The University of Chicago press, Chicago. 39-84.

Wood, T.M. and Compayo Gorcia, V. (1994). Enzymes & mechanism involved in microbial cellulolysis. Biochemistry microbial degradation, Ed. C. Ratledge. Dordrecht: Kluver 197-231.

Zabel, R.A. and Morvell, J.J. (1992). *Wood microbiology decay and its prevention.* New York Academic Press Inc. 476.

Zyska, B. (1993). Preservation of library materials. Vol 2. Factors Deteriorating Materials in Library Collection (In polish). Uniwersytet SIgski katowice.

Zyska, B. (1996) *Book Biology.* Part 1 (in Polish) Wojewodzka Biblioteka, Publiczna Katowice.

Zyska, B. (1997). Fungi isolated from Library material : A review of literature. *International Bideterioration and Biodegradation.* 40(1) : 43 – 51.

CHAPTER – 5

ROLE OF ECTOMYCORRHIZAL FUNGI AS BIOFERTILIZERS IN TREE NURSERIES

V. Mohan

Division of Forest Protection, Institute of Forest Genetics and Tree Breeding, Coimbatore - 641 002

INTRODUCTION

Microorganisms play vital role in physiological processes in the ecosystems. Beneficial rhizosphere fungi and bacteria (Biofertilizers) carryout a range of functions and are great relevance to plant growth. The most beneficial contributions of soil microorganisms to plant development are the supply of essential nutrients for plant growth, particularly those involved in nitrogen and phosphorus cycling. Nitrogen fixation is an important factor in biological productivity. More than 60% of the Nitrogen input to the plant community through fixation has a biological origin (Postgate and Hill, 1979).

Symbiotic rhizosphere microorganisms (Biofertilizes) play a critical role in the adaptation and survival of land plants. Among the micro-organisms inhabiting rhizosphere, certain species of fungi are present quite consistently on the root surface or in the tissues and cells of the root in many plant species. These are mycorrhizal, where the fungus and the host co-exist in a physiologically and reproductively active state for long periods.

BIOFERTILIZERS

The biofertilizers or microbial inoculants can be generally defined as preparations containing live or latent cells of efficient strains of nitrogen fixing, phosphate solubilizing or cellulolytic microorganisms used for the application to the seed, soil or composting areas with the objectives of increasing the numbers of such microorganisms and accelerate certain microbial processes to augment the extent of availability of nutrients in a form which can be easily assimilated by plants. Biofertilizers constitute beneficial microorganisms both symbiotic and free living forms in the soil that provide nutrients to the plants in available form through natural processes.

Symbiotic organisms like *Rhizobium* contribute a good deal to the amount of nitrogen fixed into the biosphere. The *Rhizobium* infection causes root nodules which is not merely a store house or protein but has a casual relation with the assimilation of free nitrogen. Another important group of symbiotic microbe is the mycorrhizal fungi. The fungal sheath (mantle) together with the hyphae extending to the soil absorb nutrients and other essential elements and supply to the host plants. They afford better intake of nutrients such as nitrogen, phosphorous and potassium from the surrounding soil. Metabolites produced by the mycorrhizal fungi influence the structure and morphology of root system. The mycorrhizal fungi and the host plants have mutual control over the morphological and physiological activities of the symbiotic system.

TYPES OF BIOFERTILIZERS

There are four major types of biofertilizers in use. They are as follows:

- Actinorhizal Biofertilizers
- Bacterial Biofertilizers
- Cyanobacterial Biofertilizers
- Fungal Biofertilizers

Actinorhizal biofertilizers

The microbes belonging to this group are called as *Frankia*. They are highly efficient micro aerophilic nitrogen fixers that occur within the root nodules of plants. They infect non-leguminous woody plants such as *Alnus, Casuarina,* etc. There are 24 genera from eight angiosperm families which have been reported to possess antinorhizal root nodules (Dixon and Wheeler, 1986). Among them, the genera *Casuarina* for tropical and subtropical regions and *Alnus* (*A. rubra, A. glutinosa, A. crispa, A. jorullensis, A. acuminata*) for temperate regions stand out as excellent examples for the benefits they provide to the ecosystem by way of nitrogen inputs.

Bacterial biofertilizers

There are two types of bacterial biofertilizers. They are (a) Symbiotic forms and (b) Free living forms.

(a) Symbiotic Forms

The microbes belonging to this group are called as *Rhizobium* and *Bradyrhizobium*. The details of these mirobial groups are as follows:

Rhizobium* and *Bradyrhizobium

Many soils contain native strains of *Rhizobium* capable of nodulating a given species. They fix atmospheric nitrogen symbiotically and produce nodules in the leguminous plants of both agricultural and forestry tree crops. The bacteria attack the root hairs, penetrate the root and induce proliferation of the root nodule cells.

(b) Free Living Forms

The microbes belong to this group are of different forms. They are as follows:

Azotobacter

They fix nitrogen under aerobic conditions. They are saprophytes that occur in soil, water and in plant rhizosphere. This bacteria makes nitrogen available for crops like cereals, millets, vegetables and fruits.

Azospirillum

This is an aerobic or micro aerophilic, gram negative, helical shaped, motile with polar flagella. They can occur within the roots of grasses, wheat, corn and many other plants or as a free living soil organism.

Phosphobacterium

This bacterium helps in solubilizing phosphorus and makes it available for the plants. It mainly occurs in the rhizosphere soil.

Cyanobacterial Biofertilizers

The members belong to this group are *Azolla-Anabena* and *Nostoc*. Details of the same are given below:

***Azolla-Anabena* Symbiosis**

Azolla is a small, fast growing fern that occurs in water (floating). *Anabena azollae*, a cyanobacterium lives in the cavities of *Azolla* leaves. 50% of higher yield is obtained by use of *Azolla*, that fixes nitrogen from air and excretes the nitrogenous compounds into leaf cavity of the ferns.

Fungal biofertilizers

The symbiotic microbes belong to this group are called mycorrhizal fungi. The term "Mycorrhiza" was first coined by the German forest pathologist A.B. Frank in 1885, which literally means "Fungus Root" to denote the association between certain soil fungi and plant roots where the relationship is not pathogenic. It occurs in 90% of Angiosperms, 100% of Gymnosperms and 70% of Pteridophytes (Harley and Harley, 1987). The mycorrhizal associations are found in a range of habitats. These include ecosystems ranging from aquatic to desert and from lowland tropical rain forests to high altitudes and in canopy of epiphytes. The mycorrhizal infections are known to be significantly beneficial for the nutritional uptake of host plants especially phosphorus and nitrogen from deficient soils.

TYPES OF MYCORRHIZAL FUNGI

There are different kinds of mycorrhizal fungi. Among them, the most widespread and relevant to agriculture, horticulture and forest trees are ecto-and endomycorrhizal fungi.

Ectomycorrhizal fungi

This type of symbiotic association occurs mostly on trees belonging to the family Pinaceae *(Pinus* spp.) and Cupressaceae (*Cupressus* spp.) of Gymnosperms and few Angiosperm families such as Betulaceae (*Betula* spp.), Fagaceae *(Fagus*), Juglandaceae (*Juglandus*), Myrtaceae (*Eucalyptus* spp.) etc (Molina and Trappe, 1984). The majority of forest presently grown in plantations form ectomycorrhizas; some of them also have both ectomycorrhizal and AMF associations (e.g., Acacias, Casuarinas and Eucalypts). The fungi forming ectomycorrhiza mostly belong to the higher fungi (Basidiomycetes and Ascomycetes) primarily in families like Amanitaceae, Boletaceae, Cortinariaceae, Russulaceae, Trichlomataceae, Rhzopogonaceae, Sclerodermataceae, etc., some ascomycetes and one or two phycomycetes (Trappe, 1977). Among the basidiomycetes, the hymenomycetes fungi like species of *Amanita, Boletus, Cortinarius, Laccaria, Lactarius, Leccinum, Russula, Suillus,* etc. are important. The gasteromycetes include species of *Lycoperdon, Pisolithus, Rhizopogon, Scleroderma,* etc. (Smith, 1971).

The ectomycorrhizal fungi are characterized by the presence of complete sheath of fungal tissue covering the root epidermis called mantle. The mantle serves as a barrier and prevent the entry of soil-borne pathogens and certain parasitic nematodes into the root systems of the host plants. There is an intercellular infection forming a network of fungal mycelium around the cortical cells called Hartig-net. The Hartig-net structures serve as the centers for nutrient exchange between the host plants and the fungus. Most of the ectomycorrhizal fungi can be grown in synthetic medium. The pure culture of these fungi can be mass multiplied for further inoculation in nursery and field.

Endomycorrhizal fungi

This is another group of important type of mycorrhizal fungi. It comprises three groups such as (1) Arbuscular Mycorrhizal fungi (AMF), (2) Ericoid Mycorrhizae and (3) Orchid Mycorrhizae.

Arbuscular Mycorrhizal Fungi (AMF)

The AMF fungi are the most prevalent and disturbed in varied ecosystems. They occur in agricultural, horticultural and forest trees. They occur nearly in all the families of flowering plants. Hence, plants with AMF are common in most habitats. It is easier to say where they are not found. On a global scale, AMF fungi have a broad ecological range. They are found in most ecosystems. Their occurrence within these system varies according to localized environmental conditions and plant cover. The important tree species forming AMF fungal association include species of Acacias, *Araucaria, Casuarina, Eucalyptus, Podocarpus,* etc. Among hardwoods, Sweetgum (Liquidamber), Maple (Acer) and Yellow Poplar (*Liriodendron*) are important AMF hosts. AMF fungi belong to zygomycetous group. The presence of vesicular and arbuscular structures are the characteristic features of AMF fungi. The genera such as *Acaulospora, Entrophospora, Gigaspora, Glomus* and *Scutellospora* are the major constituent of AMF fungi. The fungi occur practically in all soils but soil type, management practices, climate determine the dominant species present and its population. AMF fungi are not host specific although evidence is growing that certain endophytes may form preferential association with certain host plants. In general, majority of forest trees used in forestry programme are infected with mycorrhizal fungi, thus playing a major role in managing plant productivity.

NEED FOR MYCORRHIZAL INOCULATION IN FORESTRY

Forestry has the main objective of developing and protecting forests for their maximum productivity. With increasing demand for wood, fodder, fuel and timber, plantation forestry has gained importance in recent years. Among different plantation species, *Acacia auriculiformis, A. holosericea A. mangium, Casuarina equisetifolia, C. junghuhniana* and *Eucalyptus* spp. are being cultivated in different parts of Southern India due to its high value of pulpwood, cheap timber and fuel. Also, these tree species are well suited for afforestation of wasteland areas and other problem soils.

In natural ecosystems, trees cannot survive without mycorrhizas. In afforestation sites devoid of mycorrhizal propagules (e.g., cutover lands, treeless areas, mine spoils and other wastelands), non-mycorrhizal seedlings never fully develop and die. In these conditions, the success of the plantation depends totally on the mycorrhizas associated with the seedlings coming from the nursery. In recent years, afforestation and rehabilitation of degraded lands and mined overburdens are being done on a large scale to meet the ever increasing demand for fodder, wood and fuel. Successful afforestation programmes depend on the quick establishment of forest tree seedlings, thus assuring continuous resources supply and vigour necessary to resist insect-pests and disease pathogens and to survive climatic stresses. On mine spoils, nutrient poor, drought prone or other stressful sites, there can be only a short favourable period for growth and the seedlings that do not get well established within this critical period are unlikely to survive (Perry *et al.*, 1987).

Inoculation of mycorrhizal fungi to tree seedlings greatly improve their survival and growth (Harley and Smith, 1983). The fungal hyphae in a mycorrhizal plant can ramify a large soil volume providing greater absorptive surface area than root hairs on a non-mycorrhizal plant. Some of the plants cannot establish in nature or maintain normal growth without mycorrhizal association (Rhodes, 1980).

SIGNIFICANCE OF MYCORRHIZAL FUNGI

Tree hosts are known to derive several benefits from the mycorrhizal associations and some of them are as follows:

- Mycorrhizal fungi are highly beneficial to the plants belonging to agricultural, horticultural and forestry crops in many ways. The AMF fungi remains outside the root in the form of extramatrical mycelia which ramified in the soil beyond the depletion zone and covers a large soil mass.
- Obviously absorptive surface of the root is tremendously increased through extended hyphal extension resulting in increase in uptake of nutrients and water manifold with concomitant enhancement in plant growth.
- They mainly aid in the uptake of nutrients such as phosphorus, zinc, copper, etc. (Hayman, 1983).
- They can affect osmotic adjustment and root hydraulic conductivity.
- The improvement in plant water relations may be due to mycorrhizal fungi exploiting large soil samples.
- They provide tolerance to the host plants against drought, soil toxins, etc. (Zak, 1971; Marx and Bryan, 1971).
- They also act as a bio-control agent against soil- or root-borne pathogens (Marx, 1969; Sampangiramaiah and Bagyaraj, 1989).
- Roots colonized by AMF fungi exhibit high chitinolytic activities. Increased aggregation of silty clay loam soil by mycorrhizal fungi make the soil more porous with greater water permeability than uninoculated soil.

Because of these attributes, the mycorrhizal fungi are now considered an important biofertilizer and utilized in forestry programmes. The role of mycorrhizal fungi in man made forests in India is highlighted by Bakshi (1967). Need for inoculating tree nurseries with mycorrhizal fungi may begin with initial nursery development. Generally, new nurseries are often established on sites that have not previously or ever supported mycorrhizal hosts.

DISTRIBUTION OF ECTOMYCORRHIZAL FUNGI IN INDIA

In India mycorrhizal research did not get underway until the early fifties. Chauduri (1945) was the first to report on the mycorrhizal association in *Abies spectabilis, Cedrus deodara, Picea morina, Pinus roxburghii* and *Taxus baccata.* Later, Bakshi (1957) started the research on ectomycorrhizal association in *Abies pindrow, Cedrus deodara, Picea morina* and *Pinus roxburghii.* Bakshi (1966) recorded *Scleroderma verrucosum* as the symbiont in *Eucalyptus.* Bakshi and Thapar (1966), Bakshi *et al.* (1966) described *Cenococcum graniforme* from planted and natural stands of *Abies pindrow, Cedrus deodara, Picea smithiana, Pinus gerardiana, P. roxburghii* and *P. wallichiana* from Dehra Dun and Himalayas. Bakshi (1974) reported the association of *Amanita hemibapha, A. verna, Astraeus hygrometricus, Cantharellus cibarius, Cenococcum graniforme, Geastrum fimbriatum, Lactarius scrobiculatus, Pulveroboletus shoreae, Rhizopogon flavum, Russula delica, Scleroderma bovista, S. cepa, S. geaster, S. verrucosum, Xerocomus bakshii* with different species of *Pinus* and other trees of Dehra Dun and Himalayan regions.

Misra and Sharma (1981) have reported *Amanita muscaria, Boletus edulis, B. luteus, Cenocccum graniforme, Inocybe rimosa, Russula roesipes, Scleroderma auranticum, Suillus bovinus* were associated with *Pinus kesiya* in Shillong regions of North Eastern Himalayas. Lakhanpal and Kumar (1984) studied the appearance and disappearance of various mycorrhizal fungi with different species of tree hosts in Himachal Pradesh. Lakhanpal (1987) also studied the morphology and anatomy of ectomycorrhizal fungi.

Natarajan (1977) reported the occurrence of *Laccaria ohiensis* in *Eucalyptus* plantations in the Nilgiri Hills, South India. Later, Natarajan and his associates have reported the distribution of different ectomycorrhizal fungi such as *Amanita muscaria* and *Laccaria laccata* (Natarajan, 1977); *Scleroderma citrinum* (Natarajan and Kannan, 1979); *Thelephora terrestris* (Natarajan and Raman, 1982); *Russula parazurea, Suillus brevipes, S. pallidiceps, S. punctatipes, S. subluteus* (Natarajan and Raman, 1983) and *Lcyoperdon perlatum* (Raman, 1984; Natarajan and Purushotahama, 1987) in *Pinus patula* plantations in Pulney and Nilgiri Hills, Tamil Nadu, South India. Nataraajan *et al.* (1988) reported some new records of ectomycorrhizal fungi in Southern India. The status of ectomycorrhizal fungi in association with *Pinus patula* in the Nilgiri Hills, South India was studied and reported by Mohan (1991). Vijayakumar *et al.* (1999) reported for the first time the ectomycorrhizal fungi in association with *Acacia auriculiformis, A. holosericea* and *A. mangium* in Andhra Pradesh. Recently, Mohan (2002) reported the distribution of different ectomycorrhizal fungi viz., *Pisolithus tinctorius and Thelephora ramarioides* in association with economically important tree species in Southern India.

The success of the fungus used for the tree inoculation depends on its ability to adopt to soil and climatic conditions and compete with other fungi. A fungus, which is well suited to the conditions of the nursery, may not be well adopted to the conditions prevailing at the planting site (Molina, 1977).

The present chapter highlights a detailed account of the effect of different ectomycorrhizal inocula on growth improvement of the selected tree species in nursery conditions.

EFFECT OF ECTOMYCORRHIZAL INOCULA ON GROWTH IMPROVEMENT OF IMPORTANT TREE SEEDLINGS IN THE NURSERY

Much work in recent years has been aimed at selecting ectomycorrhizal fungal species to improve tree growth and survival. Different sources of ectomycorrhizal inoculum are available with their inherent advantages and disadvantages. Several methods of inoculating soils are currently in practice. Various types of mycorrhizal inocula are used for mycorrhization of tree seedlings in nurseries. These include soil infected by mycorrhizal fungi, colonized roots, mycorrhizal short roots, fruitbodies, and spores of mycorrhizal fungi and vegetative mycelia grown in pure cultures and used as such as or after entrapping in various types of inert material. Several types of natural and laboratory produced ectomycorrhizal inocula are used in forestry practices.

(i) Soil inoculum

Soils under the host trees/plants are rich with mycorrhizal roots, spores, hyphae, hyphal strands, rhizomorphs etc up to a depth of about 10 cm from the ground level. These soils can be used as a useful inoculum for the introduction of mycorrhizal fungi in the tree nurseries. Many researchers conducted nursery experiments to study the effect of soil inoculum on the growth improvement of different species of *Pinus* and *Eucalyptus*. Giao and Nham (1988) observed that inoculation of soil collected from *Pinus merkusii* forest showed seven to ten fold increase in growth of *Pinus elliotti* and *Pinus oocarpa* saplings at the age of 15 months when compared with uninoculated control.

Einarrsson and Kristjansson (1990) observed that the seedlings of Sitka spruce, Lodgepole pine, and Siberian larch inoculated with mycorrhiza infected soil revealed a clear superiority in growth over uninoculated seedlings. Ellis and Pennington (1992) found that the growth of seedlings was very poor in soil collected from grasslands and better growth was observed in soil collected from *Eucalyptus* stands in Australia.

In India, nursery experiment was conducted to find out the ability of *Pinus patula* seedling to acquire mycorrhizal infection when grown in unsterilized "shola" (forest soil) and grassland soils and also the ability of these seedlings to retain the acquired mycorrhizal fungi from one soil to another soil when transplanted into it (Mohan, 1991). For transplantation from one soil to another soil, a third soil type, viz., riverbed sand was also used. Among the different soil types used, the seedlings from both surface sterilized and non-surface sterilized seeds grown in "shola" soil and transplanted to "shola" soil showed better shoot biomass and root collar diameter and higher mycorrhiza formation when compared to the seedlings from both surface sterilized and non-surface sterilized seeds grown in "shola" soil and transplanted again to other soil types during the entire period of study. This is followed by the seedlings from surface sterilized seeds grown in "shola" soil and transplanted again to sand during 14th to 18th month harvests (Tables 1 and 2). The percentage of mycorrhizal tips was also highest in the seedlings from both surface sterilized and non-surface sterilized seeds grown in "shola" soil transplanted to "shola" soil as compared to the seedlings from both surface sterilized and non-surface sterilized seeds grown in "shola" soil and grassland soil, and transplanted again to other soil types during the entire period of study (Tables 3-5). Hence, the practice of using "shola" soil for raising nursery by the foresters seems to be the best method for getting better seedlings in Niligiri conditions (Mohan, 1991; Mohan and Natarajan, 1992). The influence of the type of soil on the overall seedling growth and mycorrhiza formation was also reported by other researchers (Bergemann, 1955; Parke *et al.*, 1984; Ellis and Pennington, 1992). Similarly, the differential effect of various inoculant fungi on the seedling growth and mycorrhizal production as influenced by soil type was shown by Holden *et al.* (1983).

Table 1. Mean shoot dry weight of *Pinus patula* seedlings grown in "shola" soil and transplanted again to "shola soil, riverbed sand and grassland soil (Mohan, 1991)

Months of harvesting	Non-surface sterilized seeds in "shola" soil transplanted to			Surface sterilized seeds in "shola" soil transplanted to			LSD (p=0.05)
	"Shola soil	Riverbed sand	Grassland soil	"Shola" soil	Riverbed sand	Grassland soil	
5-months	0.109 (1.023)	0.026 (0.366)	0.047 (0.631)	0.145 (1.153)	0.054 (0.697)	0.052 (0.666)	(0.184)
7-months	0.129 (1.085)	0.047 (0.649)	0.056 (0.677)	0.190 (1.268)	0.101 (0.997)	0.083 (0.891)	(0.166)
9-months	0.281 (1.432)	0.109 (1.005)	0.130 (1.101)	0.348 (1.533)	0.189 (1.269)	0.133 (1.067)	(0.146)
14-months	1.183 (2.043)	0.687 (1.825)	0.276 (1.416)	1.406 (2.132)	1.643 (2.211)	0.528 (1.609)	(0.124)
16-months	2.261 (2.353)	1.858 (2.255)	0.362 (1.523)	2.142 (2.310)	2.257 (2.342)	1.342 (2.123)	(0.105)
18-months	2.459 (2.387)	2.335 (2.345)	0.742 (1.860)	2.491 (2.388)	2.509 (2.386)	1.522 (2.174)	(0.107)

For statistical analysis, gram dry weights (W) were transformed to log (W+1)

Table 2. Mean root collar diameter of *Pinus patula* seedlings grown in "shola" soil and transplanted again to "shola soil, riverbed sand and grassland soil (Mohan, 1991)

Months of harvesting	Non-surface sterilized seeds in "shola" soil transplanted to			Surface sterilized seeds in "shola" soil transplanted to			LSD (p=0.05)
	"Shola soil	Riverbed sand	Grassland soil	"Shola" soil	Riverbed sand	Grassland soil	
5-months	0.18 (0.784)	0.11 (1.000)	0.12 (0.940)	0.26 (0.604)	0.12 (0.940)	0.15 (0.850)	(0.158)
7-months	0.26 (0.629)	0.13 (0.900)	0.17 (0.790)	0.34 (0.489)	0.18 (0.784)	0.18 (0.784)	(0.153)
9-months	0.32 (0.549)	0.18 (0.784)	0.18 (0.784)	0.46 (0.368)	0.26 (0.629)	0.22 (0.700)	(0.193)
14-months	0.54 (0.308)	0.34 (0.489)	0.26 (0.618)	0.64 (0.213)	0.60 (0.235)	0.32 (0.514)	(0.152)
16-months	0.60 (0.270)	0.40 (0.413)	0.27 (0.637)	0.76 (0.139)	0.68 (0.180)	0.37 (0.443)	(0.151)
18-months	0.72 (0.152)	0.48 (0.333)	0.36 (0.454)	0.76 (0.139)	0.74 (0.135)	0.42 (0.388)	(0.133)

Table 3. Percentage occurrence of mycorrhizal types produced by different fungi in seedlings from surface sterilized seeds raised in "shola" soil and transplanted again to "shola" soil (Mohan, 1991)

Mycorrhizal types	Months of observation					
	5-months	7-months	9-months	14-months	16-months	18-months
Thelephora terrestris	78.5	75.8	73.9	70.0	55.9	50.4
Laccaria laccata	0.0	6.0	10.0	22.0	30.3	28.0
Rhizopogon luteolus	0.0	0.0	0.0	0.0	11.0	18.0
Inocybe sp.	8.4	5.0	3.0	0.0	0.0	0.0
Hebeloma sp.	2.4	3.0	5.0	0.0	0.0	0.0
Cenococcum graniforme	0.0	3.0	0.0	4.0	0.0	2.0
Unidentified Ascomycete	4.0	0.0	2.0	0.0	0.0	0.0
Non-mycorrhizal	6.7	7.2	6.1	4.0	2.8	1.6

Table 4. Percentage occurrence of mycorrhizal types produced by different fungi in seedlings from surface sterilized seeds raised in "shola" soil and transplanted to again to grassland soil (Mohan, 1991)

Mycorrhizal types	Months of observation					
	5-months	7-months	9-months	14-months	16-months	18-months
Thelephora terrestris	0.0	0.0	0.0	25.8	36.3	44.0
Laccaria laccata	0.0	0.0	0.0	5.0	11.0	19.0
Rhizopogon luteolus	0.0	0.0	0.0	10.9	17.7	20.1
Inocybe sp.	55.2	58.0	50.5	25.0	14.0	6.0
Hebeloma sp.	13.0	10.5	16.0	15.0	10.0	4.0
Cenococcum graniforme	3.0	6.5	6.2	4.0	2.0	1.0
Unidentified Ascomycete	2.0	1.0	8.0	2.0	0.0	0.0
Non-mycorrhizal	26.8	24.0	19.3	12.3	9.0	5.9

Table 5. Percentage occurrence of mycorrhizal types produced by different fungi in *Pinus patula* seedlings from surface sterilized seeds raised in "shola" soil and transplanted again to riverbed sand (Mohan, 1991)

Mycorrhizal types	Months of observation					
	5-months	7-months	9-months	14-months	16-months	18-months
Thelephora terrestris	12.0	15.0	20.5	39.4	50.0	61.6
Laccaria laccata	0.0	0.0	0.0	0.0	0.0	0.0
Rhizopogon luteolus	0.0	0.0	0.0	0.0	0.0	0.0
Inocybe sp.	0.0	0.0	0.0	0.0	0.0	0.0
Hebeloma sp.	0.0	0.0	0.0	0.0	0.0	0.0
Cenococcum graniforme	42.8	37.6	46.0	32.0	30.0	26.0
Unidentified Ascomycete	19.8	24.3	13.6	19.4	12.1	7.7
Non-mycorrhizal	25.4	23.1	19.9	9.2	7.9	4.7

During seedling root assessments, different mycorrhizal types were also recognized from their gross morphology features and characteristic features of their mantle and extramatrical hyphae. The mycorrhizal types of the following fungi were identified (Mohan, 1991): *Thelephora terrestris, Laccaria laccata, Rhizopogon luteolus, Cenococcum graniforme, Inocybe* sp., *Hebeloma* sp., and unidentified ascomycete type (Tables 3-5). Most of the above mentioned fungi are well known ectomycorrhizal species. *Cenococcum graniforme* and the ascomycetes type have long been shown to be common mycorrhizal fungi in nursery seedlings of different tree species (Laiho, 1965; Mikola, 1965; Danielson, 1982; Wilcox *et al.*, 1983; Yang and Wilcox, 1984). The mycorrhizas of *Thelephora terrestris* were predominant in the seedlings grown in "shola" soil during the entire period of study. This fungus was retained in the seedlings of the plants transplanted from "shola" soil to other soil types. But the percentage occurrence of mycorrhizas of this type in the sand and grassland soil was less as compared to the "shola" soil. In fact, in the grassland soil, the transplanted seedlings did not have any mycorrhizas attributable to *T. terrestris* type up to 9-months. In the grassland soils, the predominant mycorrhizal types were *Inocybe* type, *Cenococcum* type and an unidentified ascomycetes type. The mycorrhizas of *Rhizopogon luteolus* occurred only in 16 and 18 months old seedlings grown in "shola" soil and transplanted to "shola" soil. The mycorrhizal tips of the seedlings grown in "shola" soil and transplanted to sand possessed abundant *Cenococcum graniforme* and other unidentified ascomycetes type mycorrhizas. The dominant mycorrhizal types were those of *Inocybe* sp., *Cenococcum graniforme, Hebeloma* sp. and unidentified ascomycetes type. Similar findings wee made by Mason *et al.* (1986) in a study on the effect of different soils on the establishment and influence of ectomycorrhizas. When seedlings were grown in mycorrhizal soil, it was found that most (68%) of the mycorrhizas that developed were attributable to *Inocybe.* In contrast, only a few mycorrhizas developed on seedlings grown in soil taken from beyond the edge of the stand (non-mycorrhizal soil) and these were attributable to *Hebeloma* type. When these seedlings were subsequently transplanted to non-mycorrhizal soil and observed for two growing seasons, the initial differences in mycorrhizal populations were still evident. The seedlings from mycorrhizal soil still possessed mycorrhizas attributable to *Inocybe* although 72% of the total also possessed a different types of mycorrhizas of seedlings propa-gated in the non-mycorrhizal soil along with *Hebeloma* type mycorrhizas, 38% were of *Lactarius pubescens* type.

Since the overall growth of the seedlings grown in "shola" soil and transplanted to "shola" soil were found to be much better than the seedlings of the other types. It is possible that the mycorrhizas produced by *T. terrestris, L. laccata* and *R. luteolus* were more efficient fungi at the early stage of the plant growth than the other fungi (Mohan, 1991; Mohan and Natarajan, 1992).

(ii) Basidiospore inoculum

Introduction of ectomycorrhizal fungi through spores and sporocarps were also common practice in Europe (Trappe, 1977). The sporocarps and spores of various fungi have been used as inocula to form specific ectomycorrhizas on tree seedlings. Puffball producing genera of ectomycorrhizal Gasteromycetes such as *Pisolithus, Rhizopogon, Scleroderma,* etc produce numerous basidiospores which can be easily collected and successfully employed as inoculum. They can even be stored for longer periods under suitable conditions (Marx *et al.*, 1979).

Seed inoculation with basidiospores is also employed with some success. Seeds of *Pinus raidata* were successfully inoculated with basidiospores of *Rhizopogon luteolus* (Theodorou and Bowen, 1973). The advantage of spore inoculation is that the spore inoculum is very light. One gram of basidiopspores of *Pisloithus tinctorius* or *Rhizopogon luteolus* contains over 1 billion basidiospores.

Different studies showed the effect of use of basidiospores as inocula on the growth improvement of important tree species in nurseries in different parts of the world. Reynaldo and Cruz (1990) used basidiospores of *Pisolithus tinctorius* and *Scleroderma cepa* as mycorrhizal tablets for inoculation to the seedlings of *Pinus, Eucalyptus, Dipterocarpus* and *Casuarina equisetifolia.* Their study revealed the height and diameter growth of *Pinus* and *Eucalyptus* species were increased by 30% to 70% due to basidiospore inoculation as compared to control.

Fakuara and Wilarso (1990) studied on the effect of inoculation of two-week old seedlings of *Pinus merkusii, Hopea mengrarawa* and *H. odorata* with three spore forms of *Scleroderma* sp. namely spore tablets, spore capsule and spore powder. The results showed that the inoculant types of tablet and capsule are not significantly different from spore powder type in all the three tree species.

In studies conducted at the Institute for Mycorrhizal Research and Development, Athens, Georgia, USA, a technique was developed involving the incorporating of *Pisolithus tinctorius* basidiospores in the external matrix of encapsulated pine seed. In this method, basidiospores of *P. tinctorius* mixed at various rates, i.e., 1, 2, and 3 mg spores per seed were used to form abundant *P. tinctorius* ectomycorrhizae on pine seedlings in container and bare root nursery studies. In certain tests, there was as much *P. tinctorius* mycorrhizas formed from spores in encapsulated seed as with vegetative inoculum. In later studies, spore pellets of *Pisolithus tinctorius* were applied at 4.05 or 8.10 g/m2 to fumigated soil in loblolly pine microplots. Pt indices of 82 to 90 for seedlings with vegetative inoculum (applied before sowing for comparison purposes) were significantly greater indices for seedlings with Pt pellets (Marx *et al.*, 1981; Marx and Bell, 1985).

In another study, Marx and Cordell (1990) found that spores of *Pisolithus tinctorius* applied by spraying formed inadequate mycorrhizae while vegetative mycelium of *P. tinctorius* placed in trenches between rows during spring formed adequate mycorrhizae in *Pinus palustris* sown during the previous autumn.

Mohan (1991) studied the effect of basidiospores of different ectomycorrhizal fungi as inocula on the growth improvement of *Pinus patula* seedlings in the nursery condition at Ootacamund, South India. He found that among seven different fungi (*Amanita muscaria, Laccaria laccata, Lycoperdon perlatum, Rhizopogon luteolus, Scleroderma citrinum, Suillus brevipes* and *Thelephora terrestris*) inoculated, the seedlings inoculated with *Thelephora terrestris* showed the highest shoot dry weight during the entire period of study (Table 6). This is followed by the seedlings inoculated with basidiospores of *Rhizopogon luteolus* and

Laccaria laccata. The least shoot dry weight was seen in the uninoculated control seedlings throughout the period of observation. Similar observations were found in other parameters like root dry weight, root collar diameter and percent mycorrhizal colonization. In the same study, Mohan (1991) also found that among the different soil samples used, the seedlings grown in basidiospores inoculated unsterilized "shola" soil showed maximum shoot dry weight and was found to be statistically significant when compared to the seedlings grown in sterilized "shola" soil and unsterilized grassland soil during the entire period of observation (Table 7).

Table 6. Effect of basidiospore inocula of different ectomycorrhizal fungi on mean shoot dry weight of *Pinus patula* seedlings grown in sterilized "shola" soil (Mohan, 1991).

Ectomycorrhizal	Months of harvesting					
fungi	5-months	7-months	9-months	14 months	16-months	18-months
Amanita	0.039	0.073	0.115	0.374	0.550	0.784
muscaria	(1.567)	(1.835)	(2.027)	(2.547)	(2.720)	(2.887)
Laccaria laccata	0.197	0.400	0.571	1.344	2.261	2.354
	(2.278)	(2.580)	(2.739)	(3.111)	(3.352)	(3.363)
Lycoperdon	0.58	0.090	0.139	0.453	0.649	0.906
perlatum	(1.721)	(1.917)	(2.122)	(2.635)	(2.798)	(2.938)
Rhizopogon	0.089	0.239	0.396	1.165	1.630	2.132
luteolus	(1.900)	(2.356)	(2.593)	(3.061)	(3.202)	(3.323)
Suillus brevipes	0.079	0.147	0.213	0.656	0.926	1.409
	(1.842)	(2.116)	(2.315)	(2.805)	(2.956)	(3.141)
Scleroderma	0.067	0.145	0.142	0.497	0.792	1.144
citrinum	(1.782)	(2.150)	(2.127)	(2.661)	(2.883)	(3.046)
Thelephora	0.243	0.529	0.790	1.829	2.264	2.688
terrestris	(2.368)	(2.716)	(2.897)	(3.258)	(3.346)	(3.424)
Uninoculated	0.022	0.037	0.056	0.168	0.240	0.346
(Control)	(1.259)	(1.523)	(1.719)	(2.206)	(2.375)	(2.525)
LSD p=0.05	(0.225)	(0.185)	(0.159)	(0.145)	(0.111)	0.127)

For statistical analysis gm dry weights (W) were transferred to log (W+1)

Table 7. Effect of different soil types inoculated with basidiospores of different ectomycorrhizal fungi on the shoot dry weight of *Pinus patula* seedlings (Mohan, 1991)

Months of harvesting	Sterilized "Shola" soil	Unsterilized "Shola" soil	Unsterilized Grassland soil	LSD p=0.05
5-months	0.099(1.840)	0.114(1.985)	0.089(1.849)	(0.080)
7-months	0.207(2.149)	0.250(2.331)	0.191(2.198)	(0.065)
9-months	0.303(2.317)	0.356(2.486)	0.277(2.356)	(0.056)
14-months	0.811(2.785)	1.084(3.002)	0.811(2.812)	(0.051)
16-months	1.164(2.954)	1.594(3.172)	1.131(2.977)	(0.039)
18-months	1.470(3.081)	1.947(3.274)	1.389(3.068)	(0.045)

For statistical analysis gm dry weights (W) were transferred to log (W+1)

An experiment was conducted at the Institute of Forest Genetics and Tree Breeding, Coimbatore, to investigate the effectiveness of basidiospore inoculum of *Pisolithus tinctorius* in forming mycorrhizal association and the growth improvement of *Eucalypus globulus* seedlings in nursery. It was observed that the inoculated plants had better growth by means of total biomass, plant height and root collar diameter. It was observed that the basidiospore inoculated seedlings had better seedling health, plant height, leaf area, collar diameter and shoot and root biomass over inoculated control seedlings. The plant height was invariably increased (3 fold) on 3-months old seedlings. Similarly, the shoot and root biomass were also increased (2-3 fold) over uninoculated control seedlings.

Table 8. Effect of biofertilizers (AMF, Ectomycorrhizal fungus and Phosphobacterium) on growth improvement of *Eucalyptus camaldulensis* in nursery (Mohan and Manoharan, 2002)

Treatment	Plant height (cm)	Shoot dry weight (gm)	Root dry weight (gm)	Collar diameter (cm)
T1-Control	7.8	0.056	0.036	0.17
T2-AMF	18.8	0.790	0.662	0.57
T3- Ectomyocrrhizal fungus (Pt)	14.5	0.656	0.524	0.37
T4-Phosphobacterium	10.3	0.400	0.213	0.27
T5-AMF + Ectomyocrrhizal fungus (Pt)	11.7	0.688	0.410	0.32
T6-AMF + Phosphobacterium	11.1	0.497	0.311	0.35
T7-Pt + Phosphobacterium	10.3	0.453	0.319	0.33
T8-AMF + Ectomyocrrhizal fungus (Pt) + Phosphobacterium	17.2	0.649	0.587	0.55
LSD p=0.05	0.107	0.159	0.183	0.249

Another experiment conducted at the Institute of Forest Genetics and Tree Breeding, Coimbatore to study the efficacy of ectomycorrhizal fungus (*Pisolithus tinctorius*) with other biofertilizers (AMF and Phsophobacterium) on growth improvement of *Eucalyptus camaldulensis* in glasshouse condition (Mohan and Manoharan, 2002). The results of the experiment revealed that AMF, Ectomycorrhizal fungus and Phosphobacterium inoculated seedlings performed better in terms of plant height, shoot and root biomass and root collar diameter. The data on plant height, shoot biomass and root biomass was significantly greater in inoculated plants as compared to control (uninoculated) seedlings (Table 8).

(iii) Vegetative/Mycelial inoculum

Mycorrhizal introduction with vegetative inoculum ensures the introduction of appropriate fungi in suitable quantities on suitable host plants. Utilization of pure cultures or vegetative mycelium do not pose the problem of bulk handling and transportation as compared to natural soil inoculum. Different studies were conducted to test the mycorrhizal association of various tree species by using vegetative or pure mycelial inoculum. Pure myclelial or vegetative inoculum of ectomycorrhizal fungi has been regularly used since several years throughout the world (Marx, 1980; Saravanan and Natarajan, 1996). Commercial preparations of vegetative mycelium and procedures for inoculating bare root and container grown seedlings are now readily available. Ectomycorrhizal inoculum of *Pisolithus tinctorius* and species of *Hebeloma, Laccaria, Scleroderma* etc are commercially available.

Danielson *et al.* (1984) inoculated 15 ectomycorrhizal fungi to the seedlings of Jack pine and found that 7 fungi viz., *Thelephora terrestris, Laccaria proxima, Hebeloma* sp., *Pisolithus tinctorius, Sphaerosporella brunnea, Cenococcum geophilum* and E-strain isolate can able to form mycorrhizal colonisation to the pine seedlings. Boyle *et al.* (1987) inoculated mycelial slurry of *Hebeloma longicandum* into the root zone of containerized seedlings resulted in consistently high colonization, though the application of the slurry to the surface of the growing medium was also effective. The slurry infectivity was dropped after it was mixed into a peat-vermiculite growing medium particularly in the presence of high levels of fertilizers. In inoculation trials, slurries of nine representative fungi were infected into the root zone of eight-week old black spruce or Jackpine seedlings. Five of the fungi consistently formed ecto-mycorrhizae.

Einarsson and Kristjansson (1990) found that the seedlings of Sitka Spruce, Lodgepole pine and Siberian larch inoculated with pure fungal cultures of different ectomycorrhizal fungi showed a clear superiority in growth over uninoculated seedlings. Saravanan and Natarajan (1996) observed that the *Pisolithus tinctorius* inoculum produced on vermiculite and inoculum entrapped in alginate gels (mycobead) significantly improved *Acacia nilotica* seedling growth and nitrogen fixation activity.

(iv) Comparison of mycobead inoculum and laboratory grown inoculum

Different studies have been conducted by the researchers to compare the effect of alginate bead inoculum and laboratory grown inoculum of ectomycorrhizal fungi on growth and mycorrhizal colonization on various tree species. Ingleby *et al.*, (1994) observed that up to 114% more Sitka Spruce (*Picea sitchensis*) seedlings were found in plots inoculated with live inoculum of *Laccaria proxima* in vermiculite-peat-carrier than in control plots with or without the same carrier. A similar, although reduced, effect was found when the live fungus was encapsulated in alginate beads. Out of the two nurseries where the experiments were conducted, 40-50% per cent less seedlings in one nursery and 66%-72% less seedlings in the other nursery were obtained in fertilized and mycorrhiza inoculated plots than in unfertilized inoculated plots.

Mortier *et al.* (1990) studied two types of inocula of *Laccaria laccata* namely mycelial inoculum grown in vermiculite peat mixture and bead inoculum produced by entrapping the mycelium grown in liquid medium in calcium alginate gel with different quantities of mycelium were compared. The experiment was conducted in a fumigated nursery bed on a sandy loam soil which was seeded with Douglas fir (*Pseudotsuga menziesii*). At the end of the first growing season, the alginate inoculum at the rate of 5 g mycelial dry weight/m^2 proved to be most efficient. The top dry weight of the seedlings in the treatment was 2.3 times that of non-inoculated fumigated controls. This inoculum treatment also ensured nearly total mycorrhizal infection by *L. laccata*.

In an other study, Gagnon *et al.* (1991) observed that seedlings of *Quercus rubra* grown in moss-vermiculite substrate and inoculated at the time of sowing, with mycelial suspension of *Laccaria bicolor* at 19 weeks had significantly more mycorrhizae but lower shoot height, root collar diameter, and shoot, root, and total dry weights than those inoculated with calcium alginate beads containing the aforesaid mycelial suspension and also when compared to controls regardless of nitrogen levels. They also found that there was no significant difference for any growth parameter between the bead and control treatments.

INOCULATION TECHNIQUES

For successful mycorrhization, the fungal inoculum must be placed in the rooting zone where the seedling roots can grow into the inoculum. Once the initial root infection takes place, extramatrical mycelium spreads rapidly to infect other roots. Successful production of ectomycorrhizal seedlings is dependant upon several factors such as type and age of inoculum used, inoculum placement in the growing medium, time of inoculating the seedlings and a number of host, fungus and soil interactions. To gain maximum benefit from mycorrhizas, seedlings should be inoculated with fungal symbionts which are best suited to specific tree species and which are well adopted to environmental conditions of the planting site. Cultural practices, biological factors, and environmental conditions influence short root development and mycorrhizal formation. Indiscriminate use of biocides (herbicides, insecticides and fungicides) and soil fumigants will also have deleterious effects on mycorrhizas (Bowen, 1984). Several types of natural and laboratory produced or commercially produced ectomycorrhizal and AMF inocula are available and these inoculum can be applied to the tree seedlings in nursery by adopting the following methods:

(i) Layering method

In this method, the inoculum should be placed in the form of layer or pad below the seed. It ensures the infection of roots immediately after seed germination, because the developing roots will penetrate the inoculum. 3-5 gm of AMF or ectomycorrhizal inocula (viable and healthy) per polybag/root trainer can be added to the size of the containers.

(ii) Banding or Side Dressing Method

AMF or ectomyocrrhizal inoculum can also be banded or side dressed next to the seeds or seedlings. This method is effective, when inoculum quantity is limited. For best results, bands should be placed in an area of root proliferation usually about 3-5 cm from seeds or seedlings.

(iii) Mixing inoculum with soil

This method is quite close to natural conditions. In this case, entire inoculum is thoroughly mixed with potting mixture and the population of inoculated soil is maintained close to the population of the AMF or ectomycorrhizal fungi existing in nature.

(iv) Pre inoculation of transplanted seedlings

A seedling can be inoculated with a selected AMF or ectomycorrhizal fungi before transplanting. This method is useful in situations where there is a risk of failure in direct inoculation in the field and hence precluding observations of mycorrhizal potential.

FUTURE PERSPECTIVES

Mycorrhizas are known to occur naturally on a majority of plants hence the beneficial effect on plant growth may be considerable, especially due to mycotropic habits of many tropical plant species. Successful reforestation depends on the capacity of tree seedlings to establish early thereby ensuring continued resource supply to resist pests and diseases and to survive climatic stress.

Mycorrhizas may be useful as biological fertilizers to soils with intrinsically low availability of phosphorous or for plants depending on mycorrhizas for normal growth and development especially on conifers and certain angiosperms and also in establishment of exotic trees. Mycorrhizas play a role as biofertilizers as partial substitutes to chemical fertilizers wherever economic or environmental considerations prohibit or restrict their use. Similarly, the suitable mycorrhizal fungi would probably most beneficial for not only production of quality planting stocks in the nursery and useful for better survival and establishment in the wasteland areas such as minespoil areas, degraded soils, eroded soils, etc.

In vitro vegetative propagation techniques are being increasingly employed for rapid multiplication of high quality healthy plants. The precocious inoculation of such seedlings with efficient mycorrhizae during their *in vitro* multiplication is a promising potential for the use of mycorrhizas.

CONCLUSION

Tree seedlings have evolved a beneficial mutual dependency upon mycorrhizal fungi for normal infections. Interactions between the host plants, mycorrhizae and soil largely determine the mycorrhizal effect on plant growth which is further conditioned by the variability in abiotic and biotic factors. A better understanding of the interrelationships among various factors is needed. Further research is aimed at finding appropriate technology for large scale commercial multiplication and inoculation techniques of mycorrhizal inoculum. Future research should endeavour at locating appropriate host – mycorrhizal combinations adopted to well defined soil and environmental conditions.

REFERENCES

Bakshi, B.K. (1957). *Mycologia.* 49: 269-272.

Bakshi, B.K. (1966). Mycorrhiza in *Eucalyptus* in India. *Indian Forester* 92: 19-21.

Bakshi, B.K. (1967). Mycorrhiza, its role in man-made forests. World Symposium on man-madeforests and their industrial importance. Canberra, Australia. 2: 1031-1042.

Bakshi, B.K. (1974). Mycorrhiza and its role in forestry. P.L. 480 Project Report, Forest Research Institute and Colleges, Dehra Dun, India. pp 89.

Bakshi, B.K. and Thapar, H.S. (1966). *Proc. Nat. Inst. Sci. India.* 32: 6-20.

Bakshi, B.K., Thapar, H.S. and Singh, B. (1966). *Proc. Nat. Inst. Sci. India.* 32: 1-5.

Bergemann, J. (1955). The different mycorrhizae of some conifers in different soils. *Z. Welforstwirt.* 18: 184-202.

Bowen, G.D. (1984). Tree roots and the use of soil nutrients. In: Nutrition of plantation forests. (Eds.) G.D. Bowen, G.D. and E.K.S. Nambiar. pp. 164-175. Academic Press, London. pp 516.

Boyle, C.D., Robertson, W.J. and Salonius, P.O. (1987). Use of mycelial slurries of mycorrhizal fungi as inoculum for commercial tree seedling nurseries. *Can. J. For. Res.* 17: 1480-1486.

Chaudhuri, H. (1945). *Proc. 32nd Indian Sci. Cong.* 3: 1-66.

Danielson, R.M. (1982). Taxonomic affinities and criteria for identification of the common ectendomycorrhizal symbiont of pines. *Can. J. Bot.* 60: 7-18.

Danielson, R.M, Visser, S. and Parkinson, D. (1984). The effectiveness of mycelial slurries of mycorrhizal fungi for the inoculation at container grown jack pine seedlings. *Can. J. For. Res.* 14: 140-142.

Dixon, R.O.D. and Wheeler, C.T. (1986). Nitrogen Fixation in Plants. Blackie and Son Ltd., Glasgow.

Einarsson, S. and Kristjansson, J.K. (1990). Ectomycorrhizal fungus inoculation by means of forest soil and pure culture. Production of mycorrhizal inoculum. In: Innovation and Hierarchial Integration. Proceedings of the Eighth North American Conference on Mycorrhizae. 5-8 September, 1990. Jackson, Wyoming. pp. 92.

Ellis, R.C. and Pennington, P.I. (1992). Factors affecting the growth of *Eucalyptus delegatensis* seedlings in inhibitory forest and grassland soils. *Plant and Soil.* 145: 93-105.

Fakuara, Y.M.T. and Wilarso, S. (1990). Effectivity test of tablet, capsule and spore powder of mycorrhizae toward the growth of several species of forest tree seedlings. In: Innovation and Hierarchical Integration. (Eds.) M.F. Allen and S.E. Williams. Proceedings of the Eighth North American Conference on Mycorrhiza. 5-8 September, 1990. Jackson, Wyoming. pp. 317.

Gagnon, J., Langlois, C.G. and Garbaye, J. (1991). Growth and ectomycorrhiza formation of container-grown red oak seedlings as a function of nitrogen fertilization and inoculum type of *Laccaria bicolour. Can. J. For. Res.* 21: 966-973.

Giao, N.S. and Niham, N.T. (1988). Study on the use of mycorrhizae in the nursery for the production of higher quality pine saplings. In: Mycorrhizae for Green Asia. Proceedings of the First Asian Conference on Myocrrhiza. (Eds.) A. Mahadevan, N. Raman and K. Natarajan. 29-31 January, 1988. University of Madras, Chennai-25, India. Pp. 274-275.

Harley, J.L. and Smith, S.E. (1983). Mycorrhizal symbiosis. Academic Press, London. 461 pp.

Harley, J.L. and Harley, E.L. (1987). A check list of mycorrhiza in the British flora. *New Phytol.*105: 1-102.

Hayman, D.S. (1983). The physiology of Vesicular-Arbuscular endomycorrhizal symbiosis. *Can. J. Bot.* 61: 944-953.

Holden, J.M., Thomas, G.W. and Jackson, R.M. (1983). Effect of mycorrhizal inocula on the growth of sitka spruce seedlings in different soils. *Plant Soil.* 71: 313-317.

Ingleby, K., Wilson, J., Mason, P.A., Munro, R.C., Walker, C. and Mason, L. (1994). Effect of five mycorrhizal inoculation and fertilizer regime on emergence of Sitka spruce seedlings in bare-root nursery seedbeds. *Can. J. For. Res.* 24: 618-623.

Laiho, O.(1965). Further studies on ectendotropic mycorrhiza. *Acta. For.Fenn.* 79: 1-35.

Lakhanpal, T.N. (1987). Morphology and anatomy of ectomycorrhizal fungi from India. In:Mycorrhiza Round Table. (Eds.) A.K. Verma, A.K. Oka, K.G. Mukerji, K.V.B.R. Tilak

Janak Raj. Proc. Workshop, Jawaharlal Nehru University, New Delhi, India. pp. 53- 83.

Lakhanpal, T.N. and Kumar, S. (1984). *J. Tree Sci.* 3: 5-9.

Marx, D.H. (1969). The influence of ectotrophic mycorrhizal fungi on the resistance of pine roots to pathogenic infections. I. Antagonism of mycorrhizal fungi to root pathogenic fungi and soil bacteria. *Phytopathology.* 59: 158-163.

Marx, D.H. and Bryan, W.C. (1971). Influence of ectomycorrhizaeon survival and growth of aseptic seedlings of Loblolly pines at high temperatures. *For. Sci.* 17: 37-41.

Marx, D.H. (1979). Synthesis of *Pisolithus* ectomycorrhizae on pecan seedlings in fumigated soil. *US Dep. Agric. For. Serv. Res. Note. SE*-283: 4

Marx, D.H. (1980). Variability in ectomycorrhizal development and growth among isolates of *Pisolithus tinctorius* as affected by source, age and reisolation. *Can. J. For. Res.* 11: 166-174.

Marx, D.H. and Bell, W. (1985). Formation of *Pisolithus* ectomycorrhizae on loblolly pine seedlings with spore pellet inoculum applied at different times. USDA Foret Service, Southeastern Forest Experiment Station Research Paper. SE-249. pp 7.

Marx, D.H., Bell, R.W. and Jari, K. (1981). Use of seed encapsulated with basidiospores of *Pisolithus tinctorius* to form ectomycorrhizae on pine. Fifth North American Conference on Mycorrhizae. 16-21 August 1981. University of Laval, Quebec, Canada.

Marx, D.H. and Cordell, C.E. (1990). Inoculation of fall and spring sown longleaf pine seedlings with *Pisolithus tinctorius. Research Note. Southeastern Forest Experiment Station, USDA Forest Service SE* 358: 5.

Mason, P.A., Last, F.T. and Wilson, J. (1986). Effect of different soils on the establishment and influence of sheathing mycorrhizas. In: *Physiological and Genetical aspects of Mycorrhizae.* (Eds. Va. Gianninazzi,-Pearson and S. Gianninazzi. INRA Press, Paris. pp. 767-7720.

Mikalo, P. (1965). Studies on the ectendotropic mycorrhiza of pine. *Acta For. Fenn.* 79: 1-55.

Misra, R.R. and Sharma, G.D. (1981). *Indian Bot. Soc.* 60: 168-171.

Mohan, V. (1991). Studies on ectomycorrhizal association in *Pinus patula* plantations in the Nilgiri Hills, Tamilnadu. Ph.D. thesis, University of Madras, Madras, India. pp 260.

Mohan, V. (2002). Distribution of Ectomycorrhizal Fungi in Association with economically important Tree Species in Southern India. In: *Frontiers of Fungal Diversity in India.* (Eds. G.P. Rao, C. Manoharachari, D.J. Bhat, R.C. Rajak and T.N. Lakhanpal). International Book Distributing Co., Lucknow, U.P., India. pp. 863-872.

Mohan, V. and Natarajan, K. (1992). Effect of soil type on the establishment of ectomycorrhizas and growth of *Pinus patula* seedlings in the nursery. In: *Programme and Abstracts. Agriculture, Horticulture and Forestry.* The International Symposium on Management of Mycorrhizas. Nedlands: University of Western Australia. pp 145.

Mohan, V. and Manoharan, P. (2002). Participated and presented a paper on "Effect of biofertilizers (AMF, Ectomycorrhiza and Phosphobacterium) on growth performance of *Eucalyptus camaldulensis* in nursery and field" in International Conference on Milestones in the Development of Mycology and Plant Pathology held at C.A.S. in Botany, University of Madras, Chennai-600 025, India on 28th to 30th March, 2002.

Molina, R. (1977). Ectomycorrhizal fungi and forestry practice.In: *Mushroom and Man: Interdisciplinary Approach to Mycology* (Walker, T. Ed.) USDA Forest Service. pp. 147-151.

Molina, R. and Trappe, J.M. (1984). Mycorrhizal management in bareroot nurseries. In: Forest nursery manual. (Eds.) M.L. Duryas and T.D. Landis. Martinus Nijhoff Publ. The Netherlands. pp. 211-223.

Mortier, F., Tacon, F. le. And Garbaye, J. (1990). Effect of dose and formulation of *Laccaria laccata* inoculum on mycorrhizal infection and growth of Douglas fir in a nursery. *Agriculture, Ecosystems and Environment.* 28: 351-354.

Natarajan, K. (1977). South Indian Agaricales. III. *Kavaka* 5: 35-39.

Natarajan, K. and Kannan, K. (1979). On the occurrence of *Scleroderma citrinum* in India. *Curr. Sci.* 48: 1086.

Natarajan, K. and Purushothama, K.B. (1987). On the occurrence of *Lycoperdon perlatum* in *Pinus patula* plantations in Tamil Nadu. *Curr. Sci.* 56: 117-118.

Natarajan, K. and Raman, N. (1982). Occurrence of *Thelephora terrestris* in India. *Curr. Sci.* 51: 483-484.

Natarajan, K. and Raman, N. (1983). South Indian Agaricales. XX. Some mycorrhizal species. *Kavaka.* 11: 59-66.

Natarajan, K., Mohan, V. and Kaviyarasan, V. (1988). On some ectomycorrhizal fungi occurring in Southern India. *Kavaka.* 16: 1-7.

Parke, J.L., Lindermann, R.G. and Trappe, J.M. (1984). Inoculum potential of ectomycorrhizal fungi in forest soils of sourthwest Oregon and northern California. *For. Sci.* 30: 300-304.

Perry, D.A., Molina, R. and Amaranthus, M.P. (1987). Mycorrhizas, mycorrhizosphere and reforestation: Current Knowledge and Research Needs. *Can. J. Res.* 17: 929-940.

Postgate and Hill. (1979). Microbial Ecology. (eds.) N.M. Lynch and N.J. Poole. Blackwell, Oxford. pp. 191-213.

Raman, N. (1984). Proc. 6th NACOM, Oregan, USA, 440.

Renaldo, E. and dela Cruz. (1990). Current status of nursery and field applications of ectomycorrhiza in the Phillippines. In: Innovation and Hierrachical Integration, Proceedings of the Eighth North American Conference on Mycorrhiza, 5-8 September, 1990. Jackson, Wyoming. pp. 75.

Rhodes, L.H. (1980). The use of mycorrhizae in crop production systems. *Outlook in Agriculture*10: 275-281.

Sampangiramaiah, K. and Bagyaraj, D.J. (1989). Root diseases and mycorrhizae - a review. *Journal of Phytopathological Research* 2: 1-6.

Saravanan, R.S. and Natarajan, K. (1996). Effect of *Pisolithus tinctorius* on the growth nodulation and nitrogen fixing potential of *Acacia nilotica* seedlings. *Kavaka* 24: 41-49.

Smith, A.H. (1971). Taxonomy of ectomycorrhizal forming fungi. In: Mycorrhizae (Ed.) E. Hackskay. Pp l. 8. USDA For. Serv. Misc. Publ. 1189. pp 255.

Theodorou, C. and Bowen, G.D. (1973). Inoculation of seeds and soil with basidiospores of mycorrhizal fungi. *Soil. Biol. Biochem.* 5: 765-771

Trappe, J.M. (1977). Selection of fungi for ectomycorrhizal inoculation in nurseries. *Ann. Rev. Phytopathol.* 15: 203-222.

Vijayakumar, R., Prasada Reddy, B. V. and Mohan, V. (1999). Distribution of ectomycorrhizal fungi in forest tree species of Andhra Pradesh, Southern India – A new record. *Indian Forester.* 125: 496-502.

Wilcox, H.E., Yang, C.S. and Lo-Bugilo, K. (1983). Responses of pine roots to E-strain ectendomycorrhizal fungi. *Plant Soil.* 71: 293-297.

Yang, C.S. and Wilcox, H.E. (1984). An E-strain ectendomycorrhiza formed by a new species *Tricharina mikolae. Mycologia.* 76: 675-684.

Zak, B. (1971). Detoxification of autoclaved soil by a mycorrhizal fungus. *USDA For. Serv. Res.* Paper PNW. 159: 1-4.

CHAPTER - 6

FUNGI IN BIOREMEDIATION OF TOXIC METALS FROM WASTE WATER

S.R. Dave

Department of Microbiology, School of Sciences, Gujarat University, Ahmedabad 380 009

INTRODUCTION

Fungi have been known either as 'friend' or 'foe' for several centuries, and many of them have been described prior to the printing era in carvings, sculptures and frescoes. During the early years, the main areas of applications of fungi were directed towards food, medicines, poisons and religious ceremonies. These uses were need based and with the availability of technologies at that time. The foundation of fungal biotechnology was laid quite early, when the Romans were credited with developing the art of wine making. The 21st century is going to be dominated by products and processes based on various microbiological systems. The explicit role of fungi is apparent when one views the vast arena of their applicability in the field of food, fodder, beverages, fermented and pharmaceutical products, environmental clean up, etc. (Molitoris, 1989).

They are ubiquitous in the natural environment and can be the dominant organisms in many soils, particularly those of low pH. It is estimated that 1 g of leaf litter on the forest floor may contain fungal hyphae as long as one kilometre. This indicates the ability of fungi to invade their environment, an attribute which is valuable in the development of bioremediation processes based on fungal biocatalysts (Bucke, 1998). Inspite of this, fungi have not yet been used on any significant scale in bioremediation.

Ability of fungi to secrete a variety of extracellular enzymes is widely known. They are natural scavengers and if it were not for them, we would be surrounded with piles of detritus and dead plant matter (Johri and Goel, 1998). Fungi in general are well known for their natural role in the decomposition and/or degradation of lignocellulose, polysaccharides, complex aromatic compounds, pulp effluents, plant polymers, etc.

Among these fields, treatment of waste and soil bioremediation has attracted attention on account of presence of high BOD; coloured compounds and toxic metals in aquatic and/or terrestrial polluted ecosystems. White rot fungi particularly are useful for the bioremediation of lignocellulose by their extracellular enzymatic reactions. *Aspergillus niger* and *Chaetomium cupreuna* have been found to be effective for treatment of tannins in the tannery effluents having high pH and elevated heavy metal concentrations (Johri and Goel, 1998). With increased awareness and changed perception towards environmental issues, the versatility of fungi has been viewed in terms of its role in environmental clean up in general and bioremediation of metallic pollutants in particular.

Table 1 . Common toxic effects on human and maximum permissible limits of some heavy metals in effluents discharged in waste water.

Metal	Toxic effect	Threshold value ($mg.l^{-1}$)*
As	Skin and nasal septum cancer, Jaundice	0.2
B	Innocuous for human consumption	2.0
Cd	Shortness of breath, Anaemia, Nercosis, Hepatic and Renal disorder	1.0-2.0
Cr	Dermatitis, Ulceration, Cancer	0.1-0.5
Cu	Uremia, Thalassemia, Hemachromatoses	2.0 - 3.0
Hg	Tremors, Gingivitis, Renal disorder, Asphyxiation, Nervous failure	0.01-0.05
Mg	Cathatic and diuretic	150.0
Ni	Lung cancer and Respiratory systosis	3.0
Os	Bronchitis, Halo around eyes	
Pb	Obesity, Colic, Anaemia, Pneumosis	0.1-1.0
Se	Gastrointestinal disturbance, Skin and eye irritation	0.05
Te	Garlic smell to sweat and breath	
V	Catarrh, Cough, Wheezing, Sore throat, Dyspnea, Dermatitis	
Zn	Bitter astringent test, cancer	15.0

*The thresold value given are for release of effluents in sewage or wast water.

Drinking water requirements are as bellow if you want to add in table make one additional column and give with table.

All valus are in mg/l.

As 0.05,Cd 0.01, Cr 0.05, Cu 0.05, Hg 0.001, Mg 30.0, Ni 0.05, Pb 0.1 Se 0.01 Zn 5.0 mg/l

Heavy metals form a part of our day to day life. Heavy metal and radioactive pollutants are known to have dangerous effects on flora and fauna. Toxicity data and permissible limits of some metals are given in Table 1. Urbanisation, industrialisation, non–judicious uses and careless disposal of toxic metals like mercury, cadmium, lead, chromium, zinc, cyanide etc. have elevated the metal pollution levels beyond acceptable safety levels. Increasing awareness among the people about the hazardous effects of such metallic pollutants has necessitated development of simple, feasible and non-polluting methods to remove toxic metals from industrial effluents, acid mine drainage, soil, sewage and sludge. Presently available conventional technologies such as chemical precipitation, use of ion exchange resins, solvent extraction, reverse osmosis etc. are successfully used at field scale only in case of wastes having sufficiently high i.e. more than 100 $mg.l^{-1}$ metal concentrations. But for large volume–low tenor waste, these technologies are not appropriate, and this technological gap needs to be bridged immediately.

Biohydrometallurgy is an emerging technology, which deals with microbial methods like bioleaching and bioremediation for metal removal, from low metal containing waste (Torma, 1977; Dave, 1980, 1995; Ehrlich, 1981; Shumate and Strandbery, 1985).

Microorganisms play a significant role in metal mobilisation by means of biosorption, bioaccumulation and bioprecipitation. Fungi and yeasts along with other microbial groups such as bacteria, cyanobacteria and algae are capable of removing heavy metals and radionuclides even from dilute solutions (Table 2) (Zajic and Chiu, 1972; Tuovinen and Kelly, 1974; Gadd and Griffiths, 1978). Fungi and yeasts are easy to grow and harvest producing high yields of biomass, and they are more tolerant to acidic conditions and even toxic metals (Gadd, 1990).

Table 2. Reported maximum metal accumulation by fungi and yeast.

Metals	Microorganism	Maximal concentration in biomass (% dry wt.)	Remarks
Cadmium	*Saccharomyces cerevisiae*	3.12	
Copper	*Rhizopus arrhizus*	1.71	
	Penicillium notatum	8.00*	
Zinc	*Penicillium notatum*	2.30*	Incubation time : 30 d
	Spinulosum	0.13	
Uranium	*Aspergillus niger*	21.50	
Thorium	*Rhizopus arrhizus*	18.50	
Lead	*Penicillium* spp.	0.61	
Silver		5.40	Expected maximum uptake capacity calculated from isotherm
Chromium		3.10	
Copper		1.60	
Cadmium	*Rhizopus arrhizus*	3.00	
Mercury		5.80	
Lead		10.40	
Uranium		19.50	
Zinc		2.10	

Fungi and yeasts are used in many industries for the production of valuable products. Thus, their biomass may be available easily in sufficient amount as a waste product. Since both living and dead biomass are efficient metal absorbents, such waste products serve to be good agents for the biotechnology of metal removal and recovery process. *Aspergillus niger, Saccharomyces cerevisiae, Rhizopus arrhizus, Penicillium chrysogenum* etc. are good examples of waste biomass products from many industries (Tsezos and Volesky, 1981; Gadd, 1990; Modi, 1996; Shah, 2000). Though both live and dead fungal biomass are good metal sorbents, there are marked differences in metal uptake mechanisms among them. Copper and nickel removal by various microbial biomass studied in our laboratory are depicted in Fig. 1 and 2. The metal uptake data clearly show the variation in the metal sorption ability of various biomass even within a genus. Even the differences in the biomass drying conditions influence metal removal ability, which is clear from the observed differences in copper and nickel removal by *Aspergillus* and *Penicillium* spp. dried under different conditions. The range of copper removal by *Aspergillus* and *Penicillium* genera is shown in Table 3. Under identical conditions, biomass of both genera showed higher copper sorption as compared to nickel. Considerable metal sorption diversity is prevalent among the fungal biomass. Thus, detailed investigations in terms of screening and optimisation for metal remediation are very essential before selecting the biomass for metal removal. Among the fungal biomass studied, *Penicillium griseofulvum* spent mycelial biomass, the waste product of the pharmaceutical industry was found to be the choice of biomass for metal remediation. *P. griseofulvum*, both in free and immobilised state was found to be a very useful biosorbent for copper and nickel removal from aqueous laboratory waste as well as industrial waste irrespective of the type of copper salt present. The maximum metal loading of as high as 110 mg copper.g^{-1} biomass was achieved with repeated cyclic exposures in column studies (Shah, 2000). Moreover, the studies and available literature reveal that living fungal cells can accumulate more metal intracellularly than dead cells and also precipitate metals by various metabolic products.

Table 3. Minimum and maximum copper and nickel sorption among different species of the genus.

Genus	% Copper sorption			
	Copper		Nickel	
	Minimum	Maximum	Minimum	Maximum
Aspergillus	13.7	88.8	8.4	24.0
Penicillium	53.6	90.0	30.7	75.8

MECHANISMS OF METAL REMOVAL

Metal removal by fungi can be mainly explained by metabolism–dependent accumulation, metabolism–independent accumulation, extracellular precipitation, complexation and removal by compounds derived from the biomass etc.

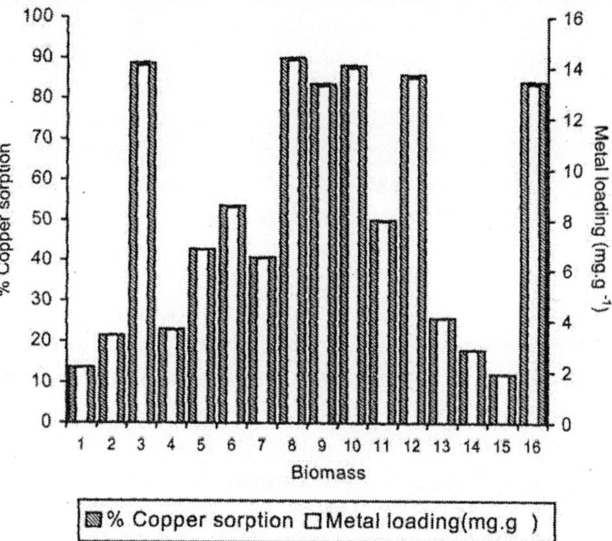

Fig. 1. Screening of biomass for copper sorption. (Copper concentration: 1000 μM aqueous solution, biomass: 100 mg, System: 25 ml.); Biomass 1. *Aspergillus* sp., 2. *Aspergillus* sp. 48, 3. *Aspergillus ochraceus*, 4. *Aspergillus niger*, 5. *Aspergillus flavus*, 6. *Penicillium* sp., 7. *Penicillium* sp. SD, 8. *Penicillium griseofulvum*, 9. *Rhizopus* sp. GNC, 10. *Rhizopus* sp. O, 11. *Streptomyces* sp. I, 12. *Streptomyces* sp. II, 13. Yeast, 14. *Saccharomyces cerevisiae*, 15. *Bacillus* sp., 16. Activated sludge

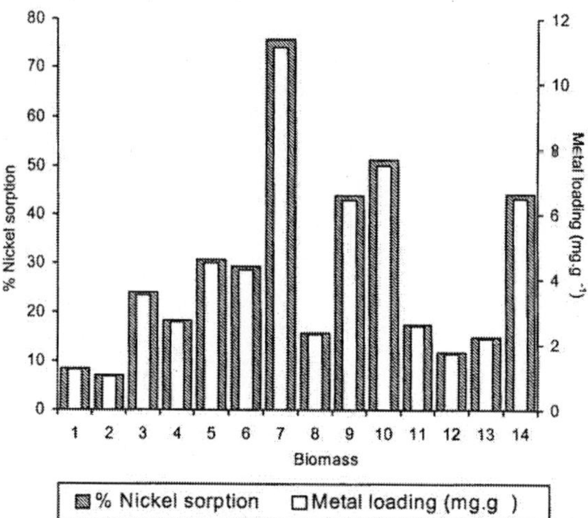

Fig. 2. Screening of biomass for nickel sorption. (Nickel concentration: 1000 μM aqueous solution, biomass: 100 mg, system: 25 ml) *Biomass 1. Aspergillus sp., 2. Aspergillus sp. 48, 3. Aspergillus ochraceus, 4. Aspergillus niger,* 5. *Penicillium* sp., 6. *Penicillium* sp. SD, 7. *Penicillium griseofulvum,* 8. *Rhizopus* sp. O, 9. *Streptomyces* sp. I, 10. *Streptomyces* sp. II, 11. *Yeast,* 12. *Saccharomyces cerevisiae,* 13. *Bacillus* sp., 14. Activated sludge.

Metabolism dependent

In case of metabolism–dependent uptake, metal ions are transported into cells across the cell membrane. This process is significantly influenced by the incubation temperature, presence of energy source and metabolic inhibitors (Norris and Kelly, 1977; Gadd, 1986). Since many metals are essential for growth and metabolism, almost all organisms intake the required metals to satisfy their physiological needs. This is comparatively a slower process than adsorption and thus in many cases of filamentous fungi, metabolism–independent biosorption hide this low rates of intracellular influx (Gadd and White, 1985). But in certain fungi and particularly in yeasts, the metabolism–dependent process is responsible for higher amounts of metal accumulation than the metabolism–independent mechanism. Energy–dependent uptake of Cu, Cd, Ni, Zn, Co, Sr has been demonstrated by fungi. The mechanisms of intracellular uptake are complex and still not fully understood (Borst–Pawels, 1981).

There are two opinions regarding metal uptake and growth phase relation. According to one group, maximum uptake is reported during exponential phase (Khovrychev, 1973; Failla and Weinberg, 1977), while others observed it during the lag period of early stages of growth and decline as the culture reached the stationary phase, for example, in the case of *A. pullulans, Debaryomyces hansonii, P. spinulosum, A. niger and T. viridae* (Gadd and Griffiths, 1978; Townsley *et. al.*, 1986)). Irrespective of the phase showing highest metal uptake, it is often influenced by medium composition, production of metabolite and even morphological changes (Gadd, 1986). Moreover, metal uptake is directly dependent on levels of intracellular K^+ and inversely on external K^+ concentration (Okorokov *et. al.,* 1979). Change in electroneutrality due to intracellular metal influx may be restored by K^+ efflux. Very often heavy metals may disrupt cell membranes and lead to release of K^+ which may be wholly or partially responsible for cell death (White and Gadd, 1987).

Many fungi have the ability to tolerate high amounts of toxic metals, which may lead to decreased metal removal. No detectable Co^{+2} uptake was noted in metal resistant strain of *N. crassa* and the same has been shown in *S. cerevisiae* for Cu^{+2}, Cd^{+2} and Zn^{+2} (Venkateswerlu and Sastry, 1979). Metal taken inside the cell may be compartmentalised, precipi-tated or bound to low molecular weight polyphosphates in the cell. Some species of yeast are known for oxidative detoxification of Ti within mitoc-hondria. Deposition of Ag^{+2} within vacuoles has been observed in *Cryptococcus albidus* (Brown and Smith, 1976). It is well studied in *S. cerevisiae* in connection with Cu^{+2}.

Metal Binding Proteins

The presence of certain metals induces the production of metal binding proteins. Copper metallothionein is induced by Cu^{+2} but not by other metals. The intracellular metal binding protein, metallothionein is responsible for detoxification and regulation of intracellular metal ions.

Presence of Cd^{+2} induces a similar metal binding protein. Metallothioneins are small cysteine – rich polypeptides produced in *Saccharomyces cerevisiae*, which are responsible for the binding of Ag, Au, Cd, Co, Cu, Zn etc. The *cup I* gene is responsible for the production of metal binding protein in *S. cerevisiae*, which has been successfully cloned in *E. coli*, resulting in expression of a functional Cu, Cd and Zn binding protein product and enhanced ability of metal accumulation in *E. coli* (Berka *et. al.*, 1988). Certain fungi and yeasts are reported to have cysteine containing γ glutamyl peptides as phytochelators. *Schizosaccharomyces pombe* and *Candida glabrata* synthesise them and show binding of Ag, Cu, Pb and Zn. *Candida glabrata* can show both, metallothionein as well as γ glutamyl peptides.

Siderophores

Siderophores are low molecular weight ferric coordinating compounds, which are excreted under iron–limiting conditions by selected bacteria and fungi, to enable accumulation of iron from the surrounding environment. They are of two types, the phenolate/ catecholate and hydroxamate type. They are mainly specific for ferric, but they also form complex with some metals such as gallium, chromium, scandium, indium, nickel, uranium, plutonium and thorium (Premuzic *et. al.*, 1985; Hausinger, 1987; Birch and Bachofen, 1990).

There seems to be relatively very little know-how, which is available for siderophore in metal remediation from industrial waste.

Cell wall constituents

A wide variety of metal binding compounds are found in the cell wall of fungi. Fungal cell wall contains chitin, chitosan, glucan, mannan, melanin and phosphomannans (Gadd, 1990). The chitin glucan complex constitutes about 30% of *A. niger* biomass. Chitin and chitosan phosphates are found to adsorb greater amounts of U than Cu, Cd, Zn, Co, Ni, Ca, Mg and Mn. For chitin phosphate, the order of adsorption efficiency was found to be $UO^{+2} > Cu^{+2} > Cd^{+2} > Mn^{+2} > Zn^{+2} > Mg^{+2} > Co^{+2} > Ni^{+2} > Ca^{+2}$, whereas for chitosan phosphate, it was $UO^{+2} > Cu^{+2} > Zn^{+2} > Mn^{+2} > Co^{+2} > Ni^{+2} > Mg^{+2} > Ca^{+2}$. Chitin can work for selective removal of UO^{+2} and Co^{+2}. Phosphorylated derivatives of chitin and chitosan are more efficient metal adsorbents than their non–phosphorylated forms (Sakaguchi and Nakajima, 1982). Insoluble chitosan–glucan complexes and glucan possessing amino or sugar acid groups from *A. niger* exhibit metal biosorption properties and show efficient remediation of transition metal ions from solution (Muzzarelli *et. al.*, 1986).

Alkali treatment extracts insoluble chitosan–glucan complexes, that exhibit efficient chelating ability for a range of metal ions (Muzzarelli and Tanfani, 1982). Alkali extracted *A. niger* biosorbent was found to be more effective in sequestering Zn and Cd from Hussain Sagar Lake water than Dowex–50. The generated biosorbent was able to remove Zn and chromium from battery manufacturing and electroplating industrial effluent respectively (Akhtar and Maruti Mohan, 1995).

However, the extraction of cellular components from the biomass sometimes show detrimental effects on its metal sequestering ability. The chitin extracted from cell mass showed decreased uranium sorption than the biomass.

Melanins

Melanins are electron–dense, granular fungal pigments located in and/or exterior to cell walls. When they are released into the external medium, they are termed as "extracellular melanin". The extracellular and cell wall associated melanin generally have identical composition (Gadd, 1992).

The presence of phenolic units, peptides, carbohydrate, aliphatic hydrocarbons and fatty acids in melanin provide many potential metal–binding sites (Sakaguchi and Nakajima, 1982). Oxygen containing groups in these substances may be particularly important for metal binding (Gadd, 1988).

A variety of melanin types occur in *Basidiomycetes, Ascomycetes* and *Deuteromycetes*. The melanin production may be induced or accelerated in fungi due to the presence of some heavy metals and a formed melanized cell exhibits high metal uptake capacity. The maximum binding ability of fungal phenolic polymers and melanins follow the sequence of Cu > Ca > Mg > Zn (Zumino and Martin, 1977; Gadd and De Rome, 1988).

Aureobasidium pullulans is reported to remove significant amount of Cu^{+2} and Fe^{+3}. The more developing interest is the removal of metals particularly from solution using fungal biomass. Such as *A. niger* waste biomass that removes metals from from citric acid fermentations.

Unicellular polymorphic and filamentous fungi have the ability to reduce metalloid oxyanions e.g. selenate, selenite, tellurate, tellurite to elemental selenium or tellurium. The formed elemental selenium or tellurium can be extensively deposited around and within cellular compartments (Gharieb *et. al.*, 1995).

Non–living biomass of *Rhizopus arrhizus* was reported for the uptake of radionuclides such as U, Pu, Am, Ce, Pm, Eu and Zr, which indicates that biomass of

Rhizopus arrhizus can be used for the treatment of radioactive effluents containing various radionuclides (Dhami *et. al.*, 1998).

Gomes *et. al.* (1998) have suggested the possible application of *A. niger* biomass for the removal and recovery of gold and silver from cyanide containing waste of gold mining industry. *S. cerevisiae* showed three different time bound mechanism of lead removal, the first, very fast process which is over in 3–5 min and is metabolism–independent cell wall binding of the metal, second, which requires about 24 h and is metabolism dependent and the third, which is metabolism–dependent or independent after 24 h of contact time (Jung *et. al.*, 1998).

Desorption of metal from biomass

The selection of metal remediating biomass depends on the recovery of the metals from the loaded biomass and the regeneration of desorbed biomass for the next cycle. Widely used desorption methods are the use of mineral acids and organic complexing agents. Generally, about 0.1 M HCl, H_2SO_4 or HNO_3 effectively remove metals from *Candida utilis, R. Arrhizus, Penicillium italicum, Cladosporium resinae* (De Rome and Gadd, 1987). EDTA 0.01 M, is reported to be effective in desorbing uranium. In case of cadmium desorption from loaded *A. niger*, the efficiency of 0.1 M NaOH as an extractant increases with increasing amounts of biomass. Carbonates have also proved to be efficient desorbing agents with good commercial potential. The selection of desorbing agents and pH varies from metal to metal and sometimes for the type of loaded biomass under the study. An efficient desorbing agent should have high concentration ratio, cause minimum damage to the biomass and require mild conditions for desorption so that the regenerated biomass could be used for many cycles.

Commercial aspects of metal removal

Both living and dead fungal cellmass remove toxic metals and radionuclides by a variety of mechanisms and some of these have commercial potential. Living cells, particularly of yeast, may have higher metal uptake capacity than dead biomass, but the bioaccumulated or precipitated metals are difficult to remove from the cells and incineration or destructive chemical treatments may be needed. On the other hand, the application of dead biomass in the process circumvents toxicity problems, eliminates the requirement of nutrient supply and withstands adverse operating conditions. Also metal and biomass recovery is often easier.

Irrespective of whether live or dead biomass is used in an industrial application, the main limitations are small particle size, low mechanical strength and low density, which raise the question for appropriate reactors to be used and how to separate the loaded biomass from the waste. The use of immobilised or pelleted fungal biomass is the choice for field level applications.

Use of immobilised biomass helps to achieve the desired particle size and strength, high flow rates, minimum clogging, easy separation, high metal loading etc. (Shumate *et. al.*, 1980; Tsezos, 1986). Commercial processes are available particularly for recovery of valuable metals. Example immobilised *Rhizopus arrhizus* for uranium recovery from bioleach solution. BIO-FIX, a potential metal biosorption process, employs yeast biomass alongwith the other microbial biomass immobilised in polysulfone for removal of metals from dilute metal containing waste stream using readily available reactors (Fergusen *et. al.*, 1989). Pellets of *A. niger* have been successfully used in fluidised bed reactors for uranium removal and it was noted that the pelleted *A. niger* biomass was 14 time more efficient than the commercial ion exchange resins IRA 400 (Yakubu and Dudeney, 1986). The fungal pellets have lower strength and it requires some development to increase the strength so that it can be used in continuous process.

From the accumulated literature, it is quite clear that metal biosorption technique employing fungal biomass has potential as a metal remediation process in wastewater treatment. But much more can be done on the path of developing biosorption technology in terms of recycling of biomass, recovery of metals, immobilisation of biomass and reactor design. Thus, the success of fungal sorption technology needs the contribution of researchers from the different fields of biology, chemistry and engineering. Of course, it also requires the co-operation and initiative from industries to adopt such new biology based technology. The combined effort of the researcher, industrialist and waste management people can result into a new eco-friendly products, that combat metal pollutants and help in keeping our planet healthy.

REFERENCES

Akhtar, M.N. and Maruti Mohan, P. (1995). Bioremediation of toxic metal ions from polluted lake waters and industrial effluents by fungal biosorbent. *Current Sci.* 69 (12): 1028 – 1030.

Berka, T., Shatzman, A., Zimmerman, J. Strickler, J. and Rosenbery, M. (1988). Efficient expression of the yeast metallothionein gene in *Escherichia coli. J. Bacteriol.* 170: 21 – 26.

Birch, L. and Bachofen, R. (1990). Complexing agents from microorganisms. *Experientia* 46: 827 – 834.

Borst–Pauwels, G.W.F.H. (1981). Ion transport in yeast. *Biochim. Biophya. Acta.* 650: 88 – 127.

Brown, T.A. and Smith, D.G. (1976). The effect of silver nitrate on the growth and ultrastructure of the yeast *Cryptococcus albidus. Microbios. Lett.* 3: 155 – 162.

Buck, C. (1998). Biochemistry of bioremediation by fungi. *J. Chem. Technol. Biotechnol.* 71: 356–357.

Dave, S.R. (1980). Microbiological and bioleaching studies on metallurgical bacteria cultured from Indian sulfidic mine waters. Ph.D. Thesis, University of Mysore, India.

Dave, S.R. (1995). Bioremediation of metals – an attractive potential microbial technology. In The Proceedings of 3rd Intl. Conf. Appropriate Waste Mgmt. Technologies for developing countries, NEERI, Nagpur, 773 – 779.

De Rome, L. and Gadd, G. M. (1987). Copper adsorption by Rhizopus arrhizus, *Cladosporium resinae* and *Penicillium italicum. Appl. Microbiol. Biotechnol.*, 26: 84 – 90.

Dhami, P.S., Gopalkrishnnan, V., Kannan, R., Ramanujan A., Salvi N. and Udupa S.R. (1998). Biosorption of radionuclides by *Rhizopus arrhizus. Biotechnol. Letts.*, 20 (3): 225 – 228.

Ehrlich, H.L. (1981). Geomicrobiology. Dekkar, New York.

Failla, M.L. and Weinberg, E.D. (1977). Cyclic accumulation of zinc by *Candida utilis* during growth in batch culture. *J. Gen. Microbiol.,* 99: 85 – 97.

Ferguson, C.R., Peterson, M.R. and Jeffers, T.H. (1989). Removal of metal contaminants from waste water using biomass immobilised in polysolfone beads. In : "Biotechnology in mineral and metal processing" (B.J. Scheiner, F.M. Doyle and S.K. Kawatra, Eds.), Society of mining engineers of AIME, Littleton Company, pp. 193 –199.

Gadd, G.M. (1986). Fungal responses towards heavy metals. In : "Microbes in extreme environments" (R.A. Herbert and G.M. Gadd, Eds.), Academic Press, London, pp. 83 – 110.

Gadd, G.M. (1988). Accumulation of metals by microorganisms and algae. In : "Biotechnology – A comprehensive treatise" Vol. 6b, Special Microbial Processes (H.J. Rehm, Ed.), Weinheim : VCH Verlagsgesellschaft, pp. 401 – 433.

Gadd, G.M. (1990). Fungi and yeasts for metal accumulation. In : "Microbial mineral recovery" (H.L. Ehrlich and C.L. Brierley, Eds.), McGraw–Hill Inc., New York, pp. 249 – 275.

Gadd, G.M. (1992). Microbial control of heavy metal pollution. In : "Microbial control of pollution" (Fry, J.C., Gadd, G. M., Herbert, R.A., Jones C.W., and Watson–Craik, I.A. Eds.), Cambridge Univ. Press, pp. 59 – 88.

Gadd, G.M. and Griffiths, A.J. (1978). Microorganisms and heavy metal toxicity. *Microb. Ecol.,* 4: 303 – 317.

Gadd, G.M. and White, C. (1985). Copper uptake by *Penicillium ochrochloron* : Influence of pH on toxicity and demonstration of energy dependent copper influx using protoplasts. *J. Gen. Micobiol.* 131: 1875 – 1879.

Gadd, G.M. and De Rome, L. (1988). Biosorption of copper by fungal melanin. *Appl. Microbiol. Biotechnol.* 29: 610 – 617.

Gharieb, M.M., Wilkinson, S.C. and Gadd,G.M. (1995). Reduction of selenium oxyanions by unicellular polymorphic and filamentous fungi : cellular location of reduced selenium and implications for tolerance. *J. Ind. Microbiol.,* 14: 300 – 311.

Gomes, N.C.M., Camargos, E.R.S. Dias, J.C.T. and Linardi, V.R. (1998). Short note : Gold and silver accumulation by *Aspergillus niger* from cyanide containing solution obtained from the gold mining industry. *World J. Microbiol. Biotechnol,* 14: 149.

Hausinger, R.P. (1987). Nickel utilisation by microorganisms. *Microbiol. Rev.* 51: 22 – 42.

Johri, B.N. and Goel, R. (1998). Fungi in biotechnology: Historical to current perceptions. In: "Fungi in biotechnology" (Prakash, A. Ed.), CBC Publishers and Distributors, New Delhi, pp. 1–16.

Jung, H.S., Kim, D.S., Yun J.W. and Song, S.K. (1998). Process of Pb^{+2} accumulation in *Saccharomyces cerevisiae. Biotechnol. Letts.* 20 (2):153 – 156.

Khovrychev, M.P. (1973). Absorption of copper ions by cells of *Candida utilis. Microbiol.,* 42: 475 – 749.

Modi, N.G. (1996). Bioremoval of mercury from waste. Ph.D. Thesis, Gujarat University, Gujarat, India.

Molitoris, H.P. (1989). Fungi in biotechnology. *Mushroom Science*. 12 (I): 445–446.

Muzzarelli, R.A.A. and Tanfani, F. (1982). The chelating ability of chitinous mateials from *Aspergillus niger, Streptomyces, Mucor rouxii, Phycomyces blakeseanus* and *Choanephora cureurbitum*. In : "Chitin and chitosan" (Mirano S. and Tokura, S. Eds.), The Japanese Society of Chitin and Chitosan, Tottori, Japan, pp. 183 – 186.

Muzzarelli, R.A.A., Bregani, F. and Sigon, F. (1986). Chelating capacities of amino acids glucans and sugar acid glucans derived from chitosan. In : "Immobilisation of ions by bio–sorption" (Eccles, H. and Hunt, S. Eds.), Ellis Horwood, Chichester, UK, pp. 173 – 182.

Norris, P.R. and D.P. Kelly (1977). Accumulation of cadmium and cobalt by *Saccharomyces cerevisiae. J. Gen. Microbiol.*, 99: 317 – 324.

Okorokov, L.A., V.M. Kadomtseva and B.L. Titovakii (1979). Transport of manganese into *Saccharomyces cerevisiae. Folia. Microbiol.*, 24: 240 – 246.

Premuzic, E.T., Francis, A.J., Lin, M. and Schubert, J. (1985). Induced formation of chelating agents by *Pseudomonas aeruginosa* grown in the presence of thorium and uranium. *Appl. Environ. Microbiol.*, 14: 759 – 768.

Sakaguchi, T. and Nakajima, A. (1982). Recovery of uranium by chitin phosphate and chitosan phosphate. In : "Chitin and chitosan" (S. Mirano and S. Tokura, Eds.), The Japanese Society of Chitin and Chitosan, Tottori, Japan, pp. 177 – 182.

Shah, M.P. (2000). Bioremediation of metallic pollutants by spent microbial biomass, Ph.D. Thesis, Gujarat University, Gujarat, India.

Shumate, S.E. and Strandberg, G.W. (1985). Accumulation of metals by microbial cells. In : "Comprehensive bitechnology" (Robinson, C.N. and Howell, J.A. Eds.) Vol. 4, Pergamon Press, New York, pp. 235 – 247.

Shumate, S.E., Strandberg, G.W., McWhirter, D.A., Parrott, J.R., Bogacki G.M. and Locke, B.R. (1980). Separation of heavy metals from aqueous solution using 'biosorbents' – development of contacting devices for uranium removal. *Biotechnol. Bioengg. Symp.*, 10: 27 – 34.

Torma, A.E. (1977). The role of *Thiobacillus ferrooxidans* in hydrometallurgical processes. In : "Advance in biochemical engineering", (Ghose, T.K. Feichter A. and Blakebrough, N. Eds.), Springer Verlag, Germany, pp. 1 – 37.

Townsley, C.C., Ross, I.S. and Atkins, A.S. (1986). Biorecovery of metallic residues from various industrial effluents using filamentous fungi. In: "Fundamental and applied biohydrometallurgy" (Lawrence, R.W. Branlon, R.M.R. and Ebner, H.G. Eds.), Elsevier, Amsterdam, 279 – 289.

Tsezos, M. (1986). Adsorption by microbial biomass as a process for removal of ions from process or waste slutions. In : "Immobilisation of ions by biosorption" (Eccles, H. and Hunt, S. Eds.), Ellis Horwood, Chichester, UK, pp. 201 – 218.

Tsezos, M. and Volesky, B. (1981). Biosorption of uranium and thorium. *Biotechnol. Bioengg.*, 23 : 583 – 604.

Tuovinen, O.H. and Kelly, D.P. (1974). Use of microorganisms for the recovery of metals. *Intl. Metall. Rev.,* 19: 21-31.

Venkateswerlu, G. and Sastry, K.S. (1979). Cobalt transport in a cobalt resistant strain of *Neurospora crassa. J. Biosci.,* 1:433-439.

White, C. and Gadd, G.M. (1987). Inhibition of H^+ efflux and K^+ uptake, and induction of K^+ efflux in yeast by heavy metals. *Toxicol. Assess.,* 2: 437-447.

Yakubu, N.A. and Dudeney, A.W.L. (1986). Biosorption of uranium with *Aspergillus niger.* In : "Immobilisation of ions by biosorption" (Eccles, H. and Hunt, S. Eds.), Ellis Horwood, Chichester, UK, pp. 183-200.

Zajic, J.E. and Chiu, Y.S. (1972). Recovery of heavy metals by microbes. *Dev. Ind. Microbiol.,* 13: 91 – 100.

Zumino, H. and Martin, J.P. (1977). Metal binding organic macromolecules in soil. I. Hypotheses interpreting the role of soil organic matter in the translocation of metal ions from rocks to biological systems. *Soil Sci.,* 123, 65 – 76.

CHAPTER – 7

RECYCLING OF AGRO-WASTES FOR PROTEIN PRODUCTION THROUGH MUSHROOM CULTIVATION

R.D. Rai and O.P. Ahlawat

National Research Centre for Mushroom,
(Indian Council of Agricultural Research), Solan-173213, India

INTRODUCTION

It is ironical that problem of malnutrition is most acute in the so-called 'agricultural-countries' where more than half of the population is engaged in food production. Widespread protein malnutrition prevails in these countries and the gap between supply and demand of protein is likely to widen further because production of alternative sources of protein has not kept pace with the population growth. Deficient intake of protein, particularly good quality protein is the main dietary defect in many countries. Many novel sources of proteins tried in the past have met with consumer resistance due mainly to nonpalatability. Mushrooms due to good quality and quantity of protein, very high productivity per unit area and time and production on worthless agrowastes have been identified as food source contributing to the protein requirement of the developing countries (Chang, 1982; Lelley, 1987; Chang and Miles, 1989).

Dry mass production on earth due to photosynthesis has been estimated at 155.2 billion tonnes per annum of which two third occurs on land and remaining in ocean. Of the terrestrial portion, 65% is in forests and a bit more than 15% is generated in grass- and cultivated lands. According to an estimate (McHale, 1970) about 75% of approximately 24 billion tonnes of bio-mass generated on the cultivated and grasslands remains waste or residue.

Of the total dry biomass of a crop plant, the quantity of straw and stalks is at least as much as the grain or the main product yield. Technological breakthroughs in crop-productivity coupled with increase in arable land area and intensive farming have led to enormous increase in the production of crop residues also. Crop-residues pose disposal problems and a sizeable portion, which may be more than 60% in some countries, is being burnt *in situ*. As a rough estimate more than one quarter of straw annually produced is burnt in most of the countries (Chang, 1982). According to a report of the Institute of Energy and Environment Research, Washington, about 5% of global biomass burning takes place in India. Contribution of methyl chloride emanating from burning of biomass to ozone-depletion is considered to be a serious problem. Allowing the biomass to decay in the field is also not without disadvantages. Biomass burning or decaying *in situ* is a poor utilisation, rather terrible waste, of utilisable energy. These can be recycled into food, feed, fuel and fertiliser through mushroom cultivation (Chang, 1982; Chang and Miles, 1989, Rai and Saxena, 1990).

Problems of unemployment, poverty and malnutrition are inextricably linked and are acute in the vulnerable sections of the society. Obviously, these can not be solved by conventional land-based agriculture. Creation of opportunities for enhancement of income of small farmers and landless labourers by diversification of opportunities of income generation through subsidiary occupations are the new challenges (Swaminathan, 1981). Mushroom production is an indoor activity and land is of little or no importance. It is labour-intensive and high-profit venture for gainful employment of the poor people. It would also provide good quality protein to bridge the 'protein gap'. If promoted properly, mushroom cultivation may go a long way in contributing significantly to the solution of the twin problems of poverty and malnutrition in the developing countries in general and India in particular.

AGROWASTES

Ever-increasing production of agrowastes is a natural corollary of the efforts to boost food production. Developing countries have predominantly agricultural economy and, in fact, produce larger quantities of agrowastes than the foodstuffs. Agro-residues are also, like the food grains, the products of labour, energy and investment and should not therefore be treated as 'wastes'. Main crop-residues in the countries like India are the cereal straws, haulms of oilseeds and pulses, sugarcane bagasse, cotton and jute stalks and leaves, etc. Natural plant wastes are the basic substrates for production of all the species of edible mushrooms, capable of utilizing lignin, cellulose and hemicellulose and producing fruitbodies with excellent nutritional and medicinal attributes (Rajarathnam and Zakia Bano, 1991; Rajarathnam *et al.*, 1992; Rajarathnam *et al.*, 1997; Poppe, 2000). Forest residues like leaves of deciduous trees, twigs, sawdust and household wastes like paper garbage etc. are also valuable for mushroom cultivation.

Though it is quite difficult to the assess production of total crop residues in the world or any country because of varied types of crops, fruits, vegetables and agro-forestry done; however, taking into account major crop groups guestimate has been made (Table 1).

Table1. Production of crop residues (1994) (x 1000 MT)

Crop group	World	India
Cereals	7,802,400	849,928
Pulses	175,941	43,608
Oil seeds	354,185	35,527
Plantation crops	1,370,065	221,580
Total	9,702,591	11,50,643

Source: Rajarathnam *et al.* (1997) based on FAO Production Year Book (1994)

Chemistry of agro-and forest residues

Cellulose, hemicellulose and lignin are the major constituents of agro- and forest residues which are therefore also termed as lignocellu-loses (LCs). Gross chemical composition of the broad categories of lignocelluloses are given in Table 2.

Table 2. Chemical composition (per cent dry weight) of lignocelluloses

Lignocellulose	Cellulose	Hemicellulose	Lignin
Leaves	15-20	80-85	-
Straws and grasses	25-40	25-30	10-30
Softwoods	45-50	25-35	25-35
Hardwoods	40-55	20-40	18-25
Cotton	89	5	-

Source: Satyanarayana and Johri (1988)

Chemical composition of the cereal straws, the most abundant and important crop-residues for cultivation of mushrooms is given in Table 3. It may vary as per the cultivar used, cultivation and harvesting practices, climate and the maturity level etc.

Table 3. Chemical composition (per cent dry weight) of cereal straws

Cereal straw	Cellulose	Hemicellulose	Lignin
Paddy straw	36	27	16
Wheat straw	43	36	14
Barley straw	40	39	13
Maize straw	18	31	19

Source: Rai and Saxena (1990).

It would be evident from the Tables 2 and 3 that cellulose, hemicellulose and lignin constitute about 80% of the dry matter of most of the crop and forest residues. soluble substances and minerals are the other important constituents.

MUSHROOMS FROM WASTES

Though many studies have been carried out on the bioconversion of lignocellulosic wastes into useful products like sugar and alcohol etc., but the efforts have not met with commercial success. The cultivation of mushrooms around the world currently represents the only large-scale commercial conversion of lignocellulosic-residues from agriculture and forestry, and the industry is now worth more than $ 6.0 billion per annum (Wood, 1989).

Lignin in the agrowastes is a recalcitrant polymer and acts as physical barrier in degradation and utilization of cellulose and hemicellulose also by the microbes most of which can not degrade it efficiently. White-rot basidiomycetes, including the cultivated mushrooms, are the only microbes known to degrade lignin, cellulose and hemicellulose efficiently and some of these degrade lignin preferentially (Zadrazil and Reiniger 1988; Rai and Saxena, 1990). It is their extensive degrading ability, discussed in detail later in the article, that has made mushrooms one of the most efficient bioconvertors of lignocelluloses into protein (fruitbodies).

Table 4. Lignocellulosic substrates for cultivation of mushrooms

Mushrooms	Substrates
Agaricus bisporus	Wheat straw, paddy straw, maize leaves, sugarcane bagasse
Lentinula edodes	Wood-logs, sawdust, corncobs, used tea leaves, cotton seed hulls, bagasse
Volvariella volvacea	Paddy straw, cotton waste, bagasse, water hyacinth, banana leaves
Pleurotus spp.	Almost all cereal straws, stalks and leaves of oilseeds and pulses, coffee pulp, used tea leaves, cotton waste, cotton stalks, pulp-mill sludge, wood-logs, forest leaves, grasses, waste paper, coconut fibre, jute stalks, corncobs, apple pomace.
Auricularia polytricha	Sawdust, wheat straw, wood logs
Flammulina velutipes	Sawdust, (hard wood + pinewood mixture)
Pholiota nameko	Sawdust
Stropharia rugoso-annulata	Bagasse, corncobs, cereal straws

About twenty types of mushrooms are being commercially cultivated now a days but eight of these are produced and traded in sizeable quantities-white button mushroom (*Agaricus bisporus*), shiitake (*Lentinula edodes*), paddy straw mushroom (*Volvariella volvacea*), oyster or 'dhingri' mushroom (*Pleurotus* spp.), black ear mushroom (*Auricularia polytricha*), winter mushroom (*Flammulina velutipes*), giant mushroom

(*Stropharia rugoso-annulata*), silver ear mushroom (*Tremella fuciformis*) and nameko (*Pholiota nameko*). Various wastes on which these mushrooms have generally been cultivated are given in Table 4.

Table 5. Pretreatments, growth substrates and average yield of mushrooms

Species	Substrate	Production time	Yield (g dry per kg dry substrate)
A. From non-axenic culture methods			
1. Without/little pretreatment			
Lentinula edodes	Wood logs	3-6 years spring/ autumn	40
Auricularia spp.	Wood logs	2-5 years spring/ autumn	-21
Tremella fuciformis	Wood logs	3-6 years, 7 months/ year	-20
Kuchneromyces mutabilis	Wood logs	2-10 years, 3 months/year	27-30
Pleurotus sajor-caju	Unfermentedcereal straw	-35 days	110
2. Some pretreatment			
Volvariella volvacea	Rice straw, cotton waste	7-10 weeks	6-10
	Pasteurized/ferment ed cereal straws	5-6 weeks	40
Pleurotus spp.	Cereal straws	8-12 weeks	100
Stropharia rugoso-annulata	Straw bales (soaked in $CaNO_3$)	10-40 weeks	-20
Coprinus fimetarius		4-5 weeks	-60
3. Long compositing process			
Agaricus bisporus	Composed cereal otraw/animal manure	14-16 weeks	65-80
Coprinus comatus	Composted straw	14-16 weeks	40-50
B. From axenic culture methods			
Flammulina velutipes	Sterilized saw dust & rice bran mixture (polypropylene bottles)	12-20 weeks	70-100
Lentinula edodes	Sterilized saw dust & rice bran mixture (polyethylene bags)	12-20 weeks	70-100
Auricularia species	Sterilized saw dust & rice bran mixture (polyethylene bags)	8-10 weeks	70-75

Source: Smith *et al.* (1987)

There are very few crop- and forest wastes, as evident from table 4, which can not be utilized for cultivation of one or the other mushroom.

Of course, selection of a mushroom type and a suitable substrate for cultivation in a region and season would be guided by many factors including the local availability of the substrate. Substrates are used singly as well as in combinations with or without elaborate pretreatments. For instance, while *Pleurotus* spp. are grown on chemically or steam pasteurized fresh cereal straws, *A. bisporus* is grown on a elaborately composted substrate. The substrates, pretreatments, production time and average yields of different mushrooms are given in Table 5.

The substrate with or without pretreatment is seeded with spawn (mushroom mycelium pregrown on a suitable substrate, mostly cereal grain) and the seeded (spawned) substrate is incubated for some time at appropriate temperature for spawn-run. Fungus colonises and degrades the substrate for profuse mycelial growth. The colonised substrate is then subjected to conditions unfavourable for mycelial growth but favourable for mushroom formation (fructification). Mushrooms appear in flushes, generally at weekly intervals, and continue to appear for a period ranging from few weeks to years. Common features and steps in cultivation of most of the commercial mushrooms are depicted in Fig. 1 and have been described in detail in many publications on the cultivation technology of mushrooms. Mushrooms have very short shelf-life and should be sold or consumed as fresh as soon as possible. If to be stored, modified atmosphere packaging (MAP), controlled atmosphere packaging (CAP), modified humidity packaging (MHP) coupled with precooling technology, refrigerated transport and storage is practised to extend the shelf-life. Some long-term storage and processing technologies for mushrooms are canning, drying, pickling, IQF and freeze-drying depending upon the market demand. Many value-added products and recipes like mushroom soup powder etc. are already in the market (Fig 2).

Yields are recorded either as per unit weight of the substrate or as per unit bed area. Average yield of *A. bisporus* varies between 15-25 kg/q compost, roughly about 40-75kg/q straw used. In other mushrooms like *Pleurotus*, yields are also expressed as Biological Efficiency (B.E.) which denotes the yield of fresh mushrooms per unit dry weight of the substrate; one quintal fresh mushroom on one quintal dry straw would be expressed as 100% B.E. Yields of various mushrooms vary as per the strain, substrate, cultivation technology, season and location. *Pleurotus* mushrooms are very high yielders with an average B.E. of 70-80%. *Pleurotus sajor-caju*, an introduction from India, has been reported to record to 200% B.E., with overall average of above 80% (Chang and Miles, 1989). Importantly, such yields are obtainable within a short period of 1-2 months and thus many crops are possible in a year. It has been estimated that one square meter of land can produce 620 kg of *A. bisporus* or 586 kg of *Volvariella* annually; values for *Pleurotus* will be still higher. In view of the above, mushroom cultivation is regarded as one of the most efficient methods of protein production per unit area and time (Table 6).

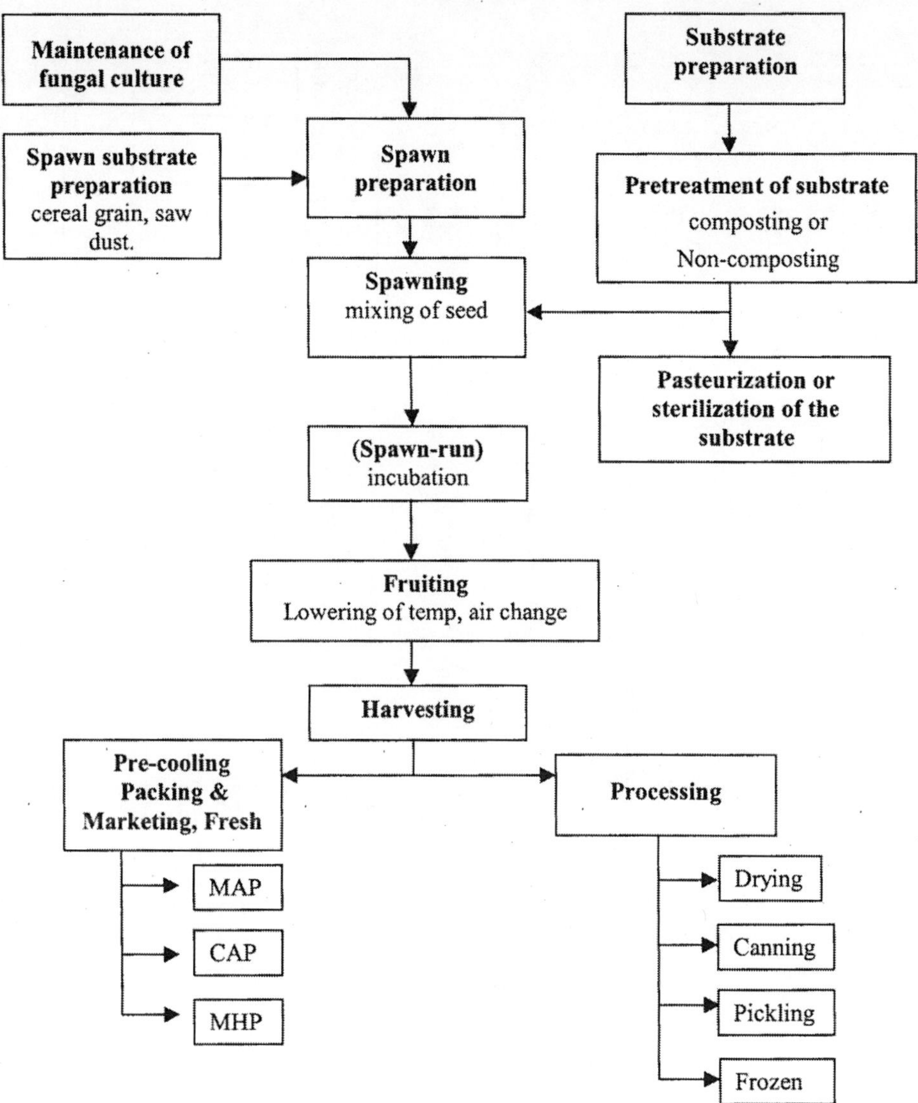

Fig. 1. Major steps in mushroom production, storage and processing.

Table 6. Protein production through various methods

System	Dry protein/ha/year
Beef cattle (conventional agriculture)	78
Fish (intensive pond-rearing)	675
Mushroom (*A. bisporus*)	65000

Source: Cooke (1977)

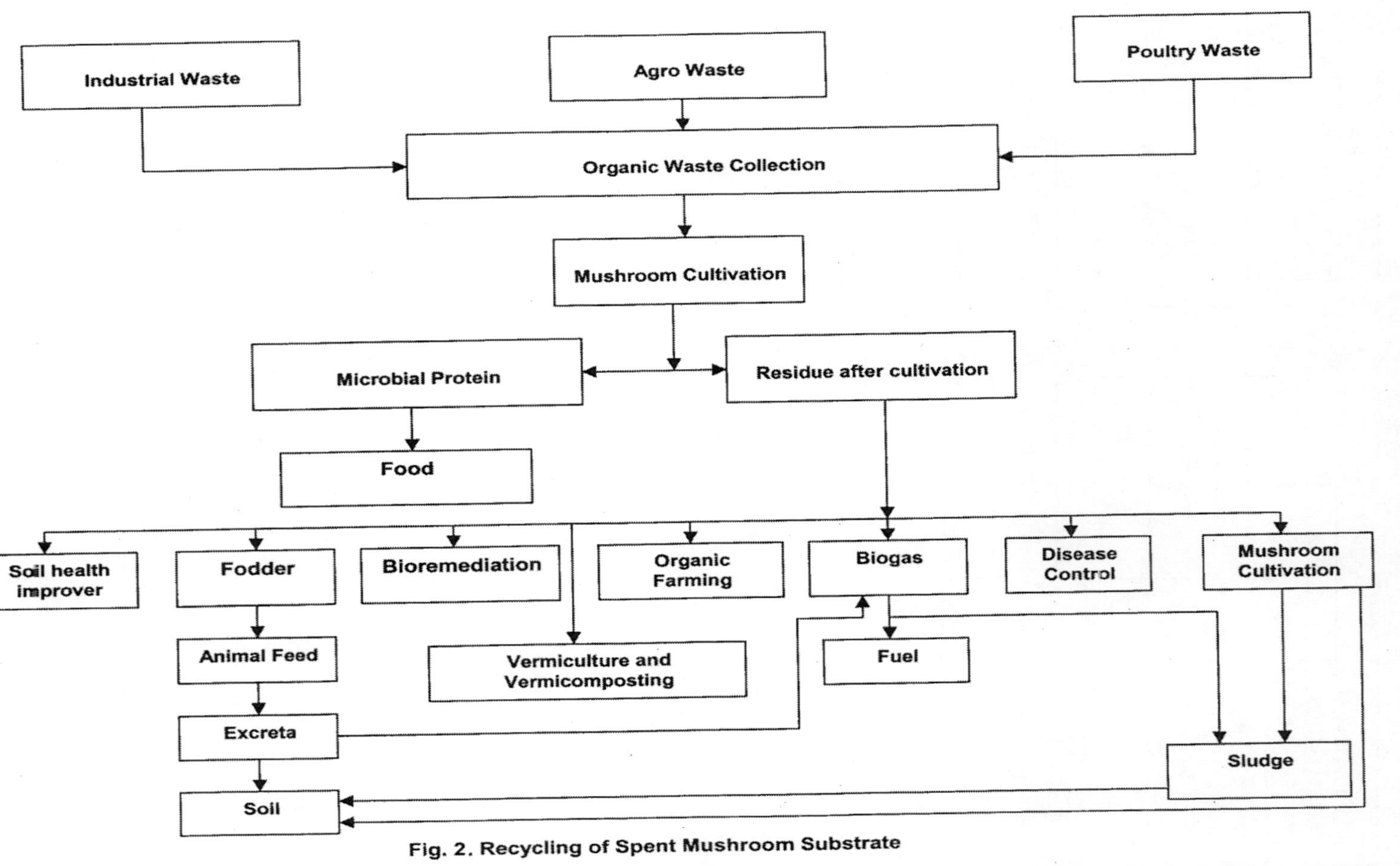

Fig. 2. Recycling of Spent Mushroom Substrate

Biodegradation of Lignocellulose degradation by the edible fungi

With a view to appreciate the ability of edible fungi to utilise a variety of lignicellulosic wastes, it would be appropriate to give a brief treatment to the biochemical mechanisms of the degradation. Due to comparatively low lignin content (~20%), the crop residues are degraded to varied extent by a number of microbes (Kirk, 1983) but the lignin basically acts as physical barrier in degradation of cellulose and hemicellulose and most of the microbes can not degrade lignin efficiently. White rot fungi (basidiomycetes) are the only microbes known to completely degrade lignin cellulose and hemicellulose, some of them degrade lignin preferentially (Janshekhar and Fiechter, 1983; Kirk, 1983, 1984).

As the major constituents of straw, namely cellulose, hemicellulose and lignin are insoluble polymers, white rot fungi elaborate a variety of extracellular enzymes to degrade these polymers outside the cell. Cellulase is a complex of many enzymes but has three main types of activities: exoglucanase (cellobiohydrolase, synonymous with C_1 cellulase), endoglucanase (Cx cellulase or CMCase) and glucosidase (cellobiase). Cellulases are inducible and act in sequential and synergistic manner (Wood and MacCrae 1979). Xylan is the most abundant hemicellulose and xylanases are most studied hemicellulases but mannanases, galactanases and arabinases have also been studied. Like cellulases, these are also inducible and exhibit exo- and biosidase type activities. It was biochemistry of lignin degradation which was not very clear till the search of ligninases and other peroxidases (Kirk and Farrel, 1987; Dahiya,. 1989). Undoubtedly,. the discovery (Tien and and Kirk, 1983) of ligninase, a peroxidase in *Phanerochaete chrysosporium*, a white-rot basidiomycete, was a breakthrough; some other enzymes Mn^{++} peroxidase, aryl alcohol oxidase, glucose oxidase have been shown to play important roles in lignin degradation (Upadhyay and Fritsche, 1997). Role of other enzymes like oxygenases and oxidases in lignin degradation has not been completely ruled out (Szklarz and Leonowicz, 1986). Laccases and peroxidases, often in cooperation with other enzymes are suggested to be involved in lignin degradation (Kelly *et al.* 1986; Szkalarz and Leonowicz, 1986) Laccase has been shown to oxidise vanillyl acetone, a model lignin compound.

Degradation of lignocelluloses by various mushrooms have been studied by many workers, but *Pleurotus* spp. have been the centre of investigations in this respect. Other mushrooms which have been investigated with respect to LC degradation are *Lentinula edodes, Agaricus bisporus* and *Volvariella volvacea* etc. *Pleurotus* spp. are reported to produce extracellular degradative enzymes for degradation of all the major polymeric constituents of LCs. (Platt *et al.* 1984; Hong *et al.* 1985; Arora and Sandhu, 1987; Sharma 1987; Bourbannais and Paice 1988). There are conflicting reports about production of the Filter paper degrading (FPase) activity; some authors have reported negligible on nil activity in the celture filtrate (Manning and Wood ,1978; Swift, 1982: Rai and Saxena, 1988). It has been suggested that initial disordering of native cellulose in these cases might be through nonenzymatic oxidation as suggested for *A. bisporus* and brown-rot fungi (Hedger and Basuki, 1982; Montgomery, 1982). However, some workers have reported C_1 cellulase in *Pleurotus* spp. (Vetter, 1987) and avicelase in *P. sajor-caju* (Hong *et al.,* 1985). All mushrooms produce activities of xylanase to degrade hemicellulose. Many mushrooms have now been reported to produce lignin degrading enzymes including ligninase, Mn^{++} peroxidase and laccase etc. In a study (Rai and Saxena, 1988) on the dynamics of appearance of extracellular degradative enzymes it was observed that during mycelial growth of *P. sajor caju* on LCs, polyphenol oxidase and laccase activities appeared and peaked earlier than endoglucanase and β-glucosidase. Similar trends have been reported in many other mushrooms including *A. bisporus* (Wood and Smith, 1988) and *L. edodes* (Leatham, 1985). These findings indicate that removal of protective lignin precedes the cellulolysis in the degradation of LCs by the white rot basidiomycetes (Hedger and Basuki, 1982; Kirk and Fenn, 1982). Lignin is reported to be degraded at a faster rate than cellulose during the mycelial colonisation (spawn-run) but trend is reversed during the fruiting (Ginterova and Lazarova, 1987). Platt *et al.* (1984) have reported higher lignin degradation in cotton straw by *P. ostreatus florida* than by *Phanerochaete chrysosporium* and also that strains with poor laccase activity showed slower degradation of cotton straw in general and lignin component in particular.

MUSHROOM AS FOOD

It would be impossible to ascertain with certainty for how long man has been eating mushrooms but it can be speculated that man started collecting and eating wild mushrooms since 'hunting and gathering' period of the human history (Cooke, 1977). Mushroom hunting is still practised in many parts of the world including India but their appearance in the wild is not only uncertain, it is declining due to deforestation. Besides, mushroom hunting is fraught with danger of chanced collection and consumption of poisonous mushrooms some of which are deadly. In the past hundred years, many mushrooms have been 'domesticated' and are being artificially cultivated. Earlier, these were being consumed mainly for their palatability, unique flavour and texture, but their nutritional and medicinal values have been recognized recently (Rai, 1994, 1997). Mushrooms, on an average, contain about 90% moisture and are basically a low-calorie food but diets which are primarily cereal based rarely lack calories. It is often the protein component which is inadequate and deficient intake of protein, especially good quality protein (mostly animal proteins), is the main dietary defect in many countries like India.

Mushrooms, due to high quantity and quality of protein, have been recognised by the Food and Agriculture Organisation (FAO) as food contributing to the protein nutrition of the countries depending largely on cereals. Expectedly, in the nutritional evaluation of mushrooms, proteins have been the focus of attention of the researchers and shall be dealt here in detail. Protein values for most popular mushroom, *A.*

bisporus as reported by various workers are given in Table 7 which shows that wide variations have been reported by various workers; similar is the case with other mushrooms also. One of the reasons for the variability is the method of quantitation and the multiplication factor. Crude protein content in the foodstuffs is traditionally determined by estimating nitrogen and multiplying with a conversion factor of 6.25 with the assumption that all proteins contain 16 per cent N and no non-protein N is present in the foodstuff. Keeping in view high non-protein N (chitin) in mushrooms, a multiplication factor of 4.38 was suggested (Crisan and Sands, 1978).

Table 7. Protein content of *Agaricus bisporus* reported by different workers

Protein (% DWT)	Source
19.4-37.1	Weaver *et al.* (1977)
26.3	Anderson and Fellers (1942)
24.9	McConnel and Esselan (1947)
33.2	Pilat (1954)
28.1	Watt and Merill (1963)
27.5	Maggioni *et al.* (1968)
23.9	FAO(1972)
26.3	Humifeld and Sugihara (1949)
32.7	Adriano and Cruz (1933)
25.5	Dommel (1964)
38.8	Hayes and Haddad (1976)
50.0	Udipa and Punekar (1980)
25.6	Haytowitz and Mathews (1984)
29.3	Rai and Sohi (1988)

Source: Beelman and Edward (1989)

Rai *et al.* (1988) determined protein in seven *Pleurotus* species by various methods of protein determination and found that true protein values obtained with the method of Lowry *et al.* (1951) were closest to N x 4.38 values (Table 8). Protein content of mushrooms has also been reported to vary from the flush to flush (Crisan and Sands, 1978) and also with the cultivation substrate (Bano and Rajarathnam, 1982).

Table 8. Protein content in *Pleurotus* spp. by various analytical methods

Species	Protein (N x 6.25)	Protein (N x 4.38)	Protein (Lowry)	Protein (dye-binding)
P. eryngii	3.10	2.18	2.18	2.12
P. flabellatus	2.80	1.97	2.01	1.89
P. florida	2.24	1.57	1.61	1.45
P. membranaceus	3.04	2.13	2.10	1.91
P. ostreatus	2.61	1.83	1.91	1.66
P. sajor-caju	3.47	2.43	2.51	2.20
P. sapidus	3.18	2.23	2.37	2.01

Source: Rai *et al.* (1988)

Protein content in *Pleurotus* spp. has been documented to range between 8.9 and 38.7 per cent on dry weight basis (Bano and Rajarathnam, 1982). Chang and Miles (1989) have suggested that total protein content of 3.5 to 4 per cent in commercial edible mushrooms may be more representative, which means that the protein in mushrooms, in general, is about twice that in asparagus and cabbage, and 4 and 12 times of those in oranges and apples, respectively. On a dry weight basis, 19-35 per cent protein in mushrooms compares well with 7.3 per cent in rice, 13.2 per cent in wheat and 25.2 per cent in milk. In terms of protein quantity, mushrooms, though below animal meats, rank well above common vegetables and fruits (Table 9).

Table 9. Protein content in some food stuffs

Food stuff	Protein in edible portion (% fresh weight)
Agaricus bisporus	3.7
Volvariella volvacea	3.9
Peas	2.6
Cabbage	1.5
Cauliflower	2.7
Potato	2.1
Apple	0.3
Banana	1.1
Chicken	20.01
Fish	19.0
Egg	12.8
Milk (cow)	3.2

Source: Kaul (1983)

It is now established that quality of protein is nutritionally more important than the quantity. Requirement of protein is, in fact, the requirement of amino acids and not of the protein *per se*. Human body is incapable of producing some (nine) amino acids in required quantities which are called essential amino acids and must be present simultaneously and in correct proportions; deficiency of any of these amino acids limits the utilisation of total protein in direct proportion to the deficit of that particular amino acid. Based on the contents of these amino acids, nutritionists have classified food proteins as of good, bad and intermediate quality. In general, as single source, vegetable proteins (cereals, pulses, vegetables) are of poorer quality than the animal proteins (milk, egg, fish, meat). It is recommended that one third of total protein intake should be of good quality (animal protein) to make up for the deficiencies inherent in the vegetable proteins but the suggestion is impracticable for the vegetarians. Quality of mushroom protein is far superior to the vegetable proteins and is as good as or just inferior to animal proteins which is because all the essential amino acids are present in mushrooms (Table 10) (Crisan and Sands, 1978; Bano and

Rajarathnam, 1982; Chang and Miles, 1989). Most abundant amino acid in mushrooms is lysine, and the cereals, the staple constituents of the diets of many countries, are deficient in lysine. Mushrooms can thus effectively supplement the cereals in terms of protein quality.

Table 10. Essential amino acids (% crude protein) in edible mushrooms

Amino acid	*Agaricus bisporus*	Pleurotus sajor-caju	*Volvariella volvacea*
Leucine	7.5	7.0	4.5
Isoleucine	4.5	4.4	3.4
Valine	2.5	5.3	5.4
Tryptophan	2.0	1.2	1.5
Lysine	9.1	5.7	7.1
Threonine	5.5	5.0	3.5
Phenyl alanine	4.2	5.0	2.6
Methionine	0.9	1.8	1.1
Histidine	2.7	2.2	3.8

Source: Bano and Rajarthnam (1982); Li and Chang (1982)

Thayumanavan and Manickam (1980) reported digestibility, biological value (BV) and net protein utilisation (NPU) of *P. sajor-caju* to be 84.1, 89.4 and 75.1 per cent respectively which are indicative of very good quality. Other indices of protein quality like chemical scores (CS), essential amino acid index (EAA), amino acid score (AAS), protein score (PS) and nutritional index (NI) have also been employed in mushrooms (Crisan and Sands, 1978; Bano and Rajarathnam, 1982) to demonstrate superior quality. *Pleurotus* mushrooms are reported to contain about 0.3 to 0.5 per cent free amino acids (Rai *et al.*, 1988). Nutritionally free amino acids are equally important and their contents should be taken into account particularly when true protein values have been given. But decline in the protein content during storage of white button mushroom, *A. bisporus*, due to elevated levels of proteases, has been observed; higher the storage duration and temperature, higher was the degradation (Rai and Saxena, 1989). Besides protein, they contain significant quantities of vitamins of B complex group and vitamin C (Crisan and Sands, 1978; Rai and Saxena, 1989) and are especially rich in thiamine, riboflavin and niacin. Folic acid and vitamin B12, normally absent in plant foods, are also present in mushrooms. Among the minerals, phosphorus, potassium and iron contents are significant. Fat content is low but it is rich in polyunsaturated fatty acids (PUFA), and cholesterol is absent – a combination suited to the heart patients (Rai, 1994). High fibre and alkaline ash content in them are desirable for those with hyperacidity and constipation. With low calorie, low sugar, nil starch and high protein, mushrooms are highly suited to those with obesity and diabetes. Besides, they are reported to possess many medicinal properties like antitumor, hypocholesterolemic, and immunomodulating activities, etc. (Rai, 1994).

MUSHROOMS FOR DEVELOPING COUNTRIES

Poverty, malnutrition and predominant agricultural economy are few common features of the developing or the Third World Countries. These are the countries where agro-wastes are also available in plenty and the demand for protein is also increasing due to population growth. Promotion of mushroom cultivation in these countries, besides utilising the agrowastes, would contribute significantly to the solution of the protein malnutrition. Being an indoor labour-intensive technology, it would also providc gainful employment to the population in general and women-folk in particular. Land is to do very little and even space above the surface is also used in a tier-system which makes it highly suited venture for the economic upliftment of the landless labourers and small farmers. Cultivation technology can be made as simple as rural farming of the tropical mushrooms in the huts or highly sophisticated big industries of *A. bisporus.* But in the developing countries it should be aimed as village- or cottage-level industry in rural areas and sophisticated industries in and around big cities. Kitchen gardening of tropical mushrooms like *Pleurotus* and *Volvariella* for fighting malnutrition should also be encouraged and promoted. Consumption of mushrooms in India is very low as compared to some other countries (Table 11) which indicates immense scope for promoting mushroom production and consumption in the country. Low consumption is mainly due to less availability, high cost and less awareness about nutritional and medicinal virtues of mushrooms which should be addressed to boost the consumption and demand.

Table 11. Consumption of mushrooms in some countries

Country	Consumption (per capita/annum) (g)
Netherlands	2800
United Kingdom	3350
France	3030
Italy	2060
India	25

Source: Kaul (1999)

It was not without any reason that cultivation of button mushroom in India first became popular in H.P., J & K and U.P. hills which are suited for cultivation of many types of mushrooms, especially the white button mushroom, *A. bisporus* under natural conditions. However, many types of mushrooms can be grown in form of a crop rotation throughout the year as per the prevailing temperature in different parts of the country (Table 12).

Table 12. Temperature requirement for production of different mushrooms

Mushroom	Spawn run	Fruiting
Agaricus bisporus	22-25	15-17
Agaricus bitorquis	28-30	22-24
Lentinula edodes	24	15-20
Pleurotus spp.	20-30	15-30
Volvariella volvacea	32-35	28-32
Auricularia polytricha	20-34	24-27
Stropharia rugoso-annulata	20-22	10-20

Pleurotus with a number of species differing in their temperature requirement (Table 13) are the most versatile and the species can be grown throughout the year in a crop rotation. Therefore, a mushroom growing unit designed exclusively for *Pleurotus* can be run throughout the year even under the natural conditions.

Table 13. Temperature for cultivation of various *Pleurotus* spp.

Species	Temperature
P. fossulatus	15-20
P. cornucopiae	15-20
P. eryngii	15-22
P. ostreatus	15-26
P. flabellatus	20-28
P. sajor-caju	20-28
P. sapidus	22-28
P. opuntiae	24-28
P. euosmus	24-28
P. citrinopileatus	25-30
P. membranaceus	25-30

Source: Upadhyay (1990)

FURTHER RECYCLING OF SPENT MUSHROOM SUBSTRATE

Compost is considered "spent" when one full crop of mushroom has been taken and further extension of cropping becomes unremunerative (Wuest and Fahy, 1991). Spent mushroom compost is still nutrient-rich and contains about 80% of the total nitrogen in bound-form with high molecular weight fractions of lignin and humic substances (Grabbe, 1982). Mushroom growing is an eco-friendly activity as it utilizes the wastes from agriculture, poultry, brewery etc.; however, piling up of 'spent' mushroom substrate after crop harvest may cause various environmental problems including ground water contamination and nuisance (Beyer, 1996). The large dumped piles of spent mushroom substrate (SMS) become anaerobic and give off offensive odours. The run-off from such piles may contaminate nearby water sources and pollute them (Beyer, 1996). In the past, efforts have been directed towards profitable utilization of the post-mushroom substrate and considerable mitigation of environmental pollution (Levanon *et al.*, 1994; Ahlawat and Sagar, 2001; Ahlawat and Rai, 2002).

Spent mushroom substrate normally contains 1.9:0.4:2.4% (N-P-K) before weathering and 1.9:0.6:1.0 (N-P-K) after weathering for 8-16 months. Nitrogen and phosphorus do not leach out easily but, potassium being more leachable is lost in significant amount during weathering. SMS contains much less heavy metals than sewerage sludge, which precludes its classification as hazardous substance (Wuest and Fahy, 1991). Weathering also causes a slow decrease in the organic matter contents (volatile solids). The SMS obtained from different commercial sites reveal that except calcium, boron, copper and zinc,coefficient of variability are less than 25% while, the conductivity and pH are consistently high (Devonald, 1987). The SMS obtained from various sources usually have conductance in the range of 1.9 to 8.3 mmhos/cm. However, the Cl content in fresh SMS varies from 1.5 to 7.5 kg/1000 kg (Gerrits, 1987) and only about 294 ppm in a well rotten SMS (Chong and Wickware, 1989). The lower chloride content in SMS has been attributed to less chicken manure incorporation in fresh mushroom substrate (Chong *et al.*, 1991). There are contradictory reports regarding the pH of fresh SMS and pH after weathering. SMS has an initial pH of around 7.3 which increases during weathering (Wuest and Fahy, 1991). However, Devonald (1987) reported that the pH of the freshly collected SMS ranged between 7.01 and 8.04 with mean ammonia and nitrate content of 174 and 39 ppm respectively. The pH of well-rotten SMS was found to be little higher (7.70) with 15 ppm NO_3 content (Chong and Wickware, 1989). In a separate experiment Chong et al. (1988) found a decrease in pH from its initial value 7.9 to 7.0 on weathering. The volume of SMS also decreases (shrinkage) over time. The fresh SMS obtained from various sources varied in density: 0.198 g/cm^3 with a range of 0.15 to 0.24 g/cm^3 in UK (Devonald, 1987); 0.475g/cm^3 in Ireland (Maher, 1991) and 0.24 to 0.62 g/cm^3 in USA (Dvorak *et al.*, 1991). The composting of SMS in semi-enclosed drum for 6 weeks led to an increase in its bulk density from 0.256 g/cm^3 to 0.293 g/cm^3 (Lohr *et al.*, 1984).

As mushroom production is increasing so is the SMS generation which calls for management of this so-called waste. Fortunately, SMS has many positive attributes still left for potential uses. The material has been found to be good nutrient source for agriculture because of its nutrient-status, a high cation exchange capacity and a slow mineralization rate, retaining its quality as organic matter. Further, SMS contains 45% water and though bulky, is light in weight (Dann, 1996). SMS can be used and utilised for various purposes (Fig 2).

ORGANIC FARMING

Horticulture

SMS made the soil suitable for raising vegetables (Kaddous and Morgans, 1986). However, suitable treatments like rapid salt leaching (Chong *et al.*, 1991) and weathering in open for two to three years make SMS more suitable for either complete or partial substitution of growing media for flowers, vegetables, fruits, saplings, ornamental shrubs and other horticultural plants of economic importance (Beyer, 1996). The SMS being rich in N, P and K acted as a good growing medium for vegetables like cucumber, tomato, broccoli, tulip, cauliflower, peppers and spinach etc. but the response of the plants varied with the level of SMS incorporation; incorporation @ 25% and 37.5% in the growing medium supported the growth of lettuce, marigold and tomato; the aged-mushroom compost was, however, preferable over the fresh compost (Lohr and Coffcy, 1987).

SMS had certain advantages over the manures from other sources or composted bark, in raising the bulbs of tulip; 50t/ha incorporation gave the maximum yield of bulbs with higher content of P, K, Ca and Mg (Wisniewska and Penkiewicz, 1989). It also acted as good substitute for peat in bag cultivation of tomatoes (Abak and Celikel, 1993; Szmidt, 1994). Some other crops including *Cucurbita pepo* (Stephens *et al.*, 1989), *Capsicum annum*, spring broccoli, autumn broccoli, aubergines (Maynard, 1994), Sweet corn cv. Seneca gold, snap beans (*Phaseolus vulgaris*) (Steffen *et al.*, 1994) and *Pennisetum glaucum* (Rhoads and Olson, 1995) were also found to give more yield in SMS incorporated plots. SMS incorporation also improved the firmness and ascorbic acid content in tomatoes besides the yield (Dundar *et al.*, 1995). However, in few cases SMS application deteriorated the quality of the produce (Manivel *et al.*, 1994). The application of SMS is not limited to the open field vegetable cultivation system only but also in greenhouse crops. The 1:1 mixture of SMS and speedel medium with osmocote (14:14:14 NPK), incorporated with root medium at 5oz/ft3 gave the highest number of flower buds in *Chrysanthemum* and *Lilium longiflorum* (Dallon, 1987). The containerized nursery crop of *Cornus sericea* and *Forsythia* x *intermedia* cv Lynwood also showed good response to SMS incorporation in the growing medium and the top dry weight of the species increased in proportion to the amount of unweathered and weathered spent mushroom compost incorporated in the medium (Chong *et al.*, 1987).

Besides vegetables, greenhouse and nursery crops, the woody ornamental and forage crops including *Cotoneaster dammeri* cv Coral Beauty, *Deutzia gracilis*, *Cornus alba*, *Argenteo marginata*, *Forsythia* x *intermedia* cv. Lynwood, *Juniperus sabina* cv. Blue Danube, *J. virginiana* cv. Hetzii, *Physocarpus opulifolius*, *Potentilla fruticosa* cv. Red Ace, *Ligustrum vulgare*, *Rosa indica* cv. John Franklin, *Weigela* cv. Bristol Ruby and *W. florida* cv. *variegata* Nana also showed good growth at different levels (33%, 67% & 100%) of SMS mixed with bark (Chong and

Rinker, 1994). In a separate experiment, four deciduous ornamental shrubs, namely Coral Beauty Cotoneaster (*Contoneaster dammeri*), Tartarian Dogwood (*Cornus alba*), Lynwood forsythia (*Forsythia intermedia*) and Variegata weigala (*Weigala florida*) were grown in trickle fertigated containers and 25 or 50% supplementation with the SMS was found to promote excellent growth of all the four ornamental shrubs (Chong and Hamersma, 1997). In addition to the above mentioned plant species, SMS also increased the total spear FW and number/m^2 in white asparagus (Pill *et al.*, 1993) and total height in mountain persimmon (*Diospyros oldhamii*) and nonastringent persimmon (*D. Kaki*) cv. Fuyu (Nee *et al.*, 1994).

FIELD CROPS

SMS application has been used to improve productivity of many major food crops. The SMS incorporation @ 100, 200 and 400 tons (Fresh weight/acre) in nutrient-poor soil like silty clay loam showed positive effect on the silage and grain yield of corn crop. The silage production increased by 33% to 68% and the grain yield varied depending upon the climatic conditions and rainfall in that particular growing season. Incorporation of SMS @ 28kg/plot in maize variety Kanchan enhanced the grain yield by 79% over the control (no manure or fresh fertilizer) in a rain-fed crop at Solan (HP) (Yadav *et al.*, 1999). In a separate study the SMS addition in corn field did not cause any toxicity problem and the quality of ground water and the run off water matched with the water from the control plots (Wuest and Fahy, 1991). Another study carried out by Wuest *et al.*, in 1995 also showed significant effect of SMS incorporation on the yield of corn; the grain as well as stover were also found to contain significantly higher levels of nitrogen and water quality was unaffected. SMS in combination with peat stimulated the diazotrophs present in the soil also leading to an increase in the number of nodules/plant in *Cytisus canariensis, Exacum affine* and *Dracaena fragrans* (Gronborg, 1990). Further, yearly incorporation of SMS upto 90 kg/m^2 area did not affect the quality of surface water and the run-off had very low amount of ammonia-N, biological oxygen demand (BOD) and chloride content.

ORGANIC FERTILIZER

Nitrogen release from the SMS is very slow and not sufficient for plant nutrition which can, however, be improved by the addition of some easily available form of nitrogen. On the other hand, phosphorus, potassium, micro-nutrients and the growth substances in SMS are present in sufficient quantity and 'available' form. Availability of nutrients from the SMS can be enhanced by the preparation of organic-mineral fertilizer mixture from the SMS. The absorption capacity of the stable 'humus' in SMS can be used for the retention of nitrogen in the top soil (Grabbe, 1978). Conversion of SMS into an organic-mineral fertilizer seems to be a good alternative using SMS for soil amelioration and plant nutrition. Process of the organic-mineral fertilizer from the SMS has been patented in USA (Robinson, 1988).

DISEASE MANAGEMENT

Due to unique chemical constituents and the microflora of the SMS, its application can be made more diversified than what is normally practised. The actinomycetes, bacteria and fungi inhabiting the compost not only play role in its further decomposition but also exert antagonism to the normal crop pathogens surviving and multiplying in the soil ecosystem. The organic amendment of soil with SMS has been found to help in restricting the root knot infestation of tomato plants caused by *Meloidogyne incognita* and efficacy was higher than carbofuran (2kg ai/ha), a common nematicide used against above organism (Verma, 1986). Under *in vitro* conditions, the anaerobically fermented aqueous extract of the SMS inhibited the conidial germination of *Venturia inaequalis*, the causal agent of the apple scab; the seedling assay also showed similar type of results but the extract did not prove its potential under *in vivo* conditions. The extract also inhibited the conidial germination of *Cochliobolus carborum* and *Sphaeropsis sapinea* (*Diplodia pinea*) causing diseases of maize and red pine (*Pinus resinosa*) respectively. The aqueous extract obtained after 5 to 9 days of incubation in the ratio of 2:1 to 4:1 (water:SMS) maintained its efficacy for about 4 months on storage at -20^{0}C, 4^{0}C and room temperature. The biological analysis of SMS extract showed that it contained a pseudononad and a *Bacillus*. SMS obtained from different growers harboured different mycoflora and showed differences in their effect on conidial germination, inhibition and disease suppression. Weekly or bi-weekly application of spreader/sticker amended SMS extract, starting from green tip to petal fall stage of apple tree reduced the scab-affected leaf area. It maintained higher population of bacteria even after 1 month of the final spray which might be the reason for less scab affected leaf area in SMS sprayed treatment (Yohalem *et al.*, 1996).

BIOREMEDIATION OF SOIL AND WATER CONTAMINATION

The uncontrolled release of industrial wastes on the open land and non availability of the effluent pretreatment facilities causing alarming pollution of the environment. The SMS has the property to absorb the organic and inorganic pollutants and, in addition, it also contains diverse category of microbes capable of degrading organic xenobiotic compounds present in the soil and water (Buswell, 1995; Semple and Fermor, 1995; Hofrichter *et al.*, 1997). In laboratory experiments, the mixing of pentachlorophenol (PCP)-contaminated sterile soil with aliquots of spent sawdust substrate of shiitake mushroom, supplemented with nutrient solution of glucose, thiamine and mineral salt, resulted in disappearance of 44.4-60.5% of PCP in 21 days of incubation; the disappearance of PCP did not have any relation with the residual polyphenol oxidase activity in the inoculum (Okeke *et al.*, 1993; Buswell, 1995). The SMS also has the decontamination potential for land sites used for disposal of hazardous wastes (Buswell, 1994). Similar studies but with diverse category of chemicals were conducted by Fermor *et al.* (2000) which demonstrated significant potential of phase II compost in remediation of lands contaminated with xenobiotics pollutants like chlorophenols (CPs), polycyclic aromatic hydrocarbons (PAHs) and aromatic monomers.

Microbes, especially actinomycetes (*Streptomyces* sp. and *Thermomonospora* sp.) present in SMS have strong PCP catabolizing capabilities (Semple and Fermor, 1995). The residues obtained after mushroom cultivation can be utilized for large scale production of phenol oxidases which can later play vital role in environment protection activities (Steffen *et al.*, 1994). The second category of chemical contaminants affecting the soil are synthetic fungicides, herbicides and insecticides, which are being indiscriminately used in agriculture. The 5% w/w SMS incorporation in soil accelerated the alachlor disappearance and protected garden pea cv. Taichung from root injury. It was also observed that SMS stimulated the microbial population and maintained it for 28 days; the stimulated microbial population of *Aspergillus* sp, *Penicillium* sp, *Trichoderma* sp, *Aeromonas* sp and *Bacillus* sp, were found to degrade alachlor within 7 days of incubation under *in vitro* conditions (Huang *et al.*, 1995, 1996).

The limestone, gypsum, organic matter and 'bulk' that constitute SMS make it an ideal choice for passive treatment of coal mine drainage. The water quality was improved by creating wet-lands with SMS. The water coming out of such wet lands had high pH, while the iron and sulfur contents went down. The water released from these wet-lands met the clean stream specifications (Pannier, 1993; Stark and Williams, 1994). This method was so effective that even after several weeks of operation there was no report of nitrogen leakage from these wet- lands (Stark *et al.*, 1995). Increasing environmental awareness and the use of bio-indicators, have made the studies related to the adverse effects caused by SMS land applications on adjacent surface water quality much more easier than before (Reed and Keil, 2000).

RECLAMATION OF SOIL

Incorporation of the SMS in the nutrient-poor soil leads to an improvement in the soil health by improving the texture, water holding capacity and the nutrient status (Kaddous and Morgans, 1986; Maher, 1991; Beyer, 1996). However, it reduces the thermal conductance, bulk density and water stable aggregates and increased organic carbon content of the soil (Kaddous and Morgan, 1986). The experiments carried out for studying the effect of SMS on the containerized nursery crops like *Cornus sericea* and *Forsythia* x *intermedia* cv. Lynwood have shown that their dry matter content increased with addition of increasing amounts of weathered or unweathered SMS (Chong *et al.*, 1987). Phosphorus and potassium requirements of the crop plants can be fulfilled by incorporating 5% of SMS by volume, while nitrogen requirement can be fully met by 25% of SMS (Maher, 1991). The mixing of carbonized SMS in soil have also shown high growth promoting activity of marigold plant (Iwase *et al.*, 2000).

RE-GROWING OF MUSHROOMS

Production of second crop of mushrooms from the SMS would, of course, prove most profitable utilization and would also ameliorate the problem of solid waste disposal in the mushroom industry (Fahy and Wuest, 1984) as only a small fraction of cellulose, hemicellulose and lignin is utilized by the mushroom in its first crop (Table 14). Re-spawning of *Agaricus* SMS after addition of spawn mate (delayed - release nutrient) and Bonaparte peat (adsorbent material) resulted in good yield from a second crop of *Agaricus*. The residual compost so obtained could be reused for raising tomato seedlings (Schisler, 1988).

Table 14. Fibrous composition of synthetic substrate before and after production of shiitake, *Lentinula edodes* for 63 days

Substrate type	Hemicellulose	Cellulose	Lignin	Fermentation efficiency (%)
Fresh	26.5	15.1	41.3	
Spent	22.4	8.4	32.0	
Difference	4.1	6.7	9.3	86.1

Source: Moyson and Verachart (1991)

The shiitake SMS supplemented with 10% wheat bran and 10% millet could be utilized for cultivation of *Pleurotus sajor-caju* after air-drying, grinding, supplementation, pasteurization and spawning. Very high yield (79% B.E.) of *P. sajor-caju* could be obtained by supplementing the shiitake SMS basal medium with 12% soya-bean and 1.0% $CaCO_3$ (Royse, 1993). Another use of SMS in mushroom cultivation is as casing material for the button mushroom. The use of aerobically fermented SMS as casing material gave mushroom yields at par with peat based casing material with the additional advantage that the fruitbodies got less bacterial blotch infection (*Pseudomonas tolaasii*) than those from peat-based casing material (Szmidt, 1994).

ANIMAL FEED

Normally the domestic animals in the developing countries are fed with wheat straw or paddy straw as staples supplemented with concentrates mostly oilseed cakes or cereal brans. The high lignin content of the cell fibres of the straws reduces the accessibility of digestive enzymes present in the animal gut to cellulose and hemicellulose (Silanikove and Levanon, 1986). *Pleurotus* spp. (oyster mushroom) have been shown to break down the lignin and make cellulose available for further degradation (Kerem, 1991). Protein value of the spent straw is also increased (Silanikove *et.al.*, 1988) and the digestibility of spent straw also gets improved after *Pleurotus* cultivation

(Table 15). The *Pleurotus* spent straw could substitute about 30% of the total feed without affecting the growth of the animals (Kerem, 1991). The spent paddy straw obtained after first flush of *Pleurotus* was found better in chemical composition as feed as compared to that obtained after third flush.

Table 15. Effect of pilot-scale SSF on the contents of SMS medium

Constituent (% DW)	Media with 35% spent compost of			
	P. ostreatus		*L. edodes*	
	Control	Fermented	Control	Fermented
Crude Protein	24.1	32.3	28.4	36.7
Crude Fibre	14.8	10.2	13.3	9.8
Crude ash	6.5	9.0	6.5	9.3
In vitro crude protein Digestibility	64.2	67.8	68.7	70.1

Source: Zhang *et al.* (1995)

BIOGAS PRODUCTION

Anaerobic fermentation of animal manures, domestic wastes and municipal garbage is carried out by a consortium of bacteria which break down the wastes to produce methane (Marchiam *et al.*, 1991). Solid residual wastes present in the fermenting tank, already decontaminated by the heat generated during the process, can be separated from the liquid and can be used as a good substitute for peat as a casing material (Levanon *et al.*, 1994) for cultivation of *Agaricus* spp. The SMS obtained as a waste from *Agaricus* cultivation when used in organic-farming would produce crop-residues which could be used for *Pleurotus* cultivation to get food (mushrooms) and feed for cattle, and the manure obtained from cattle is a good source of biogas (fuel) producing energy and the cycle repeats with the use of the solid residual waste as casing with no more waste left out (Levanon and Danai, 1997) as shown in Table 16. Use of SMS for biogas generation has further potentials and multiple benefits, such as possibility to utilize feed stocks of high moisture contents, possibility of scaling-up to suit family as well as community needs, effluent being a good source of manure which can replace chemical fertilizers in organic farming, and can give indirect economic benefits to the users (Madan, 1997). The process has thus immense scope of producing food, feed, fuel and fertilizer (manure) by complete recycling of agro-residues; farmers of Israel have adopted the technology.

In the era of growing awareness about the environment, pollution and strict legislations, the exploitation of spent mushroom substrate for the management of environment, agriculture and production of recyclable energy requires strict watch on physical, chemical and microbiological properties of SMS. Yes, its diversified uses in managing agriculture, environment and recycling energy forces us to change its name from spent mushroom substrate to "used mushroom substrate"

(Dann, 1996). Wuest rightly said, "There is no doubt about it - spent compost is a valuable natural resource" (Pannier, 1993).

Table 16. Recycling mushroom substrates and wastes

Recycling Course	Product	Waste
Cultivation cotton	Fibres	Plant residue cotton straw
Cultivation of oyster mushroom on straw	Oyster mushrooms	Spent substrate (SMS)
SMS used as cattle feed	Meat and milk	Manure
Manure fermented to produce biogas	Methane energy	Residue ("Cabutz")
Cabutz used as casing soil for champignons	Champignons	SMS
SMS composted for organic farming	Organic food crops	No further waste

Source: Levanon and Danai (1997)

CONCLUSION

Lignocellulosic substances on the earth, like forest- and crop-residues should not be termed as 'Wastes'. Edible fungi, also called mushrooms, with their extensive degrading ability of lignin, cellulose and hemicellulose, can be cultivated on these wastes to produce good quality food rich in protein quantity and quality besides other nutritional and medicinal attributes. Different types of mushrooms can be grown in form of a crop rotation throughout the year in many parts of the world even under the natural conditions. Cultivation of various mushrooms, if coupled with cattle rearing and biogas production, represents an unique system for complete recycling and utilization of agrowastes for production food (mushroom), feed, fuel (biogas) and fertilizer (manure). The venture is highly suitable for the developing countries with predominant agricultural economies----it can ameliorate the problems of poverty, malnutrition and unemployment. Burning of the residues *in situ* is a terrible waste of the utilizable energy and important source for many processes and production systems. Mushroom cultivation should be promoted in a big way in the developing countries like India.

REFERENCES

Ahlawat, O. P. and Sagar, M.P. (2001). Recycling of spent mushroom substrate for beneficial purposes. *Indian Farming.* 51(5): 24-27.

Ahlawat, O.P. and Rai, R.D. (2002). Recycling of spent mushroom waste. In *Recent Advances in Mushroom Cultivation Technology* (Verma, R.N. and Vijay, B. eds.) Pp.261-282. National Research Centre for Mushroom, Solan, India.

Abak, K. and Celikel, G. (1993). Comparison of some Turish originated organic and inorganic substrate for tomato soiless culture. In *Sec. Symposium on Protected Cultivation of Solanacea in Miled Winter Climates*, Adana, Turkey. 13-16.

Arora, D.S. and Sandhu, D.K. (1987). Decomposition of angiospermic wood saw dust and laccase production by two *Pleurotus* species. *J. Basic Microbiol.* **27**(4) : 179-184.

Bano, Z. and Rajarathnam, S. (1982). *Pleurotus* mushroom as nutritious food. In *Tropical Mushroom-Biological Nature and Cultivation Methods* (Chang, S.T. and Quimio, T.H., eds.). Pp. 363-82. The Chinese University Press, Hong Kong.

Beelman, R.B. and Edwards, C.G. (1989). Variability in the composition and nutritional value of the cultivated mushroom, *Agaricus bisporus. Mushroom News.* **37**(7): 17-26.

Beyer, M. (1996). The impact of the mushroom industry on the environment. *Mushroom News.* 44(11): 6-13.

Buswell, J.A. (1994). Potential of spent mushroom substrate for bioremediation purposes. *Compost - Science and Utilization.* 2(3): 31-36.

Buswell, J.A. (1995). Potential of spent mushroom substrate for bioremediation purposes. *Mushroom News.*43(5): 28-34.

Bourbannais, R. and Leonowicz, A. (1984). Veratryal alcohol oxidase from the lignin degradation basidiomycete *Pleurotus sajor-caju. Biochem. J.* **255**. 455-460.

Chang, S.T. (1982). Prospects for mushroom protein in developing countries. In *Tropical Mushrooms-Biological Nature and Cultivation Methods* (S.T. Chang and T.H. Quimio eds.), pp. 463-473, Chinese University Press, Hong Kong.

Chang, S.T. and Miles, P.G. (1989). *Edible Mushrooms and Their Cultivation*, CRC Press, Boca Raton.

Chong, C., Cline, R.A., Rinker, D.L. (1987). Spent mushroom compost and papermill sludge as soil amendments for containerized nursery crops. *Combined Proceedings - International Plant Propagators Society.* 37:347-353.

Chong, C. and Hamersma, B. (1997). Container growing with spent mushroom substrate. *Mushroom News.* 45(11): 12-14.

Chong, S. T.(1988). Proc. Ontario Hort. Crops Conf., Mushroom Session. 8 P.

Chong, C. and Wickware, M. (1989). *Hort. Rev.* 7(6): 10.

Chong, C., Cline, R. A. and Rinker, D. L. (1991). Rapid leaching of salts - key to successful plant growth in spent mushroom compost potting mixes. *Mushroom News.*39 (10): 12-13.

Chong, C., Rinker, D.L. and Cline, R.A. (1991). A comparison of five spent mushroom composts for container culture of ornamental shrubs. *Mush. Sci.* XIII : 637-644.

Chong, C., Rinker, D.L. (1994). Use of spent mushroom substrate for growing containerized woody ornamentals: an overview. *Compost - Science and Utilization* 2(3): 45-53.

Cooke, R.C. (1977). *Fungi, Man and his Environment*, Longman, London.

Crisan, E.V. and Sands, A. (1978). Nutritional value of edible mushrooms. In. T*he Biology and Cultivation of Edible Mushrooms* (Chang, S.T. and Hayes, W.A., eds.). pp. 137-68. Academic Press, New York.

Dahiya, J.S. (1989). Fungal degradation of lignin. *Indian J. Microbiol.* **29**. 1-14.

Dallon, J.Jr. (1987). Effects of spent mushroom compost on the production of greenhouse - grown crops. *Combined Proceedings, International Plant Propagators Society.* 37:323-329.

Dann, M.S. (1996). The many uses of spent mushroom substrate. *Mushroom News.* 44 (8): 24-27.

Devonald, V.G. (1987). Spent mushroom compost, a possible growing medium ingredient. Compost: Production, Quality and Use. 785-791. Els., April. Sci. London.

Dundar, O., Paksoy, M. and Abak, K. (1995). Quality changes during cold storage of tomato fruits grown in different substrates. In *First International Symposium on Solanacea for Fresh Market,* Malaya, Spain, 28-31 March, 1995.

Dvorak, D.H. (1991). In: Proc. 1991 Nat.l Mtg. Am. Soc. Surf. Mine Rec. 2000, Tech. for Succ. Durango, Co. P. 109-122.

Fahy, H.K. and Wuest, P.J. (1984). Best practices for environmental protection in the mushroom farm company. *Chester County Planning Commission West Chester,* PA 65 pages.

Femor, T., Wastls, N., Duncombe, T., Brooke, R., McCarthy, A., Semple, K. and Reid, B. (2000). Bioremediation: Use of composts and composting technologies. *Mushroom Sci.* XV(II): 833-839.

Gerrits, J.P.G. (1987). Compost for mushroom production and its subsequent use for soil improvement. In (Bertoldi, M.de, Ferranti, M.P. Hermite, P.L. and Zucconi, F. eds.). pp. 431-439. Els. April. Sci. London.

Ginterova, A. and Lazarova, A. (1987). Degradation dynamic of lignocellulose materials by wood-rotting *Pleurotus* fungi. *Folia Microbiol.* **32**. 434-437.

Grabbe, K. (1978). Verfahren zur Herstellung eines Dung emittels. *BRD - Patent Nr. 2831583.*

Grabbe, K. (1982). Prinzipien des mikrobiellen Stoffumsatzes bie der Aufbereitung Flussiger and fester Reststoffe und ihre Bedeutung fur die Substratherstellung im *Champignonanbau Wissenschaft and Umwelt* 4: 210-218.

Gronborg, H. (1990). Trials at the advisory centre. *Gartner Tidende.* 106(8): 208-209.

Hayes, W.A. (1969). Mushrooms, microbes as a weapon against starvation. *New Scientist.* 44: 450-452.

Hegder, J.N. and Basuki, T. (1982). The role of basidiomycetes in composts-a model system for decomposition studies. In: *Decomposer Basediomycetes-their Biology and Ecology.* (Frankland, J.C., Hedger, J.N. and Swift, M.J. eds.). Pp. 263-305. Cambridge University Press, Cambridge.

Hofrichter, M.; Scheibner, K.; Sack, U. and Fritsche, W. (1997). Degradative capacities of white-rot and litter decaying fungi for persistent natural and xenobiotic compounds. In *Advances in Mushroom Biology and Production* (Rai, R.D., Dhar, B.L. and Verma, R.N. eds.), pp. 271-280, Mushroom Society of India, NRCM, Solan, India.

Hong, J.S., Lee, J.B., Koh, M.S., Imm, J.S., Lee, K.R. and Kim, M.C. (1985). Studies on cellulolytic enzymes produced by *Pleurotus* species in synthetic medium: effect of carbon and nitrogen sources. *Korean J. Mycol.* **13**. 213-220.

Huang, J.W.; Hu, C.K. and Shih, S.D. (1996). The role of soil microorganisms in alleviation of root injury of garden pea seedlings by alachlor with spent golden mushroom compost. *Pl. Pathology Bulletin.* 5(3): 137-145.

Huang, J.W.; Hu, C.K.; Tzeng, D.D.S. and Ng, K.H. (1995). Effect of soil amended with spent golden mushroom compost on alleviating phytotoxicity of alachlor to seeling of garden pea. *Pl. Pathology Bulletin.*4(2):76-82.

Iwase, K.; Umezawa, Y. and Masuda, K.(2000). Cultivation of *Pleurotus ostrearus* with beer spent grains and utilization of carbonized waste substrate as a soil ameliorant. *Mushroom Sci.* XV(II): 819-826.

Janshakhar, J. and A. Fiechter. (1983). Lignin: biosynthesis, application and biodegradation. In: *Advances in Biochemical Engineering and Biotechnology.* pp. 119-178. (Fiechter, A. and Jefferies, Th. W. eds.)., Springer Verlag, Berlin.

Kaddous, F.G.A. and Morgans, A.S. (1986). Spent mushroom compost and deep litre fowl manure as a soil ameliorant for vegetables. In *Proc. of Surface Soil Management Rotorua*, New Zealand 138-147.

Kaul, T.N. (1983). *Cultivated Edible Mushroom.* CSIR, New Delhi. pp. 56.

Kaul, P.L. (1999). Scope of export of mushrooms from India potential market. Pp. 17-33. In: *Souvenir of Conference on Mushroom Industry in India-a Decade of Achievements and Future Perspective.* 3rd. Dec. 1999. NBPGR, IARI, New Delhi.

Kelley, R.L., Ramasamy, K. and Reddy, C.A.. (1986). Characterisation of glucose oxidase negative mutants of a lignin degrading basidiomycete *Phanerochaete chrysosporium. Arch. Micro.* 144. 254-257.

Kerem, Z. (1991). Lignocellulose degradation and improvement of cotton straw digestibility by *Pleurotus ostreatus. M.Sc. Thesis*, Hebrew University of Jerusalem (Hebrew).

Kirk, T.K. (1983). Degradation and conversion of lignocellulose. In *The Filamentopus Fungi vol IV: Fungal Technology.* (Smith, J.E., Berry, D.R. and Kristiansen, B. eds.). pp. 266-295. Edward Aronold, London.

Kirk, T.K. (1984). Degradation of ligning. In: *Microbial Degradation of Organic Compounds.* (Gibson, D.T. eds.). pp. 399-437. Marcel Dedder, New York.

Kirk, T.K. and Farrel, R.L. (1987). Enzymatic combustion the microbial degradation of lignin. *Ann. Rev. Microbiol.* 41. 465-505.

Kirk, T.K. and Fenn, P. (1982). Formation and action of the lignolytic system in basidiomycetes. In *Decomposer Basidiomycetes-their Biology and Ecology.* (Frankland, J.C., Wedger, J.N. and Swift, M.J. eds.).pp. 67-90. Cambridge University Press, Cambridge.

Leatham, G.F. (1985). Growth and development of *Lentinula edodes* on a chemically defined medium. In: *Development Biology of Higher Fungi.* (Moore, D., Casselton, L.A., Wood, D.A. and Frankland, J.C. eds.). pp. 403-428. Cambridge University Press, Cambridge.

Lelley, J. (1987). Edible mushrooms as a weapon against starvation. *Mushroom J. Tropics* **7**:135-140.

Levanon, D. and Danai, O. (1997). Recycling agricultural wastes through mushroom production. In: *Advances in Mushroom Biology and Production.* (Rai, R.D., Dhar, B.L. and Verma, R.N. eds.). pp. 305-308. Mushroom Society of India, NRCM, Solan, India.

Levanon, D. and Danai, O. (1995). Chemical, physical and microbiological considerations in recycling spent mushroom substrate. *Compost - Science and Utilization.* 3(1): 72-79.

Levanon, D. and Danai, O. (1997). Recycling agricultural residuals in Isreal. *Biocycle.* 38(6): 56-57.

Levanon, D., Hadar, Y. and Wuest, P.J. (1994). Nature and use of spent mushroom substrate. *Compost Science and Util.* 2(3) : 22-23.

Li, G.S.F. and Chang, S.T. (1982). Nutritive value of *Volvariella volvacea.* In: Tropical Mushrooms-Biology Nature and Cultivation Methods. (Chang, S.T. and Quimio, T.H. eds.). pp. 199-219. The Chinese University Press, Hong Kong.

Lohr, V.I., Brien, R.G.O. and Coffey, D.L. (1984). Spent mushroom compost in soiless media and its effects on the yield and quality of transplants. *J. Amer. Soc. Hort. Sci.* 109(5): 693-697.

Lohr, V.I.; Wang, S.H. and Wolt, J.D. (1984). Physical and chemical characteristics of fresh and aged spent mushroom compost. *Hort Science.* 19(5): 681-683.

Lohr, V.I. and Coffey, D.L. (1987). Growth responses of seedlings to varying rates of fresh and aged spent mushroom compost. *Hort Science.* 22(5):913-915.

Lowry, O.H., Roseborough, N.J., Farr, A.L. and Randall, R.J. (1951). Protein measurement with folin-phenol reagent. *J. Biol. Chem.* 193:265-75.

Madan, M. (1997). Role of mushroom cultivation in rural development in India. In *Advances in Mushroom Biology and Production.* (Rai, R.D., Dhar, B.L. and Verma, R.N. eds.), pp. 407-413. Mushroom Society of India, NRCM, Solan, India.

Maher, M.J. (1991). Spent mushroom compost (SMC) as a nutrient source in peat based potting substrate. *Mushroom Science.* XIII (Part II): 645-650.

Manning, K. and Wood, D.A. (1978). Cellulose production and regulation by *Agaricus bisporus, Proc. Soc. Gen. Microbiol.* **5**. 99.

Manivel, L., Kumar, R.R. and Marimuth, S. (1994). SMC as an amendment in tea. *Planters Chronicle.* Feb.: 69-72.

Marchiam, U., Levanon, D, Danai, O, Masaphy, S., Chen, Y., Inbar, Y. and Klinger, I. (1991). A suggested solution for slaughterhouse wastes: uses of the residual materials after anaerobic digestion. *Biores. Technol.* 37 :127-134.

Maynard, A.A. (1994). Sustained vegetable production form three year using composted animal manure. *Compost Science and Utilization* 2(1): 88-96.

McHale, M. (1970). *Ecological Context*, Braziller, New York.

Montgomery, R.A.P. (1982). The role of polysaccharidase enzymes in the decay of wood by basidiomycetes. *In Decomposer Basidiomycetes-their Biology and Ecology.* (Frankland, J.C., Hedger, J.N. and Swift, M.J. eds.). pp. 51-65. Cambridge University Press, Cambridge.

Moyson, E. and Verachtert, H. (1991). Growth of higher fungi on wheat straw and their impact on the digestibility of the substrate. *Appl. Microbiol. Biotech.* 36: 421-424.

Nee, C.C.; Chou, W.Y. and Chen, S. (1994). Study on compost of winter mushroom - a reused medium for persimmon potting. *J. Agriculture and Forestry.* 43(4):39-44.

Okeke, B.C., Smith, J.E., Paterson, A. and Watson - Craik, I.A. (1993). Aerobic metabolism of pentachlorophenol by spent sawdust culture of "shiitake" mushroom (*Lentinus edodes*) in soil. *Biotech. Let.* 15, 1077-1080.

Pannier, W. (1993). Spent mushroom compost. A natural resource that provides Solutions to Environmental Problems. *Mushroom News.* 41(11): 10-11.

Pill, W.G.; Evans, T.A. and Garrison, S.A. (1993). Forcing white button asparagus in various substrates under cool and warm regimes. *Hort. Science.* 28(10): 996-998.

Platt, M.W., Daddar, Y. and Chet, I. (1984). Fungal activities involved in lignocellulose degradation by *Pleurotus. App. Microbiol. Biotechnol.* 20: 150-154.

Poppe, J. (2000). Use of agricultural waste materials in the cultivation of mushrooms. *Mushroom Science.* 15:3-23.

Rai, R.D. (1994). Nutritional and medicinal values of mushrooms. In: *Advances in Horticulture* Vol. 13.-Mushroom. (Chadha, K.L. and Sharma, S.R. eds.) pp.537-551. Malhotra Publishing House, New Delhi.

Rai, R.D. (1997). Medicinal mushrooms. In: *Advances in Mushroom Biology and Production* (Rai, R.D., Dhar, B.L. and Verma, R.N. eds.) pp.355-368. Mushroom Society of India. NRCM, Solan, HP.

Rai, R.D. and S. Saxena (1988). How protein-rich are mushrooms. *Indian Hort.* 33(2): 2-3.

Rai, R.D. and Saxena, S. (1989a). Biochemical changes during post-harvest storage of button mushroom (*Agaricus bisporus*). *Curr. Sci.*, 58: 508-10.

Rai, R.D. and Saxena, S. (1989b). Suitability of methods of estimation for critical assessment of the vitamin C content in mushrooms. *Mushroom J. Tropics*, 9: 43-46.

Rai, R.D. and Saxena, S. (1990). Biodegradation of cereal straw by white rot fungi, *Pleurotus* spp. Recent findings. Proc. Natl. Sem. Lignin Degradation IVRI, Izatnagar (in press).

Rai, R.D., Saxena, S., Upadhyay, R.C. and Sohi, H.S. (1988). Comparative nutritional value of various *Pleurotus* spp. grown under identical conditions. *Mushroom. J. Tropics,* 8: 93-98.

Rai, R.D. and Sohi, H.S. (1988). How protein rich are mushrooms. *Indian Hort.*, 33(2): 2-3.

Rajarathnam, S., Shashirekha, M.N. and Zakia Bano. (1992). *Adv. App. Microbiol.* 37: 233.

Rajarathnam, S. and Zakia Bano. (1991). In: *Advances in Applied Mycology. 3 Food and Feeds* (Arora, D.K. and Marth, E.H. eds.) p.241. Marcel Dekker Press, New York.

Rajarathnam, S., Shashirekha, M.N., Zakia Bano and Ghosh, P.K. (1997). Renewable lignocellulosic wastes - the growth substrates for mushroom production : National strategies. In: *Advances in Mushroom Biology and Production.* Pp. 291-304 (Rai, R.D., Dhar, B.L. and Verma, R.N. eds.). Mushroom Society of India, NRCM, Solan, India.

Reed, W. and Keil, C. (2000).The effect of spent mushroom substrate and application on adjacent surface water using aquatic macroinvertebrates as bio-indicators. *Mushroom News.* 48(11): 4-13.

Rhoads, F.M. and Olson, S.M. (1995). Crop production with mushroom compost. Proceedings - Soil and Crop Science. Society of Florida. 54: 53-57.

Robinson, E.C. (1988). Fertilizer and Method. United - States - Patent, U.S. 4, 743, 287.

Royse, D. J. (1993). Recycling of spent shiitake substrate for production of the oyster mushroom (*Pleurotus sajor-caju*). *Mushroom News.* 41 (2): 14-18.

Satyanarayana, T. And Johri, B.N. (1988). Cellulases and xylanases of thermophilic moulds. In *Perspectives in Mycology and Plant Pathology* (eds. Agnihotri, V.P., Sarbhoy, A.K. and Kumar, D.) pp. 24-60. Malhotra Publishing House, New Delhi.

Schisler, L.C. (1988). Why mushroom production declines with each successive break and the production of a second crop of *Agaricus* mushrooms on "Spent" Compost. *Mushroom News.* 36(9): 6-11.

Semple, K.T. and Fermor, T.R. (1995). The bioremediation of xenobiotic - contamination by composts and associated microflora. *Mushroom Sci.* XIV (Part II): 917-924.

Sharma, H.S.S. (1987). Comparative study of the degradation of flax shive by strains of *Pleurotus* . *Appl. Microbiol. Biotechnol.* 25: 542-546.

Silanikove, N.; Danai, O. and Levanon, D. (1988). Composted cotton straw silage as a substrate for *Pleurotus* spp. cultivation. *Bio. Wastes*, 25:219-26.

Silanikove, N. and Levanon, D. (1986). Cotton straw: Composition, variability and effect of anaerobic preservation. *Biomass.* 2: 101-112.

Smith, J.F., Fermor, T.R. and Zadrazil, F. (1987). In: *Treatment of Lignocellulosics with White Rot Fungi* (Zadrazil, F. and Reiniger, P. eds) Elsevier, New York.

Stark, L. and Williams, F. M. (1994).The roles of spent mushroom substrate for the mitigation of coal-mine drainage. *Compost Science and Utilization.* 2(4): 84-94.

Stark, L.M., Wenerick, W.R., Williams, F.M. and Wuest, P.J. (1995). The effects of pH, flow rate and carbon supplementation on manganese retention in mesocosm wetlands. *J. Environ. Quality.* 24(5): 816-826.

Steffen, K.L., Dann, M.S., Fager, K., Fleischer, S.J. and Harper, J.K. (1994). Short-term and long-term impact on an initial large scale SMS soil amendment on vegetable crop productivity and resource use efficiency. *Compost Science and Utilization* 2(4): 75-83.

Stephens, J.M., Henry, G.C., Castro, B.F. and Bennett, D.L. (1989). Mushroom compost as a soil amendment for vegetable gardens. *Proceedings of the Florida State Horticultural Society.* 102: 108-111.

Swaminathan, M.S. (1981). *Building a National Food Security System.* Indian Environment Society, New Delhi.

Swift, M.J. (1982). Basidiomycetes as components of forest ecosystems. In. *Decomposer Basidiomycetes: their Biology and Ecology.* (Frankland, J.C., Hedger, J.N. and Swift, M.J. eds.). pp. 307-388. Cambridge University Press, Cambridge.

Szklarz, G., and Leonowicz, A. (1986). Cooperation between fugal laccase and glucose oxidase in degradation of lignin derivatives. *Phytochem.* 25: 2537-2539.

Szmidt, R.A.K. (1994). Recycling of spent mushroom substrates by aerobic composting to produce novel horticultural substrates. *Compost Science and Utilization.* 2(3): 63-72.

Szmidt, R.A.K. and Chong, C. (1995). Uniformity of spent mushroom substrate (SMS) and factors in applying recommendations for the use. *Compost Science and Utilization.* 3(1): 64-71.

Szmidt, R.A.K. and Chong, C. (1997). Uniformity of spent mushroom substrate (SMS) and factors in applying recommendations for usage. *Mushroom News.* 45(11): 16-27.

Tien, M. and Kirk, T.K. (1983). Lignin degrading enzyme from the hymenomycete, *Phanerochaete chrysoporium. Science.* 221: 661-663.

Thayumanavan, B. and Manickam, A. (1980). Protein quality of the sporophore of the fungus *Pleurotus sajor-caju* (Fr.) Singer. *Indian J. Nutr. Dietet.*, 17:

Upadhyay, R.C. (1990). Cultivation of oyster mushroom. Tech. Bull. 1, NCMRT Solan.

Upadhyay, R.C. and Fritche, W. (1997). Ligninolytic enzymes of *Pleurotus* species. In: *Advances in Mushroom Biology and Production.* (Rai, R.D., Dhar, B.L. and Verma, R.N. eds.). Pp. 281-290. Mushroom Society of India, NRCM, Solan, India.

Verma, R.R. (1986). Efficacy of organic amendments against *Meloidogyne incognita* infesting tomato *India J. Nematology* 16(1): 105-106.

Vetter, J. (1987). Extracellular cellulase and xylanase production of *Pleurotus* species. *Acta Bot. Hung.* 32:285-294.

Wisniewska, G.H. and Pankiewicz, T. (1989). Evaluation of the suitability of spent mushroom substrate for tulip cultivation. Prace - Instytute - Sadownictwa-i-Kwiaciarstwa- w.- Skierniewicach. Seria-B,- Rosliny - Ozdobne, 14: 7 -14.

Wood, D.A. 1989. Mushroom biotechnology, *Intern. Industl. Biotechnol.* 9: 5-8.

Wood, T.M. and McCrae, I. (1979). Ynergim between enzyme involved in the solubilisation of native cellulose. *Adv. Chem.* 181. 181-209.

Wood, D.A. and Smith, J.F. (1988). The cultivation of mushroom (Part III). *Mushroom J.* 189: 688-691.

Wuest, P.J. and Fahy, H.K. (1991). Spent mushroom compost traits and uses. *Mushroom News.* 39 (12): 9-15.

Wuest, P.J.; Fahy, H.K. and Fahy, J. (1995). Use of spent mushroom substrate (SMS) for corn (maize) production and its effect on surface water quality. *Compost Science and Utilization.* 3(1):46-54.

Yadav, M.C., Verma, R.N., Vijay, B. and Singh, S.K. (1999). Utilization of spent mushroom compost as manure for maize cultivation. *ICAR News (a Science and Technology Newsletter)* 5(1):13.

Yohalem, D.S.; Nordheim, E.V. and Andrews, J.H. (1996). The effect of water extracts of spent mushroom compost on apple scab in the field. *Phytopathology* 86(9): 914- 922.

Zadrazil, F. and Reiniger, P. (eds.). (1988). *Treatment of Lignocellulosics with White-Rot Fungi.* Elsevier Applied Science, Barking.

Zang, C.K. Gong, F. and Li, D.S. (1995). A note on the utilization of spent mushroom composts in Animal feeds. *Bioresource Technology.* 52:89-91.

CHAPTER – 8

ROLE OF FUNGI IN BIOREMEDIATION

Manoj Kumar, S.C. Agrawal*, G. Prakash**

*Department of Applied Microbiology and Biotechnology,
Dr. H.S. Gour University, Sagar (M P)
**Department of Biochemical Engineering and Biotechnology,
Indian Institute of Technology, New Delhi

INTRODUCTION

Bioremediation is defined as the use of living organisms or their products to metabolize or detoxify environmental pollutants. There are innumerable strains of microbes belonging to bacteria, yeast or fungus, which biodegrade xenobiotic through digestion of harmful chemicals and compounds present into simpler less toxic or non-toxic substances. Microorganisms like all living organisms needs nutrients, carbon and energy for survival. These beneficial microbes break down a variety of organic compounds found in waste to obtain energy for their growth. Sometimes bioremediation removes the pollutant from the environment without degrading pollutants (Alexander, 1994). Fungi and bacteria in particular take up large amounts of metals and minerals to ensure adequate resources for binary fission. Algae and plants are also very good at absorbing nitrogen, phosphorus, sulfur and many minerals and metals from the environments. For example, plants like locoweed remove large amounts of the toxic element selenium. The selenium is stored in plant tissues where it poses no harm until/unless the plant is eaten.

The demand for engineered use of microbes to treat waste is increasing and significant reasons for this increase includes societal pressure to treat waste more thoroughly, improved performance of bio-processes arising from better scientific understanding environment pressure, societal and technical limitations of alternative techniques. Technologies using bioremediation treatment include bioaugmentation, biofilters, biostimulation, bioventing, composting and land farming.

Biostimulation

The use of nutrients or substrates to stimulate the naturally occurring organisms that can perform bioremediation. Fertilizer and growth supplements are the common stimulant. The presence of small amounts of the pollutant can also act as a stimulant by turning on operons for the bioremediation enzymes. Bioventing is similar to biostimulation. It involves the venting of oxygen through soil to stimulate the growth or natural and introduced bioremediation organisms. This is used predominantly for soils contaminated with petroleum products. It is not suitable for removing halogenated gases that contribute to ozone layer damage.

Bioaugmentation

This is a general term describing the addition of organisms or enzymes to a material to remove unwanted chemicals. Bioaugmentation is used to remove byproducts from raw materials and potential pollutants from waste. Bacteria are the most common bioaugmentation organisms. Many applications are accomplished using vegetation to remove excess nutrients, metals and pathogenic bacteria. Waste water from human and agricultural effluent is cleaned this way using wetland plants.

Biofilters

The removal of organic gases by passing air through compost or soil containing microorganisms capable of degrading the gases. It has been used to remove volatile organic compounds from air.

Land farming

The use of farming tilling and soil amendment techniques to encourage the growth of bioremediation organisms in a contaminated area. It has been used successfully to remove large petroleum spills in soil.

These technologies are classified as either *in situ* or *ex situ*. *In situ* technologies involve the use of organisms or enzymes to remove pollutants in the location that is polluted. *Ex situ* technologies involve the removal of the contaminated material where it can be treated using bioremediation. The process of developing bioremediation techniques may involve the following common steps:

- Isolating and characterizing naturally occurring micro-organisms with desired degradation/adsorption potential Laboratory cultivation to develop viable populations.
- Studying the catabolic activity of these microorganisms in contaminated material through bench-scale experiments.
- Monitoring and measuring the progress of bioremediation through analytical and chemical analysis and toxicity testing.
- Field applications of bioremediation techniques.

Advantage of bioremediation

Bioremediation provides a cost-effective solution for providing a sustainable clean environment for future generations by building upon the many capabilities that microorganisms and plants have in dealing with toxic materials: they "eat up" the contamination as a food substrate, hence providing a natural way to cleanse the soil and water. Bioremediation—both above and below ground—promises to be a premier technology for restoring the terrestrial subsurface and may be the only solution to many of our deepest toxic "plumes." However, bioremediation is limited in the number of toxic materials it can handle but where applicable, it is cost-effective (Atlas and Unterman, 1997). Bioremediation's popularity is further enhanced because it is perceived as being more "green" than other remediation technologies. Use of microbes offers several advantages relative to alternative waste treatment technologies:

- Bioremediation is an ecologically sound, natural process.
- The residues from biological treatment are usually harmless products (CO_2, H_2O, fatty acids, etc.).
- Biodegradation is usually a destructive process (Nature's incinerator), which instead of merely transferring contaminants from water to air or land, destroy the target chemicals.
- Biodegradation occurs at nearly ambient conditions, which minimize energy and equipment requirement
- Biodegradation is usually cost efficient than other technologies that are often used to clean up hazardous waste (such as incineration, thermal desorption, extraction) and it is a gentle non-aggressive approach.
- Bioremediation can often be accomplished where the problem is located (*in-situ*). This eliminates the need to transfer large quantities of contaminated waste off-site, and the potential threats to human health and the environment that can arise during such transportation.
- Biodegradation can be coupled with other treatment techniques to treat a host of environmental problems.

Bioremediation potential of fungi

In nature, fungi are among the major decomposers, particularly with respect to plant polymers such as cellulose and lignin. The ability of fungi to transform a wide variety of hazardous chemicals (Table 1) has aroused interest in using them in bioremediation (Alexander, 1994). The fungi are unique among microorganisms in that they secrete a variety of extracellular enzymes. The role of fungi in the degradation of complex compounds such as starch, cellulose, pectin, lignin, lignocellulose, inulin, xylan, araban, etc. is well known. The basidiomycetous, mucoraceous fungi and arbuscular mycorrhizal fungi are potentially used in bioremediation. Aquatic fungi and anaerobic fungi are the other candidates for bioremediation. Several fungal enzymes that are released such as laccases, polyphenol oxidases, lignin peroxidases, etc. play a role in the degradative process. In addition, a variety of intracellular enzymes such as reductases, methyl transferases and cytochrome oxygenases are known to play a role in xenobiotic degradation (Barr and Aust, 1994).

Table 1. Xenobiotics metabolized/adsorbed by fungus

Xenobiotic	Common fungi for bioremediation	Reference
Heavy metals Cadmium, Nickel Copper, Lead, Chromium	*Aspergillus niger, Penicillium passturianum* *Nocardia amarae, Candida tropicalis* *Candida glabrata, Candida guilliermondii* *Saccharomyces cerevisiae,* *Fusarium oxysporum,* *Cryptococcus laurentii, Cryptococcus luteolus, Sporobolomyces alborubescens, Rhizopus* sp., *Trichoderma* sp., *Byschoclamyss* sp. and *Mucor* sp.	Kim *et al.*, 2002 Price *et al.*, 2001 Jianlong *et al.*, 2001 Zielinski *et al.*, 2001 Koch *et al.*, 2001 Dias *et al.*, 2000 Bayhan *et al.*, 2001 Hefnawy & Razak 1998 Elinov, 1999 Mogollon *et al.*, 1998
Polyphenolics	*Abortiporus biennis, Dichomitus squalens Inonotus hispidus, Irpex lacteus, Lentinus tigrinus, Lentinus edodes, Panellus stipticus, Pleurotus ostreatus, Trametes hirsuta, Phanerochaete chrysosporium*	Aggelis *et al.*, 2002 Aiken and Logan 1996 Mileski *et al.*, 1985
Diuron, metalaxyl, atrazine or terbuthylazine	*Coriolus versicolor, Hypholoma fasciculare* and *Stereum hirsutum*	Bending *et al.* 2002
Lignin	*Pleurotus ostreatus, Cyathus stercoreus, Phanerochaete chrysosporium*	Ha *et al.*, 2001 Baldrian *et al.*, 2000 Sethuraman, 1999
TNT	*Phanerochaete chrysosporium, Clitocybula dusenii, Stropharia rugosa-annulata, Penicillum corylophilum, Diachromitus squalens, Clitocybula dusenii, Stropharia rugosa- annulata*	Hodgson *et al.*, 2000 Jackson *et al.*, 1999
Petroleum hydrocarbons and Polynuclear aromatic hydrocarbons (PAHs).	*Merasmeiellus troyanus, Pleurotus* spp., *Phanerochaete* spp., *Trametes versicolor, Aspergillus niger, Penicillium* spp., *Cunninghamella elegans, Candida* spp., *Cladophialophora* sp.	Ismailov *et al.*, 1985 Higson 1991 Yateem *et al.*, 1998 Prenafeta-Boldu *et al.*, 2002
Azo and phthalo-cyanine dyes	*Bjerkandera adusta, Phanerochaete chrysosporium, Pleurotus ostreatus*	Heinfling *et al.*, 1998 Spadaro *et al.*, 1992
Carbendazim and other enzimidazole compounds	*A. alternata,Phoma eupyrena, Penicillium janthinellum, Trichoderma harzianum*	Yarden *et al.*, 1990
Plastics and their monomers styrene, polystrene, nylon	*Pleurotus ostreatus* (oyster mushroom), *Tramates versicolor, Bjerkandera adjusta, Phanerochaete chrysosporium*	Deguchi *et al.*, 1997
Insecticide, Pesticide	*Rhizopus arrhizus, Penicillium notatum, Coriolus versicolor, Hypholoma fasciculare,* and *Stereum hirsutum.*	Flashinski and Lichtenstein, 1975 Bending *et al.*, 2002

The white rot fungi are most studied and used organisms in fungal bioremediation. White-rot fungi are physiological rather than taxonomic grouping comprising fungi that are capable of degrading lignin within lignocellulosic substrates. The name white-rot derives from the appearance of wood attacked by the fungi in which lignin removed resulting in a bleached appearance to the substrates. Most known white-rot fungi are basidiomycetes, although a few ascomycete genera within the Xylariaceae are also capable of white rot decay (Eaton and Hale, 1993). Lignin degradation by white rot fungi has been extensively studied, and studies revealed that three kinds of extracellular phenoloxidases, namely, lignin peroxidase, manganese peroxidase and laccasse are responsible for initiating the depolymerisation of lignin. The expression pattern of these enzyme depends on the organisms: some secrete lignin peroxidase, manganese peroxidase (no laccasse), whereas other secrete manganese peroxidase, laccasse (lignin peroxidase). In addition to lignin, white rot fungi are able to degrade a variety of environmentally persistent pollutants, such as, polyaromatic hydrocarbons (PAHs), chlorinated aromatic hydrocarbons (CAHs), polycyclic aromatics, polychlorinated biphenyls, polychlorinated dibenzo (p) dioxins, the pesticides DDT and lindane, and some azo dyes. The white rot fungi degrade chemicals nonspecifically without using them for their growth. Probabily, this degradability of white rot fungi is due to the strong oxidative activity via a free redical mechanism and low substrate specificity of their ligninolytic enzymes.

The white-rot fungus *Phanerochaete chrysosporium* most versatile and studied among ligninolytic white rot fungi that mineralize a broad range environmental pollutants. Degradation of PAHs, which include benzo (a) pyrene, pyrene, fluorene, chrysene, naphthalene and phenanthrene, is favored at nitrogen-limiting conditions and low pH (about 4.5). Bumpus (1989) showed that at least 22 of the PAHs, including all of the more abundant PAHs in the anthracene oil, undergo 70-100% breakdown in 27 days in nitrogen limiting culture of *P. chrysosporium.* A cell-associated cytochrome P450 monooxygenase mediated PAH transformation reaction is also thought to occur in certain white rot fungi (Bezalel *et al.,* 1997). White rot fungi other than *P.chrysosporium* also appear to be effective in mineralizing PAHs (Sutherland, 1992). For example, *Trametes versicolor* and *Chrysosporium lignorum* have been reported to mineralize phaenanthrene (Morgan *et al.,* 1991). *Pleurotus ostreatus* and *T. versicolor* and *Coriolopsis polyzona* have shown to oxidize anthracene to anthraquinone (Vyas *et al.,* 1994). Davis *et al.* (1993) showed that *P. sordida* was capable of degrading efficiently the three ring PAHs, but less efficiently the four-ring PAHs. The ability of the white-rot fungi to degrade PAHs has led to its use in the treatment of coal – and creosote –contaminated soils augmented with wood chips, corn cobs or sawdust (Bumpus, 1989; Aust, 1990). In a study *Pleurtus ostreatus* removed 48% of the 5 ring PAH from creosote –contaminated soil, in addition to 87-89% of 3 – and 4 –ring PAH (Eggen, 1999).

White-rot fungi are able to mineralize tri-, tetra-, and pentachlorophenol (PCP). Laboratory soil studies have shown that PCP concentration in *P. chrysosporium* and *P. sordida* decrease by 96% and 82%, respectively (Lamar *et al.,* 1990). Alleman *et al.* (1995) have used rotating tube bioreactors to study PCP degradation, demonstrating 99% PCP removal within one day with the white rot fungi *P. chrysoporium, T. versicolor* and *Inonatus dryophilus* when grown in nitrogen sufficient media. The highest percentage of dehalogenation of PCP is observed with *T. versicolor* (62%), followed by *P. chrysosporium* (38%) and *I. dryophilus* (21%). Lamar (1993) has investigated the ability of white rot fungi to deplete PCP in two different sites contaminated with wood preservatives. Fungal inocula, consisting of wood chips heavily colonized by *P. chrysosporium* or *P. sordida,* were added to soil at 3.35 and peat (carbon source) was added at 1.9%. The result shows that PCP and low molecular weight PAH concentrations are highly reduced by the end of 6.5 weeks of fungal activity. *Trametes versicolor* mineralized 29% of the added PCP during 42 days of growth in soil (Tuomela *et al.,* 1999).

Degradation of cyclodiene insecticides, including chlordane, by white-rot fungus has been demonstrated. White –rot fungi have also been shown as capable of mineralizing DDT (Bumpus and Aust, 1987). The chlorinated triazine herbicide 2- chloro- 4- ethylamine-6-isopropyl-amino-1, 3,4-triazine (atrazine) is recalcitrant in environment, although the white rot fungi *P. chrysosporium* (Mougin *et al.,* 1994) and *Pleurotus pulmonarius* (Masaphy *et al.,* 1993) have both been demonstrated to transform atrazine, yielding hydroxylated and N-dealkylated metabolites. Numerous studies have shown that white –rot fungi including *Coriolopsis polyzona, P.chrysosporium, Pleurotus ostreatus* and *Trametes versiocolor* are capable of significant polychlorinated biphenyls removal (Yadav *et al.,* 1995; Novotny *et al.,* 1997). White-rot fungus has been observed to degrade TNT in laboratory-scale studies (using pure cultures), however, factors that limit their effectiveness may retard widespread use in the field. Results from bench-scale studies of mixed fungal and bacterial systems indicate that most of the degradation of TNT is attributable to bacteria and that most of the losses of TNT is due to adsorption onto the fungus and soil amendments. Recent studies showed the brown rot fungi *Gleophyllum* spp. transforms TNT by reduced oxygen species produced by fenton –like reactions (Newcombe *et al.,* 2002).

P. chrysosporium efficiently degrades benzene, toluene, ethylbenzene and ortho-, meta-, and para-xylene (BTEX) individually or in mixtures, both in liginolytic and non-liginolytic culture conditions (Zheng and Obbard, 2002). Deguchi *et al.* (1997) has demonstrated the nylon degradation by lignin-degrading fungi. Yateem *et al.* (1998) have evaluated three strains of white rot fungi, namely *Phanerochaete chrysosporium, Pleurotus ostreatus* and *Coriolus versicolor* for their ability to degrade oil in contaminated soil. *Coriolus versicolor* showed the highest degradation rate as the total petroleum hydrocarbon concen-tration decreased from 32 g/kg to reach 7 g/kg within year. The studies conducted by

Prenafeta–Boldu *et al.* (2002) shows that soil fungi *Cladophialophora* sp can contribute significantly to bioremediation of BTEX pollution. *Phanerochaete chrysosporium* has been shown to effect the bioleaching of organic dyes (Nigam *et al.,* 1995; Sami and Radhaune, 1995) *Phanerochaete chrysosporium* and microbial consortia were effective in color removal from textile dye effluents (Asoka *et al.,* 2002). The fungus caused 80% decolorization in broth containing 2.5% of effluent. The basidiomycetous fungus *Pleurotus ostreatus* has been shown to produce an extracellular hydrogen peroxide dependent lignolytic enzyme that removes the color due to remozol brilliant blue.

Among other fungi used in bioremediation, the yeasts, e.g., *Candida tropicalis, Saccharomyces cerevisiae, S. carlbergensis* and *Candida utilis* are important in clearing industrial effluents of unwanted chemicals. *Agaricus bisporus* and *Lentinus edodes* are important in lignocellulose decomposition. *Corius versicolor* is important in cleaning up pulp and paper mill wastes. Consortia of fungi and bacteria (usually uncharacterized) are used in composting, the most useful waste disposal practice. Phenolic azo dyes have been shown to be oxidized by the enzyme laccase produced by *Pyricularia oryzae* (Chivukula and Renganathan, 1995). Beside that number of fungi are well known to degrade keratinous materials (Deshmukh and Agrawal, 1982; Deshmukh *et al.,* 2000)

Fungi are good in the accumulation of heavy metals such as cadmium, copper, mercury, lead and zinc. Systems using *Rhizopus arrhizus* have been developed for treating uranium and thorium. The fungus *Aspergillus niger* has been used to remove lead, cadmium, copper and nickel ions from wastewaters (Jianlong *et al.,* 2001; Price *et al.,* 2001). There are several promising fungi that can degrade zinc, cyanide and chromium and radioactive metals (Gray, 1998) in effluents. The filamentous fungus *Penicillium ochro-chloron* (ATCC 36741) as a biological trap (biotrap) for remediation of heavy metal contaminants in industrial wastewaters. Mogollon and co-workers (1988) studied nickle uptake capability from aqueous solutions in *Rhizopus* sp., *Penicillum* sp., *Aspergillus* sp., *Trichoderma* sp., *Byschoclamyss*, and *Mucor* sp. The nickle uptake of *Rhizopus* sp. was found highest. Young and Banks (1998) had shown bio-sorption of lindane to non-viable biomass of *Rhizopus oryzae.* The adsorption of the negatively charged lindane to negatively charged mycelium was facilitated by hydrogen ions acting as bridging ligands. Fungal degradation of polyurethane has also been reported; a species of *Curvularia senenegalensis* was shown to degrade the polymer *via* secretion of an extracellular esterase (Crabbe *et al.,* 1994)

The *Pleurotus* sp. has been used for de-caffeination of coffee wastes paired with oyster mushroom production. Caffeine contamination of watersheds represents a major ecological disaster in coffee growing regions of the world. Oyster mushrooms can be grown on a wide variety of coffee wastes, producing commercial crops, destroying the majority of the caffeine in the process, and rendering the substrate into secondary products and the decomposition of hydrocarbon-based contaminants, including petroleum-based products such as creosote, oil and pesticides as well as for the concentration and removal of heavy metal based contaminants.

Advantage from Fungal bioremediation

In nature, fungi are among the major decomposers, particularly with respect to plant polymers such as cellulose and lignin. Whenever bioremediation figures as the topic of discussion, bacterial agents come into focus and fungi are much less studied. One reason is that classical microbiological enrichment techniques favor the isolation of bacteria since fungi are slowe growing and often require co-metabolic substrates for growth. When supplemental carbon sources and bacteriostatic agents are incorporated into enrichment schemes, fungi are readily isolated. These organisms have the following advantages:

(1) The aggressive growth of mycelia allows rapid colonization of substrates and extensive hyphal reach in soil as well as generates greater biomass production.

(2) Diverse and versatile fungal flora.

(3) The extracellular nature of the degradative enzymes allows fungi to tolerate high concentration of toxins.

(4) Many fungal enzymes attack a variety of substrates in a nonspecific fashion.

(5) Fungi offers specific advantages with respect to stability and activity of populations, especially under conditions of reduced water activity and low pH.

CONCLUSIONS

Most of the bioremediation currently conducted on a commercial scale utilizes prokaryotes, with comparatively few and recent attempts to use fungi. Fungi however, offer advantages over bacteria in diversity of compounds they are able to oxidize. The extracellular nature of fungal enzyme and low molecular mass mediators enhance bioavailability of pollutants to fungi in situations where bacteria, with their cell associated pollutant catabolism, may not have access. The high surface-to-cell ratio of filamentous fungi makes them better degraders under certain niches like contaminated soils. This further makes these organisms unique and attractive for bioremediation of polluted sites. There are certain limitations with use of fungi in bioremediation. A major limitation of certain fungus is its sensitivity to biological process operations. It has been observed that the fungus does not grow well in suspended cell systems, enzyme induction is negatively affected by mixing action; and the ability of the fungus to effectively attach itself to fixed media is poor. The transformations by fungus are known to be slow in the field. Moreover, the degree of success achieved in the laboratory in the mineralization of major pollutants, however, been realized in field scale studies. Much research is needed to study the factors involved in limiting the effectiveness of the fungus in field scale applications. Also, further research is needed to understand the molecular mechanism of fungi to enable the design and application of optimal bioremediation strategies for treating contaminated sites.

REFERENCES

Aggelis, G., Ehaliotis, C., Nerud, F., Stoychev, I., Lyberatos, G., and Zervakis, G.I. (2002). Evaluation of white-rot fungi for detoxification and decolorization of effluents from the green olive debittering process. *Appl Microbiol Biotechnol* 59(2-3): 353-360.

Aiken, B.S. and Logan B.E. (1996). Degradation of pentachlorophenol by the white rot fungus *Phanerochaete chrysosporium* grown in ammonium lignosulphonate media. *Biodegradation* 7(3): 175-182.

Alexander, M. (1994). *Biodegradation and Bioremediation.* Acad. Press, San Diego, New York.

Alleman, B.C., Logan, B.E. and Gilbertson, R.L. (1995). Degradation of pentachlorophenol by fixed films of white- rot fungi in rotating tube bioreactors. *Wat Res* 29: 61-67.

Ashoka, C., Geetha, M.S. and Sullia, S.B. (2002). Bioleaching of composite textile dye effluent using bacterial consortia. *Asian Jr Microbial Biotech Env Sci.* 4: 65-68.

Ashoka, C., Manjunath, K. and Sullia, S.B. (2000). Biological treatment of combined textile dye effluent. In: Ecology of Fungi, D.J. Bhat and S. Raghukumar (eds.), pp 63-67.

Atlas, R.M. and Unterman, R. (1999). Bioremediation. In: Industrial Microbiology & Biotechnology, 2nd Ed., ASM Press, Washington, 666-681

Baldrian, P., Der Wiesche, C., Gabriel, J., Nerud, F., and Zadrazil, F. (2000). Influence of cadmium and mercury on activities of ligninolytic enzymes and degradation of polycyclic aromatic hydrocarbons by *Pleurotus ostreatus* in soil. *Appl Microbiol Biotechnol* 53(3): 328-334.

Barr, D.P. and Aust, S.D. (1994). Mechanisms the white rot fungi use to degrade pollutants. *Env Sci Technol* 28: 79A-87A.

Bayhan, Y.K., Keskinler, B., Cakici, A., Levent, M. and Akay, G. (2001). Removal of divalent heavy metal mixtures from water by *Saccharomyces cerevisiae* using crossflow microfiltration. *Water Res* 35(9): 2191-2200

Bending, G.D., Friloux, M. and Walker, A. (2002). Degradation of contrasting pesticide by white –rot fungi and its relationship with ligninolytic potential. FEMS *Microbiol Lett* 212: 59-63.

Bennett, J.W. and Faison, B.D. (1997). Use of Fungi in Biodegradation. In: *Environmental Microbiology,* ASM Press, Washington.

Bezalel, L., Hadar, Y., and Cerniglia, C.E. (1997). Enzymatic mechanisms involved in phenanthrene by the white rot fungus. *Pleurotus ostreatus. Appl Environ Microbiol* 63: 2495-2501.

Bumpus, J..A. (1989). Biodegradation of polycyclic aromatic hydrocarbon by *Phanerochaete chrysosporium. Appl Environ Microbiol* 55: 3946-3950.

Bumpus, J.A., and Aust, S.D. (1987). Biodegradation of DDT (1,1,1-tricholro –2,2 bis (4-cholrophenyl ethane) by the white rot fungus *Phanerochaete chrysosporium. Appl Environ Microbiol* 53: 2001-2008.

Bumpus, J.A., Kakar, S.N. and Coleman, R.D. (1993). Fungal degradation of organophosphorus insecticides. *Appl Biochem Biotechnol* 39:715-726.

Bumpus, J.A., Fernando, T., Jurek, M., Mileski, G.J and Aust, S.D. (1989). Biological treatment of hazardous wastes by *Phanerochaete chrysosporium.* In biotechnology applications in hazardous in hazardous waste treatment. Edited by Lewandowski G, Armenante P, Baltzis B New York: Engineering Foundation (US): 167-183.

Cohen R, Persky., L, and Hadar, Y. (2002). Biotechnological applications and potential of wood degrading mushrooms of the genus *Pleurotus. Appl Microbiol Biotechnol* 58: 582-594.

Crabbe, J.R., Campbell, J.R,, Thompson, L, Walz, S.l., and Scultz, W. (1994). Biodegradation of a colloidal ester-based poyurethane by soil fungi. *Appl Environ Microbiol* 56: 1114-1118.

Chivukula, M. and Renganathan, V. (1995). Phenolic azo dye oxidation by laccase from *Pyricularia oryzae. Appl. Environ. Microbiol.* 61: 4374-4377.

Davis, M.W., Glaser, J. A., Evans, J. W. and Lamar, R, T. (1993). Field evolutions of the lignin –degrading fungus *Phanerochaete sordida* to treat creosote contaminated soil. *Environ Science technol* 27: 2572-2576.

Deguchi, T, Kakezawa, M. and Nishida, (1997). Nylon biodegradation by lignin degrading fungi. *Appl Environ Microbiol* 63: 329-331.

Deshmukh, S.K., Agrawal, S.C. (1982). *In vitro* degradation of human hair by some keratinophilic fungi. *Mycoses* 25: 454-458.

Deshmukh, S.K., Agrawal, S.C., and Jain, P.C. (2000). Isolation of dermatophytes and other keratinophilic fungi from soil of Mysore India. *Mycoses* 43: 55-57.

Dias, J.C., Rezende, R.P., Rosa, C.A., Lachance, M.A.,and Linardi, V.R. (2000). Enzymatic degradation of nitriles by a *Candida guilliermondii* UFMG-Y65.*Can J Microbiol* 46(6): 525-531.

Eaton, R.A. and Hale, M.D.C. (1993). Wood, decay, pests and prevention. Chapman and Hall, London.

Eggen, T. (1999). Application of the fungal substrate from commercial mushroom prediction – *Pleurotus ostreatus* – for bioremediation of creosote contaminated soil. *Int Biodeter Biodegr* 44: 117-126.

Elinov, N.P., Anan'eva, E.P. and Iaskovich, G.A. (1999). Activity of exoglycans as sorbents of heavy metal ions. *Prikl Biokhim Mikrobiol* 35(2): 190-193.

Flashinski, S.J. and Lichtenstein, E.P. (1975). Environmental factors affecting the degradation of Dyfonate by soil fungi. *Can. J. Microbiol* 21(1): 17-25.

Gray, S.N. (1998). Fungi as potential bioremediation agents in soil contaminated with heavy or radioactive metals. *Biochem. Soc. Trans.* 26: 666-670.

Hefnawy, M.A. and Razak, A.A. (1998). Alteration of cell-wall composition of *Fusarium oxysporum* by copper stress. *Folia Microbiol.* 43(5): 453-458.

Heinfling, A., Martinez, M.J., Martinez, A.T., Bergbauer, M. and Szewzyk, U. (1998). Transformation of industrial dyes by manganese peroxidases from *Bjerkandera adusta* and *Pleurotus eryngii* in a manganese-independent reaction. *Appl Environ Microbiol* 64(8): 2788-2793.

Higson, F.K. (1991). Degradation of xenobiotics by white rot fungi. *Rev. Environ. Contam Toxicol.* 122: 111-152.

Hodgson, J, Rho, D., Guiot, S.R., Ampleman, G., Thioboutot, S. and Hawari, J. (2000). Tween 80 enhanced TNT mineralization by *Phanerochaete chrysosporium. Can J Microbiol* 46: 110-118.

Ismailov, N.M. (1985). Biodegradation of petroleum hydrocarbons in soil inoculated with yeasts. *Microbiologiia,* 54(5): 835-841.

Jianlong, W., Xinmin, Z., Decai, D. and Ding, Z. (2001). Bioadsorption of lead (II) from aqueous solution by fungal biomass of *Aspergillus niger. J Biotechnol* 87(3): 273-277.

Kim, D.W., Cha, D.K., Wang, J. and Huang, C.P. (2002). Heavy metal removal by activated sludge: influence of *Nocardia amarae. Chemosphere* 46(1): 137-142

Ha, H.C., Honda, Y., Watanabe, T. and Kuwahara, M. (2001). Production of manganese peroxidase by pellet culture of the lignin-degrading basidiomycete, *Pleurotus ostreatus. Appl Microbiol Biotechnol* 55(6): 704-711.

Koch, K.A., Allard, S., Santoro, N., Cote, J. and Thiele, D.J. (2001). The *Candida glabrata* Amt1 copper-sensing transcription factor requires Swi/Snf and Gcn5 at a critical step in copper detoxification. *Mol Microbiol* 40(5): 1165-1174.

Lamar, R.T. (1993). On-site remediation of soils contaminated with wood-preservatives using lignin-degrading fungi. In Proceedings FEMS symposium on lignin biodegradation and transformation: biotechnological applications. Edited by Duarte JC, Costa_Ferreira M, Ander P. Lisbon, FEMS: 259-263.

Lamar, R.T., Larsen, M.J. and Kirk, T.K. (1990). Sensitivity to and degradation of pentachlorophenol by *Phanerochaete spp. Appl Environ Microbiol* 58: 4048-4050.

Masaphy, S., Levanon, D., Vaya, J. and Henis, Y. (1993). Isolation and characterization of a novel atrazine metabolite produced by the fungus *Pleurotus pulmonarius* 2 cholro-4-ethylamino-6-91-hydroxyisopropyl)amino-1,3,5 triazine. *Appl Environ Microbiol* 59: 4342-4346.

Mogollon, L, Rodriguez, R., Larrota, W., Ramirez, N. and Torres, R. (1998). Biosorption of nickel using filamentous fungi. *Appl Biochem Biotechnol* 70-72: 593-601.

Morgan, P., Lewis, S.T. and Watkinson, R.J. (1991). Comparison of abilities of white rot fungi to mineralize selected xenobiotic compounds. *Appl Microbiol Biotechnol* 34: 693-696.

Mougin, C., Laugero, C., Asther, M., dubroca, J. and Frasse, P. (1994). Biotransformation of the herbicide atrazine by the white - rot fungi *Phanerochaete chrysosporium. Appl Microbiol Biotechnol* 134: 693-696.

Newcombe, D., Paszczynski, A., Gajewska, W., Kroger, M., Fies, G. and Crawford, R. (2002). Production of small molecular weight catalysts and the mechanism of trinitrotoluene degradation by several *Gleophyllum* species. *Enzy. Microbiol. Tech.* 30: 506-517.

Nigam, P., Banat, I.M., McMullan, G., Singh, D. and Marchant, R. (1995). Microbial degradation of textile effluent containing Azo, Diazo and reactive dyes by aerobic and anaerobic bacterial and fungal cultures. 36th Annu. Conf, AMI, Hisar, 37-38.

Novotny, C., Vyas, B.R.M., Erbanova, P., Kubatova, A. and Sasek, V. (1997). Removal of various PCBs by various white –rot fungi in liquid cultures. *Folia Microbiol* 42; 136-140.

Ollikka, P. Alhonmaki, K., Leppanen, V., Glumoff, T., Timo Raigola and Hari Suominen (1993). Decolorization of Azo Triphenyl Methane Heterocyclic and Polymeric dyes by lignin peroxidase, isoenzymes from *Phanerochaete chrysosporium. Appl Environ Microbiol* 59: 4010-4016.

Prenafeta –Boldu, F.X., Vervoort, J. Grotenhuis, J.T. and Van Groenestijin, J.W. (2002). Substrate interactions during the biodegradation of benzene, toluene, ethylbenzene, and xylene (BTEX) hydrocarbons by the fungus *Cladophialophora* sp. strain T1. *Appl Environ Microbiol* 68: 2660-2665.

Price, M.S., Classen, J.J. and Payne, G.A. (2001). *Aspergillus niger* absorbs copper and zinc from swine wastewater. *Bioresour Technol* 77(1): 41-49.

Price, M.S., Classen, J.J. and Payne, G.A. (2001). *Aspergillus niger* adsorbs copper and zinc from swine wastewater. *Bioresour Technol* 77: 41-49.

Sami and Radhaune (1995). Role of lignin peroxidase and manganese peroxidase from *Phanerochaete chrysosporium* in the decolorization of olive mill waste water. *Appl. Envron. Microbiol* 61: 1098-1103.

Spadaro, J.T., Gold, M.H. and Renganathan, V. (1992). Degradation of azo dyes by the lignin-degrading fungus *Phanerochaete chrysosporium. Appl Environ Microbiol* 58: 2397-2401.

Sutherland, J.B. (1992). Detoxification of polycyclic aromatic hydrocarbons by fungi. *J Ind Microbiol* 9: 53-62.

Tuomela, M. Lyytikainen, M., Oivanen, P. and Hatakka, A. (1999). Mineralization and conversion of pentachlorophenol (PCP) in soil inoculated with white rot fungus Trametes versicolor. *Soil Biol Biochem* 31: 65-74.

Vyas, B.R.M., Bakowski, S., Sasek, V. and Matucha, M. (1994). Degradation of anthracene by selected white –rot fungi. FEMS *Microbiol Ecol* 14: 65-70.

Yadav, J.S., Quensen, III J.F., Tiedje, J.M. and Reddy, C.A. (1995). Degradation of benzene, toluene, ethylbenzene and xylenes (BTEX) by the lignin – degrading basidiomycetes *Phanerochaete chrysosporium. Appl Environ microbiol* 59: 756-762.

Yarden, O., Salomon, R., Katan, J. and Aharonson, N. (1990). Involvement of fungi and bacteria in enhanced and nonenhanced biodegradation of carbendazim and other benzimidazole compounds in soil. *Can J Microbiol* 36(1): 15-23.

Yateem, A., Balba, M.T., Al-Awadhi, N. and El-Nawawy, A.S. (1998). White–rot fungi and their role in remediating oil-contaminated soil. *Environment Int* 24: 2560-2565.

Young, E. and Banks, C.J. (1998). The removal of lindane from aqueous solution using a fungal biosorbent: influence of pH, temperature, biomass concentration and culture age. *Environ Technol* 19: 619-625.

Zheng, Z. and Obbard, J.P. (2002). Removal of surfactant solubilized polycyclic aromatic hydrocarbons by *Phanerochaete chrysosporium* in a rotating biological contact reactor. *J biotechnol* 96: 241-249.

Zielinski, R., Kubinski, K., Stefaniak, E. and Szyszka, R. (2001). Heavy metals biosorption from aqueous solution by simple eucaryotic organisms. *Folia Histochem Cytobiol* 39: 154-155.

CHAPTER – 9

Curvularia lunata: A VERSATILE ORGANISM FOR BIOTRANSFORMATION OF ORGANIC COMPOUNDS

R.Z. Sayyed, A.H. Jobanputra and S.B. Chincholkar*

Department of Microbiology, School of Life Sciences,
North Maharashtra University, PB. 80, Jalgaon 42500;
E-mail : chincholkar_sb@indiatimes.com
*Corresponding author

INTRODUCTION

Curvularia lunata is a versatile organism, which has fascinated the scientific and commercial world because of its ability to biotransform steroidal substrates viz. cortexolone to hydrocortisone, a wonder drug for anti-inflammatory activity against rheumatoid arthritis and other inflammatory diseases. The organism is also known for its ability to biotransform antibiotics like rifamycin- B to rifamycin-S, a potent antituberculosis drug. Because of this biotransformation potential, it has been exploited for the development of many commercial processes (Shull *et al.,* 1953; Chincholkar, 1994).

Members of *Curvularia* are septate and mostly curved at the third cell from the base (hence the name *Curvularia*), which is also much larger and darker in colour than the rest of the cells. *C. lunata* species is known for crescent or half moon shaped spores. The conidiophore forms a sympodula bearing a cluster of conidia, which may be compact. The detached conidium leaves a knob like scar on the conidiophore on the 'bends' or geniculations formed due to the characteristic growth as a sympodula. The perfect stage of *C. lunata* is *Cochliobolus lunata*. Species of *Curvularia* are common leaf-spot fungi and also occur in soil. Some are active degraders of cellulose and damage textile (Shull *et al.,* 1953).

ULTRA-STRUCTURE OF *C. lunata*

Sukhodol'skaya *et al.* (1994) have illustrated the ultrastructure of *C. lunata* with electron micrograph. Based on their studies it has been established that the polymorphism in culture is independent of the time and cultivation device. The mycelium was described as separate vegetative hyphae, round oval cells and bi or tri septate cellular forms with one central morphologically intact cell. The cytoplasm was characterized by high density and rich in ribosomal content. The cytoplasm of 48-96 h age hyphae lost its density due to the reduction of the ribosomes, accumulation of glycogen like material and increase in the number and size of lipid inclusions. Forms of asexual propagation of the chlamydospores and arthospore type were described for the first time in the life cycle of this fungus (Sukhodol'skaya *et al.*, 1994).

BIOTECHNOLOGICAL EXPLOITATION OF POTENTIAL OF *C. lunata*

Antibiotic biotransformation

There are scanty reports available on antibiotic biotransformations. Modification(s) in the parent structure of antibiotics may lead to an enhanced activity of such antibiotics. Chemical or biochemical transformation of naturally produced antibiotics can potentially be used to alter some of the characteristics of natural substances. The products of such transformations can be used as starting material for the synthesis of another potent and pharmacologically superior derivative drugs. Changes in antimicrobial activity, spectrum of action, oral absorption, toxicity and allergenic responses may also be achieved (Banerjee, 1994a).

A commercially used example of antibiotic biotransformation is the conversion of rifamycin B to S (Rasalkar *et al.*, 2002).

Rifamycin : chemical properties and therapeutic importance

Rifamycins belong to ansamycin group of antibiotics (White *et al.*, 1973; Lancini, 1986). Rifamycin and its derivatives have been used successfully over the years for the treatment of tuberculosis and leprosy (Fox, 1985). Rifamycins are basically yellow to orange compounds, soluble in polar solvents and insoluble in non-polar solvents with acidic properties (Patil and Chincholkar, 1997).

Rifamycin B has been reported to be microbiologically inactive (Banerjee *et al.*, 1992); its apparent activity depends on its transformation to active product like rifamycin O, SV and S (Maggi *et al.*, 1965) which have been widely used against *Mycobacterium tuberculosis*. Rifamycin SV, rifamide and rifampicin have already been employed in the therapy of bacterial infections (Riva and Silvestri, 1972).

Recently, rifamycin derivatives like rifabutane and rifapentane have been found to be effective against MAIS (*Mycobacterium avium/ intracellulare/ scrofalucumm* (MAC) complex, which is associated with the AIDS complex (Murali-krishna *et al.*, 1999).

Rifamycin S is of considerable pharmacological and economic importance because it is the precursor for majority of semisynthetic rifamycins and is highly active against Gram positive and Gram negative bacteria. It inhibits DNA dependent RNA synthesis (Umezawa *et al.*, 1968). Open ansa chain of rifamycin S derivative exhibited inhibitory activity against HIV-1 reverse transcriptase (Bartolucci *et al.*, 1995).

Caruso (1997) observed that intra-auricular rifamycin was effective against active synovitis and could be used as antirheumatic drug. Antimicrobial and chemical properties of some of novel semisynthetic rifamycin derivatives are summarized in Table 1.

Table 1. Chemical properties and some important activities of novel semisynthetic rifamycins.

Rifamycin derivatives	Chemistry	Derivatives Obtained	Important activities
Rifamycin B	With amines, hydrazine and alcohol, yields amides, hydrazides and esters respectively	Rifamycin O/S/SV and Rifamide	Biologically inactive, derivatives active against bacteria, lower toxicity
Rifamycin O	With aromatic amines and hydrazine, yields quinonimine derivatives	Rifamycin AG	Broad spectrum, Antimicrobial
Rifamycin S	With O amino-phenols, produce phenazine	Rifazine, rifamycin SV, Rifaldehyde rifampicin	Broad spectrum, antimicobial. Inhibits polymerase of RNA tumor virus.
Rifampicin (Rifadine/ Rimactane)	Obtained from rifaldahyde	—	Antibacterial, used in tuberculosis drug regime, leprosy treatment, non-tuberculosis infection and *S. aureus* and *S. epididermis*, inhibits viral reverse transcriptase
Rifabutin (Mycobutin)	—	—	Anti MAC complex and TB bacilli, bactericide, inhibits RNA polymerase

Chemical conversion of rifamycins

Existing conventional chemical conversion process of rifamycin B to S has warranted use of strong oxidizing agents like MnO_2, tetrahydrofuran and H_2SO_4 for oxidation and subsequent acid hydrolysis. Such stringent chemicals significantly reduce the yield (50-60%) and generates large amounts of hazardous acidic water. Overall, the process suffers severe set back due to requirement for low pH, acid resistant equipment, vigorous foaming during reaction and low yield (Seong *et al.*, 1982).

Biotransformation of rifamycin

Biotransformation of rifamycin B to S was reported in 1969 (Lancini *et al.*, 1969). The efficiency with biological system was found to be 85%

vis-a-vis 60% yield with chemical process (Seong and Han, 1982). Biotransformation of rifamycin B to S has been reported with an intracellular rifamycin oxidase from *Humicola* sp. ATCC 20620 and ATCC 20621 (Han *et al.*, 1983) and with an extracellular rifamycin oxidase from *C. lunata* var. *aeria* (Vohra and Dube, 1989). First real biotransformation of rifamycin B to S was reported by Han *et al.* (1983) using rifamycin oxidase form *Monocillium* sp. ATCC 20621. The organism was found to convert rifamycin B to O which was spontaneously hydrolyzed to rifamycin S in neutral aqueous milieu.

Rifamycin oxidase: A rifamycin biotransforming enzyme

The enzyme system reported for the bioconversion belongs to dioxygenase type. It was classified under the class E.C. 1.10.3.36 (Han *et al.*, 1983). It has been predicted that the enzyme might be requiring co-factors like NADH, flavoproteins and cytochrome P_{450} (Han *et al.*, 1983; Patil, 1998).

(a) *Humicola* sp.

The catalytic properties of rifamycin oxidase of *Humicola* sp. showed that the enzyme is most specific for rifamycin B than other substrate analogues, viz. polyol, P-hydroxyphenoxyacetic acid and p-hydroquinone, the corresponding structural analogues in the quinonoid moiety was reported to significantly decrease the enzyme activity. The Km value of the enzyme from *Humicola* sp. was strongly inhibited by Fe^{2+} and Hg^{2+}, but not other ions. Presence of chelating agent such as EDTA did not affect the enzyme activity (Seong *et al.*, 1985; Banerjee *et al.*, 1992).

(b) *Monicillium* sp.

Another report of intracellular rifamycin oxidase from *Monocillium* sp. showed that the enzyme had quite different properties. It oxidized rifamycin B, rifamycin SV, 3-formyl rifamycin SV, hydroquinone, pyrogallol and catechol. The enzyme activity remains unaffected by Ag^{+} or Hg^{2+}, Cu^{2+}, Ca^{2+}, Mg^{2+}, Fe^{3+}, Fe^{2+}, Co^{2+}, Mn^{2+}, Zn^{2+} and Mo^{2+}. It showed pH and temperature optima of 7.8 and 40^{0}C respectively (Han *et al.*, 1983).

(c) *C. lunata*

Reports on the characteristics of rifamycin oxidase from *C. lunata* showed that a neutral pH was most favorable for growth as well as enzyme production and enzyme activity. It has been shown that elevated concentration of glucose in the reactor decreased the enzyme yield per unit mass of glucose (Banerjee, 1993b). The temperature optima of the enzyme was 50^{0}C; at 40^{0}C and 45^{0}C the enzyme exhibited 75% and 85% of its maximum activity respectively (Vohra *et al.*, 1989 a). The kinetic studies of the enzyme from *C. lunata* showed that the enzyme had a conversion efficiency of 85% as compared to 60% in the chemical process

(Banerjee and Patnaik, 1996). The Km and Vmax values, with rifamycin B as substrate were found to be 0.67 mM and 11 IU/ml, respectively (Banerjee *et al.*, 1992). No inhibition of enzyme activity was noted upto 10-mM rifamycin B (Banerjee *et al.*, 1992). The activation and deactivation energies of the partially purified enzyme were 5.80 and 35.10 kcal/mole respectively (Banerjee *et al.*, 1992). The enzyme activity was inhibited by Fe^{2+}, Ag^{+} and Hg^{2+}, but not by EDTA (Banerjee *et al.*, 1992).

Mechanism of biotransformation of rifamycin B to S

In the reaction mechanism one mole of rifamycin B reacted with one mole of oxygen to produce one mole each of rifamycin O and hydrogen peroxide. Further hydrolysis of rifamycin O yielded one mole each of rifamycin S and glycolic acid, thus 100 mg of rifamycin gave 92 g of rifamycin S. From the action of rifamycin oxidase, it appeared that substituted hydroquinone moiety of rifamycin B was converted to quinone form which was rifamycin S (Fig. 1) (Banerjee *et al.*, 1992).

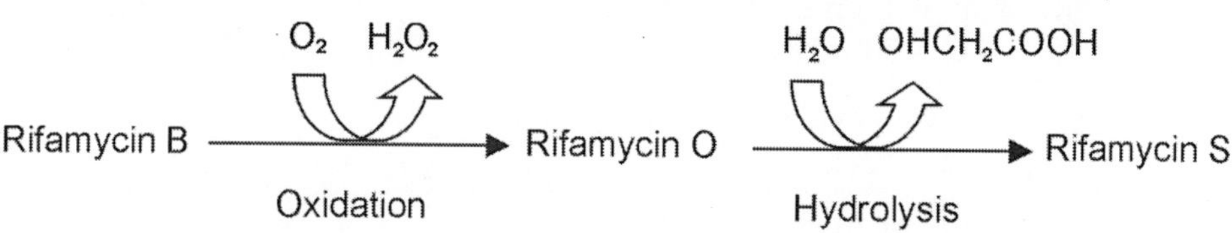

Fig.1. Mechanism of biotransformation of rifamycin B to S

Vohra *et al.* (1989) reported highly active rifamycin oxidase from *Curvularia lunata* var. *aeria*. The time course of this conversion showed that 90 % of rifamycin B was consumed within the first twenty minutes of incubation. This led to a quick synthesis of rifamycin and its spontaneous hydrolysis to S. Entire rifamycin B was converted to O in an hour and most of it was hydrolyzed to S (Vohra and Dube, 1989).

METHODS OF BIOTRANSFORMATION

Biotransformation process can be carried out in different ways by using free and immobilized whole cells, free and immobilized enzymes and by performing process in organic solvents.

Biotransformation with whole cells

Whole cell preparations of *Humicola* sp. as enzyme source showed a time lag in the conversion time profile in the initial stages of reaction due to diffusional resistance to the substrates and products through the cell membrane. Acetone treatment and increased temperature were reported to reduce the lag and hence diffusional limitation (Seong *et al.*, 1985). Delipidation of whole cells of *Humicola* sp. has been emphasized for attaining maximal yield of rifamycin B to S (Han *et al.*, 1983).

Banerjee (1993b) has reported the transformation of rifamycin B to S using growing and resting cells of *C. lunata*. Addition of 1% (w/v) of rifamycin B in the medium inoculated with growing cells affected the growth however; transformation was completed within 20 hours of incubation. Resting cells in both exponential and stationary phases of growth also transform rifamycin B to S. Biotransformation of rifamycin B with growing and resting cells did not give stoichiometric yields; i.e. rifamycin S was obtained in less quantity than expected. Whole resting cells or stationary phase cells and culture filtrate of *C. lunata* (source of enzyme) also exhibited this bioconversion potential (Patil, 1998).

Biotransformation with immobilized whole cells

The whole cells of *Humicola* sp. ATCC 20620 with rifamycin oxidase activity immobilized on acrylamide and defatted with acetone showed the maximum activity at pH 7.5 for both free and immobilized forms. The apparent Km values of rifamycin oxidase for the free and immobilized acetone-defatted cells were 0.3 mM and 0.6 mM respectively. Substrate inhibition was observed for immobilized cells. Compared to free enzyme, the whole cell immobilized preparations were less sensitive to temperature and pH changes. A temperature of 50°C was reported as optimal for polyacrylamide immobilized whole cells (Lee *et al.*, 1984).

Biotransformation with soluble enzymes

Extracellular, soluble rifamycin oxidase from *C. lunata* has been used for the transformation of rifamycin B to S. Optimum temperature reported was 28°C; agitation speed, 200 rpm; enzyme concentration, 5 % (v/v/m) and aeration rate 1-1.25 (v/v/m). Polypropylene glycol was found to suppress the foam without affecting the enzyme activity (Banerjee, 1993c).

Conceptually, this enzyme route avoids unwanted side reactions, offers process flexibility and simplicity; however, large scale preparation of such systems becomes difficult. Hence, the use of cell-free soluble enzymes is less practicable as compared to other procedures.

Biotransformation with immobilized enzymes

Rifamycin oxidase from *C. lunata var. aeria* immobilized on kappa carrageenan gel was reported to show excellent catalyzing activity and operational stability. Thermostability and reusability of immobilized enzyme had increased as compared to free enzyme (Banerjee, 1993e). Rifamycin oxidase from *C. lunata* have been immobilized on polyacrylamide gel. Optimization of acrylamide percentages, pH and temperature were the mainly focused lines of study. Reusability was found to impart colour to beads after 3-4 times and transformation was found to be diffusionally limited (Banerjee, 1993f).

Biotransformation in organic solvents

Biotransformation using organic solvents can be applied for antibiotic biotransformation with minimized problems of substrate solubility and mixed crystal formation etc. It was reported that several organic solvents unexpectedly gave much higher rates than were obtained in aqueous systems (Patil and Chincholkar, 1997).

FACTORS INFLUENCING BIOTRANSFORMATION

Series of attempts have been made to improve rifamycin oxidase activity and hence biotransformation of rifamycin B to S. various physiological and physicochemical parameters are known to regulate the activity of rifamycin oxidase. These include -

(i) Temperature and pH

Rifamycin oxidase from *C. lunata* is shown to be a thermostable enzyme requiring an optimum temperature of 50^{0}C and pH of 6.5 value (Banerjee, 1994b). Whereas intracellular rifamycin Oxidase of *Humicola* sp. and *Monocillium* sp. is reported to function optimally at 45^{0}C and 28-30^{0}C and pH of 7.8 and 7.5-8.5 value respectively (seong *et al.*, 1985).

(ii) Agitation

Mixing of nutrient medium was reported to have a significant effect on biotransformation in *C. lunata*. As the shaking speed increased upto a threshold level of 200 rpm, time required for transformation of rifamycin B to S decreased (Banerjee, 1993g). However, in a batch reactor with varying stirrer speeds, it was revealed that the production of rifamycin oxidase increased upto 400 rpm and then remained constant with further increase of speed (Banerjee, 1993g). Production of extracellular rifamycin oxidase was reported from *C. lunata* var. *aeria* grown in YPD medium at 28^{0}C. Mycelial form of growth gave higher enzyme activity than the pellet form (Banerjee and Srivastava, 1993). The organism was found to form black pigment when the glucose level was almost zero and the medium became alkaline. Formation of black pigment was considered as a indication of reduced enzyme activity (Banerjee, 1993). Metal ions (10-75 mgL^{-1}) like Mg^{2+} Co^{2+} Zn^{2+}, and Mn^{2+} had no stimulatory effect where as Fe^{3}, Ca^{2+} and Cu^{2+} stimulated rifamycin oxidase production (Banerjee, 1994b). Method of substrate preparation in different solvents revealed that ethanol was best suited for maximum secretion of enzyme (Patil and Chincholkar, 1997).

(iii) Aeration

Aeration has been reported to have profound effect on biotransformation and production of rifamycin oxidase in *C. lunata*.. 1 vvm aeration required 3 h while 0.5 vvm of aeration required more than 3 h to complete biotransformation. Although rate of disappearance of rifamycin B increased with aeration rate, complete transformation occurred only after 3 h incubation in case of reactions with 1 and 1.5 vvm of aeration (Banerjee, 1993g). In batch reactor it was reported that the enzyme activity increased with increase in air flow rate upto 0.75-1 vvm and activity remained constant (12 IU mL^{-1}) in range from 1-1.25 vvm of aeration (Banerjee, 1993g).

(iv) Antifoam

For biotransformation with *C. lunata* use of polypropylene glycol has been reported to control foam without causing an interference in the isolation of rifamycin S from the spent medium (Banerjee, 1993b). 'Neorin 202' (1%) was used as an antifoam agent during production of rifamycin oxidase in *Humicola* sp. ATCC 20620 (Kim *et al.*,1984).

(v) Substrate concentration and method of substrate preparation

Substrate concentration and method of substrate preparation are the most critical parameters in biotransformation of rifamycin B to S in *C. lunata.* Patil and Chincholkar (1997) have claimed that ethanol is the best solvent for dissolution of rifamycin B, it also resulted in increased enzyme activity and transformation of rifamycin B to S. The substrate rifamycin B has been found to promote the reaction upto a critical concentration and inhibit thereafter i.e. self inhibitory above 3.4 gl^{-1} concentration (Banerjee and Patnaik, 1996).

The enzyme exhibited a Km value of 0.67 mM and Vmax of 11 $mMh^{-1}ml^{-1}$. The substrate inhibition was apparent at substrate concentration higher than 2 mM. Kinetic studies of the enzyme revealed uniform enzyme activity upto 9 days at 30^{0}C. The activation and deactivation energies of partially purified enzymes, calculated from Arrhenius plot, were 5.80 and 35.10 kcal mol^{-1} respectively (Banerjee *et al.*, 1992).

(vi) Biomass

Specific activity of enzyme was found to increase with increase in cell biomass. In *C. lunata,* 5% inoculum size was optimal for growth and enzyme secretion. It also shortened the lag phase. Further increase in inoculum size did not reduce the lag phase or increase the enzyme activity (Banerjee, 1994b).

(vii) Carbon source

C_3 precursors namely glycerol, glucose and ribose could be used for glycolic acid moiety in rifamycin B. Ghisalba and co-investigators (1982) claimed carboxy-methyl cellulose (CMC) as an effective carbon source for rifamycin oxidase synthesis. Since CMC enhances viscosity of medium and thereby promotes better mycelial growth than pellet form, such conditions were conducive for secretion of enzyme and hence biotransformation.

(viii) Nitrogen source

Peptone, peanut meal and ammonium sulphates were found to be the most effective nitrogen source to secrete maximum enzyme. Compounds like ammonium acetate and ammonium oxalate were found to be inhibitory to enzyme synthesis and biotransformation. *C. lunata* could neither grow nor secrete enzyme in synthetic medium without peptone and yeast extract (Banerjee, 1994a).

SOLID SUBSTRATE CULTIVATION

It is well recognized that solid state cultivation has advantages not only on soluble and immobilized enzymes, but also over whole cell system in terms of total capital investment (10-18% of submerged process) and economics of process (Ghildayal *et al.*, 1985).

Several studies using *C. lunata* showed that the organism produces the enzyme extracellularly with high specific activity (Vohra and Vyas, 1982; Han *et al.,* 1983). Growth characteristics of *C. lunata*, production of extracellular rifamycin oxidase and bioransformation capabilities have been studied (Vohra and Vyas, 1992; Banerjee and Patnaik, 1996; Patil and Chincholkar, 1997).

Biotransformation of rifamycin B to rifamycin S using two strains of *C. lunata* namely NCIM 716 and NMU grown on various solid substrates viz., grass, paper, jowar/wheat straw, bran and bagasse has been reported by Rasalkar *et al* (2002). Almost complete biotransformation efficiency of rifamycin B (0.06 mM) was observed within 24 hr. However, *C. lunata* NMU strain showed higher bioransformation efficiency (90 %) vis-à-vis reference strain (85 %). Column bioreactor studies with bagasse revealed further improvement in biotransformation efficiency of *C. lunata* NMU strain (Rasalkar *et al.*, 2002).

C. lunata NCIM 716 could grow in presence of rifamycin B and could convert it to rifamycin S as evidenced by celllulose utilization (Table 2). *C. lunata* NMU maximally utilized the cellulose from each substrate except paper. *C. lunata* NCIM 716 poorly utilized cellulose from bagasse and paper. Whereas *C. lunata* NCIM 716 poorly utilized cellulose from baggase and paper. Both strains showed maximum utilization of cellulose on grass. These results have also been supported by cellulase activity of both strains on solid substrates (Table 3). *C. lunata* NMU showed higher exoglucanase activity on jowar/wheat straw while reference strain exhibited maximum exogluconase activity on grass and paper (Rasalkar *et al.*, 2002).

Table 2. Profile of cellulose utilization by *C. lunata* NCIM 716 and *C. lunata* NMU at 27°C for 24 h (Rasalkar *et al.*, 2002)

Solid substrate	Total cellulose (g %)	Cellulose utilization (%)	
		C. lunata NMU	*C. lunata* NCIM 716
Baggase	46.0	35.0	28.0
Grass	57.0	54.4	51.0
Paper	57.0	26.3	30.0
Jowar / wheat straw	56.0	43.0	39.3

Table 3 Pattern of cellulase activity by *C. lunata* NMU and *C. lunata* NCIM 716 on matrices at 27°C

Solid substrate	Filter paper activity (Uml^{-1})	Exoglucanase activity (Uml^{-1})	Endoglucanase activity (Uml^{-1})	β-glucosidase activity (Uml^{-1})
C. lunata NMU				
Bagasse	0.052	0.61	0.074	0.0065
Grass	0.066	0.66	0.059	0.0070
Jowar/wheat straw	0.064	1.11	0.040	0.0035
Paper	0.066	0.77	0.083	0.0072
C. lunata NCIM 716				
Bagasse	0.072	0.61	0.129	0.006
Grass	0.070	0.88	0.144	0.007
Jowar/wheat straw	0.085	0.77	0.055	0.007
Paper	0.088	0.88	0.20	0.007

Biotransformation in petri plate

Both strains showed typical growth behaviour on each solid substrate. *C. lunata* NMU strain appeared as black mycelial lawn, while *C. lunata* NCIM 716 showed white mycelial growth on each solid substrate containing petri plate after 4 days of incubation. In petri plate, shallow depth of substrate (0.5-1.0 cm) eliminated the critical requirement like agitation, heat and oxygen transfer for growth and transformation of rifamycin. The headspace of 1-1.5 cm was enough to articulate the conditions. Both strains on each solid substrate showed almost complete transformation of rifamycin B at 27°C within 24 h. As illustrated in Table 4, 10 days old grown cells of both strains of *C. lunata* converted rifamycin B to rifamycin S within 24 h of incubation. In the experiments performed with 50 mgl^{-1} of rifamycin B in petri plate, 45 mgl^{-1} of rifamycin S (90 % biotransformation) was obtained with *C. lunata* NMU strain grown on bagasse, while *C. lunata* NCIM 716 produced 42.5 mgl^{-1} of rifamycin S (85% biotransformation)

Column bioreactor studies

Column bioreactor studies revealed increase in efficiency. *C. lunata* NMU strain showed 95% of biotransfromation efficiency against 87.5 % biotransformation efficiency obtained with reference strain under similar conditions on bagasse after 24 h. Bioactivity of biotransformed product was confirmed with increased zone of growth inhibition (38 mm) of *Staphylococcus aureus* compared to rifamycin B (35 mm) (Rasalkar *et al.*, 2002).

Table 4. Biotransformation of rifamycin B on solid substrate (bagasse) in petri plate and column bioreactor at 27°C

	C. lunata NCIM 716		*C. lunata* NMU	
	12 h	24 h	12 h	24 h
Petri plate Assay Rifamycin S (mgl^{-1})	30.0	42.5	35.0	45.0
% Transformation efficiency	60.0	85.0	70.0	90.0
Column Reactor Assay Rifamycin S (mgl^{-1})	30.0	43.8	35.0	47.5
% Transformation efficiency	60.0	87.6	70.0	95.0

Thin layer chromatography (TLC) during the course of column bioreactor operation confirmed the identity of reaction products. Rifamycin S (dark red, R_f 0.9) was identified along with unconverted rifamycin B (yellow, R_f 0.1) (Rasalkar *et al.*, 2002).

Rifamycin oxidase activity in the culture filtrate with strain on bagasse, grass, paper, jowar / wheat straw was on an average 0.0107 U. Poor enzyme activity may be due to production of some metabolites, which inactivated the enzyme, or it may be due to higher *p*H (Rasalkar *et al.*, 2002).

CURRENT STATUS OF THE DEVELOPMENT OF RIFAMYCIN DERIVATIVES

A large number of rifamycin derivatives have been synthesized/developed over the years. For semisynthetic rifamycins, naphthalene ring modifications is preferred since ansa chain modification reduces the biological activity of rifamycins (White *et al.*, 1973).

Rifamycin S is a commercially significant intermediate for the synthesis of semisynthetic rifamycins. Rifampicin, a physiological derivative of rifamycin S, has been used clinically for the treatment of tuberculosis and leprosy. Now, rifabutin is approved as a preventive drug for MAC infection in AIDS patients. Rifalzil (KRM - 1648) possesses a promising potent antituberculosis (Hidaka, 1999).

Preliminary experiments on solid substrate cultivation of *C. lunata*, a local isolate, for biotransformation of rifamycin in our laboratory also offered another possibility in this regard.

Steroid biotransformation

Steroid transformation is a second only to antibiotic production as a vital contribution of microbes to the production of pharmaceuticals. The critical role of microbes in steroid synthesis has been found in connection with the synthesis of the adrenocortical hormones, corticosterones, cortisone and their therapeutically superior derivatives such as prednisone, prednisolone, triamicolone etc. It centers on the introduction of oxygen atom at C11 of the steroid molecule. These steroids are important because they serve as anti-inflammatory and anti-allergic action and as a starting material for other drugs (Chincholkar, 1994).

As total synthesis of sex hormones and corticosteroids by chemical route is lengthy and expensive, natural steroids are in great demand for their partial synthesis. Therefore, plant materials such as hecogenin, solasodine and diosgenin were exploited as starting material for the synthesis of corticosteroids (Chincholkar, 1994).

Today, the steroid industry consumes more than 2,000 tons of diosgenin equivalent to make corticoids and contraceptives, which represents 85 and 5% of the market in volume or 55 and 5 % in value. The current areas of research interest include sterol degradation, use of immobilized cells and enzymes and conversions at high substrate levels in organic solvents (Patil and Chincholkar, 1997).

MICROBIAL FEATURES

For the hydroxylation of steroids number of cultures have been reported. Few examples of such organisms are cited in Table 5. The common microbiological reactions of steroids include – dehydrogenation, epoxidation, esterification, isomerization and hydrolysis of acetal.

Table 5. Microbes involved in steroid biotransformation

Microbes	Substrate	Reaction	Reference
Bacteria			
A. simplex	Hydroxycortisone	1-dehydrogenation	Kondo and Masuo,1961
Bacillus sphaericus	Androstenedione	1-dehydrogenation	Ringgold *et al.*, 1963
Ps. testosterone	A-3-oxo-β steroid	Dehydrogenation	Dixon and Webb,1979
Fungi			
S. argenteolus	Progesterone	16α-hydroxylation	Perlman *et al.*, 1952
Cochliobolus lunatus	Norethisterone acetate	Hydroxylation	Horhold *et al.*, 1986.
Paecilomyces lilacinus	Progesterone	Side-chain Cleavage	Vitas *et al.*, 1995.
Cunninghamella blacksleeana	Progesterone	Hydroxylation	Chincholkar *et al.*, 1995
Algae			
Chlorella sp.	Progesterone	Hydroxylation	Pollio *et al.*, 1996

MICROBIAL HYDROXYLATION OF STEROIDS

Out of common microbiological reactions of steroids, hydroxylation reactions is of more practical importance. These reactions are unique in their diversity and importance in that they offer access to otherwise inaccessible sites in the steroid molecule. The broad specificity of hydroxylase system is a great advantage for these processes, allowing the selection of the optimal intermediate for hydroxylation in a sequence of chemical and microbial steps to desired product. All microbial hydroxylation involves direct replacement of the hydrogen atom on a given carbon by a hydroxyl group. The oxygen atom in the hydroxyl group is derived from molecular (gaseous) oxygen. At present, from the practical point of view, more interest is focused on the microbial hydroxylations at 11 α, 11 β 16 and 21 positions of steroidal nucleus (Miller, 1986).

11 β hydroxylase

11 β-hydroxylation of steroid mediated by 11β hydroxylase was first reported in *C. lunata* (Shull *et al.*, 1953) and *C. blacksleeana*. The enzyme required molecular oxygen and NADPH and was inducible with compound S and other substituted pregnanes lacking functionality at the C_{11} or C_{14} position serving as inducers. The purified enzyme is quite stable at 30 °C and at pH value of about 8 (Zuidweg, 1968).

FACTORS AFFECTING ENZYME ACTIVITY / BIOTRANSFORMATION

11β hydroxylase is dependent on conditions during enzyme synthesis. To increase the steroid hydroxylase activity and stability and at the same time to decrease the activity of other enzymes responsible for by-product formation, number of alternative routes were applied in recent years (Sukhodol'skaya *et al.*, 1994).

(a) Carbon source

Dextrin and starch have been reported to increase 11β hydroxylase activity. In presence of glucose (0.5 %), 11 β -hydroxylase of androstene-4 en-3, 17-dione (AD) by *R. nigricans* was completed within 24 h where as in absence of glucose it required 46 h. Similarly incorporation of glucose in the growth medium stimulated steroid hydroxylation activity of immobilized microorganisms (Sedlaczek *et al.*, 1979).

(b) Nitrogen source

Of the different nitrogen sources used in steroid hydroxylation, corn steep liquor has been reported to be satisfactory and economical substrate. Amino acids like glycine, glutamic acid, histidine, valine, methionine, aspartic acid and threonine were found to support growth of *C. lunata* whereas formic acid was found to create a marked effect on transformation. Certain vegetable oils also seem to encourage better steroid yields when added to the medium (Dulaney and Stapley, 1959). Number of reports has shown that inorganic nitrogen source affects growth and negligible effect on biotransforming ability of *C. lunata, C. blacksleeana, A. ochraceus*. However, with the use of ammonium salts, increase in 11 β hydroxylase activity was reported with *C lunata* (Dulaney and Stapley, 1959).

(c) Oxygen

It has been reported that microbial hydroxylation of steroids proceeds only in presence of oxygen (Zuidweg, 1968). Oxygen has been claimed to serve as an inducer for steroid hydroxylase. In case of *C. lunata*, DO concentration for a maximum hydroxylation rate was shown to be much higher (40-60%) than required for its growth (20%) (Chincholkar *et al.*, 1992).

Sukhodol'skaya *et al.* (1994) established that at low pO_2 values (20-40%) the phase of exponential growth was absent and the phase of acceleration was prolonged in *C. lunata* VKM F-644. Increase in the pO_2, concentration (95%) in the medium caused the inhibition of initial rate of mycelial growth, which appeared as extended phase of growth acceleration, namely 72 h. transition to stationary growth phase also approached after 96 h of cultivation.

(d) Growth stage- growing and resting cells

11 β hydroxylation of steroids is carried out with growing and grown (resting stage) microorganisms. It was found that 11β hydroxylation by resting cells was rapid than catalyzed by growing cells. Biotransformation with fully grown washed mycelium offer numerous advantageous like -

- (i) It regulates enzyme activity in desired direction more efficiently.
- (ii) Rate of aeration is increased with the increase in the effectiveness.
- (iii) Ease of isolation and purification of the products due to the absence of any contamination.
- (iv) Decrease in time required for hydroxylation, transformation being completed in 6 h
- (v) Resting whole cell suspension could be changed with substrate several times, yielding product until endogenous reserves of nutrients and cofactors were exhausted.

Washed free cells transform cortexelone to hydrocortisone even after 13 days of incubation in the nutrient medium. Interestingly, using 13 days old mycelium as a biocatalyst did not result in any significant change in the activity of enzyme as compared to 3-5 days old mycelium. However, significant changes in the character of the process and quantitative and qualitative composition of the transformation products have been reported with 33 days old rnycelium as a catalyst. This also resulted in the formation of numerous by-products not characteristics of 11β hydroxylation, which is indicative of active enzymatic degradation of the product.

(e) Substrate solubility, method of introduction and concentration

In the last few years, lot of attention was given to the poor solubility of steroidal substrate in water. When steroid substrates are added in crystalline form, another problem of binding of steroids comes in to picture. It has been reported that in 11 β hydroxylation of progesterone (50 gl^{-1}), practically complete substrate binds to the mycelium of *A. ochraceus*. Binding and biotransformation of steroids by thermophillic moulds has been reported. β cyclodextrins have been reported to be were helpful in decreasing the substrate absorption by mycelium, increasing substrate and product solubility in aqueous medium, accelerating velocity of reaction and enhancing yield of hydrocortisone (Chincholkar *et al.*, 1989).

To improve the hydroxylation process through increase in the solubility of steroids, following methods have been reported :

(i) Introduction of steroid substrates in suspension with surface active agents. Surface active agents are claimed to increase the synthesis and activity of enzymes as they improve the intensity of respiration

(ii) Introduction of steroid substrates dissolved in organic solvents.

(iii) Introduction of water soluble steroid complexes.

(iv) Use of two phase aqueous systems based en water soluble polymers.

(v) Use of β cyclodextrin in 11β hydroxylation of cortexelone has been reported to be the most effective for the improvement of hydroxylation (Chincholkar *et al.*, 1989).

Substrate concentration has also been reported to influence 11β hydroxylation of progesterone. Upon using, substrate concentration in a range of 2 gl^{-1} to 20-50 gl^{-1}, hydroxylation apparently occurred rapidly in fermentation at low substrate concentration (Weaver *et al.*, 1960). However, at higher concentrations, the substrate was known to exert toxic effect on the microorganisms (Chincholkar, 1991).

(f) Cultivation device

The mycelium produced by cultivation in flasks on a rocker exhibited steroid 11 β hydroxylase activity in all phases of growth. However, highest enzyme activity was reported in stationary and delayed growth phase. The specific steroid 11 β hydroxylase activity of such mycelium comprised 23 μM of product g^{-1}. 60% yield of hydrocortisone was obtained with 2 gl^{-1} of cortexelone (Angelova *et al.*, 1985).

IMMOBILIZATION

Although fungi were among the few living things, which were immobilized, there are very few reports about the growth and physiology of fungi in immobilized states (Sonomoto *et al.*, 1981). Investigations on the influence of carrier structure on the growth and 11β hydroxylase activity of *C. lunata* were carried out by Sonomoto *et al.* (1981). For the immobilization of spores, photocross linkable polymers with prepolymer chains of varying length (10-60 nm) were used. High level activity was in the carriers with lengthy chain of monomers (40-50 nm) correlated with the intensive growth of microorganisms (Sonomoto *et al.*, 1981).

In 11β hydroxylation of cortexelone, growth of fungal mycelium and enzyme activity were found to be dependent upon the nature and chemical characteristics of the matrix used for immobilization. A correlation between 11β hydroxylase activity with the cellular ATP levels, indicative of viability of cells in immobilized state was established (Chincholkar *et al.*, 1993).

Immobilized enzymes

There are few reports on steroid biotransformation with immobilized enzymes entrapped in crosslinked polyacrylamide gel (Mosbach and Larsson, 1970). The conversion of hydrocortisone to predriisolone has been demonstrated when the substrate dissolved in ethanol along with methosulfate was passed through a column packed with the immobilized enzyme (1-dehydrogenase). It has been shown that it is much easier to entrap whole cells instead of enzyme in the gel (Larsson *et al.*, 1976).

Morphogenesis and enzyme activity in different carriers

Sukhodol'skaya *et al.* (1989) and Gulevskaya *et al.* (1994) have investigated the morphogenesis and its relation to 11β hydroxylase activity of growing free *C. lunata* VKM F-644 during the transformation of cortexolone into hydrocortisone. Pronounced sporogenesis and conidiogenesis was reported in polyacrylamide immobilized cells as well as starved cells.

Use of polyacrylamide gel (PAG)

Immobilization of spores in polyacrylamide gel and the subsequent incubation of carrier in a nutrient medium led the formation of vegetative mycelium in a carrier, eliminated the mechanical damage of the gel entrapped mycelium and decreased the toxic effect of the reagent. The most pronounced polymorphism was observed in the culture immobilized in PAG particles while that entrapped in small particle was the most homogenous. PAG immobilized mycelium hydroxylates cortexelone less efficiently than free mycelium. The yield of the product did not exceed 40-50% which was found to be 20% less than the free mycelium process (Gulevskaya *et al.*, 1994).

Use of Ca-alginate

11β hydroxylase activity of Ca alginate immobilized biocatalyst after different inoculation times showed that, after 20 days of growth, despite the preservation in the gel, the enzymatic activity of the biocatalyst was extremely low (Gulevskaya *et al.*, 1994).

Biotransformation in organic solvents

Biotransformation using organic solvents has been reported mainly in steroids (Antonie *et al.*, 1981; Miller, 1986). Buckland *et al.* (1975) has successfully converted cholesterol to cholest-4-en-3-one in such systems using strain of *Nocardia*. It was reported that several organic solvents unexpectedly gave much higher rates than were obtained in aqueous systems.

Pseudocrystallo fermentation

Biotransformation of cortisol (powderized from) by *A. simplex* also reported to give high yields of 1-dehydrogenated product. The finely powdered solid cortisol (1-50%) of crystalline prednisolone in high yields (>93%) directly in the fermentation beers was obtained within 5 days (Kondo and Masuo, 1961)

FUTURE PERSPECTIVES

Worldwide importance of rifamycin has increased tremendously, but efforts to improve the yield of rifamycin derivatives are still scanty. Further improvements in the process could be achieved by optimizing the cultural conditions, concentration of rifamycin B, solid substrate and bioreactor design for high yield of rifamycin S. Further exploration of biotransformation of rifamycin B will lead to new and highly potent rifamycin derivatives with significant therapeutic value. For future novel antituberculosis therapy, it is therefore prerequisite and imperative to develop a new rifamycin derivatives with (i) a preferable antimicrobial activity, (ii) shortened treatment period and (iii) potent activity against MDR bacilli.

Although technical know-how of conventional steroid conversions by microorganisms is available, there is a great interest in alternative routes for improvement of steroid hydroxylations (Chincholkar, 1994).

CONCLUSION

The overview of literature focuses that *C. lunata* is a biotechnologically potential organism for biotransforming industrially viable compounds like antibiotics and steroids. Either cell free enzyme extracts or whole cells secreting such enzymes played a key role in the biotransformation of antibiotics desired by the industry.

Free enzymes are now being replaced by immobilized enzymes, which offers distinct advantages in terms of greater stability, reusability, productivity, and exclusion of possible allergic reactions.

11β hydroxylation catalysed by 11β hydroxylase seems to be important microbiological reaction amongst different steroid biotransformation reactions. Optimization of cultural conditions, use of β cyclodextrin for increasing the solubility of steroidal substrates (Chincholkar *et al.,* 1989) and use of organic solvents for dissolving steroidal substrates (Bukhland *et al.*, 1975; Patil and Chincholkar, 1997) have been suggested for improving the hydroxylation activity/ biotransformation.

REFERENCES

Angelova B.A., Sukhodol'skaya G.V. and Koscheenko K.A. (1985). *Microbiologiya,* 54 (5) : 704-709.

Antonie E., Carrea C and Cremonesi P. (1981). Enzyme catalyst function in water-organic solvent two-phase system. *Enz Microbial Technol,* 3: 291-296.

Banerjee U.C. (1993a). Transformation of rifamycin B with growing and resting cells of *C. lunata,* Enz *Microbial Technol,* 15: 1037-1041.

Banerjee U.C. (1994a). Optimization of culture conditions for the production of rifamycin oxidase by *Curvularia lunata. World J Microbiol Biotechnol* 10: 462-464.

Banerjee U.C and Patnaik P.R. (1996). A kinetic model for enzymatic conversion of rifamycin B to S by rifamycin Oxidase from *Curvularia lunata, Ind Chem Eng, A,* 38(1): 28-30.

Banerjee U.C. (1993e). Studies on rifamycin oxidase immobilized on nylon fibers, *Biotechnol Tech,* 6 (1): 61-64.

Banerjee U.C. (1993b). Transformation of rifamycin B with soluble rifamycin oxidase from *Curvularia lunata, J Biotechnol,* 29: 137-143.

Banerjee U.C. (1993c). Effect of glucose and carboxy methylcellulose on growth and rifamycin oxidase production by *Curvularia lunata. Current Microbiol,* 26: 261-265.

Banerjee U.C. (1993d). Transformation of rifamycin B with soluble rifamycin oxidase from *Curvularia lunata, J Biotechnol,* 29: 137443.

Banerjee U.C. (1993f). Studies on rifamycin oxidase immobilized on agar gel, *J Gen Appl Microbiol,* 39: 251-255.

Banerjee U.C. (1993g). Transformation of rifamycin with immobilized rifamycin oxidase of *C. lunata, Biotechnol Tech,* 7 (5): 339-345.

Banerjee U.C. (1994b). Growth and production of rifamycin oxidase by *Curvularia lunata. Folia Microbiol,* 39 (1): 49-52.

Banerjee U.C. and Srivastava J.P. (1993). Effects of pH and glucose concentration on the production of rifamycin oxidase by *Curvularia lunata* in a batch reactor, *J Biotechnol,* 29: 229-236.

Banerjee U.C., Saxena B. and Chisti Y (1992). Biotransformation of rifamycins: process possibilities, *Biotech Adv,* 10: 577-595.

Bartolucci C, Cellai L.E., Mannina L, Marzano M., Brufani M. Filocamo L., Messina S., Brizzi V. Benedetto A. and Di Carro A. (1995). Synthesis of nucleocidal rifamycins on inhibitors of human immunodeficiency virus type I, *Antiviral Chem Chemother,* 8: 215-221.

Buckland B. C., Dunhill P. and Lily M. D. (1975). The enzymatic transformation of water insoluble in non-aqueous solvents. Conversion of cholesterol to cholest-4-ene-3-one by a *Nocardia* sp. *Biotechnol Bioeng,* 17: 815-826.

Caruso I. (1997). Twenty years of experience with intraauricular rifamycin for chronic arthritides, *J. Int. Med. Res.,* 25: 307-317.

Chincholkar S B. (1991). Ph. D. Thesis, Institute of biochemistry and physiology of microorganisms, Pushchino. Moscow, Russia.

Chincholkar S. B. (1994). Microbial steroid hydroxylation with free and immobilized cells, Microbes for better living, MICON- 94 and 35th International conference, CFTRI, Mysore, India.

Chincholkar S. B., Sukhodol'skaya G. V., Baklashova T. G. and Koshchyenko K. A (1989). Influence of cyclodextrins on 11β hydroxylation of cortexsolone by free and Ca-alginate immobilized *C. lunata,* Int. Conf Proc IV, Varna, Bulgaria, 311-319.

Chincholkar S.B., Sukhodol'skaya G.B. and Koshchyenko K.A. (1993). Physiology of immobilized cells with special reference to steroid biotransformation, paper presented in *62nd SBC (I) meeting.* 5.

Chincholkar S.B., Seeta Laxman and Wakharakar, R.D. (1995). Hydroxylation biotransformation of progesterone by *C. blakesleeana* NCIM 687, *World J. Microbiol.* and *Biotech.,* 11: 75-83.

Dixon M. and Webb H. C. (1979). Enzymes, 3 edition, Academic Press, New York, 218-219, 698.

Dulaney E. L. and Stapley E. O. (1959). Studies on the transformation of 11-deoxy 17α-hydroxy-corticosterone to hydrocortisone with a strain of *Curvularia lunata, Appl Microbiol.* 7: 276-284.

Fox W. (1985). Short course chemotherapy for pulmonary tuberculosis and some problems of its program application with particular reference to India, *Bull Jut Union Tuberc,* 60: 40-49.

Ghildyal N.P., Lonsane B.K., Srikantiahya, K.R. and Murthy, V.S. (1985). Economics of submerged and solid state fermentations for the production of amyloglucosidase, *J. Food Teechnol,* 22: 171.

Ghisalba, O., Rooss Schuup T. and Nuesh J. (1982). Transformation of rifamycin S into the rifamycin B and L, A revision of the current biosynthetic hypothesis, *J Antibiot,* 35: 74-80.

Gulevskaya S.A. Sukhodol'skaya G.V. Koshchyenko K.A., Joshi, A.K., Bari, V, Bajpai, V.K. and Basu S.K. (1994). Morphogenesis and 11β hydroxylase activity of *Curvularia lunata* grown in free and immobilized states, *J. Microbiol and Biotechnol,* 9 (1):1-21.

Han M.H., Seong B.L, Son, H.J., Mheen T.I. (1983). Rifamycin B oxidase from *Monocillium* sp., A new type of diphenol oxidase, *FEBS Lett,* 151: 36-40.

Hidaka T. (1999). Current status and perspectives on the development of rifamycin derivative antibiotics, *Aidsline Med,* 74 (1): 53-61.

Horhold C., Undisz K., Groh H., Sahn R., Schada W. and Komel R. (1986). Bioconversion of steroids by *Cochliobolus lunatus.* Transformation of Reichstein's compound S with cell preparation of *Cochliobolus lunatus, Basic Microbiol,* 26: 335-339.

Kim E. K., Choi C. Y., Park J. M., Han M. H. and Park Y. H. (1984). Effects of pH and glucose concentration on production of rifamycin oxidase, *J Ferment Technol*, 62 (2): 117-121.

Kondo E. and Masuo E. (1961). A new process for preparing prednesolone by microorganisms, *J. Gen. Appl. Microbiol.,* 7: 113-117.

Lancini G. C. (1986). Ansamycins in: Biotechnology, 4, Ed. Pape H and Rehm H. J., VCH, Weinhem. Germany, 431-463.

Lancini G. C., Gallo G. G., Sartori C. and Sensi P. (1969). Isolation and structure of rifamycin L and its biogenetic relationship with other rifamycin, *J. Antibiot,* 22 (8) : 369-377.

Larsson P.O., Ohlson S. and Mosbach K. (1976). New approach to steroid conversion using activated immobilized microorganisms, *Nature* (London), 263: 796-797.

Lee G M., Choi Y., Park J.M. and Han M.H. (1984). The properties of immobilized whole cell of *Humicola* sp. with rifamycin oxidase activity, *Biotechnol Lett,* 6 (3): 143-148.

Maggi N., Arioli V. and Sensi P. (1965). Rifamycins XLI. A new class of active semisynthetic rifamycins. N-substituted aminomethyl derivatives of rifamycin SV, *J. Med Chem*, 8: 790-793.

Miller T. L. (1986). Steroid fermentations in: Comprehensive Biotechnology, Ed. Murray-Moo Young, Univ Waterloo, Canada, 297-318.

Mosbach K and Larsson P.O. (1970). Preparation and application of polymer-entrapped enzymes and microorganisms in microbial transformation processes with special reference to steroid 11β hydroxylation and 1-dehydrogenation, *Biotethnol Bioeng.* 12: 19-27.

Muralikrishna P.S.M., Venkateswarlu G., and Rao L.V. (1999). Production of rifamycin SV using mutant strains of *Amycolatopsis mediterranei* MTCC 17, *World J. Microbiol Biotechnol,* 15: 741 -743.

Patil B. B. (1998). Studies on the role of iron metabolism in biotransformations with special reference to hydroxylation of progesterone and oxidation of rifamycin B, Ph. D. Thesis, North Maharashtra University, Jalgaon, (MS), India.

Patil B.B. and Chincholkar S.B. (1997). Influence of substrate preparation method on rifamycin oxidase activity of *C. lunata, Ind. J. Expt Biol.,* 35: 917-919.

Perlman D., Titus E. and Fried J. (1952). Microbiological hydroxylation of progesterone, J *Am. Chem. Soc.*, 74: 2126.

Polio A, Pinto G, Della G. M., Fiorentiono A. and Previtera L. (1996). Biotransformation of progesterone by *Chlorella* sp., *Phytochem*, 42: 685-688.

Rasalkar A. A., Salunke B. K., Sayyed R. Z. Chaudhari A. B. and Chincholkar S. B. (2002). Solid state cultivation of *Curvularia lunata* for transformation of rifamycin B to S. *Ind. J. Expt. Biol.*, 40: 930-933.

Ringgold H. J., Hayano M. and Stefanavic V. (1963). Concerning the stereochemistry and mechanism of the bacterial C-12-dehydrogenalion of the steroids, *J. Biochem.,* 238: 1960-1965.

Riva S. and Silvestri J.G. (1972). Rifamycins: A general view, *Ann. Rev. Microbiol.,* 26: 199-219.

Sedlaczeck L., Zarowski A., Sasiak A and Zajaczkowska, E. (1979). *Acta Microbiol. Polon.,* 28:111-121.

Sensi P. and Thiemann J. E. (1967). Production of rifamycins, Progress *Ind. Microbiol.,* 6: 21-59.

Sensi P., Ballota R, Greco A. M. and Gallo G. C. (1961). Rifamycin XV-Activation of rifomycin B and rifomycin 0 - Production and properties of rifomycin S and rifomycin SV, *Fannaco* (Ed. Sc.), 16:165.

Seong B. L, Son H. J., Mheen F. I., Park V. F. and Han M. H. (1985). Enzymatic oxidation of Rifamycin by microorganism of genus *Humicola, J. Ferment. Technol.,* 63 (6): 515-522.

Seong B. L. and Han H (1982). A Facile preparation of rifamycin derivatives by use of Manganese Dioxide, *Chemistry Lett.,* 627-628.

Shull G. M., Kita D. A. and Davidsonn J. W. (1953). Oxygenation of steroids, US patent, 2 658 023.

Sonomoto K., Hoq M. M. Tanaka A. and Fukui S (1981). Growth of *C. lunata* spores into mycelial forms within various gels and steroid 11 β hydroxylation by the entrapped mycelia, *J. Ferment. Technol.,* 59: 465-469.

Sukhodol'skaya G. V. Chincholkar S.B., Baklashova T.C., Angelova B. A. and Koshcheyenko K.A. (1989). Physiological-biochemical features of immobilized *Curvularia lunata,* VKM F-644 mycelium with stero-11β hydroxylase activity, in : *Physiology of immobilized cells,* International Symp, de Bont, J. A. M., Visser, Ed. J. Mattinsson B and Tramper J. Wageningen, The Netherlands, 591 -595.

Sukhodol'skaya G. V. Gulevskaya S. A. Chincholkar S. B., Satrutdinov A. D. and Kochceyenko K. A. (1994). Influence of cultivation conditions on growth, development cycle and steroid 11β hydroxylase activity of *Curvularia lunata,* VKM F-644, *Appl Biochem Microbiol,* 30 (1) : 85-93.

Umezawa H., Mıjuno S., Yamazaki H. and Nitta K. (1968). Inhibition of DNA dependent RNA synthesis by rifamycins, *J. Antibiot.,* XXI (3): 234-236.

Vitas M., Rozman D., Komel P. and Kelly S. L. (1995). Progesterone side chain cleavage by *Paecilomyces lilacinus, Phytochemistry,* 40: 73-76.

Vohra R.M. (1992). Novel assay for screening rifamycin B producing mutants, *Appl Environ Micrbiol,* 403-404.

Vohra R.M. and Vyas V.V. (1992). Characterization of rifamycin oxidase immobilized on nylon fibers, *Biotechnol Techniques.* 6 (1): 61-64.

Vohra R.M. and Dube S. (1989). Identification and quantititation of rifamycins by reverse phase high performance liquid chromatography, *J Chrornatogr,* 477: 463-466.

Vohra R.M., Banerjee U.C., Das S. and Dube S. (1989). Microbial transformation of rifamycin B: A new extracellular oxidase from *C. lunata. Biotechnol Lett,* 2: 851-854.

Weaver F. A Kinney H. E. and Wall M. E. (1960). Effect of substrate concentration on the microbiological hydroxylation of progesterone. *Appl Microbiol,* 8: 345-348.

White R.J., Martinelli E, Gallo G.G. and Lancini G. (1973). Rifamycin biosynthesis studied with ^{13}C enriched precursors and carbon magnetic resonance, *Nature*, 243: 273-277.

Zudweg M.H.J. (1968). Hydroxylation of compound S with cell free preparations from *Curvularia lunata, Biochem Biophys Acta,* 152:144-158.

CHAPTER - 10

DISTRIBUTION OF ARBUSCULAR MYCORRHIZAL FUNGI IN NURSERY AND PLANTATION IN TROPICAL FOREST RESEACH INSTITUTE JABALPUR, MADHYA PRADESH

Pratiksha Chaturvedi and Jamaluddin

Forest Pathology Division, Tropical Forest Research Institute, P.O. - R.F.R.C, Jabalpur — 482 021, M.P.

INTRODUCTION

Mycorrhizae are symbiotic association between beneficial soil fungi and plant roots. They have an important role in increasing plant uptake of phosphorus and other poorly mobile nutrients (O' Keefe and Sylvia, 1991). The arbuscular mycorrhizal (AM) fungi have a very broad host range. The members of more than 85% of all vascular plant families are colonized by them.

Arbuscular mycorrhizae are obligate endo-symbiotic fungi occurring under varying climates. Some of the important studies on AM fungi associated with forest tree species were made by Thapar and Khan (1985), Kandaswamy *el al.* (1987), Mohan *el al.* (1995). Most of the surveys showed that endomycorrhizae of the arbuscular type (AM) are found to be the predominant mycorrhizal association in the natural forest of humid tropics (Thapar and Khan, 1973; Tupas and Sajise, 1976; De Alwis and Abeynayake, 1980; Gunatilieke and Maheswaran, 1987).

Influence of ecological factors on distribution of arbuscular mycorrhizae is well documented in temperate forests (Bakshi, 1980). Mukerji and Kapoor (1986) recorded the occurrence of AM on sixteen species of trees and shrubs in a natural forest stand in the old Delhi ridge. Mohan Kumar and Mahadevan (1987) investigated the frequency of mycorrhizal association in the Kalakad Reserve Forest, Tamil Nadu. Rano Jagpal and Mukerji (1988) screened 62 dicot plants belonging to 20 different families and reported that all were mycorrhizal though the extent of infection varied. Thapar *et al.* (1991) while studying the frequency of occurrence of AM fungi in saline soil in Saraswati Range of Haryana State reported sixteen fungi of which thirteen were represented.

The present study on AM fungi was undertaken with a view to study the AM flora, occurring in a centralized nurseries of Tropical Forest Research Institute, Jabalpur, Madhya Pradesh.

EXPERIMENTAL

Study area

The study area selected was Tropical Forest Research Institute, which is located near village Neemkheda about 10 km. away from Jabalpur town. The Institute is spread over an area of 108 ha which lies between 230 15' 37" to 230 6' 10" N latitude and 790 59' 49" to 790 591 4211 E longitude. The climate of Jabalpur is fairly pleasant except later part of summer season. There are three distinct seasons in the year, viz., summer, rainy season and winter. The annual rainfall is about 358.43 mm the mean winter temperature is 18.64^{0}C and mean summer temperature is 29.14^{0}C.

Three soil series are found to occur in this area of study. They belong to red and black soil group area has scanty vegetation. Dry deciduous trees are of common occurrence of which *Butea monosperma* is prominent. The area falls under forest type 5A /C3 i.e. southern mixed dry deciduous forest.

METHOD OF SAMPLE COLLECTION AND SPORE EXTRACTION FROM SOIL

Under study area the soil samples and root samples were taken from seven different tree species such as *Tectona grandis, Azadirachta indica, Pongamia pinnata, Albizzia procera, Dendrocalamus strictus, Sesbania grandiflora* and *Eucalyptus* species. The samples were also taken in different seasons i.e. summer, rainy and winter season from the plants grown in polybags, nursery beds and in plantations.

The soil samples were collected up to a depth of 10 to 15 cm. from the rhizosphere soil. Each sample was collected in 500 gm. Then the samples were collected in fresh polythene bags and there opening were tied with rubber bends. AM spores were extracted from soil samples by Daniel and Skipper (1982) centrifugation technique. The % frequency of AM species and the occurrence of AM species in each site separately are calculated by the formula as below:

$$(1)\ \%\ \text{frequency of species} = \frac{\text{No. of site in which a species is present}}{\text{Total no. of sites}} \times 100$$

$$(2)\ \%\ \text{frequency of all species} = \frac{\text{No. of species present in the site}}{\text{Total number species}} \times 100$$

RESULTS AND DISCUSSION

In all, 28 species of AM fungi on 7 host species were collected during the study (Table 1). The different species of AM fungi vary in all the three conditions of plant species viz., poly bag (PB), nursery bed (NB) and plantation (P) sites respectively *T. grandis, D. strictus, A. indica, P. pinnata, A. procera* and *S. grandiflora* were found associated with a number of AM fungi while *Eucalyptus* showed comparatively less per cent of mycorrhizal fungi.

Table 1. Host wise distribution of AM fungi in the study area

S.No.	Name of AM species	Name of host species	Place/Site		
			Polybag	Nursery bed	Plantation
1	*Glomus aggregatum*	*Tectona grandis*	+	+	+
		Dendrocalamus	+	+	+
		Azadirachta indica	+	+	+
		Pongamia pinnata	+	+	+
		Albizia procera	+	+	+
		Sesbania grandiflora	+	+	+
		Eucalyptus	+	+	+
2	*Glomus albidum*	*Tectona grandis*	—	—	—
		Dendrocalamas strictus	—	—	—
		Azadiarachta indica	+	+	+
		Pongamia pinnata	—	—	—
		Albizia procera	—	—	—
		Sesbania grandiflora	—	—	—
		Eucalyptus	—	—	—
3	*Glomus deseticola*	*Tectona grandis*	+	—	—
		Dendrocalamus strictus	—	—	+
		Azadirachta indica	—	—	—
		Pongamia pinnata	—	—	—
		Albizia procera	—	—	—
		Sesbania grandiflora	—	—	—
		Eucalyptus	—	—	—
4	*Glomus diaphanum*	*Tectona grandis*	—	—	—
		Dendrocalamus strictus	—	—	—
		Azadirachta indica	—	—	—
		Albizia procera	—	—	—
		Pongamia pinnata	—	—	—
		Sesbania grandiflora	—	—	—
		Eucalyptus	—	—	—
5	*Glomus dimorphicum*	*Tectona grandis*	—	—	—
		Dendrocalamus strictus	—	—	—
		Azadirachta indica	—	—	—
		Pongamia pinnata	—	—	—
		Albizia procera	—	—	—

Contd....

S.No.	Name of AM species	Name of host species	Place/Site		
			Polybag	Nursery bed	Plantation
		Sesbania grandiflora	—	—	—
		Eucalyptus	+	—	—
6	*Glomus etunicatum*	*Tectona grandis*	+	—	+
		Dendrocalamas strictus	—	+	+
		Azadirachta indica	—	—	+
		Pongamia pinnata	—	—	—
		Albizia procera	—	—	+
		Sesbania grandiflora	—	—	—
		Eucalyptus	—	—	+
7	*Glomus geosporum*	*Tectona grandis*	—	—	—
		Dendrocalamus strictus	+	—	+
		Azadirachta indica	—	—	—
		Pongamia pinnata	+	+	—
		Albizia procera	—	—	—
		Sesbania grandiflora	—	—	+
		Eucalyptus	—	—	—
8	*Glomus intraradices*	*Tectona grandis*	+	—	—
		Dendrocalamus strictus	—	—	—
		Azadirachta indica	—	+	—
		Albizia procera	—	—	—
		Pongamia pinnata	—	—	—
		Sesbania grandiflora	—	—	—
		Eucalyptus	—	—	—
9	*Glomus macrocarpum*	*Tectona grandis*	—	—	—
		Dendrocalamus strictus	—	—	—
		Azadirachta indica	—	—	—
		Pongamia pinnata	—	—	—
		Albizia procera	+	—	+
		Sesbania grandiflora	—	—	—
		Eucalyptus	—	—	—
10	*Glomus microaggregatum*	*Tectona grandis*	—	—	—
		Dendrocalamas strictus	—	—	—
		Azadiarachta indica	—	—	—
		Pongamia pinnata	—	—	+
		Albizia procera	—	—	—
		Sesbania grandiflora	—	—	—
		Eucalyptus	—	—	—
11	*Glomus invermim*	*Tectona grandis*	—	—	—
		Dendrocalamus strictus	+	—	—

Contd....

S.No.	Name of AM species	Name of host species	Place/Site		
			Polybag	Nursery bed	Plantation
		Azadirachta indica	—	—	—
		Pongamia pinnata	—	—	—
		Albizia procera	—	—	—
		Sesbania grandiflora	—	+	—
		Eucalyptus	—	—	—
12	*Glomus monosporum*	*Tectona grandis*	—	—	—
		Dendrocalamus strictus	+	—	+
		Azadirachta indica	—	—	—
		Albizia procera	—	—	—
		Pongamia pinnata	—	—	—
		Sesbania grandiflora	—	—	—
		Eucalyptus	—	—	—
13	*Glomus mosseae*	*Tectona grandis*	+	+	+
		Dendrocalamus strictus	+	+	+
		Azadirachta indica	+	+	+
		Pongamia pinnata	+	+	+
		Albizia procera	+	+	+
		Sesbania grandiflora	+	+	+
		Eucalyptus	+	+	+
14	*Glomus occultum*	*Tectona grandis*	—	—	—
		Dendrocalamas strictus	—	—	—
		Azadiarachta indica	—	—	—
		Pongamia pinnata	—	—	+
		Albizia procera	—	—	—
		Sesbania grandiflora	—	—	—
		Eucalyptus	—	—	—
15	*Glomus reticulatum*	*Tectona grandis*	—	—	—
		Dendrocalamus strictus	—	—	—
		Azadirachta indica	—	+	—
		Pongamia pinnata	—	—	—
		Albizia procera	—	—	—
		Sesbania grandiflora	—	+	—
		Eucalyptus	—	—	—
16	*Acaulospora delicata*	*Tectona grandis*	—	—	—
		Dendrocalamus strictus	+	—	—
		Azadirachta indica	—	—	—
		Albizia procera	—	—	—
		Pongamia pinnata	—	—	+
		Sesbania grandiflora	—	—	—
		Eucalyptus	—	—	—

Contd....

S.No.	Name of AM species	Name of host species	Place/Site		
			Polybag	Nursery bed	Plantation
17	*Acaulospora longula*	*Tectona grandis*	—	—	—
		Dendrocalamus strictus	—	—	
		Azadirachta indica	—	+	—
		Pongamia pinnata	—	—	—
		Albizia procera	—	—	
		Sesbania grandiflora	—	—	—
		Eucalyptus	—	—	—
18	*Acaulospora foveata*	*Tectona grandis*	—	—	—
		Dendrocalamas strictus	—	—	+
		Azadirachta indica	—	—	—
		Pongamia pinnata	—	—	+
		Albizia procera	—	—	—
		Sesbania grandiflora	—	—	+
		Eucalyptus	—	—	—
19	*Acaulospora scrobiculata*	*Tectona grandis*	—	—	+
		Dendrocalamus strictus	+	+	+
		Azadirachta indica	—	—	+
		Pongamia pinnata	—	+	—
		Albizia procera	—	+	+
		Sesbania grandiflora	+	+	+
		Eucalyptus	+	—	—
20	*Acaulospora spinosa*	*Tectona grandis*	+	+	—
		Dendrocalamus strictus	+	—	—
		Azadirachta indica	—	—	—
		Albizia procera	—	—	—
		Pongamia pinnata	—	—	
		Sesbania grandiflora	—	—	+
		Eucalyptus	—	+	—
21	*Gigaspora gilmori*	*Tectona grandis*	+	—	—
		Dendrocalamus strictus	—	—	—
		Azadirachta indica	—	—	—
		Albizia procera	—	—	—
		Pongamia pinnata	—	—	—
		Sesbania grandiflora	—	—	—
		Eucalyptus	—	—	—
22	*Gigaspora margareta*	*Tectona grandis*	+	+	+
		Dendrocalamus strictus	—	+	+
		Azadirachta indica	+	—	—
		Albizia procera	+	+	+

Contd....

S.No.	Name of AM species	Name of host species	Place/Site		
			Polybag	Nursery bed	Plantation
		Pongamia pinnata	+	+	—
		Sesbania grandiflora	+	+	—
		Eucalyptus	+	—	—
23	*Gigaspora rosea*	*Tectona grandis*	—	—	—
		Dendrocalamus strictus	+	—	—
		Azadirachta indica	—	—	—
		Albizia procera	—	—	—
		Pongamia pinnata	—	—	+
		Sesbania grandiflora	+	+	—
		Eucalyptus	—	+	—
24	*Gigaspora gigantia*	*Tectona grandis*	—	+	+
		Dendrocalamus strictus	+	+	—
		Azadirachta indica	—	+	+
		Albizia procera	—	—	+
		Pongamia pinnata	+	—	+
		Sesbania grandiflora	+	+	—
		Eucalyptus	—	+	—
25	*Scutellospora pellucida*	*Tectona grandis*	—	—	—
		Dendrocalamus strictus	—	—	+
		Azadirachta indica	+	+	+
		Pongamia pinnata	—	—	—
		Albizia procera	—	—	—
		Sesbania grandiflora	—	—	+
		Eucalyptus	—	—	—
26	*Sclerocystis microcarpum*	*Tectona grandis*	—	+	—
		Dendrocalamas strictus	—	—	—
		Azadirachta indica	—	—	—
		Pongamia pinnata	—	—	+
		Albizia procera	—	—	—
		Sesbania grandiflora	—	—	—
		Eucalyptus	—	—	—
27	*Sclerocystis sinuosa*	*Tectona grandis*	—	—	—
		Dendrocalamus strictus	—	—	—
		Azadirachta indica	—	—	—
		Pongamia pinnata	—	—	+
		Albizia procera	—	—	—
		Sesbania grandiflora	—	—	—
		Eucalyptus	—	—	—

Contd....

S.No.	Name of AM species	Name of host species	Place/Site		
			Polybag	Nursery bed	Plantation
28	*Sclerocystis rubiformis*	*Tectona grandis*	—	—	—
		Dendrocalamus strictus	—	—	—
		Azadirachta indica	—	—	—
		Albizia procera	—	—	—
		Pongamia pinnata	—	—	+
		Sesbania grandiflora	—	—	—
		Eucalyptus	—	—	—

Perusal of data in Table 3 reveals that the *Glomus mosseae and* G. *aggregatum* were two species (having 100% frequency) present in all seasons in study area, whereas *Glomus albidum, G. diaphanum, G. monosporum, G. occultum, G. reticulatum, Acaulospora longula, A. foveata, Gigaspora gilmorei, Sclerocystis sinuosa and S. rubiformis* (having 11.11% frequency) were present only in a single season. The Maximum number of species (53.57% frequency) was found in winter season, whereas summer season was least represented by these fungi (21.42% frequency).

Table 2. Occurrence of AM fungi in different season in study area.

S.N.	Name of AM species	Poly bag			Nursery bed			Plantation			% of frequency of species
		S1	S2	S3	S1	S2	S3	S1	S2	S3	
1	*Glomus aggregatum*	+	+	+	+	+	+	+	+	+	100
2	*Glomus albidum*	—	+	—	—	—	—	—	—	—	11.11
3	*Glomus deserticola*	—	—	+	—	+	—	—	—	—	22.22
4	*Glomus diaphanum*	—	—	—	—	—	—	—	—	+	11.11
5	*Glomus diamorphicum*	—	+	—	—	+	—	—	—	—	22.22
6	*Glomus etunicatum*	—	—	+	+	—	—	+	—	—	33.33
7	*Glomus geosporum*	—	+	—	—	—	+	—	+	—	33.33
8	*Glomus intraradices*	+	—	+	—	—	+	—	—	—	33.33
9	*Glomus macrocarpum*	—	+	—	—	—	—	—	+	—	22.22
10	*Glomus microaggregatum*	—	—	+	—	—	—	—	+	—	22.22
11	*Glomus invermium*	+	—	—	—	+	—	—	—	—	22.22
12	*Glomus monosporum*	—	+	—	—	—	—	—	—	—	11.11
13	*Glomus mosseae*	+	+	+	+	+	+	+	+	+	100
14	*Glomus occultum*	—	—	—	—	—	—	—	—	+	11.11
15	*Glomus reticulatum*	—	—	—	—	—	—	—	+	—	11.11
16	*Acaulospora delicata*	—	—	—	—	—	—	—	+	+	22.22
17	*Acaulospora longula*	—	—	—	—	+	—	—	—	—	11.11
18	*Acaulospora foveata*	—	—	—	—	—	—	—	+	—	11.11

Contd....

S.N.	Name of AM species	Poly bag			Nursery bed			Plantation			% of frequency of species
		S1	S2	S3	S1	S2	S3	S1	S2	S3	
19	*Acaulospora scrobiculata*	—	+	+	+	+	+	+	+	—	77.77
20	*Acaulospora spinosa*	—	+	—	—	+	—	—	+	—	33.33
21	*Gigaspora gilmorei*	—	+	—	—	—	—	—	—	—	11.11
22	*Gigaspora margarita*	+	+	+	+	+	—	+	—	—	66.66
23	*Gigaspora rosea*	+	+	—	+	+	—	—	+	—	55.55
24	*Gigaspora gigantea*	—	—	—	—	+	+	+	+	+	55.55
25	*Scutellospora pellucida*	—	—	+	—	+	—	—	+	—	33.33
26	*Sclerocystis microcarpum*	—	—	—	—	+	—	—	+	—	22.22
27	*Sclerocystis sinuosa*	—	+	—	—	—	—	—	—	—	11.11
28	*Sclerocystis rubiformis*	—	—	—	—	—	—	—	+	—	11.11
	Percentage frequency of species in the site	21.42	46.42	32.14	21.42	42.85	2142	21.42	53.57	21.42	

Note : S1= Summer, S2 = Winter, S3 = Rainy season.

It is evident from Table 3 that *Glomus aggregatum, Gl. dimorphicum, G. geosporum, G. mosseae, Acaulospora scrobiculata, A. spinosa, Gigaspora gilmorei, G. margarita, G. rosea, G. gigantea* and *Scutellospora pellucida* were the 10 species (having 100% frequency) present in all the sites under study. However, *Glomus albidum, G. dimorphicum, G. microaggregatum, G. occultum, G. reticulatum, Acaulospora longula, A. foveata, Gigaspora gilmorei* and *Sclerocystis microcarpum* (having 33.33% frequency) were present only in a single site. The maximum number of species (75% frequency) was found in poly bags. while plantation showed least representation by these fungi (57% frequency). The distribution of AM fungi was governed by the host site and season. AM fungi were found to be wide spread in the study area.

Some AM fungi were specific for some particular host, site and season. Some species were restricted to a particular host species, while others could infect a wide range of hosts, sites and seasons. This ability can be attributed to the ability of AM fungi for colonization, which differ from species to species (Jamaluddin *el al.,* 1997). *Glomus mosseae and* G. *aggregatum* exhibited a wide range of host plants belonging to hardwoods and conifers. Both of these species i.e. *Glomus mosseae and G. aggregatum were* recorded in all the seasons in the study area (Table 2). Hamikumar and Bagyaraj (1988) studied the effect of annual seasons on mycorrhizal colonization and sporulation by native mycorrhizal fungi and found out the maximum colonization and sporulation during winter, (November to January). Present study also showed maximum number of species in their higher frequency (53.57 percent) during winter season. However, summer months i.e. April to June were unfavourable for the proliferation of AM fungi.

Table 3 clearly shows that *Glomus aggregatum, G. etunicatum, G. geosporum, G. mosseae, G. acaulospora, Acaulospora scrobiculata, A. spinosa, Gigaspora margarita, G. rosea, G. gigantea* and *Scutellospora pellucida* were the 10 species, which were categorized as cosmopolitan species. Nine species viz., *Glomus deserticola, G. diaphanum, G. intraradices, G. macrocarpurn, G. invermium, G. monosporum, Acaulospora delicata, Sclerocystis microcarpum* and *S. rubiformis* were second in rank in distribution. While *Glomus albidum, G. dimorphicum, G. microaggregatum, G. occultum, G. G. reticulatum, Acaulospora longula, A. foveata, Gigaspora gilmorei* and *Scierocystis sinuosa* were represented the third category in order of occurrence in different sites.

Table 3. Occurrence of AM fungi in different site in study area

S.N.	Name of AM species	Poly bag	Nursery Bed	Plantation site	Percentage of frequency of species in study area (%)
1	*Glomus aggregatum*	+	+	+	100.00
2	*Glomus albidum*	+	—	+	33.33
3	*Glomus deserticola*	+	+	—	66.66
4	*Glomus diaphanum*	—	+	+	66.66
5	*Glomus diamorphicum*	+	—	—	33.33
6	*Glomus etunicatum*	+	+	+	100.00
7	*Glomus geosporum*	+	+	+	100.00
8	*Glomus intraradices*	+	—	+	66.66
9	*Glomus macrocarpum*	+	+	—	66.66
10	*Glomus microaggrigatum*	—	+	—	33.33
11	*Glomus invermim*	+	—	+	66.66
12	*Glomus monosporum*	+	+	—	66.66
13	*Glomus mosseae*	+	+	+	100.00
14	*Glomus occultum*	—	+	—	33.33
15	*Glomus reticulatum*	—	—	+	33.33
16	*Acaulospora delicata*	+	+	—	66.66
17	*Acaulospora longula*	—	—	+	33.33
18	*Acaulospora foveata*	—	+	—	33.33
19	*Acaulospora scrobiculata*	+	+	+	100.00
20	*Acaulospora spinosa*	+	+	+	100.00
21	*Gigaspora gilmorei*	+	—	—	33.33
22	*Gigaspora margarita*	+	+	+	100.00
23	*Gigaspora rosea*	+	+	+	100.00
24	*Gigaspora gigantea*	+	+	+	100.00
25	*Scutellospora pellucida*	+	+	+	100.00
26	*Sclerocystis microcarpum*	—	+	+	66.66
27	*Sclerocystis sinuosa*	+	—	—	33.33
28	*Sclerocystis rubiformis*	+	+	—	66.66
	Percentage of frequency of species (%)	75	67.85	57.14	

The distribution of AM fungi as in Table 3 shows that they were in abundance in polybags. As moisture is the most important factor affecting the growth, occurrence and distribution of AM fungi, more favorable conditions of moisture were met in polybags regularly. In the plantation of teak dry conditions were prevailing due to dry deciduous conditions because of less species cover and unirrigated conditions. The number of AM species counted in this site was lower than those recorded in polybags. However, the position of AM fungi is medium in nursery beds. The reason behind this was due to less irrigation schedule, over crowding of plants and soil compactness in the nursery beds.

REFERENCES

Bakshi, B.K. (1980). The status and future of mycorrhiza research in India. In : Mikola P. (Ed.) *Tropical Mycorrhiza Research.* Clarenden Press, Oxford. pp 102-106

Daniel, B.A. and Sldpper, H.D. (1982). Methods for the recovery and quantitative estimation of propagules from soil. In. Schenck, N.C. (Ed.) Methods and principles of mycorrhizal research. *Amer. Phytopatho Soc. St. Paul.* pp 29-35.

De Alwis, D.P. and Abeynayake (1980). A survey of mycorrhizae in some forest trees of Sri Lanka. *Tropical Mycorrhiza Research* (Ed. by P. Mikola), Clarendon Press, Oxford. pp 187-202.

Gunatilleke, I.A.U.N. and Maheshwaran, J. (1987). Some microbiological in a tropical rain forest and deforested area in Sri Lanka. Regional Workshop on the Ecology and Conservation of the tropical humid forests of the Indomalayan Realm, Colombo, Sri Lanka (mimeo).

Harinikumar, K.M. and Bagyaraj, D.J. (1988). *Acid soil Research and Rehabilitation.* 2: pp 139-143.

Jamaluddin, Chandra, K.K. and Chaturvedi, P. (1997). Development of VA-mycorrhizal fungi in different bamboos in bambusetum. *Indian Phytopath.* 50 (4); pp 552-556.

Kandaswami, D., Rangarajan, M., Oblisami, G., Narayan, R. and Ponniah, C. (1987). Mycorrhizae occurrence and their effect on certain forest species. In : *Nat. Workshop on mycorrhiza, JNU,* New Delhi, 13-15 March.

Mohankamar, V. and Mahadeven, A. (1987). Ecological distribution of vesicular-arbuscular mycorrhizae in tropical forest In : Mycorrhizal Round Table, Proc. of a workshop, (Eds : A.K. Varma, A.K. Oak., K.G. Mukerjl, K.V.B.R. Tilak and Janakraj), IDRC - CRO- CIID, New Delhi, pp. 238-256.

Mohan, V., Neelam, Verma and Singh, Y. P. (1995). Distribution of VAM fungi in nurseries and plantations of Neem *(Azadirachia indica)* in Arid Zone of Rajasthan. *Indian Forester.* 11 121.

Mukerji, K.G. and Kapoor, A. (1986). *Forest Ecology and Management* 16: pp. 117-126

O' Keefe, D.M. and Sylvia, D.M. (1991). Mechanisms of the vesicular- arbuscular mycorrbizal plant-growth response In D.K. Arora *et al.* (Ed.) Handbook of applied mycology. Marcel Dekker, New York, pp.34-35.

Rano Jagpal and Mukerji, K.G. (1988). *Jour. Phytopath. Research.* I : 31-35

Thapar, H.S. and Khan, S.N. (1973). Studies on Endomycorrhiza in some forest species. Proc. *Indian Nat. Sci Acad. B39*: pp. 687-694

Thapar, H.S. and Khan, S.N. (1985). Distribution of AM-mycorrhizal fungi in forest soils of India. *Indian. J. For. 8*: pp. 5-7

Thapar, H.S.; Uniyal, K. and Venna, R. K. (1991). Survey of native AM fungi of sodic soils of Hariyana State. *The Indian Forester.* 117 (2).

Tupas, G.L. and Sajise, P.E. (1976). Mycorrhizal associations in some Savannah and reforestation trees *Kalikasas.* 5: pp. 235-240.

CHAPTER - 11

STARCH-HYDROLYSING ENZYMES OF THERMOPHILIC MOULDS

Sanjeev Kumar and T. Satyanarayana

Department of Microbiology, University of Delhi South Campus, Benito Juarez Road, New Delhi-110 021

INTRODUCTION

Biopolymers such as cellulose, starch, pectin, xylan and lignin constitute a major part of living matter. The capability to utilize these substances as carbon and energy sources is widely distributed among animals, plants and microorganisms. After cellulose, starch is the second most abundant polysaccharide produced by plants. A wide variety of bacteria, fungi and yeasts produce a large array of extra-cellular enzymes to degrade these substances in different environmental niches (Antranikian, 1992). The great diversity of polysaccharide-hydrolyzing enzymes have fulfilled the requirement of enzyme industry over the last few decades and also led to interest in further screening programmes to look for enzymes with novel properties. Starch is a major storage product of many economically important crops such as wheat, rice, maize, tapioca and potato.

Natural starch is insoluble and is present in plant cells in the form of microscopic granules. The size and shape of starch granules are characteristic of the plant. Heating in water weakens the hydrogen bonds that bind the granules and causes swelling and gelatinization. This macromolecule is composed of two high molecular weight components: Amylose (15-20%) and Amylopectin (75-85%). Amylose consists of a linear chain of glucose residues linked by α-1,4-glycocidic bonds. It is twisted into a helical coil with about six glucose residues per turn. It complexes with iodine and forms an intense blue colour complex. One end of the chain has a free C1 hydroxyl group and is called the reducing end. Amylopectin, on the other hand, is a branched polymer containing α-1,4 linked glucan with α-1,6 linked branch points at every 17-26 glucose residues.

Unlike amylose, amylopectin is relatively stable in aqueous solutions and it does not form compact aggregates. Due to the nature of the branched structure, the iodine binding power is less pronounced and this gives a red-violet colour. Amylose and amylopectin show different physical properties due to difference in their structure. Their ratios vary in different natural starches. The waxy starches contain no amylose whereas amylomaize starch contains upto 80% of amylose. In general, amylose to amylopectin ratios are in the range between 1:3 and 1:4. These variations in structure and composition of native starches give rise to differences in their degradability by various amylolytic enzymes.

Starch is the chief storage polysaccharide whereas cellulose is the structural component of plants. Both polysaccharides are composed exclusively of glucose units and are bound by glycosidic linkages. Starch is the principal food reserve in the plant kingdom and serves as a bulk nutrient and energy source in animal and human diets. The glucose residues in starch are linked in the alpha orientation i.e. α- glucan. On the other hand cellulose is composed of β-D glucose units. Due to the different anomeric conformations of these polymers, different enzymatic systems are required for their degradation. Glycogen is the counterpart of starch in animals. Starch can be extracted commercially from many raw materials including maize, wheat, barley, potato, rice, oat, cassava, sorghum and others (Antranikian, 1992; James and Lee, 1997).

A large-scale starch processing industry has emerged in the last century. In the past decades, we have seen a shift from the acid hydrolysis of starch to the use of starch-hydrolyzing enzymes in the production of maltodextrin, modified starches, and maltose, glucose and fructose syrups. Amylases constitute about 15% of the world's enzyme sales. Besides the use in starch hydrolysis, starch-hydrolyzing enzymes are also used in a number of other industrial applications, such as laundry and porcelain detergents and as anti-staling agents in baking.

STARCH - HYDROLYZING ENZYMES

Amylolytic enzymes (α-glucanases) hydrolyze the glycosidic linkages in various α-glucans. The action of these enzymes can be divided into two categories: Endoamylases and Exoamylases. Endoamylases: The dextrinogenic or liquefying amylases act randomly on the α–1,4 linkages and the branch points (α–1, 6 linkages) remain unattacked. Their action results in the formation of linear and branched oligosaccharides of varying chain lengths (dextrins). The α–amylases are a type of endoamylases. Exoamylases: The saccharifying or saccharogenic amylases hydrolyze polysaccharides from the non-reducing end successively resulting in short end products. One type cleaves each bond to produce solely α–glucose (glucoamylases) and another type breaks every alternate bond to produce maltose (β-amylases).

Several thermophilic moulds have been reported to secrete amylolytic enzymes and some of these have been purified and characterized (Jensen *et al.*, 1988; Fagerstrom and Kalkkinen, 1995; Mishra and Maheshwari, 1996; Duo Chan *et al.*, 1998; Parvinder Kaur and Satyanarayana, 2001). In this chapter an attempt has been made to review various aspects of amylolytic enzymes of thermophilic moulds including their potential applications.

α-Amylases (1,4-α-D-Glucan Glucanohydrolase, E.C.3.2.1.1.)

α-Amylases are extra-cellular enzymes, which hydrolyze α-1,4 glycosidic linkages of starch randomly and are capable of bypassing the branching points. They are widely distributed in microbes, which liberate linear and branched oligosaccharides of varying lengths as well as glucose from starch. Such branched oligosaccharides are also called α-limit dextrins. The end products have an α–configuration at C1 (Antranikian, 1992). The action of α–amylases is accompanied by a rapid loss of viscosity and a rapid reduction in iodine staining power due to the formation of oligosaccharides. α-Amylases are divided into two categories according to degree of hydrolysis of substrate. Saccharyfying α-amylases produces free sugars whereas liquefying α-amylases break down the starch polymer but do not produce free sugars. Liquefying α–amylases cause more rapid reduction in viscosity of starch pastes, whereas the saccharyfying α–amylases reduce the viscosity less rapidly relative to the amount of reducing sugars released. Commercial applications of α–amylases are numerous. Probably the largest volume is used for thinning of starch in the liquefaction process in the sugar, alcohol and brewing industries. They are also used in the desizing of fabrics, in the baking industry, in production of adhesives, pharmaceuticals, detergents, in sewage treatment and in animal feeds.

β-Amylases (1,4-α–D-Glucan Maltohydrolase, E.C.3.2.1.2)

β-Amylases are usually of plant origin, but some microbial producers are also known. This exoenzyme hydrolyzes the alternate α-1,4 linkage from the non-reducing end, causing inversion of the anomeric configuration of the liberated maltose to β, thus the name of β-amylase. This enzyme is incapable of bypassing α–1,6-linkages of branched substrates such as amylopectin. The degradation of such substrates is therefore, incomplete. Consequently, the end products are maltose and high molecular weight β-limit dextrins.

In some cases, microorganisms form similar exoacting enzymes, but instead of maltose, other oligosaccharides with defined size (maltotetrose, maltohexose etc.) are formed, the reducing ends of which have α–configuration. Food and beverage industries employ β-amylase to convert starch into maltose (Vihinen and Mantsala, 1989).

Glucoamylase (1,4-α-D-Glucan Glucanohydrolase, E.C.3.2.1.3)

Glucoamylase, also known as amyloglucosidase and χ–amylase, is a typical fungal enzyme. It attacks α–1,4 linkages of α-glucans from the non-reducing ends; and also α–1,6 linkages but at a very slow pace. It preferentially degrades polysaccharides with a high molecular weight. These enzymes hydrolyze starch to give glucose in theoretically 100% yield. The reaction rate decreases with the decreasing chain length of the dextrin substrate, and maltose is attacked most slowly (Gerhartz, 1990). Prolonged incubation time and high concentration of starch usually cause reversion of the reaction catalyzed by glucoamylase. This results in the formation of maltose, isomaltose and some oligosaccharides (Antranikian, 1992). Glucoamylases are commercially important in the hydrolysis of starch for the production of glucose and fructose syrups for use in confectionery and ethanol production (Vihinen and Mantsala,1989; Jensen and Olsen, 1999).

Cyclodextrin Glycosyl Transferase (CGTase) {1,4-α-D-Glucan 4-α- D – (1,4 α-D-Glucano)- Transferase (Cyclizing), E.C.2.4.1.19}

CGTase, an enzyme found only in bacteria, produces a series of α, β, χ cyclodextrins (rings made up of 6,7 and 8 glucose units respectively, bound by α–1,4 bonds) from starch, amylose and other polysaccharides. In addition, CGTases also catalyze coupling reaction by which the rings are opened and transferred to cosubstrates like glucose, maltose or sucrose. The enzyme also catalyzes disproportionation reaction in which one or more glycosyl units are transfered between linear oligosaccharides. Finally, a hydrolysis reaction is catalyzed by the enzymes producing dextrins.

α-Glucosidase (α-D-Glucoside Glucohydrolase, E.C.3.2.1.20)

α–Glucosidase hydrolyzes α–1,4-and/or α–1,6-linkages of saccharides, that are formed by the action of other amylolytic enzymes and liberates α–D glucose units from the non-reducing end. This appears to be the final enzyme involved in the breakdown of starch. Most enzymes show high affinity towards maltose and so they are referred to as maltases.

Isoamylase (Glycogen 6-Glucanohydrolase, E.C.3.2.1.68)

Isoamylase hydrolyses the α–1,6-linkages in polysaccharides like amylopectin, glycogen and branched dextrins in an exo fashion. However, it cannot degrade α–1,6 linkages of pullulan. Isoamylases are used to debranch starch in the production of glucose and maltose syrups (Vihinen and Mantsala, 1989).

Pullulanase

Pullulanases are the debranching enzymes that hydrolyze α–1,6-linkages of amylopectin and pullulan, but their activity on glycogen is very weak.

Pullulanase type I (α-dextrin 6 – glucanohydrolase, E.C.3.2.1.4)

It is a typical bacterial enzyme that is specific for α–1,6 linkages in amylopectin or it's degradation products and is unable to attack α–1,4-linkages in α–glucans. It attacks exclusively the α–1, 6-linkages in pullulan in a random endofashion and causes its complete conversion to maltotriose.

Pullulanase type II (amylase–pullulanase/amylopullulanase),

An enzyme distributed mostly among anaerobic bacteria, and is capable of hydrolyzing α-1, 4 linkages in addition to the branch points (α-1,6 linkages), causing the complete conversion of starch to small sugars without the requirement for other enzymes.

AMYLASE PRODUCING THERMOPHILIC MOULDS

Amylolytic enzymes are produced by an extremely wide variety of microorganisms. α-Amylase and glucoamylase are the most widely reported in microorganisms, whereas, β-amylase are generally of plant origin, and this has been reported in only a few microbes. Properties of these enzymes vary and are somewhat linked to the natural environment i.e. niche occupied by the producing organisms (Table 4). Degradation of starch usually requires cooperation of several enzymes because only a few enzymatic activities can degrade it wholly alone, and thus, starch-degrading microbes usually have several amylolytic activities. The enzymes of thermophilic microorganisms are of potential industrial importance due to their thermostability. There are many advantages in the use of thermostable enzymes in the starch processing industry, such as increased reaction rates and decreased contamination risk (Kelly *et al.*, 1986). High activity and stability at high temperatures are the requisite features for an ideal amylolytic enzyme.

Thermophilic microorganisms are, therefore, the first choice when looking for a new enzyme. The thermophilic moulds that are known to produce amylolytic enzymes are listed in Table 1.

Amylolytic enzymes are of widespread occurrence in fungi, which often produce several activities into the medium. Several thermophilic moulds i.e. *Humicola lanuginosa, H. griseus, Malbranchea pulchella, Rhizomucor pusillus, R. meihei* and others have been reported to produce extracellular amylolytic enzymes in both surface and submerged cultivations. From three thermophilic fungi viz., *Humicola grisea, Talaromyces emersonii* and *Thermomyces lanuginosus,* the amylolytic systems have been identified, and in all three cases the amylolytic enzymes contained α–amylase, glucoamylase and α-glucosidase (Jeckson and Seidman, 1985; Vanacker *et al.*, 1990; Bunni *et al.*, 1992).

Table 1. Thermophilic moulds which are known to produce amylolytic enzymes.

Sources	References
α-Amylase	
Chaetomium thermophilum	Fergus 1969; Flannigan & Sellars 1977
Humicola grisea var. *thermoidea*	Jackson and Seidman 1985
H. insolens	Gupta and Gautam 1992, 1995
H. lanuginosa	Arnesen *et al.*1998, Duo-Chuan *et al.* 1997, Jensen *et al.*1988, Jensen and Olsen 1992, Mishra & Maheshwari, 1996, Singh *et al.* 1996
H. stellatus	Fergus 1969, Subrahmanyam *et al.*1977
Malbranchea sulfurea	Gupta and Gautam 1992, 1995
Myceliophthora thermophila	Sadhukhan *et al.* 1992, 1993
Rhizomucor miehei	Thomas *et al.* 1984
R. pusillus	Kanlayakrit *et al.* 1987
Taloromyces emersonii	Bunni *et al.* 1992
Thermoascus aurantiacus	Vanacker *et al.* 1990
Thermomyces stellatus	Fergus 1969,
Torula thermophila	Fergus 1969, Subrahmanyam *et al.*, 1977
Scytalidium thermophilum	Roy *et al.*, 2000
Papulospora thermophilum	Adams, 1985
Glucoamylase	
Chaetomium thermophilum	Dey, 1978
Humicola grisea var. *thermoidea*	Allison *et al.*, 1992, Campos and Felix, 1995, Jackson and Seidman, 1985, Tosi *et al.*, 1993
H. insolens	Basaveswara *et al. 1979*
H. lanuginosa	Rao *et al.* 1979, Rao *et al.* 1981a, Duo-Chuan *et al.* 1998, Haansum *et al.* 1991, Jensen *et al.*, 1988, Mangallam *et al.* 1977, Mishra and Maheshwari, 1996, Singh *et al.*, 1986, Tamura *et al.*, 1981, Taylor *et al.*, 1978
Malbranchea pulchella	Rao *et al.*, 1979
Rhizomucor miehei	Thomas *et al.*, 1984
R. pusillus	Kanlayakrit *et al.*, 1987
Taloromyces dupontii	Tamura *et al.*, 1981
Thermomucor indicae-seudaticae	Parvinder Kaur and Satyanarayana, 2001
Thermoascus aurantiacus	Rao *et al.*,1979, Fergus, 1969
T. crustaceus	Vanacker *et al.*, 1990
Torula thermophila	Mangallam *et al.*, 1997
α-glucosidase	
Thermoascus aurantiacus	Obert, 1981
Malbranchea sulfurea	Gupta and Gautam, 1993

PRODUCTION OF AMYLASES

Solid state fermentation (SSF) involves the growth and fermentation by microorganisms especially fungi on moist, water insoluble, solid substrate in the absence of free flowing water. The water content is quite low with the necessary moisture existing in absorbed or complex form within the solid material and the microorganism is almost in contact with gaseous oxygen in the air. Lonsane and Ramesh (1990) have drawn the following general guidelines for SSF:

Inoculum preparation

Substrate preparation includes particle size, nutrient and pH

Sterilization

Inoculation to the moist medium

Incubation under optimum conditions for growth

Product extraction and drying of the solid substrate

Filtration, concentration and purification

Fungal enzymes (Fungal amylase–Taka–Koji) were extensively produced by SSF technique in United States, pioneered by Takamine (1914). The solid substrate not only supplied the nutrients to the culture but also serves as an anchorage for the microbial cells. The ideal solid substrate is one that provides all the necessary nutrients to the microorganism for their optimum function. A variety of substrate have been employed for the production of amylases: wheat bran (Pandey 1990, Jaleel *et al.*, 1992), wheat straw, Cassava (Soccol *et al.*, 1994), rice bran (Pandey, 1993) and Copra waste (Pandey, 1995).

The main advantages of SSF are the ease with which fungi can grow on complex natural solid substrates like agro-industrial residues without any chemical transformation. There are some very important factors (macromolecular structure, particle size and shape, porosity, particle consistency) that should be considered for SSF. Complexity of the solid substrate arises from the presence of macromolecular compounds like starch, cellulose, pectins, lignin and some small sugar molecules. Several agroresidues have been used for solid state cultivation of moulds and bacteria (Table 2).

Table 2. Different agro-residue used for solid state cultivation of fungi

Starchy residues	cellulosic residues	Pectin residues	Oily residues
Wheat bran	Wheat Straw	Citrus pulp	Mustured oil seedcake
Rice	Rice Stover	Citrus peel	Cotton oil seed cake
Cassava	Corn Stover	Apple pulp	Linseed seed cake
Corn	Sugar beet pulp	Sunheamp	
Banana meal	Feedlot waste	Peer pulp	
Buckwheat seeds	Wood		
Rye meal			

Fungi can not directly utilize these macromolecules, and therefore, they are induced by low molecular weight compounds to synthesize and secrete the enzymes to hydrolyze the macromolecules into smaller metabolizable compounds (Mitchell *et al.*, 1992). There are, however, certain limitations such as heat build up, difficulty in growth estimation, uncontrolled pH and moisture content, problems associated with scale up etc. Although this technique has been in practice since ancient times for some fermented foods, little progress has been made in this field as compared to SmF processes. Gowthaman *et al.* (2001) reviewed fungal solid state fermentation in detail highlighting the importance of nutritional factors in the solid substrate and the cultivation techniques like tray system, packed column reactor, rotating drum bioreactor, and several other configurations.

There has been a surge of interest in SSF in the last two decades for the production of variety of microbial products. *Thermomucor indicae-seudaticae* produced amylolytic enzymes optimally in solid state fermentation at 45°C with substrate to moisture ratio of 1:2.5 using wheat bran as the substrate (Bhattacharya, 1996). There was no production of the enzyme below aw of 0.92. The enzyme titres were high when 5/10 g wheat bran was taken in 250 ml Erlenmeyer flask; at high substrate level, the enzyme production declined. The secretion of amylolytic enzymes by *H. lanuginosus* in SSF commenced after two days, and it reached a peak in 4 days at 45°C (Chawla, 1996). The substrate to moisture ratio of 1:2.5 and aw in the range between 0.92 to 0.8 favored enzyme production; it declined steeply at aw of 0.75. The enzyme titres under optimal conditions in trays containing wheat bran were slightly lower than in flasks.

Submerged Fermentation

The thermophilic fungus *Thermomyces lanuginosus* (syn *Humicola lanuginosa*) has been shown to produce several amylases. Chadha *et al.* (1996) optimized the cultural conditions in submerged fermentation for the production of amylases by *T. lanuginosus*. This mould produced α-amylase (18.4 U/ml) and glucoamylase (11.2 U/ml) optimally, when the medium contained rice flour (2% w/v) and corn steep liquor as carbon and nitrogen sources at pH 5.5, 50 °C and 150 rpm in 72 h.

The α-amylase enzyme synthesis was higher when *Myceliophthora thermophila* D-14 was grown in culture medium incorporated with starch or other carbohydrates containing maltose units (Sadhukhan *et al.*, 1992). Maximum enzyme production was attained with 1% starch followed by a gradual decline with increase in concentration. Marked decrease in α-amylase synthesis occurred with the addition of glucose to the culture medium, and this was concentration dependent. The enzyme synthesis was resumed as soon as the glucose concentration fell below a critical level. The addition of cAMP did not eliminate the repressive effect of glucose, suggesting that extracellular α-amylase synthesis in *M. thermophila* D-14 was inducible and subject to catabolite repression (Sadhukhan *et al.*, 1992).

The production of extracellular starch hydrolyzing enzyme from *T. lanuginosus* was compared in four different cultivation systems (Jensen *et al.*, 1987), and the highest extracellular amylase production was attained in static cultures. The static conditions promoted synthesis of α–amylase, while glucoamylase synthesis was supported in shake flasks.

Rosenberg (1975) reported that nearly half of the thermophilic and thermotolerant species required yeast extract (0.01%) for growth. Yeast extract was employed as a nitrogen source for many thermophilic moulds. The organic nitrogenous sources supported conidial germination (Haasum *et al.,* 1991). Some inorganic nitrogen sources have also been used frequently without any negative impact. Sadhukhan *et al.* (1993) investigated the effect of different metal ions on amylase production by *Myceliophthora thermophila.* Ca^{2+} and Mn^{2+} favored amylase yield, where as Cu^{2+}, Cd^{2+}, Co^{2+}, Zn^{2+} inhibited growth as well as amylase production. The addition of Tween-80 enhanced secretion of α-amylase by 2.2 fold in *H. lanuginoisa* (Arnesen *et al.,* 1998).

Gupta and Gautam (1995) examined the effect of a detergent Triton X-100 on extracellular, mycelial and intracellular α-glucosidase in *Malbranchea sulfurea.* The extracellular fraction of the α-glucosidase was 90.7% and 40.4% in the presence and absence of Triton X-100 respectively. The data suggested that supplementation of the medium with the detergent caused partial solubilization of the cell-bound α-glucosidase and affected its release into the growth medium. *Thermomucor indicae-seudaticae* secreted glucoamylase optimally in a synthetic medium containing starch in shake flasks as well as in laboratory fermenter (Parvinder Kaur, 2000).

Rubinder *et al.* (2002) studied the effect of different parameters on amylase production by a mutant strain [III(51)] of *T. lanuginosus* using a Box-Behnken design. The regression models computed showed significantly high R squared value for α-amylase and glucoamylase activities, suggesting that they are appropriate for predicting relationships between corn flour, soybean meal and pH.

PURIFICATION AND CHARACTERIZATION

Isolation of a protein from the biological environment requires a series of purification steps, each step removing some of the impurities and bringing the product closer to the final specification (Kaul and Mattiasson, 1993). The choice of procedure for enzyme purification depends on their location. Initial processes including crude fractionation, clarification, concentration of crude enzyme using processes such as centrifugation, ultrafiltration (Grootegoed *et al.*, 1973, Tsvetkov and Emanuilova, 1989) and salt precipitation (Yamane *et al.*, 1973; Krishnan and Chandra, 1983). The concentrate is then further purified using high resolution techniques based on chromatographic and electrophoretic separations. It is designed to remove aggregates, degradation products and to prepare a solution suitable for the final formulation of the purified enzyme (Kaul and Mattiasson, 1993). Fungal amylases are extracellular enzymes, which have pH optimum near neutrality or in the acidic range. Temperature optima ranged between 25 and 60°C. Properties of fungal α–amylases are quite similar to bacterial enzymes. They are usually less thermostable than bacterial enzymes, but have a wider range of pH stability. Most enzymes were found to require Ca^{2+} for activity and/ or stability (Vihinen and Mantsala, 1989). Properties of amylases from thermophilic moulds are summarized in Table 3.

Table 3. Properties of amylolytic enzymes of thermophilic moulds

Sources	Forms	Molecular mass (Da)	Optimum pH	Optimum Temp. (°C)	References
α-Amylase					
Rhizopus sp	2	64,000	4.0-6.0	60-65	Siqueria *et al.* 1997
Scytalidium sp.	1	87,000	6.5	50	Odibo *et al.* 1992
Thermomyces lanuginosus	1	—	4.6-5.2	60	Jensen and Olsen, 1992
Glucoamylase					
Humicola lanuginosa	2		4.9-6.6	65,70	Taylor *et al.* 1978
Thermomyces lanuginosus	1	37000		70	Rao *et al.* 1981b
Thermomucor indicae-seudaticae			7.0	50	Parvinder Kaur & Satyanarayana, 2001
α-Glucosidase					
Thermoascus auraniticus	4	140-14500			Bedino *et al.* 1983a,b
Malbranchea sulfurea	1	110000	4.8	60	Gupta & Gupta, 1993

α-Amylase

Maarel *et al.* (2002) have reviewed properties and applications explaining three-dimensional structures and X-ray crystallography in detail of α-amylase family. A large number of amylases have been purified and characterized from mesophilic and thermophilic fungi and from bacterial sources that show varying physico–chemical properties. These include pH and temperature optima, thermostability, effect of substrate concentration and effect of various cations. Azevedo *et al.* (2000) isolated *Trichoderma harzianum,* that produced substantial amounts of α-amylase. The purified α-amylase had a molecular mass of about 68700 Da showing maximal activity against soluble starch at pH 4.0 and 60 degrees C. The K_M and V_{max} values were 3.5 mg ml^{-1} and 1.67 mg min^{-1} of reducing sugar. The enzyme also hydrolyzed glycogen, amylopectin, maltotriose, and maltotetraose, but not pullulan or cellobiose. The purified amylase exerted a discrete hydrolytic effect on the C. *perniciosa* cell wall in vitro as observed by scanning electron microscopic analysis. The cations Fe^{3+}, Al^{3+}, Zn^{2+}, and Cu^{2+} inhibited the purified amylase, while Mn^{2+} considerably stimulated the activity, and Ca^{2+}, Mg^{2+}, and Co^{2+} did affect the enzyme activity.

The α–amylase of *Thermomyces lanuginosus* has a temperature and pH optima of 60°C and 4.0-5.5, respectively (Jensen and Olsen, 1992), while for *Humicola insolens,* these are 50°C and 6.0 (Ogundero, 1979). *Rhizomucor pusillus* α-amylase had pH optima of 3.5–4.0 and temperature optima of 65-70°C (Adams, 1990). α-Amylases are normally regarded as calcium metallo enzymes and the Ca^{2+} is required for catalytic activity as well as for stability. Ca^{2+} exhibited a stimulatory effect on amylase activity of *T. lanuginosus* (Jensen and Olsen, 1992) but inhibited the enzyme of *Aspergillus oryzae* (Kundu and Das, 1970). The stability of α–amylase in presence of EDTA varied greatly according to its origin. The negative role of EDTA was reported by Srivastava *et al.* (1980), Brown *et al.* (1990) and Ivanova *et al.* (1993). An α-amylase of 45 0000 Da was purified using fluidized bed chromatography from the thermophilic fungus *Scytalidium thermophilum,* the isoenzyme binding to Con A was stable for at least 3 h at 80°C (Roy *et al.*, 2000), Ca^{2+} played an important role in stabilizing the enzyme activity, A-agarose further improved higher thermal stability. A comparison of properties of microbial α-amylases are summarized in Table 3.

Glucoamylase

Fagerstrom and Kalkkinen (1995) characterized glucoamylase of *Trichoderma reesei* having pH and temperature optima of 5.5 at 70°C. The enzyme was slightly inhibited by the action of β-cyclodextrin, and no effect with EDTA and metal ions. K_M and V_{MAX} values were 0.11 mg ml^{-1} and 28.5 s^{-1} respectively. The amino acid sequencing showed about 60% identity with extensively studied *A. niger* glucoamylase. The molecular mass of glucoamylases ranged between 45000 to 77000 Da (Romanelli *et al.*, 1975; Moloney *et al.*, 1985; Jensen *et al.*, 1988; Duo-Chaun *et al.*, 1998). The thermostability was almost same in all the cases.

T. lanuginosus produced both α-amylase and glucoamylase, and they did not exhibit any synergistic effect like those of *Aspergillus sp.* (Abe *et al.*, 1988). Glucoamylase of *T. lanuginosus* effected up to 76% conversion of soluble starch or raw potato starch to glucose in 24 h, indicating that it was independent from any catabolite repression (Mishra and Maheshwari, 1996).

A thermostable glucoamylase of 57000 Da was purified from the thermophilic fungus *Thermomyces lanuginosus* (Rao *et al.,* 1981a). It was optimally active at 70°C. Amylolytic enzymes contain the starch-binding domain (SBD) positioned almost exclusively at the C-terminus. Detailed analysis of all available SBD sequences from 43 different amylases revealed its independent evolutionary behaviour with regard to the catalytic domains (Janecek and Sevcik 1999). The only known N-terminal SBD from *Rhizopus oryzae* glucoamylase is on the longest branch separated from all C-terminal SBDs. The 3-dimensional (3-D) structures of fungal glucoamylase and bacterial CGTase SBDs were compared and used to discuss the interesting SBD evolution. An informative data on structure and function of starch-degrading enzymes was provided by Gottschalk *et al.* (1998).

Toshi *et al.* (1993) produced a 2.5 to 3.0 fold higher glucoamylase production was observed when *Humicola grisea* var. *thermoidea* grown in a rich medium containing maltose as the principal carbon source after replacing starch. The *Humicola* glucoamylase of 63000 Da exhibited pH and temperature optima of 5.0 and 55°C. The enzyme had a single catalytic site for starch and maltose, as both were hydrolyzed by the same catalytic site at the same rate. Another form of glucoamylase (74000 Da) produced by the same fungus was remarkably insensitive to end product inhibition; it retained 65% activity in presence of 950 mM glucose (Campos and Eelix, 1995). The glucoamylase of *Thermomucor indicae-seudaticae* was optimally active at 60 °C and at pH 7.0 (Parvinder Kaur and Satyanarayana, 2001).

Muthukumaran and Dhar (1983) purified a glucoamylase from *Rhizopus nodosus* and was separated from the acid lipase by DEAE-cellulose chromatography at pH 8.0 It was purified by Concanavalin A Sepharose 4B affinity chromatography followed by CM-Sephadex chromatography 387 fold with 30.7% yield. The homogeneity of the enzyme was confirmed by polyacrylamide gel electrophoresis and immunological studies. The different physico-chemical properties of the enzyme were studied. The molecular weight of the enzyme was 71,000 Da. Ethylenediamine tetraacetic acid had no effect on the enzyme, whereas Hg^{2+} partially inhibited the enzyme activity. Tryptophan residues were found to be essential for the enzyme activity.

Takahashi *et al.* (1985) found different behaviour of glucoamylase isozymes from *Rhizopus* sp., Gluc1 (M.W. 74,000), Gluc2 (M.W. 58,600), and Gluc3 (M.W. 61,400). The pH optima for raw starch digestion were different, i.e. 4.5 for Gluc1 and 5.0 for both Gluc2 and Gluc3. All the enzymes digested raw starch almost completely but at quite different rates; Gluc2 and Gluc3, which lacked N-terminal portions of Gluc1, were 22 and 25 times less effective, respectively, for raw starch digestion than Gluc1. Of the three enzymes, only Gluc1 was tightly bound to raw starch. Binding of Gluc1 to raw starch occurred pH-dependently with a broad pH optimum of 4.5-5.5, but temperature and ionic strength affected it only slightly. The binding constant of Gluc1 for raw starch at pH 5.0 and 4 °C was estimated as 1.2 X 10(5) M-1. Fragment H (M.W. 16,700), presumably released from the N-terminal part of Gluc1, not only bound to raw starch itself but also inhibited the binding of Gluc1 to raw starch. pap-Gluc (M.W. 57,000) and chymo-Gluc (M.W. 64,000), which were papain- and chymotrypsin-modified Gluc1, respectively, and lacked the N-terminal portions of Gluc1. Gluc2 and Gluc3 resembled in raw starch binding as well as digestion.

Pazur *et al.* (1990) compared some properties of the glucoamylase from *Rhizopus niveus* and compared with *Aspergillus niger*. The enzymes from these organisms possessed the following common properties: quantitative conversion of starch to glucose, molecular weights in the range 95,500 to 97,500, and glycoprotein structures with many oligosaccharide side chains attached to the protein moieties of the enzymes. Differences in the glucoamylases exist in electrophoretic mobility, amino acid composition, nature of carbohydrate units, and types of glycosidic linkages. Lysine, threonine, serine, glutamic acid, tyrosine, and phenylalanine differed in the two glucoamylases by 25 to 50%. Antibodies directed against the two glucosamylases have been isolated by affinity chromatography and found to be specific for the carbohydrate units of glucoamylases.

IMPROVEMENT IN THE PRODUCTIVITY OF AMYLASES BY GENETIC AND MOLECULAR APPROACHES

Besides sexual recombination, genetic manipulations have made possible the introduction of DNA in unrelated species, so that the sequence can be integrated and expressed while stably transferred to the next generation. Introduction of a known sequence to be expressed in an eukaryotic cell is probably the main interest for genetic transformation of filamentous fungi. Even new advantage can be explored in fungal transformation using protoplasts (Case *et al.*, 1979; Tilburn *et al.*, 1983; Buxton and Radford, 1984; Pedersen, 1984; Orbach *et al.*, 1986). Considering that the activity of thermosatbility of *Humicola* glucoamylase depends upon a native glycosylation pattern of this enzyme (Olsen and Thomsen, 1991), Allison *et al.* (1992) developed a transformation system for this thermophile to achieve hyper production of *Humicola* glucoamylase. *Humicola grisea* var. *thermoidea* was transformed using bleomycin-phleomycin resistance gene linked to regulatory sequence from *A. nidulans*. Transformations were obtained at a frequency of 0.52 per μg of plasmid DNA. Vector DNA was integrated into the genome of *Humicola* in varying pattern and copy number, and the transformants were mitotically stable. Extra copies of *Humicola* glaI gene were introduced into genome of several *Humicola* strains by transformation; some transformants produced almost 3-fold more glucoamylase as compared to the parent strain.

Bunni *et al.* (1992) synthesized cDNA using the intact amylase mRNA as template, isolated from *Talaromyces ermersonii* following growth on starch containing media. The cDNA was cloned into *E. coli* expression vector system pUC18, and this was used to transform *E. coli*. One of the transformants were found to contain a plasmid pMH1 that haboured a 1.2 kb insert. After subcloning experiments, it was reported that the amylase phenotype was encoded for by this fragment. Northeren and Southeren blotting hybridization analysis confirmed that the cloned amy+ fragment was of *T. emersonii* origin.

Gupta and Gautam (1992) isolated mutants of *Malbranchea sulfurea* that were less sensitive to catabolite repression. Mutants were selected after screening for amylase in the presence of translator inhibitor actinomycon D and puromycin. Puromycin immediately inhibited amylase synthesis, where as, actinomycin D arrested enzyme synthesis after 48-72 h of incubation. Glucose repression resistant mutants (100mM) were selected on non metabolizable glucose analogue (2-Deoxy glucose). Some mutants showed higher α-amylase and α-amyloglucosidase production as compared to the parent strain.

Rubinder *et al.* (2000) constructed a hyper-producing, catabolite-repression-resistant recombinant strain of *Thermomyces lanuginosus* by protoplast fusion using well-characterized, morphological, and 2-deoxy-D-glucose resistant markers. The fusant heterokaryons exhibited enhanced amylase activities as compared to the parent strain (T2). Diploids derived from heterokaryons segregated to stable haploid recombinant strains. In the haploid strain (Tlh 4q), approximately 5-fold higher specific activities of alpha-amylase and glucoamylase in the culture filtrate were observed as compared to the wild-type strain.

Table 4. Comparison of properties of microbial amylase

Properties	Bacteria		Fungi		Others	
	Liquefying	Saccharolytic	*H. grisea*	*M. sulfurea*	Endomycopsis	Oospora
Heat stability °C	65-90	55-70	55-60	55-65	35	50-70
pH stability	4.8-10.6	4.0-7.8	6.0	4.8	6.0-7.5	6.0-10.5
Optimum pH	5.4-6.0	4.8-5.2	6.0	4.8	5.4	5.6
Activity/ mg protein	1800	1190	170	—	760	970
Adsorption on starch	++	—	+	+	+	+
Action on starch Maximum hydrolysis Ratio (%)	35	70	—	—	90	37
Main product	Dextrin Maltose	Glucose, maltose, Maltotriose	Glucose	Maltose	Glucose	Dextrin Maltose
Action on Maltose	—	—	+	+	+	—
Action on Phenylmaltose	—		—	—	+	—

Thermomyces lanuginosus was subjected to three cycles of mutagenesis (UV/NTG) and a selection procedure to develop amylase-hyperproducing, catabolite-repression-resistant and partially constitutive strains. One of the selected derepressed mutant strain III(51), produced approximately 7- and 3-fold higher specific activity of alpha-amylase (190 U/mg protein) and glucoamylase (105 U/mg protein), respectively, as compared to the wild-type parent strain (Rubinder *et al.*, 2002).

INDUSTRIAL APPLICATIONS

Starch hydrolyzing enzymes are among the most important enzymes that are produced on industrial scale. An important application is in the production of various sugar syrups. The manufacture of sugars from starch is a multistage process, that involves different microbial enzymes in successive steps (Fig. 1). The variation of parameters (pH and temperature) in this process cause many handicaps in the starch industry including extensive use of ion exchangers for salt removal (used for pH adjustments), reaction reversal and lower yields due to increased time consumed during cooling. The discovery of various amylases that are active at high temperatures i.e. thermozymes that function optimally between 60°C and 125°C (Vieille and Zeikus, 1996), and in the narrow pH range will significantly lower the sugar production costs. Various applications of starch hydrolyzing products are listed in Table 5.

Table 5. Properties and industrial applications of hydrolyzed starch products

Type of syrup	DE	Composition	Properties	Applications
Low-DE Maltodextrins	15-30	1-20 D-glucose 4-13 maltose 6-22 maltotriose 50-80 higher oligomers	Low osmolarity	Clinical feed formulations, raw materials for enzyme saccharification, thickness, filters, stabilizers, glues, pastes
Maltose syrups	40-50	16-20 D-glucose 14-44 maltose 36-43 higher oligomers	High viscosity Reduced crystal-lization, moderately sweet	Confectionery, soft drinks, brewing and fermentation, jams, jellies, ice cream, sauces
High-maltose syrups	48-55	2-9 D-glucose 48-55 maltose 15-16 maltotriose	Increased maltose content	Hard confectionary, brewing and fermentation
High-DB syrups	50-68	23-35 D-glucose 40-48 maltose	Increased moisture holding, increased sweetness, reduced viscosity, higher fermentation	Confectionary, soft drinks, brewing and fermentation, jams, sauces
Glucose syrups	96-98	95-98 D-glucose 1-2 maltose 0.5-2 isomaltose	Commerical liquid ``dextrose"	Soft drinks, carmel, backing, brewing and fermentation, new materials
Fructose syrups	98	48 D-glucose 52 D glucose	Alternative indust-rial sweetner to sucrose	Soft drinks, jams, sauces, yoghurts, canned fruit

Maltodextrins: The degree of starch hydrolysis is expressed as dextrose equivalents (DE). The DE values for starch and dextrose or glucose are 0 and 100 respectively. The DE values for the limit dextrin are in between 3 and 30 depending on the enzyme saccharifcation.

Glucose manufacturing: The dextrins are saacharified with thermostable glucoamyase in order to produce high conversion of glucose syrup. The syrup either can be converted to crystallized glucose or can be converted to high fructose syrup (HFS) with the action glucose isomerase. For the HFS the glucose concentration should be very high up to 95%. The sugar syrups have found wider applications in the manufacturing of soft drinks, dairy products, baked products and canned foods.

Maltose syrup: In the industrial process small amounts of maltose and isomaltose are produced. Syrups with a higher maltose content can be produced by aiding α-amylase. Maltose syrups are being used in confectionary, soft drinks, jams, ice creams etc., for the production of food grade maltose and high purity maltose for intravenous injection.

Cyclodextrins: Cyclodextrins are at present used in the food industry to remove unwanted taste and odours and for stabilization of fragrances and colouring compounds. Cyclodextrin may have some applications in medicine as drug carrier.

Baking: A number of enzymes, mostly produced by bacteria and fungi, are commercially available for use by the baking industry. These include α-amylase, β-amylase and protease, glucoamylase, pentosanase and phytase (Ledford and Ledford 1992). α- and β-amylases produce dextrins and maltose, respectively.

Textile Desizing: in fabric manufacture, the wrap is highly stressed mechanically during weaving. Application of starch paste, or sizing, reinforces the wrap and prevent loss of string by friction, as well as generation of static electricity on the string. Continuous desizing procedures that depends on the stability of α-amylase at high temperatures are a principal approach to the removal of sizing. After being prewashed in hot water at 90°C for about 10 s, the cloth is passed one or more times through a bath containing 0.5-1% w/v solution of α-amylase, a 0.5% w/v solution of $CaCl_2$ (stabilization of the enzyme), and a wetting agent (0.2% w/v nonionic surfactant) at temperature between 60-80°C for about 20 s each time. The cloth is then rinsed.

FUTURE PROSPECTS AND CONCLUSIONS

Present studies on thermophilic moulds gained importance by the prospects of finding new fungi capable of secreting high levels of thermostable enzymes with a long shelf life. Thermostability and long shelf life are the desired characteristics for commercial purposes in the industries. The extracellular enzymes of thermophilic moulds are appreciably thermostable and are comparable to bacterial enzymes. Amylases are the most important enzymes used for industrial purposes and are considered very useful for biopharmaceutical applications as a useful tool in medicine and clinical chemistry. Since thermostable amylases are useful in starch hydrolysis, it is desirable to isolate and screen thermophilic fungi for α-amylases and glucoamylases to further improve the process. Utilization of agro-residues for the production of these valuable products should be emphasized. It is worthwhile to design novel strains for hyper production of amylases, and to improve thermostability using protein engineering. The forgoing account indicates that a few attempts have been made to understand genetic and molecular aspects of thermophilic moulds that produce starch-hydrolyzing enzymes, and many such attempts are expected in the near future.

REFERENCES

Abe, J., Nakajiima, K., Nagano, H., Hizkuri, S. and Obata, K. (1988). Properties of the raw starch digesting amylase of *Aspergillus* sp. K-27: a synergistic action of glucoamylase and α-amylase. *Carbohydr. Res.* 175: 85-92

Adams, P.R. (1985). Amylase and growth characteristics of *Papulaspora thermophilia. Mycopathologia.* 90(2): 81-83

Adams, P.R. (1990). Mycelial amylase of thermophilic species of *Rhizomucor, Humicola* and *Papulaspora. Mycopathologia* 112: 35-37.

Allison, D.S., Rey, M.W., Berka, R.M., Armstrong, G. and Dunncoleman, N.S. (1992). Transformation of the thermophilic fungus *Humicola grisea* var. *thermoidea* and overproduction of *Humicola* glucoamylase. *Curr. Genet.* 21: 225-229

Antranikian, G. (1992). Microbial degradation of starch. In: Microbial Degradation of Natural Products. Winkel, G. ed., VCH, Weinhein (Germany). pp. 27-51

Arnesen, S., Erikson, S.H., Olsen, J. and Jensen, B. (1998). Increased production of α-amylase from *Thermomyces lanuginosus* by the addition of Tween-80. *Enz. Microbial. Technol.* 23: 249-252.

Azevedo, A.M., De Marco, J.L. and Felix, C.R. (2000). Characterization of an amylase produced by a *Trichoderma harzianum* isolate with antagonistic activity against *Crinipellis perniciosa,* the causal agent of witches' broom of cocoa. *FEMS Microbiol. Lett.* 15; 188(2): 171-175

Bedino, S., Testore, G. and Obert, F. (1983a). Multiple forms of intracellular α-glucosiadse from *Thermoascus aurantiacus* Miehe: Purification and properties. *Ital. J. Biochem.* 32: 371-384.

Bedino, S., Testore, G. and Obert, F. (1983b). Mechanism of action and structural properties of an intracellular α-glucosidase from *Thermoascus aurantiacus* Miehe. *Ital. J. Biochem.* 32: 408-417.

Bhattacharya, R. (1996). Amylolytic enzymes of *Thermomucor inducae-seudaticae.* M. Sc. Dissertation, University of Delhi South Campus.

Brown, S., Constantino, H. and Kelly, R. (1990). Characterization of amylolytic enzymes associated with the hyper thermophilic archaebacterium *Pyrococcus furiosus. Appl. Environ. Microbiol.* 56: 1985-1991.

Bunni, L., Coleman, D.C., Mc Hale, L., Hackett, T.J. and Mc Hale, A.P. (1992). cDNA clone and expression of a *Talaromyces emersonii* amylase encoding genetic determinant in *E. coli. Biotech. Lett.* 14: 1109-1114.

Buxton, F.P. and Radford, A. (1984). Transformation by integration in *Podospora auserina.* 1. Methodology and phenomenology. *Mol. Gen. Genet.* 196: 339-344.

Campos, L. and Felix, C.R. (1995). Purification and charcterization of a glucoamylase from *Humicola grisea. Appl. Environ. Microbiol.* 61: 2436-2438.

Case, M.E., Schweizer, M., Kushner, S.R. and Giles, N.H. (1979). Efficient transformation of Neurospora crassa by utilizing hybrid plasmid DNA. *Proc. Natl. Acad. Sci. USA.* 76: 5529-5263.

Chadha, B.S., Singh, S., Vohra, G. and Saini, H.S. (1996). Shake culture studies for the production of amylases by *Thermomyces lanuginosus. Mycologia.* 67: 608-615.

Chawla, M. (1996). Amylolytic enzymes of the thermophilic mould *Humicola lanuginosa.* M. Sc. Dissertation, University of Delhi South Campus.

Dey, D.F. (1978). A thermophilic glucoamylase from *Cepahalosporium eichhorniae. Cur. Microbiol.* 1: 181-184

Duo-Chaun, L., Yi-Jung, Y., You-Liang, P. and Chong-Yao, S. (1998). Purification and characterization of extracellular glucoamylase from the thermophilic *Thermomyces lanuginosus. Mycol. Res.* 102: 568-572.

Duo-Chaun, L., Yi-Jung, Y., You-Liang, P., Chong-Yao, S., Peijin, Z. and Yicum, H. (1997). Purification and properties of a thermostable α-amylase from the thermophilic fungus *Thermomyces lanuginosus. Weishengwu Xuebao.* 37: 107-114.

Fagerstrom, R. and Kalkkinen (1995). Characterization, subsite mapping and partial amino acid sequence of glucoamylase from the filamentous fungus *Trichoderma resei. Biotechnol. Appl. Biochem.* 21: 223-231.

Fergues, C.L. (1969). The production of amylase by some thermophilic fungi. *Mycologia* 61: 1171-1177.

Flannigan, B. and Sellars, P.N. (1977). Amylase, beta glucosiadse and beta xylosidase activity of thermotolerant and thermophilic fungi isolated from barley. *Trans. Brit. Mycol. Soc.* 69: 316-317.

Gerhartz, W. (1990). Enzymes in Industry–Production and Applications. V.C.H. Weinheim.

Gottschalk, T.E., Fierobe, H.P. Mirgrodskaya, E., Clarker, A.J., Tull, D., Sigurskjold, B.W., Christensen, T., Payre, N., Frandsen, T.P., Juge, N. McGuire, K.A., Cottaz, S., Roepstroff, P., Driguez, H., Williamson, G. and Svensson, B. (1998). Structure, function and protein engineering of starch-degrading enzymes. *Biochem Soc. Trans.*26: 198-204.

Gowthaman, M.K., Chundakkadu, K. and Moo-Young, M. (2001). Fungal solid state fermentation – an overview. In Applied Mycology and Biotechnology Vol. 1. Khachatourians GG and Arora DK eds., Agriculture and Food production, Elsevier Science, The Netherlands, pp. 305-352

Grootegoed, J.A., Lauweres, A.M. and Heinen, W. (1973). Separation & partial purification of extracellular amylase & protease from *Bacillus caldolyticus. Arch. Microbiol.*90: 223-232.

Gupta, A.K. and Gautam, S.P. (1992). Catabolite repression of amylase synthesis of the thermophilic mould *Malbranchea sulfurea* and isolation of repressed mutants. In DAE symposium on Molecular biology of microorganism, NCL, Pune, India. Jan. 28-30, 1992. pp. 101-106.

Gupta, A.K. and Gautam, S.P. (1993). Production of extracellular alpha–glucosidse from the thermophilic fungus *Malbranchea sulfurea. Cryptogamic Botany* 3: 303-306.

Gupta, A.K. and Gautam, S.P. (1995). Effect of detergent Triton X-100 on growth and alpha glucosidase production by the thermophilic fungus *Malbranchea sulfurea,* following protoplast fusion. *W. J. Microbiol. Biotechnol.* 11: 193-195.

Haasum, I., Erickson, S.H., Jensen, B. and Olesen, J. (1991). Growth and glucoamylase production by a thermophilic fungus *Thermomyces lanuginosus* in a synthetic medium. *Appl. Micobiol. Biotechnol.* 34: 656-660.

Ivanova, V.N., Dobreva, E.P. and Emanuilova, E.I. (1993). Purification and characterization of a thermostable α-amylase from *Bacillus licheniformis. J. Biotechnol.* 28: 277-289.

Jackson, L. E. and Seidman, M. (1985). Enzymatic hydrolysis of granular starch directly to glucose, European Patent Application 0171218.

Jaleel, S.A., Srikanta, S. and Karanth, N.G. (1992). Production of fungal amyloglucosidase by solid state fermentation-influence of some parameters. *J. Microb. Biotechnol.* 7(2): 1-8.

James, J.A. and Lee, B.H. (1997). Glucoamylase: Microbial sources, industrial application and molecular Biology-A review. *Food Biochem.* 21: 1-52.

Janecek, S, and Sevcik, J. (1999). The evolution of starch-binding domain. *FEBS Lett.* 30; 456(1):119-25

Jensen, B. and Olsen, J. (1992). Physicochemical properties of α-amylase from thermophilic fungus *Thermomyces lanuginosus. Enz. Microb. Technol.* 14: 112-116.

Jensen, B and Olsen, J. (1999). Amylases and their inudstrial potential. Thermophilic moulds in Biotechnology. Johri, B.N., Satyanarayana, T. and Olsen, J. eds. Kluwer Academic Publishers, Netherlands, pp 115-137

Jensen, B., Olesen, J. and Allermann, K. (1987). Effect of media composition on the production of extracellular amylases from the thermophilic fungus *Thermomyces lanuginosus. Biotech. Lett.* 9: 313-316.

Jensen, B., Olsen, J. and Allermann, K. (1988). Purification of extracellular amylolytic enzymes from the thermophilic fungus *Thermomyces lanuginosus. Can. J. Microbiol* 34: 218-223

Kanlayakrit, W., Ishimatsu, K., Nakao, M. and Hayashida, S. (1987). Characteristics of raw starch digesting glucoamylase from thermophilic fungus *Rhizomucor pusillus. J. Ferment. Technol.* 65: 379-385.

Kaul, R. and Mattiasson, B. (1993). Secondary purification. *Bioseparation* 3:1-26.

Kelly, C.T., Giblin, M. and Fogarty, W.M. (1986). Resolution, purification & characterization of two extracellular glucohydrolases, α-glucosidase & maltase of *Bacillus licheniformis. Can .J. Microbiol.* 32 : 342-347.

Krishnan, T. and Chandra, A. K. (1983). Purification & characterization of α-amylase from *Bacillus licheniformis* CUMC 305. *Appl. Environ. Microbiol.* 46: 430-437.

Kundu, A.K. and Das, S. (1970). Production of amylase in liquid culture by strain of *Aspergillus oryzae. Appl. Microbiol* 19:598-603

Ledford, R.A. and Ledford, M.W. (1992). Encyclopedia of Microbiology, Vol. I, Academc Press, London. pp. 401-408.

Lonsane, B.K. and Ramesh, M.V. (1990). Production of a bacterial thermostable α -amylase by solid state fermentation. A potential tool for achieving economy in enzyme production and starch hydrolysis. *Adv. Appl. Microbiol.* 35: 1-56.

Maarel, M.J.E.C., Veen, B.V.D., Uitdehaag, J.C.M., Leemhuis, H. and Dijkhuijen, L. (2002). Properties and applications of starch converting enzymes of the α-amylase family. *J. Biotechnology.* 94: 137-155.

Mangallam, S., Subrahamanyam, A. and Gopalkrishna, K.S. (1977). Prelimenry observations on the production of amyloglucosidase by thermophilic fungus. *Curr. Sci.* 46: 16.

Mishra, R.S. and Maheshwari, R. (1996). Amylases of thermophilic fungus *Thermomyces lanuginosus*: Their purification, properties, action on starch and response to heat. *J. Biosci.* (Bangalore) 21: 653-672.

Mitchell, D.A., Targonski, Z., Rogalsk, J. and Leonowic, A. (1992). In Solid substrate cultivation. Doelle, H.W., Mitchell, D.A. and Rolz, C.E. eds., Elsevier App. Sci. London. pp. 29.

Moloney, A.P., McCrae, S.I., Wood, T.M. and Coughlan, M.P. (1985). Isolation and charcaterization of 1, 4-Dglucan glucanohydrolase of *Talaromyces ermersonii. Biochem J.* 225: 365-372.

Muthukumaran, N. and Dhar, S.C. (1983). Purification and properties of a glucoamylase fraction from the culture filtrate of *Rhizopus nodosus. Ital. J. Biochem.* 32(4): 239-53.

Obert, F. (1981). Suuna α-glucosidase del mycelio di *Thermoascus auranticus. Allionia.* 27 : 79-84.

Odibo, F.J.C., Okafor, N., Tom, M.U. and Oyeka, C.A. (1992). Purification and characterization of b-amylase from *Hendersonula toruloidea. Starch/starke.* 44:192-195.

Ogundero, V.W. (1979). Amylase and Celulase activities of tobacco products in Nigeria. *Mycopathalogia* 69:131.

Olsen, O. and Thomsen, K.K. (1991). Improvement of bacterial b-glucanase thermostability by glycosylation. *J. Gen. Microbiol.* 137:579-585.

Orbach, M.J., Porro, E.B. and Yanofsky, C. (1986). Cloning and characterization of the gene for β-tubulin from benomyl-resistance mutant of *Neurospora crassa* and its use as a dominant selectable marker. *Mol. Cell. Biol.* 6: 2452-2461.

Pandey, A. (1990). Improvements in solid state fermentation for glucoamylase production. *Biological wastes* 34: 11-19.

Pandey, A. (1993). The production of glucoamylase by *Aspergillus niger* NCIM 1245 in solid state fermentation. *Process Biochem.* 28: 305-309.

Pandey, A. (1995). Glucoamylase research: an overview. *Starch/Starke.* 42: 439-445.

Parvinder Kaur (2000). Production and characterization of glucoamylase of thermophilic mould *Thermomucor indicae-seudaticae.* M.Sc. Dissertation. University of Delhi South Campus.

Parvinder Kaur and Satynarayana, T. (2001). Partial purification and characterization of glucoamylase of thermophilic mould *Thermomucor indicae-seudaticae. Ind. J. Microbiol.* 41: 195-199.

Pazur, J.H., Liu B.L. and Miskiel, F.J. (1990). Comparison of the properties of glucoamylases from *Rhizopus niveus* and *Aspergillus niger*. *Biotechnol. Appl. Biochem.* 12(1): 63-78.

Pederson, S. (1984). *Escherichia coli* ribosome translate in vivo with variable rate. *EMBO J.* 3: 2895-2898.

Rao, B.V., Sastri, N.V.S. and Subba Rao, P.V. (1979). A thermostable glucoamylase from the thermophilic fungus *Thermomyces lanuginosus*. *Biochem. J.* 193: 379-387.

Rao, B.V., Sastri, N.V.S. and Subba Rao, P.V. (1981a). Purification and characterization of a thermostable glucoamylase from the thermophilic fungus *Thermomyces lanuginosus. Biochem. J.* 193: 379-387

Rao, V.B., Sastri, N.V.S. and Subba Rao, P.V. (1981b). Thermal stabilization of glucose oxidase and glucoamylase by physical entrapment. *Biochemical J.* 193: 389-394.

Romanelli, R.A., Houstan, C.W. and Barnett, S.M. (1975). Studies on thermophilic cellulolytic fungi. *Appl. Microbiol.* 30: 276-281.

Rosenberg, S.L. (1975). Temperature and pH optima for 21 species of thermophilic and thermotolerant fungi. *Can. J. Microbiol.* 21: 1535-1540.

Roy, I., Sastry, M.S., Johri, B.N. and Gupta, M.N. (2000). Purification of alpha-amylase isoenzymes from *Scytalidium thermophilum* on a fluidized bed of alginate beads followed by concanavalin A-agarose column chromatography. *Protein Exp. Purif.* 20(2):162-8

Rubinder, K, Chadha, B, Singh, N, Saini, H and Singh, S. (2002). Amylase hyper-production by deregulated mutants of the thermophilic fungus *Thermomyces lanuginosus. J. Ind. Microbiol. Biotechnol.* 29(2) : 70-74

Rubinder, K, Chadha, B.S., Singh, S. and Saini, H.S. (2000). Amylase hyper-producing haploid recombinant strains of *Thermomyces lanuginosus* obtained by intraspecific protoplast fusion. *Can J Microbiol.*46 (7):669-73

Sadhukhan, R., Roy, S.K. and Chakraborty, S.L. (1993). Immobilization of α-amylase trom *Myceliophthora thermophila* D-14 (ATCC 48104). *Enz. Microbial. Technol.* 15: 801-804.

Sadhukhan, R., Roy, S.K., Raha, S.K. and Chakraborty, S.L. (1992). Induction and regulation of α-amylase synthesis in cellulolytic fungus *Myceliophthora thermophila* D 14 (ATCC 48 104). *Ind. J. Exp. Biol.* 30: 692-696.

Singh, C.B., Arvind, S.S. and Singh, S.H. (1996). Mutagenesis of hyperproduction of extracellular amylases by *Thermomyces lanuginosus. Acta Microbiol. Polonica.* 45:31-36

Siqueira, E.M.D., Mizuta, K. and Giglio, J.R. (1997). a-amylase of *Pycnoporus sanguineus. Mycol. Res.* 101: 188-190.

Soccol, C.R., Iloki, I., Marin, B. and Raimbault, M. (1994). Comparative production of α-amylase, glucoamylase & protein enrichment of raw & cooked cassava by *Rhizopus* strains in submerged & solid state fermentations. *J. Food Sci. Technol.* 31(4) : 320-323

Srivastava, R.A.K., Nigam, J.N., Pillai, K.R. and Baruch, J.N. (1980). Purification, properties & regulation of amylases produced by a thermophilic *Bacillus* sp. *Indian J. Experimental Biol.* 18: 972-976.

Subramanyam, A., Mehrotra, B.S. and Thirumalchar, M.J. (1977). *Thermomucor*, a new genus of Mucorales. *Georgia J. Sci.* 35: 106.

Takahashi, T., Kato, K., Ikegami, Y. and Irie, M. (1985). Different behavior towards raw starch of three forms of glucoamylase from a *Rhizopus* sp. *J. Biochem. (Tokyo).* 98(3): 663-71.

Takamine, J. (1914). Enzymes of *Aspergillus oryzae* and the applications of its amylolytic enzyme to the fermentation industry. *Ind. Eng. Chem.* 6: 824 - 828

Tamura, M.K., Shimizu, M.H. and Tago, M. (1981). Highly thermostable glucoamylase and process for its production. United State patent No. 4, 247, 637

Taylor, P.M., Napier, E.J. and Fleming, L.D. (1978). Some properties of a glucoamylase produced from the thermophilic fungus *Humicola lanuginosa. Carbohydr. Res.* 61 : 301- 308.

Thomas, A.A., Drozd, T.P. and Walter, T.J. (1984). Amylase activity in fungal rennets and whey protein concentration powder. *J. Agric. food. Chem.* 32: 825-828.

Tilburn, J., Scazocchio, C., Taylor, G.G., Zabicky-Zissima, J.H., Lockington, R.A. and Davis, R.W. (1983). Transformation by integration in *Aspergillus nidulans. Gene.* 26 : 205-221.

Toshi, L.R.O., Terenzi, H.F. and Jorge, J.A. (1993). Purification and characterization of an extracellular glucoamylase from thermophilic fungus *Humicola grisea* var. *thermoidea. Can. J. Microbiol.* 39 : 846-852.

Tsvetkov, V.T. and Emanvilova, E.I. (1989). Purification & properties of heat stable α-amylase production from *Bacillus brevis. Appl. Microbiol. Biotechnol.* 31 : 246-248.

Vanacker, P., Bacle, B., Vidal, G. and Lacoste, L. (1990). Recherche de nouvelles activites saccharifiantes thermostable chez les champignons filamenteux. *Can. J. Microbiol.* 36 : 625-630.

Vieille, C. and Zeikus, J.G. (1996). Thermozymes: identifying molecular determinants of protein structural and functional stability. *TIBTECH.* 14 : 183-189

Vihinen, M. and Mantsala, P. (1989). Microbial amylolytic enzymes. *Crit. Rev. Biochem. Mol. Biol.* 24 : 321-418

Yamane, K., Yamaguchi, K. and Maruo, B. (1973). Purification and properties of a cross reacting material related to α-amylase and biochemcial comparison with the parent α-amylase. *Biochim. Biophy. Acta.* 295 : 323-340.

CHAPTER - 12

PRODUCTION AND APPLICATION OF FUNGAL XYLANASES

Naveen Kango and P.C. Jain

Department of Applied Microbiology and Biotechnology,
Dr. Hari Singh Gour Vishwavidyalaya, Sagar- 470 003 (M.P.)

INTRODUCTION

Microbial enzymes are gaining importance because of their varied applications in industries, bioprocesses and solid waste management. Numerous explorations of the microbial world for identifying enzyme producers have yielded a number of potential strains for production of amylases (Fogarty and Kelly, 1980; Fogarty, 1983; Crabb and Mitchingson, 1997; Singhai and Jain, 2003), cellulases (Durand *et al.*, 1988; Gilbert and Hazlewood, 1993) and xylanases (Dekker and Richards, 1976; Davis *et al.*, 1992; Haltrich *et al.*, 1996; Sunna and Antranikian, 1997; Archana *et al.*, 1999; Beg *et al.*, 2001) and the heated habitats rich in organic matter have been identified as reservoirs for such potential strains (Singhai and Jain, 1992; Jain, 1999; Kango *et al.*, 2003b).

An annual yield of plant biomass on land is estimated to be more than 120 billion tones comprising approximately 50% in the complex forms of lignocelluloses (Smith, 1996). For complete decomposition of lignocellulosics three major groups of enzymes are required which include lignolytic enzymes, cellulases and hemicellulases. The hemicellulose fraction of plant includes arabans, mannans, galactans, glucans, fructans and xylans of which xylans constitute a major fraction (up to 30%). Degradation of xylan is significant in plant biomass recycling, upgradation of fodder and pulp manufacture. Development of an efficient process for xylan hydrolysis using microbial xylanases opens up novel prospects for xylan utilization where the hydrolysate can be utilized for the production of fuel, SCP and other value-added products

(Woodward, 1984). A number of microorganisms including fungi, actinomycetes and bacteria are known to elaborate xylanolytic enzymes but there has been an increasing interest in use of fungi for xylanase production because of their greater potential to produce higher levels of these enzymes. The present chapter highlights production and properties of fungal xylanases and their applications.

XYLANS : OCCURRENCE AND STRUCTURE

Xylans are major hemicelluloses of plant cell wall (Northcote *et al.*, 1989). Xylans, having backbone of β-1,4 glycosidically linked xylose units with arabinose, acetyl and methyl glucuronic acid side groups are reported to be present in all terrestial plants including hardwoods, softwoods and grasses (Timmel, 1964; Whistler and Richards, 1970) while β-1,3 linked xylose backbones have been witnessed in marine algae (Dekker and Richards, 1976). Xylans from annual plants are more heterogenous than those from woody plants. Xylans with both sorts of linkages are reported from some seaweeds such as *Palmeria palmata* (Barry and Dillon, 1940). Unsubstituted xylan, composed of only D-xylose residues is considered to be a homoxylan e.g. esparto grass (*Stipa tenacissima*) xylan and tobacco stalk xylan containing 95-96% xylose (Chanda *et al.*, 1950; Eda *et al.*, 1976). Xylan content in hardwoods is 15-30% of total hemicelluloses while it is upto 7-10% in softwoods (Sjostrom, 1993). The xylan of dicots is essentially an acetylated methyl glucuronoxylan in which 4-O-methyl glucuronic acid residues are the main substituents at C-2 in approximately one in ten of the β-D-xylopyranose units, via α-1,2-glycosidic linkage while about 10-70% of xylopyranose units are acetylated at C-2 or C-3 positions (Puls and Schuseil, 1993). The xylan of softwood is mainly arabino-4-O-methyl glucuronoxylan. In addition to 4-O-methyl glucuronic acid, it also contains α-arabinofuranosyl units which are α-1,3 linked to the xylan backbone (Fig. 1). Graminaceous xylans are arabinoxylans having predominantly arabinofuranosyl units as substituents at C-3 positions (Wilkie, 1979).

XYLANASES

As the structure of xylan suggests, its complete breakdown needs hydrolysis of different sorts of covalent linkages and thus several enzymes are required. Xylanolytic enzymes include endoxylanase, β-xylosidase, arabinofuranosidase, acetyl xylan esterase, glucuronidase etc. (Fig. 2). Endoxylanase (EC 3.2.1.8) is understood to be the chief enzyme hydrolyzing the backbone of xylan macromolecule and therefore plays a central role in xylan degradation and other enzymes act upon side groups of the heteropolysaccharide (Colowick and Kaplan, 1988; Christov and Prior, 1993). The major xylanolytic enzymes are as follows:

Endoxylanase (β-1,4- D- Xylan Xylanohydrolase, EC 3.2.1.8)

These enzymes hydrolyse β-1,4 glycosidic bonds of xylan skeleton in a random fashion yielding a mixture of linear and branched xylooligosaccharides. Endoxylanases have been observed to follow different modes of action liberating a mixture of xylose and xylooligos (McCleary and Matheson, 1981). The degree of hydrolysis depends upon kind, solubility and degree of polymerism of xylan and kind and degree of substitution. Degree of substitution on xylan side chains has been found to create steric hindrance for endoxylanases. Endoxylanases have been reported to catalyze intermolecular transglycosylation in presence of high concentrations of xylooligomers (Komerlink *et al.*, 1993). Endoxylanases are the best characterized and revealed enzymes among xylanolytic enzymes. Several of these have been purified, cloned and solved for structure (Jeffries, 1996; Kulkarni *et al.*, 1999; Natesh *et al.*, 1999; Teixeira *et al.*, 2001).

β- Xylosidase (β-1,4-D-Xylan Xylohydrolase, EC 3.2.1.37)

This enzyme catalyzes hydrolysis of xylobiose or xylooligosaccharides to form D-xylose as the end product. Its efficiency of hydrolysis increases with decrease in polymerization.

Arabinofuranosidase (α-L-arabinofuranosidase, EC 3.2.1.55)

This enzyme catalyzes hydrolysis of α-L-arabinofuranosidic linkage in xylan with arabinose substitution (arabinoxylan).

Glucuronidase (α- 4-O-methyl-D-glucuronidase, EC 3.2.1.1)

This enzyme catalyzes cleavage of α-4-O-methyl-D-glucuronic side chains from the xylopyranose substituted with α-4-O-methyl-D-glucuronic acid

Acetyl xylan esterase (EC 3.1.1.6)

They bring about hydrolysis of ester linkages between xylopyranose and acetyl group, releasing acetic acid form acetylxylans.

ASSAY OF XYLANOLYTIC ENZYMES

Xylans being heteropolysaccharides are acted upon by a repertoire of enzymes hydrolyzing a variety of linkages. Detection and measurement of xylanolytic enzymes thus imply need of sensitive and specific substrates (Table 1). Detection of xylanase activity in plates can be achieved by incorporating xylan in agar or agarose and subsequently staining it by congo red (Srivastava and Srivastava, 1993) or by using synthetic substrates, RBB-xylan (Remazol brilliant blue dyed xylan), 4-methyl umbelliferyl-beta-D-xylobioside (MU-X2) and 5-bromo-3-indolyl-beta-D-xylobioside (BI-X2) in place of xylan (Kaneko *et al.*, 2000). Kaneko *et al.*

(2000) could detect a *Streptomyces* xylanase more sensitively with MU-X2 as substrate whereas it was not detected by RBB- xylan and BI-X2.

Table 1. Substrates used for assay of Xylanolytic enzymes

Enzyme	Substrates
Endoxylanase	4-methyl umbelliferyl- beta-D- xyobioside (MU- X2) 5- bromo- 3-indolyl – beta-D- xyobioside (BI- X2) Remazol brilliant blue-xylan (RBB- xylan) Xylans (Oat spelt, beech wood, birchwood, Larchwood), Carboxymethylxylan
β- Xylosidase	p- or o- nitrophenyl β- D- xylopyranoside (ONP- or PNP- xyl), 2,5- and 3,4- dinitrophenyl β- xylobiosides (DNP- xyl) 4-chloro- 2- nitrophenyl β- D- xylopyranoside (CNP- xyl)
α- L- arabinofuranosidase	pnp- α-L – arabinofuranoside
α- glucuronidase	2-O- (4- methyl -α-D- glucopyranosyluronic acid) xylobiose 2-O- (4-methyl α-D – glucopyranosyluronic acid)-xylitol 4-nitrophenol –2- (4- O- mehtyl - α- D- glucopyranosyl)-beta -D-xylanopyranoside
Acetyl xylan esterase	pnp- acetate, naphthyl acetate or 4-methyl umbelliferyl acetate, acetyl xylans

Xylans are water insoluble thus endoxylanase activity is quantified by measuring reducing sugar (xylose) liberated from suspensions of purified xylans (oat spelt xylan, beech wood xylan, birch wood xylan). However, while using these substrates, substituent non-xylose sugars liberated by the activity of other xylanolytic enzymes may also get into account. Carboxymethylation of xylan modifies substituent sugars and carboxymethylxylan (CMX) can be used to assess endoxylanase activity exclusively (Khowala *et al.*, 1988). The use of larchwood xylan, being unbranched and partially soluble in water, has also been suggested (Reilly, 1981). Certain soluble substrates such as RBB-xylan, 4-O-methyl glucuronoxylan from alderwood and o-hydroxyethyl xylan may also be used for assaying endoxylanase activity (Biely *et al.*, 1984). Considerable variations in endoxylanase activity are observed towards xylans from different sources.

β-xylosidase and α-L-arabinofuranosidase are assayed by measuring liberation of p-nitrophenol from their respective synthetic p-nitrophenol substrates (Ziser *et al.*, 1995; Li *et al.*, 2000). Activity of α-glucuronidase is assayed by using substrates such as 2-O-(4-methyl-α-D-glucopyranosyluronic acid) xylobiose and 2-O-(4-methyl-α-D-glucopyranosyluronic acid)-xylitol and the products are analyzed by TLC or HPLC. Recently, Biely *et al.* (2000) have reported use of 4-nitrophenol-2-(4-O-methyl-α-D-glucopyranosyl)-β-D-xylanopyranoside as an excellent substrate for assay of α-glucuronidase activity. Acetyl esterases can be assayed using pnp-acetate, naphthyl acetate or 4-methyl umbelliferyl acetates.

However, to ascertain xylan specific acetyl xylan esterases (AXE), liberation of acetate from acetylxylans must be measured (Bailey *et al.*, 1992).

XYLANASE PRODUCTION

Fungi are well-recognized producers of a wide variety of extracelluar hydrolases. Several fungi have been investigated for the production of xylanolytic enzymes and many of them are known to produce high levels of these. During last three decades several fungi having varied physiological characteristics have been investigated for xylanase production by different workers. These include mesols (John *et al.*, 1979; Durand *et al.*, 1984; Belancic *et al.*, 1995), thermotolerants (Mendicuti Castro *et al.*, 1997), thermophiles (Matsuo *et al.*, 1975; Kitpreechavanich *et al.*, 1984; Anand *et al.*, 1990; Dusterhoft *et al.*, 1997), alkalophiles (Raj and Chandra, 1995; Kang *et al.*, 1996; Taneja *et al.*, 2002) and plant pathogens (Christakopoulos *et al.*, 1997; Gomez-Gomez *et al.*, 2002). Due to natural competence of moulds to survive and colonize on different lignocellulosics, most of the studies carried out for xylanase production employ natural lignocellulosic substrates under different modes of cultivation. A review of literature indicates that xylan, cellulose, solkafloc, beet pulp and β-methyl xyloside are certain other substrates used by workers for production of xylanolytic enzymes (Table 2). Co-produced cellulases tend to be a major limitation for application of fungal xylanase in pulp bleaching. Attempts have been also made for obtaining cellulase-free xylanases for specific use in pulp bleaching. This has been accomplished by some workers either by using a strain that does not produce cellulase or by using xylan as substrate so that cellulase production is not induced during the growth of fungi (Gomes *et al.*, 1993a).

Table 2. Production of xylanase by various fungal forms

Microorganism	Max. yield (IU/ ml)	Sp. act. (IU/ mg)	Temp (°C)	pH	Substrate	Culture mode	Culture age	Reference
Aspergillus flavipes	5.1	—	30	6.0	Xylan	LF	96 h	Mukhopadhyay *et al.*, 1997
A. fischeri	41.8	—	30	10	Wheat bran	LF	48 h	Raj and Chandra, 1995
A. fumigatus	10	—	45	—		SSF	4 d	Kitpreechavanich *et al.*, 1984
A. nidulans	40	—	37	10	Wheat bran	LF	144 h	Taneja *et al.*, 2002
A. niger	—	5484	—	—	Rice straw	SSF	5 d	Park *et al.*, 2002
A. niger	6	—	30	5.5	Xylan	SmF	120 h	John *et al.*, 1979
A. ochraceous NG - 13	120.0	—	28	6.5	xylan	LF	14 d	Biswas *et al.*,1990
A. tamarii	—	220	30	—	β- Methyl xyloside	LF	10 d	Simao *et al.*, 1997

Microorganism	Max. yield (IU/ ml)	Sp. act. (IU/ mg)	Temp (°C)	pH	Substrate	Cult-ure mode	Cul-ture age	Reference
Allescheria terrestris	266	—	45	6.5	Cellulose	LF	96 h	Grajek, 1987
Aureobasidium pullulans	398.0	620	28		Xylan	LF		Leathers *et al.*, 1986
Bipolaris sorokiniana	228 *	—	28	6.83	Barley leaf	LF	12 d	Peltonen *et al.*, 1994
Cephalosporium sp. *RYM 202*	—	7.8	30	10	Wheat bran	SmF	3 d	Kang *et al.*,1996
Cephalosporium sp.	18	—	28	10	Wheat bran	LF	96 h	Bansod *et al.*, 1993
Emericella nidulans NK-62	362	275	45	6.5	Wheat bran	LF	7d	Kango *et al.*, 1993c
Heterobasidium annosum	4.6 *	—	23	—	Xylan	LF	28 d	Maijala *et al.*, 1995
Humicola grisea var. *thermoidea*	.55	—	—	—	Wheat straw	LF	36 h	Rana *et al.*, 1996
Malbranchea pulchella var. *sulfurea*	18.4	—	42	6.5	xylan	LF	3–4 d	Matsuo *et al.*, 1975
M. pulchella var. *sulfurea*	726 $	—	45	—	Wheat straw hemicellulose	LF	7 d	Banerjee *et al.*,1995
Melanocarpus albomyces	22.5	—	40	6	Bagasse	LF	3-4 d	Maheshwari and Kamalam, 1985
Mucor pusillus	2	—	45	6.5	cellulose	LF	3 d	Grajek, 1987
Poria- medula panis	0.9	—	28	28	Wheat bran	LF	20 d	Milagres *et al.*, 2001
Theilavia terrestris	18.8	—	48	4.0	Solkafolk BW 200	LF	1.4 d	Merchant *et al.*, 1988
Thermoascus aurantiacus	—	705	55	—	Wheat bran	SSF	7 d	Alam *et al.*, 1994
Thermoascus aurantiacus	9.67	—	45	6.5	Sugar beet	SSF	6 d	Grajek, 1987.
Thermomyces lanuginous	9168 *	—	50	6.0		LF	5-6 d	Gomes *et al.*, 1993b
Thermomyces lanuginosus SSBP	—	4200	50	—	Oatspelt xylan	LF	5-7 d	Lin *et al.*, 1999
Thermomyces lanuginosus	—	1899.6	55	—	Wheat bran	SSF	7 d	Alam *et al.*, 1994
Thermomyces lanuginosus RT 9	26170 *	—	50	6.5	Xylan	LF	7 d	Hoq and Dekwer, 1995
Thermomyces lanuginosus MH 4	11618 *	—	50	6.5	Xylan	LF	7 d	Hoq and Dekwer, 1995

(* nkat/ ml; $ IU/ L; # IU/g; h-hours; d-days)

Among Aspergilli, *Aspergillus ochraceous* NG-13 (a mutant) has been reported to yield 120 IU/ ml of xylanase on xylan containing medium (Biswas *et al.*, 1990) while xylanase specific activity of 220 IU/ mg protein was noticed in cultures of *Aspergillus tamarii* (Simao *et al.*, 1997). Several strains of *Aspergillus niger* have also been studied for xylanase production (John *et al.*, 1979; Frederick *et al.*, 1985; Qy *et al.*, 1996). Recent investigations by Park *et al.* (2002) indicated production of 5484 IU/mg of xylanase from *A. niger* using rice straw as substrate under solid state fermentation (SSF). The results are of greater importance and suggest that many more fungal strains be explored for xylanase production. An alkalitolerant fungus *Aspergillus fischeri* has been reported to produce 41.8 IU/ml of alkalistable cellulase-free xylanase on wheat bran in shake culture (Raj and Chandra, 1995). Mendicuti Castro *et al.* (1997) found 2.5 fold enhancement in xylanase specific activity of thermotolerant *Aspergillus nidulans* when grown on wheat bran at 45°C while Taneja *et al.* (2002) have reported an alkalophilic *Aspergillus nidulans* KK-99 to produce an alkaline, thermostable xylanase (40 IU/ml) in a basal medium supplemented with wheat bran at pH 10. Many more species of *Aspergillus such* as *A. fumigatus, A. terreus, A. sydowii, A. kawachi, A. sojae* have also been found to secrete xylanolytic enzymes *in vitro* (Kitpreechavanich *et al.*, 1984; Hrmova *et al.*, 1989; Ito *et al.*, 1992; Ghosh and Nanda, 1994; Kimura *et al.*, 1995). In order to understand physiological nature of xylanase producing strain, Lemos *et al.* (2001) investigated influence of nitrogen sources on xylanase production and found increased levels of endoxylanase and β-xylosidase when sodium nitrate was used as N-source during SSF of milled sugarcane bagasse. Alkali tolerant strains of *Cephalosporium* have been reported to elaborate xylanase at an alkaline pH. *Cephalosporium* RYM 202 produced 7.8 IU/mg xylanase (specific activity) at 30°C in a 5 lit. fermentor in 3 days (Kang *et al.*, 1996) while *Cephalosporium* sp. produced 18 IU/ml of xylanase in shake culture in a time course of 4 days (Bansod *et al.*, 1993).

Several strains of *Trichoderma* have been characterized for production of xylanase (Defaye *et al.*, 1985; Durand *et al.*, 1988; Wong *et al.*, 1988). Gomes *et al.* (1992) reported 188.1 IU/ml xylanase activity from *Trichoderma viride*. Relatively much higher yield (960 IU/ml) of xylanase has been noticed from *Trichoderma reesei* (Bailey *et al.*, 1993). Production of very high titres of xylanase has been noticed in *Schizophyllum commune* (Haltrich *et al.*, 1992). Steiner *et al.* (1987) reported 1244 IU/ml of xylanase along with co-secreted cellulase (65 IU/ml) from *Schizophyllum commune*.

Like thermophilic prokaryotes, thermophilic moulds also elaborate thermostable bioactive molecules (Johri *et al.*, 1999). Several thermophilic fungal forms are reported to yield thermostable cellulases and amylases (Adams and Deploey, 1976; Jain and Singhai, 2001). Thermophilic strains of *Thermomyces lanuginosus* (=*Humicola lanuginosa*) have been found to produce high titres of thermostable xylanolytic enzymes (Anand and Vithayathil, 1990; Gomes *et al.*, 1993a; Alam *et al.*, 1994; Hoq *et al.*, 1994; Purkathofer and Steiner, 1995; Lin *et al.*, 1999). Chadha *et al.* (1999) have compared several strains of *T.*

lanuginosus for xylanase production and found *T. lanuginosus* BS1 to produce 1018.1 IU/ ml of xylanase while an induced colour mutant of the same fungus could produce 2506 IU/ml. Solid state fermentation of wheat bran by *T. lanuginosus* yielded 1899.6 IU xylanase per gram of fermented wheat bran (Alam *et al.*, 1994). While screening thermophilic isolates for xylanase production, Kango *et al.* (2003a) have observed 1418 IU/ml xylanase activity in the culture filtrates of *Thermomyces lanuginosus* NK-2 grown on Czapek's mineral salt solution containing wheat bran (2% w/v). Among other thermophiles *Thermoascus aurantiacus* has been shown to produce 67 IU/ml of xylanase on sugar beet under SSF while *T. aurantiacus* BJT 190 when grown on wheat bran could produce 705.7 IU/gram of infested bran (Alam *et al.*, 1994). *Malbranchea pulchella var. sulfurea, Talaromyces byssochlamydoides, Paecilomyces variotii, Thielavia terrestris, Sporotrichum thermophile, Melanocarpus albomyces* are certain other thermophilic moulds reported to produce various levels of thermostable xylanase using different lignocelluloses (Matsuo *et al.*, 1975; Yoshioka *et al.*, 1981; Krishnamurthy and Vithayathil, 1989; Gilbert *et al.*, 1993; Saraswat and Bisaria, 1997). *Rhizomucor pusillus*, a mucoraceous fungus was found to produce 998 IU $l^{-1}hr^{-1}$ of xylanase when grown on Mandel's medium at 55°C (Hoq and Deckwer, 1995).

Xylanase production has also been noticed in basidiomycetous fungi (Elisashvili *et al.*, 1999; Inglis *et al.*, 2000). Milagres *et al.* (2001) have studied xylanase production by two basidiomycetes, *Poria-medula panis* and *Wolfporia cocos* on wheat bran. Xylanase has also been characterized from certain plant pathogens such as *Fusarium oxysporum* (Gomez-Gomez *et al.*, 2002), *Chochliobolous carbonum* (Apel-Birkhold and Walton, 1996), Wood rot fungi, *Phanerochaete chrysosporium* (Khalil, 2002) and *Heterobasidion annosum* (Maijala *et al.*, 1995). Among yeasts, *Aureobasidium pullulans* has been reported to produce 398 IU/ ml of xylanase on xylan and β-methyl xyloside (BMX) containing medium (Leathers *et al*, 1986). BMX was also found to induce higher xylanase yield (25 IU/ml) in case of a xylan assimilating yeast, *Cryptococcus flavus* (Yasui *et al.*, 1984). Morosoli *et al.* (1992) have performed cloning and expression of a xylanase from *Cryptococcus albidus* in two other yeasts, *Saccharomyces cerevisiae* and *Pichia stipitis.*

Production of β-xylosidase, an important enzyme of xylanolytic consortia has been reported from several moulds incluidng *A. niger, A. pulverulentus, Trichoderma reesei, T. virde, T. koningii, Penicillium wortmanni, Neocallimastix frontalis, Chaetomium trilaterale* etc. (John *et al.*, 1979; Deleyn *et al.*, 1982; Poutanen and Puls, 1988; Herrmann *et al.*, 1997; Li *et al.*, 2000). Among thermophilic moulds, β-xylosidase has been characterized from *Humicola grisea var. thermoidea* and *Talaromyces emersonii* (Tuhoy *et al.*, 1993; De Almedie, 1995). Curotto *et al.* (1994) found *Penicillium janthinellum* to produce 85 IU/L of β-xylosidase on oat spelt xylan while Lemos *et al.* (2001) reported production of 3.5 IU/L of β-xylosidase by *Aspergillus awamori* on milled sugarcane bagasse under SSF.

Enzymes releasing L-arabinose from arabinose containing polysaccharides are of different nature and thus classified on the basis of their specificities (Kaji, 1984). Presence of arabinose side groups in arabinoxylans can restrict its enzymatic hydrolysis by creating steric hindrance.Enzymes cleaving α-L- arabinofuranosidic linkages have been produced by *A.niger, T. reesei, A. awamori,* (Rombouts *et al.*, 1988; Poutanen, 1988; Komerlink *et al.*, 1991). The *A. awamori* arabinosidase is shown to be very specific for arabinoxylan and does not release arabinose from arabinan (Komerlink *et al.*, 1991). Multiple forms of arabinosidases have been noticed e.g. *A. niger* produced three such enzymes, α-L-arabinofuranosidae A, α-L-arabinofuranosidase B and endo-α-1,5-arabinase with varied activities towards p-nitrophenyl arabinofuranoside (PNPA), oligo and polymeric substrates (Rombouts *et al.*, 1988). Luonteri *et al.* (1995) found *Aspergillus terreus* to produce three arabinosidases liberating upto 80% of arabinose from polymeric arabinoglucuroxylan while Filho *et al.* (1996) have observed presence of two such enzymes in *Penicillium capsulatum*. Arabinosidases have been found to release more of arabinose in presence of other xylanases (Poutanen, 1988; Sinha and Sengupta, 1995). Cloning and expression of the β-xylosidase and α-L-arabinofuranosidase of *T. reesei* in *Saccharomyces cerevisiae* has been successfully achieved by Margolles-Clark *et al.* (1996).

Presence of 4-O-methyl glucuronic acid side groups confers stability and has a limiting effect on hydrolysis of neighbouring xylosidic bonds (Puls *et al.*, 1987). The first acidophilic α-glucuronidase (107 kDa) was purified from *Thermoascus aurantiacus* by Khandke *et al.* (1989). Later on these have been characterized from *Trichoderma reesei, T. reesei* RUT-30, *Phanerochaete chrysosporium* etc. (Puls *et al.*, 1987; Siika-aho *et al.*, 1994; Castanares *et al.*, 1995).

During alkaline leaching of hemicelluloses most of the esterified side groups are removed by saponification and thus acetyl xylan esterases (AXE) are not required. However, while processing lignocelluloses in near neutral pH, acetyl esters remain intact and their removal is done with the help of these enzymes. Acetyl esterases have been detected in the cultures of *Aspergillus niger, Aureobasidium pullulans, Trichoderma reesei, T. viride, Schizophyllum commune, Penicillium purpurogenum* etc. (Biely *et al.*, 1988; Halgasova *et al.*, 1994; Linden *et al.*, 1994; Egana *et al.*, 1996). Sundberg and Poutanen (1991) showed presence of two acetyl xylan esterases in *T. reesei* having similar hydrolytic properties but different isoelectric points (pI). Total removal of acetic acid from acetylated beechwood xylan was possible with these enzymes and their activity was found increased in presence of other xylanolytic enzymes (Poutanen *et al.*, 1990).

PROPERTIES OF FUNGAL XYLANASES

Pulp bleaching being the chief application of xylanases, it is desirable that the enzyme remains optimally active at higher pH (alkaline) and temperature for long durations. Thus, while evaluating feasibility of enzymatic bleaching of pulp it is necessary to look for xylanase producer that elaborates alkalistable and thermostable xylanase. Xylans are heterogenous and considerably vary in their structure as according to source. Activity of same xylanase preparation has been reported to vary significantly with the xylan-type (such as arabinoxylan, beechwood xylan, oat spelt xylan etc). Purified xylans used in laboratory testing of xylanase are much more exposed for enzyme action as compared to the 'natural state' of xylan, where binding with lignin and cellulose fibrils offers hindrances for xylanase action.

Occurrence of multiple forms of xylanases with varied specificities is a characteristic of this class of enzyme. Several moulds have been found to elaborate xylanases with varying characteristics such as pH and temperature optima, Km, Vmax, pI etc. (Table 3). John *et al.* (1979) have reported of 5 endoxylanases in *A. niger* culture filtrates. Among alkali-tolerant, thermostable and cellulase-free xylanase producing fungi, *A. fischeri* produced a 31 kDa xyalnase with pH and temperature optima at 6 and 60°C, respectively (Raj and Chandra, 1996). *Cephalosporium* sp. RYM 202 has been reported to produce two forms of 35 kDa and 24 kDa having pH optima 7.5 and 8.0 respectively, while both exhibited a temperature optima of 50°C (Kang *et al.*, 1996). Another *Cephalosporium* sp. is described to produce two xylanases of 30 kDa and 70 kDa with pH and temperature optima as 8 and 40°C, respectively (Bansod *et al.*, 1993).

Table 3. Characteristics of xylanases of different fungal origin

S.N.	Microorganism	Molecular weight (kDa)	Optimum		Stability (pH)	Km (mg/ ml)	Vmax (mM/min /mg)	Reference
			pH	Temp (°C)				
1	*Aspergillus fischeri*	31	6	60	5-9.5	4.88	58.8 x 10	Raj and Chandra, 1996
2	*A. oryzae*	21	6	58		5.1	123	Kimura *et al.*, 2000
3	*A. kawachii*	26-35	2-5.5	50-60	1-10			Ito *et al.*, 1992
4	*A. nidulans*	34	6	56	4.0-6.7	0.97	1091	Fernandez-Espinar *et al.* , 1994a
5	*A. niger*	31-50	4.0-6.5					John *et al.*, 1979
6	*A. ochraceous*	48.0	6.0	50	5-10	$1x10^{-3}$	19.6	Biswas *et al.*, 1990
7	*A. sojae*	32.7, 35.5	5, 5.5	60, 50	5-8, 5-9			Kimura *et al.*, 1995
8	*A. sydowii*	30	5.5	60				Ghosh & Nanda, 1994
9	*Acrophialophora nainiana*	17	6	50		.731		Ximenes *et al.*, 1999

Continue...

S.N.	Microorganism	Molecular weight (kDa)	Optimum		Stability (pH)	Km (mg/ ml)	Vmax (mM/min /mg)	Reference
			pH	Temp (°C)				
10	*Aureobasidium pullulans*	25	4.4	54		7.6	2.650	Li and Ljungdahl, 1994
11	*Bipolaris sorokiniana*	30	5.5	70	5 -10			Peltonen *et al.,* 1994
12	*Cephalosporium* sp. *RYM 202*	24, 35	7.5 – 8.0	50	5.5 – 12.0	3.18, 2.20	1.402x10, 1.756x10	Kang *et al.,* 1996
13	*Cephalosporium* sp.	30, 70	8	40	8-9.5	.15		Bansod *et al.,* 1993
14	*Claviceps purpurea*	21.5, 33.8						Giesbert *et al.,* 1998
15	*Cochliobolus carbonum*	22						Apel-Birkhold *et al.,* 1996
16	*Cryptococcus albidus*	48	5	25		5.7		Morosoli, 1985
17	*Fusarium oxysporum*	20.8, 23.5	6	60, 55	7 - 10	9.5, 8.45	.41x 10^{-3} .37 x 10^{-3}	Christakopolous *et al.,*1997
18	*Geotrichum candidum*	60 – 67	4	50	3- 4.5	-	-	Radionova *et al.,* 2000
19	*Humicola insolens*	6, 21	6 - 6.5	55 - 60	6 - 9	-	-	Dusterhoft *et al.,* 1997
20	*Humicola. grisea* var. *thermoidea*	13, 95	6, 6-7	60	-	3.3, 2	2 x 10, 1.8 x 10	Rana *et al.,* 1996
21	*Malbranchea pulchella* var. sulfurea *No. 48*		6.0 – 6.5	70				Matsuo and Yasui, 1985
22	*Melanocarpus albomyces*	22.9-34.3						Saraswat & Bisaria, 2000
23	*Melanocarpus albomyces*	38, 24	6.6, 5.6	65	5 - 10	.33, 1.69	311, 500	Prabhu & Maheshwari,1999
24	*Paecilomyces varioti*	20	4	50		49.5		Kelly *et al.,* 1989
25	*Penicillium purpurogenum*	33 23	7 3.5	60 50	6-7.5 4.5-7.5			Belancic *et al.,* 1995
26	*Penicillium chrysogenum*	35	6	40	4 – 8			Haas *et al.,* 1993
27	*Penicillium* sp. 40	25	2		2 –5			Kimura *et al.,* 2000
28	*Scizophyllum commune*		5.5	52	5- 8.9			Steiner *et al.,* 1987
29	*Talaromyces emersonii*	54.2, 131						Grajek, 1987
30	*Theilavia terrestris*	26	6.5	50				Gilbert *et al.,* 1993

Continue...

S.N.	Microorganism	Molecular weight (kDa)	pH	Temp (°C)	Stability (pH)	Km (mg/ ml)	Vmax (mM/min /mg)	Reference
31	*Thermomyces lanuginous SSBP*	23.6	6.5	70-75	5-12	3.26	6.3	Lin *et al.,* 1999

S.N.	Microorganism	Molecular weight (kDa)	Optimum		Stability (pH)	Km (mg/ ml)	Vmax (mM/min /mg)	Reference
			pH	Temp (°C)				
32	*Thermomyces lanuginous BS1*	25, 54						Chadha *et al.*, 1999
33	*Thermomyces lanuginosus DSM*	25.5	7	60-70	5-9	7.3		Cesar and Mrsa, 1996
34	*Thermomyces lanuginosus*	32	5.1	80				Gomes *et al.*, 1993a
35	*Trichoderma reesei*	20 19	5.5 4 – 4.5	45 40	3-8.5, 2.5- 8.5	3-6.8 14.8-2	2.3	Tenkanen *et al.*, 1992
36	*Trichoderma harzianum*	20, 29	5	50		.58	0.106 x 10^{-3}	Tan *et al.*, 1985
37	*Trichoderma harzianum*	18	5	45				Silveria *et al.*, 1999

Humicola insolens secreted two xylanases of molecular weight equal to 6 kDa and 21 kDa Dusterhoft *et al.* (1997), (while *H. grisea var. thermoidea* secreted one low molecular weight xylanase (13 kDa) and one high molecular weight xylanase (73 kDa) (Rana *et al.*, 1996). Strains of *Thermomyces lanuginosus*, a potent xylanolytic fungi elaborating high levels of xyalanse enzyme have been shown to produce single form of xylanase in the range of 20-30 kDa (Lin *et al.*, 1999) while *T. lanuginosus* BS1 strain has been reported to secrete two xylanases of 25 kDa and 54 kDa (Chadha *et al.*, 1999). Several species of Aspergilli have also been reported to produce xylanases of distinct molecular weights (Table 3). Belancic *et al.* (1995) found *Penicillium purpurogenum* to produce several xylanases, two of which were characterized as xyl A (33 kDa, pI 8.6) and Xyl B (23 kDa, pI 5.9). Some characteristics of β-xylosidases and other side group cleaving enzymes are summarized in Table 4.

Table 4. Properties of some β-xylosidases and side group cleaving enzymes

Enzyme	Organisms	Mol.wt (kDa)	pH optima	pI	Reference
β-xylosidase					
	Trichoderma koningii G-39	104	3.5-4.0	4.6	Li *et al.*, 2000
	T. reesei RUT 30	100	4	4.7	Hermann *et al.*, 1997
	T. reesei	103	4.0	4.7	Poutanen and Puls, 1988
Arabinofuranosidase					
	Aspergillus niger	83.0	3.4	3.3	Rombouts *et al.*, 1988
		67.0	4.0	3.5	Kaneko *et al.*, 1993

Continue....

	A. nidulans	36.0	5.5	4.3	Fernandez-Espinar *et al.*, 1994b

Enzyme	Organisms	Mol.wt (kDa)	pH optima	pI	Reference
	A. sojae	34.3	5.0	3.9	Kimura *et al.*, 1995
	Termimyces clypeatus		5.5		Sinha and Sengupta, 1995
	Trichoderma reesei	53.0	4.5	7.5	Poutanen, 1988
	Penicillium capsulatum	64.5	4.0	4.1	Filho *et al.*, 1996
		62.5	4.0	4.5	
α- Glucuronidase					
	Thermoascus aurantiacus	107.0	4.5		Khandke *et al.*, 1989
	Phanerochaete chrysosporium	112.0	3.5	4.6	Castanares *et al.*, 1995
	T. reesei	91	4.5-6.0	5.0-6.2	Siikaaho *et al.*, 1994
Acetyl xylan esterase					
	A. niger	30.0	5.0-5.5	3.0-3.2	Komerlink *et al.*, 1993
	Schizophyllum commune	31.0	7.7	3.4	Biely *et al.*, 1988
	T. reesei	34	5-6	7.0	Sundberg and Poutanen,
		34	5.9-6.0	6.8	1991; Poutanen *et al.*, 1990

Molecular Biology of Fungal Xylanases

According to similarities of conserved sequences of catalytic domains, xylanases have been classified into two families of glycosyl hydrolases, F/10 and G/11 (Henrissat and Bairoch, 1993). The relatedness of the enzymes within these families can be ascertained either by pairwise alignments of the protein sequence or by basic local alignment search tool (BLAST) to make out sequence similarities (Jeffries, 1996). Xylanases belonging to different families show characteristic differences in mode of action and substrate specificities (Biely *et al.*, 1997). The two xylanases produced by *Trichoderma reesei* belong to family G/11 (Torronnen *et al.*, 1992). Two family G/11 xylanases have also been observed in *Chaetomium gracile* (Yoshino *et al.*, 1995) while *Fusarium oxysporum* xylanase F3 has been assigned to F/10 (Christakopoulos *et al.*, 1997). Xylanase from thermophilic *Thermoascus aurantiacus* is grouped into F/10 (Srinivasa *et al.*, 1991). In a study comparing the efficacy of xylanases in bleaching, xylanases from family G/11 were found to be more effective at increasing final paper brightness (Clarke *et al.*, 1997)

Possibilities of improvement in their hydrolyzing efficiency using protein engineering and recombinant DNA technology have triggered many expeditions in structure elucidation and cloning of fungal xylanases. Understanding the structure of xylanases and its correlation with function is important in achieving desirable manipulations. To predict pH and thermal stability and substrate specificity, exact details of 3D structures at very high resolution are required (Davies and Henrissat, 1995). The 3-dimensional detailed structures of *Thermoascus aurantiacus* xylanase revealed that it has a TIM barrel fold (Bα_8) and its thermostability is a result of hydrophobic packing, favourable interactions between charged side chains and presence of prolines at the N-terminal helices (Leggio *et al.*, 1999, Natesh *et al.*, 1999). Teixeira *et al.* (2001) have illustrated anisotropically refined structure of *T. aurantiacus* xylanase I (TAXI) interpreting thermal properties of the enzyme. Schmidt *et al.* (1998) have given the structure of *Penicillium simplicissimum* xylanase. Studies on changes in xylanase conformation with respect to alkaline pH indicated that higher pH optimum is influenced by presence of Asn residue (Krengel and Dijkstra, 1996) while Nath and Rao (1998) explained it on the basis of microenvironment and spatial distribution of charged residues on the enzyme.

Table 5. Commercial xylanase preparations and their manufacturers

Manufacturer	Trade name	Application
Clariant, U.K.	Cartazyme HS 10, Cartazyme HT, Cartazyme SR 10	Pulp bleaching
Enzyme Development, USA	Enzeko xylanase	Baking, Food, Animal feed
Finnfeeds International, UK	Avizyme	Animal feed
Genencor International, Finland	Irgazyme 40-4X, Liftase Exp A40,	Pulp bleaching
Novo Nordisk, Denmark	Pulpzyme HA, Pulpzzyme HB	Pulp bleaching
	Biofeed Beta, Biofeed Plus	Animal feed
	Ceremix	Brewing
Rhon Enzyme, QY	Ecopulp X-100	Pulp bleaching
Solvay Interox, USA	Optipulp L-8000	Pulp bleaching
Thomas Swan, U.K.	Ecozyme	Pulp bleaching
Vocsi Alpino, Austria	VAI xylanase	Pulp bleaching
Ultra Bio-Logics Inc., Canada	Bio-Glucanase 10x	Animal feed
Dyadic International, Inc., USA	Xylanase XP Concentrate	Animal feed
RAP posieux, Germany	Roxazyme, Avizyme 2300, Econase Wheat L Plus,	Animal feed

Recombinant gene expression of fungal xylanase in other microorganisms is an important tool in metabolic engineering and commercial production of xylanases. In case of hyperxylanolytic fungi that co-secrete cellulase, cloning of xylanase genes in appropriate host holds a solution. The *Trichoderma* recombinant xylanase II has been cloned and expressed in a multicopy plasmid in *Saccharomyces cerevisiae* (La Grange *et al.*, 1996). Kimura *et al.* (2002) have reported cloning of a xylanase F3 gene from *A. oryzae* KBN 616 genomic library and its overexpression in *A. oryzae* while De-faria *et al.* (2002) could clone xylanase gene from thermophilic *Humicola grisea var. themoidea* in *T. reesei.* Dalboge (1997) has made a successful attempt in expression cloning of fungal hemicellulases including xylanase from *Aspergillus aculeatus.* Cloning and expression of a *Myceliophthora thermophila* xylanase in *A. oryzae* for use in food and pulp industry has been described by Sandal *et al.* (1997). The recombinant enzyme had a mol. wt. of 43 kDa, pI of 4.5 and temperature optima of 65°C.

APPLICATIONS

Enzymes have been at the centre stage of intense worldwide research and last few decades have witnessed an unprecedented expansion in the knowledge in use of microbial enzymes. Microbial xylanases are one such group of enzymes receiving worldwide attention. They have found several novel industrial applications while their major application has been realized in saccharification of lignocellulosic substrates and pulp bleaching. Production and application of xylanase at commercial scale has reached new vistas as evidenced by its worldwide marketing (Table 5). The details of different applications of xylanase enzymes are as follows:

Enzymic Saccharification

Hydrolysis of hemicellulosic residues, which form approximately 35% of plant material yields a mixture of sugars, primarily D-xylose. Complete hydrolysis of xylan to xylose by xylanolytic saccharification of agricultural residues, woody biomass, hemicelluosic waste from municipal and industrial sources, provides an efficient method yielding sugar syrups (Biely, 1985; Duff and Murray, 1996; Christov *et al.*, 1999). These can be further utilized for animal or human consumption or can be used in fermentation industries for microbial growth. Detroy *ct al.* (1982) have reported of production of ethanol on crude wheat straw hydrolysate containing xylose using *Pachysolen tannophilus* strain NRLL 2460. Synergistic action of xylanase and cellulase mixtures of fungal origin could result in complete and efficient method of yielding sugars from lignocelluloses (Wong *et al.*, 1988). Wang *et al.* (1980) recommended addition of exogenous isomerases to achieve xylose fermentation by yeast. Degradation of lignocellulosic biomass by xylanase in synergism with other enzymes can be used for generation of biofuels (ethanol), xylitol (sweetner) and organic acids via fermentation (Kuhad and Singh, 1993; Olson and Hann-Hagerdel, 1996; Dominguez, 1998). Moreover, the saccharified liquor can also be used as substrate to produce protein rich biomass or SCP (Ericksson *et al.*, 1990). Garcia-Kirchner *et al.* (2002) have shown that cross-synergistic action of enzymes, produced in a co-culture of *Penicillium* sp. and *A. terreus* results in enhanced hydrolysis of lignocellulsoes.

Biobleaching

One of the most attractive biotechnological uses of xylanases is biobleaching of pulp in paper industries. Several patents disclosing enhancement of bleachability of pulp by these enzymes have been filed (Fangerstrom *et al.*, 1997; Vollmond, 1997). Weighty benefits of xylanase pretreatment of pulp include higher brightness ceilings and significant reductions in the amounts of chlorine based bleaching chemicals, thus avoiding consequential release of organochlorine compounds in bleach plant effluents.

Residual xylan and lignin get covalently linked to degradation products generated during pulping and are trapped in the cellulose fibre pulp matrix imparting it brownish colour (Kantelinen *et al.*, 1991). These colour imparting chromophores are hard to extract and elemental chlorine and chlorine dioxides are used to bleach these for enhancing brightness of pulp (Gierger and Wannastrom, 1984). Chlorine and chlorine based chemicals form toxic and recalcitrant organochloro compounds that are released in environment by way of effluent and present dangers of toxicity and bioaccumulation. Environmental agencies and consumers are increasingly becoming discreet regarding use of chlorine in bleaching. Reports however indicate that xylanase pretreatment can reduce chlorine requirements in bleaching of both softwood and hardwood kraft pulps (Tolan and Canovas, 1992).

The underlying mechanism of effect of xylanase treatment is not fully understood yet. Xylanase pretreatment brings about depolymerization of xylans and this helps in release of otherwise occluded chromophoric material. The loosening of pulp structure by xylanase treatment enhances lignin extraction with minimum chemicals (Viikari *et al.*, 1990). Jeffries *et al.* (1998) have studied mechanisms for kappa reduction and colour removal by xylanase. They employed commercial enzyme preparations containing xylanase *viz.* Pulpzyme HC (3,734 IU/ml), Ecozyme (4,606 IU/ml), SP342 (2,319 IU/ml) and found significant reduction in kappa number induced by xylanase. They concluded that the mechanism involves liberation of degraded carbohydrates rather than hexeneuronic acid (HexA) or lignin liberation.

For effective biobleaching it is desirable to have cellulase-free xylanase, otherwise celluolse hydrolysis may result in lowering pulp viscosity while cellulase-free xylanase pretreatment enhances pulp viscosity as low molecular weight xylans are removed. Since alkaline pH and high temperature regimes are part of pulp cooking the xylanase should be stable at alkaline pH and must have good thermal stability (Viikari *et al.*, 1992). Xylanases with varying properties impart different effects and therefore effective molecular weight, net ionic properties and specific action patterns of xylanase suiting to the microenvironment of pulp are essential. Buchart *et al.* (1992) have shown that two *Trichoderma reesei* xylanases having different pI of 5.5 and 9.0 behave differently. At pH 7.0, former liberated more of reducing sugar while the latter hydrolyzed small amounts of xylan but conferred more brightness to the pulp. At the given pH, xylanase with pI 9.0 is positively charged and gets well absorbed onto negatively charged pulp fibres while xylanase with pI 5.5 may not get absorbed at all. Presence of metal ions in pulp has also been reported to positively affect xylanase action while pulps devoid of metal ions are poorly hydrolysed (Senior *et al.*, 1990).

Ragauskas *et al.* (1994) obtained optimal bleaching results with ozone while Yang *et al.* (1992) found that xylanase pretreatment can activate kraft pulp toward bleaching with another non chlorine reagent, H_2O_2. In general, it has been noticed that xylanase pretreatment enhances bleaching efficiency of non-chlorine reagents such as ozone, peracetic acid, peroxides etc. Use of xylanases has also been indicated in deinking or toner removal of recycled paper waste. Jeffries *et al.* (1996) have examined use of commercial enzyme preparations, Novozyme 342TM, Pulpzyme HBTM (Novo Industries), LiftaseTM Exp A40 (Genencor) etc. containing cellulase and xylanase in removal of toner from office waste paper and found much reduced levels of residual ink in enzyme treated test sheets (26 ppm) as compared to controls (258 ppm)

Food and Feed Upgradation

Xylanase pretreatment of food and feed with significant xylan content brings improvement in their nutritional and textural properties (Wong *et al.*, 1988). Treatment of xylanase upgrades poultry feed and increases its digestibility (Vanparidon *et al.*, 1992). Improved digestibility of feed due to reduced intestinal viscosity has been found to support weight gain in rye-fed broiler chicks (Bedford and Classen, 1992). Use of xylanase has been indicated in clarification of fruit juices and wines, liqufication of fruit pulp and vegetables (Wong and Saddler, 1992). Increased concentration of xylooligomers have been shown to impart better rheological properties to liquids.

Since arabinoxylans of wheat flour are hydrophilic and can absorb water ten times its weight, its hydrolysis brings a loss in water absorption of wheat flour thus changing loaf texture and helps bread to rise in volume (Krishnaru and Hoseney, 1994). Maat *et al.* (1992) have suggested quality improvement of bread by synergistic employment of amylase and xylanase. Xylanase pretreatment has been shown to enhance starch recovery from wheat flours (Simpson, 1956). Use of xylanase in extraction of coffee and oil (Wong and Saddler, 1992), pretreatment of forage crops for increased digestibility of ruminant feeds (Gilbert and Hazlewood, 1993) and maximum conversion of lignocelluloses in liquid feedstock (Jeffries, 1985) are also noteworthy.

Fibre Upgradation

A potential application of xylanases in combination with pectinolytic enzymes lies in debarking of bast fibres such as flax, juite and hemp (Wong and Saddler, 1992; Sharma, 1997; Puchart *et al.*, 1999). Efficacy of xylanse treatment on quality parameters of low quality jute fibres such as feel and touch, hemicellulose content and pressly index (fibre strength) was tested by subjecting fibres to xylanase preparations of two thermophilic moulds *Thermoascus aurantiacus* and *Thermomyces lanuginosus.* Xylanase treatments imparted softness to fibres and brought reduction in hemicellulose content while fibre strength was diminished to some extent (Alam *et al.*, 1994).

FUTURE PERSPECTIVES

Past few decades have witnessed a steady increase in the sale of industrially important enzymes and is likely to occupy 1500-2000 million market by the year 2005. Xylanases are one such group of enzymes whose demand is likely to increase many folds because of their varied industrial applications. Enumeration and exploitation of potent microbial strains coupled with advanced bioprocess technology shall certainly be able to meet out these demands.

High resolution crystallography has revealed the basis of thermostability in xylanase I (TAXI) of *Thermoascus aurantiacus* (Leggio *et al.,* 1999; Teixeira *et al.,* 2001). Such investigations on xylanases of different origin shall allow a better understanding about their mode of action and desirable site-directed manipulations for improvement of their pH and temperature stability. Attempts made for parametric optimization of xylanase production from moulds under SSF and mixed culture are quite encouraging while more investigations are needed. Cloning of fungal xylanases in appropriate host may provide us novel systems with enhanced levels of cellulase-free xylanase(s) with desirable specificities. Xylanase pretreatment of pulp enhances its bleachability by mild non-chlorine chemicals and more of such alternatives should be worked out. Comprehensive studies to establish importance of xylanases in feed upgradation and consequent benefits conferred to livestock are required.

CONCLUSIONS

Xylanases have been identified to play a key role in lignocellulose biotechnology. They find major application in biobleaching of pulp, saccharification of woody biomass, feed and fodder upgradation, fibre retting, besides their several applications in food industires. Various strains of *Thermomyces lanuginosus* have been identified as potential source for xylanase production with a wide range of temperature and pH for activity while certain strains of mesophilic fungi, *Cephalosporium* and *Aspergillus* yield highly alkalistable xylanase. Fungal xylanases are inducible enzymes and hence for their practical application their substrate requirements be identified at large. Occurrence of multiple forms of xylanases with varied specificities is a characteristic of these enzymes. An approach to clone genes in appropriate host has also been practiced and more successful attempts are needed to develop strains with desired specific activities. Commercial preparations of fungal xylanases with specified use in biobleaching, feed upgradation etc. signify that usage of xylanase has been realized and further investigations are required for their varied industrial applications.

REFERENCES

Adams, P.R. and Deploey, J.J. (1976). Amylase production by *Mucor pusillus* and *Mucor mihei*. *Mycologia* 68: 934-938.

Alam, M.M., Gomes, I., Mohiuddin, G. and Hoq, M.M. (1994). Production and characterization of thermostable xylanase by *Thermomyces lanuginosus* and *Thermoascus aurantiacus* grown on lignocellulose. *Enz. Microb. Technol.* 16: 298-302.

Anand, L. and Vithayathil, P.J. (1990). Production of extracellular xylanase by thermophilic fungus *Humicola lanuginosa* Bunce. *Ind. J. Exp. Biol.* 25: 434-437.

Anand, L., Krishnamurthy, S. and Vithayathil, P.J. (1990). Purification and properties of xylanases from thermophilic fungus, *Humicola lanuginosa* (Griffon and Maublanc) Bunce. *Arch. Biochem. Biophys.* 276: 246-553.

Apel-Birkhold, P.C. and Walton, J.D. (1996). Cloning, disruption, and expression of two endo-beta 1,4-xylanase genes, XYL2 and XYL3, from *Cochliobolus carbonum*. *Appl .Environ. Microbiol.* 62: 4129-4135.

Archana, A., Sharma, A. and Satyanarayana, T. (1999). Xylanolytic enzymes. In : Thermophilic Moulds in Biotechnology, Johri, B.N. Satyanarayana, T. and Olsen, J. Eds, Kluwer Academic Publishers, Dordrecht, 169-190.

Biely, P. (1985). Microbial xylanolytic systems. *Trends Biotechnol.* 3: 288-290

Bailey, M.J., Biely, P. and Poutanen, K. (1992). Interlaboratory testing of methods for assay of xylanase activity. *J. Biotechnol.* 23: 257-270.

Bailey, M.J., Buchert, J. and Viikari, L. (1993). Effect of pH on production of xylanase by *Trichoderma. reesei* on xylan and cellulose based media. *Appl. Microbiol. Biotechnol.* 40: 224-229.

Banerjee, S., Archana, A. and Satyanarayana, T. (1995). Xylanolytic of activity and xylose utilization by thermophilic molds. *Folia Microbiol.* 40: 279-282.

Bansod, S.M., Dutta-Choudhary, M., Srinivasan, M.C. and Rele, M.V. (1993). Xylanase active at high pH from an alkalotolerant *Cephalosporium* sp. *Biotechnol. Lett.* 15: 965-970.

Barry, V.C. and Dillon, T. (1940). The occurrence of xylan in marine algae. *Nature* (London) 146: 620.

Bedford, M.R. and Classen, H.L. (1992). The influnence of dietery xylanase on intestinal viscosity and molecular weight distribution of carbohydrates in rye-fed broiler chicks. In: Xylans and Xylanases, Visser, J., Beldman, G., Someren, M.A.K. and Voragen, A.G.J. Eds., Elsevier, Amsterdam, 361-370.

Beg, Q.K., Kapoor, M., Mahajan, L. and Hoondal, G.S. (2001). Microbial xylanases and their applications: A review. *Appl. Microbiol. Biotechnol.* 56: 326-338.

Belancic, A., Scarpa, J., Peirano, A., Diaz, R., Steiner, J. and Eyzaguirre, J. (1995). *Penicillium purpurogenum* produces several xylanases: purification and properties of two of the enzymes. *J. Biotechnol.* 41: 71-79.

Biely, P., Hirsch, J., La Grange, D.C., Van Zyl, W.H. and Prior, B.A. (2000). A chromogenic substrate for a beta-xylosidase coupled assay of alpha-glucuronidase. *Anal. Biochem.* 286: 289-94.

Biely, P., Mackenzie, C.R. and Schneider, H. (1988). Production of acetyl xylan esterase by *Trichoderma reesei* and *Schizophyllum commune. Can. J. Microbiol.* 34: 767-772.

Biely, P., Mislvicova, D. and Toman, R. (1984). Soluble chromogenic substrates for the assay of endo-1,4- β-xylanases and endo-1,4-β- glucanases. *Anal. Biochem.* 144: 142-146.

Biely, P., Vranska, M., Tenkanen, M. and Kluepfel, D. (1997). Endo-B-1,4-xylanase families: differences in catalytic properties. *J. Biotechnol.* 57: 151-166.

Biswas, S.R., Jana, S.C., Mishra, A.K. and Nanda, G. (1990). Production, purification and characterization of xylanase from a hyperxylanolytic mutant of *Aspergillus ochraceus. Biotechnol. Bioeng.* 35: 244-251.

Buchart, J., Ranua, M., Kantelinen, A. and Viikari, L. (1992). The role of two *Trichoderma reesei* xylanases in bleaching of kraft pulp. *Enzyme Microb. Technol.* 37: 825-829.

Castanares, A., Hay, A.J., Gordon, A.H., McCrae, S.L. and Wood, T.M. (1995). D-xylan-degrading enzyme system from *Phanerochaete chrysosporium*: isolation and partial characterization of an α- (4-O-methyl)-D-glucuronidase. *J. Biotechnol.* 43: 183-194.

Cesar, T. and Mrsa, V. (1996). Purification and properties of xylanase produced by *Thermomyces lanuginosus. Enz. Microb.Technol.* 19: 289-296.

Chadha, B.S., Jaswinder, K., Rubinder, K., Saini, H.S. and Singh, S. (1999). Xylanase production by *Thermomyces lanuginosus* wild and mutant strains. *World J. Microbiol. Biotechnol.* 15: 217-221.

Chanda, S.K., Hirst, E.L., Jones, J.K.N. and Percival, E.G.V. (1950). The constitution of xylan from esparto grass (*Stipa tenacissima*). *J. Chem. Soc.* 50: 1287-1289.

Christakopoulos, P., Nerinekx, W., Kekos, D., Marcis, B. and Claeyssens, M. (1997). The alkaline xylanase III of *Fusarium oxyspourm* belongs to family F/10. *Carbohydr. Res.* 302: 191-195.

Christov, L.P. and Prior, B.A. (1993). Esterases of xylan-degrading micro-organisms: Production, properties and significance. *Enz. Microb. Technol.* 15: 462-475.

Christov, L.P., Myburgh, J., O'Neill, F.H., VanTonder, A. and Prior, B.A. (1999). Modification of the carbohydrate composition of sulfite pulp by purified and characterized beta-xylanase and beta-xylosidase of *Aureobasidium pullulans. Biotechnol. Prog.* 15: 196-200.

Clarke, J.H., Rixon, J.E., Ciruela, A., Gilbert, H.J. and Hazlewood, G.P. (1997). Family-10 and family-11 xylanases differ in their capacity to enhance the bleachability of hardwood and softwood paper pulps. *Appl. Microbiol. Biotechnol.* 48:177-183.

Colowick, S.P. and Kaplan, N.O. (1988). Cellulase and hemicellulase. In: Methods in Enzymology. Wood, W.A. and Kellog, S.T. Eds., Academic Press 160: 511-649.

Crabb, W.D. and Mitchingson, C. (1997). Enzyme involved in the processing of starch of sagurs. *Trends Biotechnol.* 15: 349-352.

Curotto, E., Concha, M., Campos, V., Milagres, A.M. and Duran, N. (1994). Production of extracellular xylanases by *Penicillium janthinellum*. Effect of selected growth conditions. *Appl. Biochem. Biotechnol.* 48: 107-116.

Dalboge, H. (1997). Expression cloning of fungal enzyme genes; a novel approach for efficient isolation of enzyme genes of industrial relevance. *FEMS Microbiol. Rev.* 21: 29-42.

Davis, S., Gysin, B., Casimir, J. and Zimmermann, W. (1992). Thermostable xylanases from actinomycetes for pulp bleaching. In: Xylans and Xylanases. Visser, J., Beldman, G., Someren, M.A.K. and Voragen, A.G.J. Eds., Elsevier, Amsterdam, 551-554.

Davies, G. and Henrissat, B. (1995). Structure and mechanism of glycosyl hydrolases. *Structure* 3: 853-859.

De Almedie (1995). Purification and characterization of β-Xylosidase from *Humicola grisea var. sulfurea*. *FEMS Microbiol. Lett.* 130: 171-176.

De Faria, F.P., Te'O, V.S., Bergquist, P.L., Azevedo, M.O. and Nevalainen, K.M. (2002). Expression and processing of a major xylanase (XYN2) from the thermophilic fungus *Humicola grisea var. thermoidea* in *Trichoderma reesei*. *Lett. Appl. Microbiol.* 34: 119-123.

Defaye, J., Driguez, H., John, M., Schmidt, J. and Ohleyer, E. (1985). Induction of D-xylan-degrading enzymes in *Trichoderma lignonum* by non metabolizable inducers. A synthesis of 4-thioxylobiose. *Carbohydr. Res.* 139: 123-132.

Dekker, R.F.H. and Richards, G.N. (1976). Hemicellulases: Their occurrence, purification, properties and mode of action. Adv. Carbohydr. *Chem. Biochem.* 32: 277-352.

Deleyn, F., Claeyssens, M. and De Bruyne, C.K. (1982). Beta-D-xylosidase from *Penicillium wortmanni*. *Methods Enzymol.* 83: 639-644.

Detroy, R.W., Cunningham, R.L., Bothast, R.J., Bagby, M.O. and Herman, A. (1982). Bioconversion of wheat straw cellulose/ hemicellulose to ethanol by *Saccharomyces uvarum* and *Pachysolen tannophilus*. *Biotechnol. Bioengr.* 24: 1105-1113.

Dominguez, J.M. (1998). Xylitol production by free and immobilized *Debaryomyces hansenii*. *Biotechnol. Lett.* 20: 53-56.

Duff, S.J.B. and Murray, W.D. (1996). Bioconversion of forest products, industry waste cellulosics to fuel ethanol- A review. *Bioresource Technol.* 55: 1-33.

Durand, H., Clanet, M. and Tiraby, G. (1988). Genetic improvement of *Trichoderma reesei* for large-scale cellulase production. *Enz. Microb. Technol.* 10: 341-346.

Durand, H., Soucaille, P. and Tiraby, G. (1984). Comparative study of cellulase and hemicellulases from four fungi: Mesophiles *Trichoderma reesei* and *Penicillium* sp. and thermophiles *Thielavia terrestris* and *Sporotrichum cellulophilum*. *Enz. Microb.Technol.* 6: 175-180.

Dusterhoft, E.M., Linssen, V.A.J.M., Voragen, A.G.J. and Beldman, G. (1997). Purification, characterization and properties of two xylanases from *Humicola insolens*. *Enzyme Microb. Technol.* 20: 437-445.

Eda, S., Ohnishi, A. and Kato, K. (1976). Xylan isolated from the stalk of *Nicotiana tabacum. Agric. Biol. Chem.* 40: 359-364.

Egana, L., Gutierrez, R., Caputo, V., Peirano, A., Steiner, J. and Eyzaguirre, J. (1996). Purification and characterization of two acetyl xylan esterases from *Penicillium purpurogenum. Biotechnol. Appl. Biochem.* 24:33–39.

Elisashvili, V.I., Khardziani, T.S., Tsiklauri, N.D. and Kachlishvili, E.T. (1999). Cellulase and xylanase activities in higher basidiomycetes. *Biochemistry (Mosc)* 64: 718-22.

Ericksson, K.E.L., Blanchette, R.A. and Ander, P. (1990). Microbial and enzymatic degradation of wood and wood components. Springer-Verlag, New York.

Fangerstrom, R., Paloheimo, M., Lantto, R., Lahtinen, T. and Suominen, P. (1997). Novel xylanases and their uses. (Assignees: Primalco, Finland) Patent No. WO 9722691.

Fernandez-Espinar, M.T., Pinaga, F., deGraff, L., Visser, J., Ramon, D. and Valles, S. (1994a). Purification, characterization and regulation of the synthesis of an *Aspergillus nidulans* acidic xylanase. *Appl. Microbiol. Biotechnol.* 42: 555-559.

Fernandez-Espinar, M.T., Pena, J.L., Pinaga, F. and Valles, S. (1994b). α-L-arabinofuranosidase production by *Aspergillus nidulans. FEMS Microbiol. Lett.* 1151: 107-112.

Filho, E.X.F., Puls, J. and Coughlan, M.P. (1996). Purification and characterization of two arabinofuranosidases from solid-state cultures of the fungus *Penicillium capsulatum. Appl. Environ. Microbiol.* 62: 168-173.

Fogarty, W.M. (1983). Microbial amylases. In: Microbial Enzymes in Biotechnology. Fogarty, W.M., Ed., Applied Science Publishers, London. 1-92.

Fogarty, W.M. and Kelly, C.T. (1980). Amylases and amyloglucosidases and related glucanases. In: Economic Microbiology, Microbial enzymes and Bioconversions. Vol.5, Rose, E.H. Ed., Academic Press, New york. 115-170.

Frederick, M.M., Chin-Hen, K., Frederick, J. R. and Reilly, P. (1985). Purification and characterization of endoxylanases from *Aspergillus niger* I: Two isozyme active on xylan backbone near branch points. *Biotechnol. Bioeng.* 27: 525-532.

Garcia-Kirchner, O., Munoz-Aguilar, M., Perez-Villalva, R. and Huitron-Vargas, C. (2002). Mixed submerged fermentation with two filamentous fungi for cellulolytic and xylanolytic enzyme production. *Appl. Biochem. Biotechnol.* 98: 1105- 1114.

Ghosh, M. and Nanda, G. (1994). Purification and some properties of xylanase from *Aspergillus sydowii* MG-49. *Appl. Environ. Microbiol.* 60: 4620-4623.

Gierger, J. and Wannstrom, S. (1984). Formation of alkali stable C-C bonds between lignin and carbohydrate fragments during kraft pulping. Holforschung 38: 181-184.

Giesbert, S., Lepping, H., Tenberge, K.B. and Tudzynski, P. (1998). The xylanolytic system of *Claviceps purpurea* : Cytological evidence for secretion of xylanases in infected rye tissue and molecular characterization of two xylanase genes. *Phytopathol.* 88: 1020-1030.

Gilbert, H.J. and Hazlewood, G.P. (1993). Bacterial cellulases and xylanases. *J. Gen. Microbiol.* 139: 187-194.

Gilbert, M., Yoguchi, M., Watson, D.C. and Wong, K.K. (1993). A comparison of two xylanases from thermophilic fungi *Thielavia terrestris* and *Thermoascus crustacens. Appl. Microbiol. Biotechnol.* 40: 508-514.

Gomes, I., Gomes, J., Steiner, W. and Esterbauer, H. (1992). Production of xylanolytic enzymes by a wild strain of *Trichoderma viride. Appl. Microbiol. Biotechnol.* 36: 701-707

Gomes, J., Gomes, I., Kreiner, W., Esterbauer, H., Sinner, M. and Steiner, W. (1993a). Production of high level of cellulase-free xylanase from *Thermomyces lanuginosus* using beechwood xylan. *J. Biotechnol.* 30: 283-297.

Gomes, J., Purkarthofer, H., Hayn, M., Kappimuller, J., Steiner, M. and Steiner, W. (1993b). Production of high level of cellulase free xylanase by the thermophilic fungus *Thermomyces lanuginosus* in laboratory and pilot scale using lignocellulosic materials. *Appl. Microbiol. Biotechnol.* 39: 700-707.

Gomez-Gomez, E., Ruiz-Roldan, M.C., Di Pietro, A., Roncero, M.I. and Hera, C. (2002). Role in pathogenesis of two endo-beta-1,4-xylanase genes from the vascular wilt fungus *Fusarium oxysporum. Fungal Genet.* Biol. 35: 213-22.

Grajek, W. (1987). Production of D-xylanases by thermophilic fungi using different methods of culture. *Biotechnol. Lett.* 9: 353-356.

Haas, H., Friedlin, E., Stoffler, G. and Redl, B. (1993). Cloning and structural organization of a xylanase-encoding gene from *Penicillium chrysogenum. Gene* 126: 237-242.

Halgasova, N., Kutejova, E. and Timko, J. (1994). Purification and some characteristics of the acetyl xylan esterase from *Schizophyllum commune. Biochem. J.* 298: 751–755.

Haltrich, D., Nidetzky, B., Kulbe, K.D., Steiner, W. and Zupancic, S. (1996). Production of fungal xylanases. *Biores. Technol.* 58, 137-161.

Haltrich, D., Prenner, E., Prieb, M. and Steiner, W. (1992). Production of extremely high values of xylanase activity by *Schizophyllum commune.* In: Biotechnology in the Pulp and Paper Industry, Kuwahara, M. and Shimada, M., Eds., Uni Publishers Co. Ltd., Tokyo, 123-128.

Henrissat, B. and Bairoch, A. (1993). New families in the classification of glycosyl hydrolases based on amino acid sequence similarities. *Biochem. J.* 293: 781-788.

Herrmann, M.C., Vrsanska, M., Jurickova, M., Hirsch, J., Biely, P. J. and Kubicek, C.P. (1997). The b-D-xylosidase of *Trichoderma reesei* is a multifunctional b-D-xylan xylohydrolase. *Biochem. J.* 321: 375-381.

Hoq, M.M. and Deckwer, W. (1995). Cellulase-free xylanase by thermophilic fungi: a comparision of xylanase production by two *Thermomyces lanuginosus* strains. *Appl. Microbiol. Biotechnol.* 43: 604-609.

Hoq, M.M., Hempel, C. and Deckwer, W. (1994). Cellulase-free xylanase by *Thermomyces lanuginosus* RT9: Effect of agitation, aeration, and medium components on production. *J. Biotechnol.* 37: 49-58.

Hrmova, M., Beily, P. and Vrsanka, M. (1989). Cellulose and xylan degrading enzymes of *Aspergillus terreus. Enz. Microbiol.Technol.* 11: 610-616.

Inglis GD, Popp AP, Selinger LB, Kawchuk LM, Gaudet DA, ands McAllister TA. (2000) Production of cellulases and xylanases by low-temperature basidiomycetes. *Can. J. Microbiol* 46: 860-865

Ito, K., Ogassawara, J., Sugimoto, T. and Ishikawa, T. (1992). Purification and properties of acid stable xylanases form *Aspergillus kawachii. Biosci. Biotechnol. Biochem.* 56: 547-550.

Jain, P.C. (1999). Spoilage of stored products. In: Thermophilic Moulds in Biotechnology. Johri, B.N., Satyanarayana, T. and Olsen, J. Eds., Kluwer Academic Publishers, Dordrecht, The Netherlands, 369-395.

Jain, P.C. and Singhai, S. (2001). Cellulolytic activity of amylolytic strains of *Humicola grisea* and *Myceliophthora vellerea* isolated from saw dust and compost. *Asian Jr. Microbiol. Biotech. & Env. Sc.* 3: 197-200.

Jeffries, T.W. (1985) Emerging technology for fermenting D-xylose. *Trends Biotechnol.* 3: 208-212

Jeffries, T.W. (1996). Biochemistry and genetics of microbial xylanases. *Curr. Opin. Biotechnol.* 7: 337-342.

Jeffries, T.W., Davis, M., Rosin, B. and Landucci, L. (1998). Mechanisms for kappa reduction and color removal by xylanases. In: 7th international conference on Biotechnology in the Pulp and Paper Industry, Vol. C, Vancouver, Canada, 41-43.

Jeffries, T.W., Sykes, M.S., Rutledge-Cropsey, K., Klungness, J.H. and Abubakr, S. (1996). Enhanced removal of toners from office waste papers by microbial cellulases In: Biotechnology in the Pulp and Paper Industry: Recent Advances in Applied and Fundamental Research, Rebotnik, E.S. and Messner, K. Eds., Vienna, Austria, 141-144

John, M., Schmid, B. and Schmid, J. (1979). Purification and some properties of five endoxylanases and β-xylosidase produced by a strain of *A. niger. Can. J. Biochem.* 57: 125-134.

Johri, B.N., Satyanarayana, T. and Olsen, J. (1999). Thermophilic Moulds in Biotechnology. Kluwer Academic Publishers, Dordrecht, The Netherlands.

Kaji, A. (1984). L-Arabinases. *Adv. Carbohydr. Chem. Biochem.* 42:383–394.

Kaneko, S., Kitaoka, M., Kuno, A. and Hayashi, K. (2000). Syntheses of 4- mehtylumbelliferyl -beta-xylobioside and 5-bromo-3-indolyl-beta-D-xylobioside for sensitive detection of xylanase activity on agar plates. *Biosci. Biotechnol. Biochem.* 64: 741-745.

Kaneko, S., Shimasaki, T. and Kusakabe, I. (1993). Purification and some properties of intracellular α-L-arabinofuranosidase from *Aspergillus niger* 5-16. *Biosci. Biotech. Biochem.* 57: 1161-1165.

Kang, M.K., Maeng, J.P. and Rhee, Y. (1996). Purification and characterization of two xylanases from alkalophilic *Cephalosporium* sp. strain RYM-202. *Appl. Environ. Microbiol.* 62: 3480-3482.

Kango, N., Agrawal, S.C. and Jain, P.C. (2003a). Xylanase production by thermophilic fungi from soil and decomposing organic matter. In: Microbiology and Biotechnology for Sustainable Development. Jain, P.C., Ed., CBS Publishers & Distributors, New Delhi.

Kango, N., Agrawal, S.C. and Jain, P.C. (2003b). Isolation of thermophilic fungi from soil and decomposing organic matter. *Asian Jr. of Microbiol. Biotech & Env. Sc.* 6 (1): (In press)

Kango, N., Agrawal, S.C. and Jain, P.C. (2003c). Production of xylanase by *Emericella nidulans* NK-62 on low-value lignocellulosic substrates. *World J. Microbiol. Biotechnol.* 19z: (In press)

Kantelinen, A., Sundquist, J., Linko, M. and Viikari, L. (1991). The role of reprecipitated xylan in the enzymatic bleaching of kraft pulp. Proc. 6th Int. Symp. Wood Chem. Technol. 493-500.

Kelly, C.T., O'Mahony, M.R. and Fogarty, W.M. (1989). Extracellular xylanolytic enzymes of *Paecilomyces variotii*. *Biotechnol. Lett.* 9: 125-130.

Khalil, A.I. (2002). Production and characterization of cellulolytic and xylanolytic enzymes from the ligninolytic white-rot fungus *Phanerochaete chrysosporium* grown on sugarcane bagasse. *World J. Microbiol. Biotechnol.* 18: 753-759.

Khandke, K. M., Vithayathil, P.J. and Murthy, S.K. (1989). Purification and characterisation of an α-glucuronidase from a thermophilic fungus, *Thermoascus aurantiacus*. *Archiv. Biochem. Biophys.* 274: 511-517.

Khowala, S., Mukherjee, M. and Sengupta, S. (1988). Carboxymethyl xylan - a specific substrate directly differentiating backbone-hydrolysing and side chain reacting β-D-(1→4)-xylanases of the mushroom *Termitomyces clypeatus*. *Enz. Microb. Technol.* 10: 563-567.

Kimura, L. Sasahara, H. and Tajima, S. (1995). Purification and characterization of two xylanases and an arabinofuranosidase from *A. sojae*. *J. Ferment. Bioeng.* 80: 334-339.

Kimura, T., Ito, J., Kawano, A., Makino, T., Kondo, H., Karita, S., Sakka, K. and Ohmiya, K. (2000). Purification, characterization, and molecular cloning of acidophilic xylanase from *Penicillium* sp.40. *Biosci. Biotechnol. Biochem.* 64: 1230-1237.

Kimura, T., Suzuki, H., Furuhashi, H., Aburatani, T., Morimoto, K., Sakka, K. and Ohmiya, K. (2002). Molecular cloning, characterization, and expression analysis of the xynF3 gene from *Aspergillus oryzae*. *Biosci. Biotechnol. Biochem.* 66: 285-292.

Kimura, T., Suzuki, H., Furuhashi, H., Aburatani, T., Morimoto, K., Karita, S., Sakka, K. and Ohmiya, K. (2000). Molecular cloning, overexpression, and purification of a major xylanase from *Aspergillus oryzae*. *Biosci Biotechnol Biochem* 64: 2734-2738.

Kitpreechavanich, V., Hayashi, M. and Nagai, S. (1984). Production of xylan degrading enzymes by thermophiolic fungi *A. fumigatus* and *Humicola lanuginosa*. *J. Ferment. Technol.* 62: 415–420.

Komerlink, F.J.M., Gruppen, H., Vietov, R.J. and Voregan, A.G. (1993). Mode of action of the xylan-degrading enzymes from *Aspergillus awamori* on alkali-extractable cereal arabinoxylans. *Carbohydr. Res.* 249: 355-365.

Kormelink, F.J.M., Searle-van Leeuwen, M.F., Wood, T.M., and Voragen, A.G.J. (1991). Purification and characterization of a (1,4)-b-D-arabinoxylan arabinofuranohydrolase from *Aspergillus awamori. Appl. Microbiol. Biotechnol.* 35: 753–758.

Krengel, U. and Dijkstra, B.W. (1996). Three dimensional structure of endo-1,4-beta-xylanase I from *Aspergillus niger.* Molecular basis for its low pH optimum. *J. Mol. Biol.* 263: 70-78.

Krishnamurthy, S. and Vithayathil, P.J. (1989). Purification and characterization of endo-1,4-L-xylanase from *Paecilomyces variotii* banier. *J. Ferment. Bioeng.* 67: 77-82.

Krishnaru, L. and Hoseney, R.C. (1994). Enzymes increase loaf volume of bread supplemented with starch tailings and insoluble pentosanses. *J. Food Sci.* 59: 1251-1254.

Kuhad, R.C. and Singh, A. (1993). Lignocellulose biotechnology: current and future prospects. *Crit. Rev. Biotechnol.* 13: 151-172.

Kulkarni, N., Shendye, A. and Rao, M. (1999). Molecular and biotechnological aspects of xylanases *FEMS Microbiol. Rev.* 23: 411-456.

La Grange, D.C., Pretorius, I.S. and Van Zyl, W.H. (1996). Expression of a *Trichoderma reesei* β-xylanase gene (xyn 2) in *Saccharomyces cervisiae. Appl. Env. Microbiol.* 62: 1036-1044.

Leathers, T. D., Detroy, R.W. and Bothast, R.J. (1986). Induction and glucose repression of xylanase from a color variant strain of *Aureobasidium pullulans. Biotechnol. Lett.* 8: 867-872.

Leggio, L., Kalogiannis, S., Bhat, M.K. and Pickersgill, R.W. (1999). High resolution structure and sequence of *T. aurantiacus* xylanase I: Implications for the evolution of thermostability in family 10 xylanases and enzymes with βα-barrel architecture. *Proteins* 36: 295-306.

Lemos, J.L., Fontes, M.C. and Pereira, N. Jr. (2001). Xylanase production by *Aspergillus awamori* in solid-state fermentation and influence of different nitrogen sources. *Appl. Biochem. Biotechnol.* 91: 681-689.

Li, X.L. and Ljungdahl, L.G. (1994). Cloning, sequencing, and regulation of a xylanase gene from the fungus *Aureobasidium pullulans* Y-2311-1. *Appl. Environ. Microbiol.* 60: 46-47.

Li, Y., Yao, H. and Cho, Y. (2000). Effective induction, purification and characterization of *Trichoderma koningii* G-39 β-xylosidase with high transferase activity. *Biotechnol. Appl. Biochem.* 31: 119-125.

Lin, J., Ndlovu, L.M., Singh, S. and Pillay, B. (1999). Purification and biochemical characteristics of b-D-xylanase from a thermophilic fungus, *Thermomyces lanuginosus*-SSBP *Biotechnol. Appl. Biochem.* 30: 73-79.

Linden, J., Samara, M., Decker, S., Johnson, E., Boyer, M., Pecs, M., Adney, W. and Himmel, M. (1994). Purification and characterization of an acetyl esterase from *Aspergillus niger. Appl. Biochem. Biotechnol.* 45: 383–393.

Luonteri, E, Siika-aho, M., Tenkanen, M. and Viikari, L. (1995). Purification and characterization of three α-arabinosidases from *Aspergillus terreus. J. Biotechnol.* 38: 279 - 291.

Maat, J., Roza, M., Verbakel, J., Stam, H., Satnos da Silva, M.J., Bosse, M., Egmond, M.R., Hagemans, M.L.D., Gorcom, R.F.M.V., Hessing, J.G.M., Van der Hondel, C.A.M.J.J. and Rotterdam, C.V. (1992). Xylanases and their application in bakery. In: Xylans and Xylanases, Visser, J., Beldman, G., Someren, M.A.K. and Voragen, A.G.J. Eds., Elsevier, Amsterdam, 349-360.

Maheshwari, R. and Kamalam, P.T. (1985). Isolation and culture of a thermophilic fungus *Melanocarpus albomyces* and factors influencing the production and activity of xylanase. *J. Gen. Microbiol.* 131: 3017-3027.

Maijala, P., Raudaskaski, M. and Viikari, L. (1995). Hemicellulolytic enzymes in P- and S- strains of *Heterobasidion annosum. Microbiol.* 141: 743-750.

Margolles-Clark, E., Tenkanen, M., Nakari-Seta¨La¨, T. and Penttila, M. (1996). Cloning of genes encoding a-L-arabinofuranosidase and b-xylosidase from *Trichoderma reesei* by expression in *Saccharomyces cerevisiae. Appl. Environ. Microbiol.* 62: 3840–3846.

Matsuo, M. and Yasui, T. (1985). Properties of xylanase of *Malbranchea pulchella var. sulfurea* No. 48. *Agric. Biol. Chem.* 49: 839-841.

Matsuo, M., Yasui, T. and Kobayashi, T. (1975). Production and saccharifying action for xylan of xylanase from *Malbranchea pulchella var. sulfurea* No. 48. *Nippon Nagaikagaku Kaishi* 49: 263-270.

McCleary, B.V. and Matheson, N.K. (1981). Enzymic analysis of polysaccharide structure. In: Methods in Enzymology. Tipson, R.S. and Horton, D. Eds., Academic Press, 44: 147-276.

Mendicuti Castro, L.P., Trejo-Aguilar, B.A., Aguilar Osorio, G. (1997). Thermostable xylanases produced at 37°C and 45°C by a thermotolerant *Aspergillus* strain. *FEMS Microbiol. Lett.* 146 : 97-102.

Milagres, A.M.F. and Sales, R.M. (2001). Evaluating the basidiomycetes *Poria medula-panis* and *Wolfporia cocos* for xylanase production. 28: 522-526.

Morosoli, R. (1985). Molecular expression of xylanase genes in *Cryptococcus albidus. Biochim. Biophys. Acta.* 826: 202-207.

Morosoli, R., Zalce, E., Moreau, A. and Durand, S. (1992). Secretion of a xylanase from *Cryptococcus albidus* by *Saccharomyces cerevisiae* and *Pichia stipitis.* In: Xylans and Xylanases, Visser, J., Beldman, G., Someren, M.A.K. and Voragen, A.G.J. Eds., Elsevier, Amsterdam, 247-258.

Mukhopadhyay, S.K., Paul, S., Roy, A. and Chatterjee, S.P. (1997). Xylanolytic enzyme production by *Aspergillus flavipes. Ind. J. Microbiol.* 37: 77-80.

Nath, D. and Rao, M. (1998). Structural and functional role of tryptophan in xylanase from an extremophilic *Bacillus*: Assessment of active site. *Biochem. Biophys. Res. Commun.* 249: 207-212.

Natesh, R., Bhanumoorthy, P., Vithayathil, P.J., Sekar, K., Ramkumar, S. and Viswamitra, M.A. (1999). Crystal structure at 1.8 A resolution and proposed aminoacid sequence of a thermostable xylanase from *Thermoascus aurantiacus. J. Mol. Biol.* 288: 999-1012.

Northcote, D.H., Davey, R. and Lay, J. (1989). Use of antisera to localize callose xylan and arabinogalactan in the cell plate, primary and secondary walls of plant cells. *Planta* 178: 353-360.

Olson, L. and Hahn-Hagerdal, B. (1996). Fermentation of lignocellulosic hydrolysates for ethanol production. *Enz. Microb. Technol.* 18: 312-331.

Park, Y.S., Kang, S.W., Lee, J.S., Hong, S.I. and Kim, S.W. (2002). Xylanase production in solid state fermentation by *Aspergillus niger* mutant using statistical designs. *Appl. Microbiol. Biotechnol.* 58: 761-766.

Prabhu, K.A. and Maheshwari, R. (1999) Biochemical properties of xylanases from a thermophilic fungus, *Melanocarpus albomyces* and their action on plant cell walls. *J. Biosciences* 24: 471-476

Peltonen, S., Karjalainen, R. and Niku-Paavola, M.L. (1994). Purification and characterization of a xylanase from *Bipolaris sorokiniana. Mycol. Res.* 98: 67-73.

Poutanen, K. (1988). An a-L-arabinofuranosidase of *Trichoderma reesei. J. Biotechnol.* 7: 271–282.

Poutanen, K. and Puls, J. (1988). Characteristics of *Trichoderma reesei* b-xylosidase and its use in the hydrolysis of solubilized xylans. *Appl. Microbiol. Biotechnol.* 28: 425-432.

Poutanen, K., Ratto, M., Puls, J. and Viikari, L. (1990). Deacetylation of xylans by acetyl xylan esterases of *Trichoderma reesei. Appl. Microbiol. Biotechnol.* 33: 506-510

Puchart, V., Katapodis, P., Biely, P., Kremnicky, L., Christakopolous, P., Vranska, M., Kekos, D., Marcis, B.J. and Bhat, M.K. (1999) Production of xylanases, mannanases and pectinases by the thermophilic fungus *Thermomyces lanuginosus.* Enz. Microb. Technol. 24: 355-361

Puls, J. and Schuseil, J. (1993). Chemistry of hemicelluloses: relationship between hemicellulose structure and enzymes required for hydrolysis. In: Hemicellulose and Hemicellulases, Coughlan, M.P. and Hazlewood, G.P. Eds., Portland Press, London, 1-28

Puls, J., Schmidt, O. and Granzow, C. (1987). α-glucuronidase in microbial xylanolytic systems. *Enz. Microb. Technol.* 9: 83-88.

Purkathofer, H. and Steiner, W. (1995). Induction of endo-b-xylanase in the fungus *Thermomyces lanuginosus. Enz. Microb. Technol.* 17: 114-118.

Qy, Y., Gao, P., Wang, D., Zhao, X. and Zhang, X. (1996). Production, characterization and application of the cellulase-free xylanase from *Aspergillus niger. Appl. Biochem. Biotechnol.* 57: 375-381.

Radionova, N.A., Dubovaya, N.V. Eneiskaya, E.V., Matrinovich, L.I., Gracheva, I.M. and Bezborodov, A.M. (2000). Purification and characterization of endo-(1→4) -β-xylanase from *Geotrichum candidum* 3C. *Appl. Biochem. Microbiol.* 36: 460-465.

Ragauskas, A.J., Poll, K.N. and Cesternino, A. (1994). Effects of xylanases pretreatment procedures on non-chlorine bleaching. *Enz. Microb. Technol.* 16: 492-495.

Raj, K.C. and Chandra, T.S. (1995). A cellulase-free xylanase from alkali-tolerant *Aspergillus fischeri* Fxn1. *Biotechnol. Lett.* 17: 309-314.

Raj, K.C. and Chandra, T.S. (1996). Purification and characterization of xylanase from alkali-tolerant *Aspergillus fischeri* Fxn1. *FEMS Microbiol. Lett.* 145: 457-461.

Rana, B.K., Johri, B.N. and Thakur, I.S. (1996). Formation and activities of xylan-hydrolysing enzymes of *Humicola grisea var. thermoidea. World J. Microbiol. Biotechnol.* 12: 12-15.

Reilly, P.J. (1981). Xylanases: Structure and function. *Basic Life Sci.* 18: 111-129.

Rombouts, F. M., Voragen, A. G. J., Searle-van Leeuwen, M. F., Geraeds, C. C. J. M., Schols, H. A. and Plinik, W. (1988). The arabinases of *Aspergillus niger* - purification and characterization of two a-L-arabinofuranosidases and an endo-1,5-a-L-arabinase. *Carbohydr. Polym.* 9: 25–47.

Sandal, T., Kofod, L.V., Kauppinen, M.S., Andersen, L.N. and Dybdal, L. (1997). Enzyme with xylanase activity. (Assignees: Novo Nordisk, Denmark) Patent No. WO 9727292.

Saraswat, V. and Bisaria, V.S. (1997) Biosynthesis of xylanolytic and xylan-debranching enzymes in *Melanocarpus albomyces* IIS 68. *J. Ferm. Biotechnol.* 84: 352-357

Saraswat, V. and Bisaria, V.S. (2000). Purification, characterization and substrate specificities of xylanase isoenzymes from *Melanocarpus albomyces* IIS 68. *Biosci. Biotechnol. Biochem.* 64: 1173-1180.

Schmidt, A., Schlacher, A. Steiner, W., Schwab, H. and Kratty, C. (1998). Structure of the xylanase from *Penicillium simplicissimum. Protein Sci.* 7: 2081-2088.

Senior, D.J., Mayers, P.R., Breuil, C. and Saddler, J.N. (1990). The interaction of xylanase with pulps: Non-selective adsorption and inactivation of xylanase. In: Biotchnology in Pulp and Paper Manufacture, Kirk, T.K. and Chang, H.M. Eds., Butterworth-Heinemann, Boston, 169-182.

Sharma, H.S.S. (1997). Enzymatic degradation of residual non-cellulosic polysaccharides present on dew-retted flax fibers. *Appl. Microbiol. Biotechnol.* 26: 358-362.

Siika-aho, M., Tenkanen, M., Buchert, J., Puls, J. and Viikari, L. (1994). An α-glucuronidase from *Trichoderma reesei* RUT-30. *Enz. Microb. Technol.* 16: 813-819.

Silveira, F.P., Sousa, M., Ricart, C., Milagres, A., Medeiros, C. and Filho, E. (1999). A new xylanase from a *Trichoderma harzianum* strain. *J. Ind. Microbiol. Biotechnol.* 23: 682-685.

Simao, R.C.G., Souza, C.G.M. and Peralta, R.M. (1997). The use of methyl β- D-Xyloside as a substrate for xylanase production by *Aspergillus tamarii Can. J. Microbiol. 43:* 56-60.

Simpson, F.J. (1956). Microbial pentosanases, II Some factors affecting the production of pentosanases by *Bacillus pumilus* and *Bacillus subtilis. Can. J. Microbiol.* 2: 28-38.

Singhai, Sharmeela and Jain, P.C. (1992). Thermophilic fungi from decomposing organic matter: compost and saw dust. *Geobios new reports.* 11: 41-44.

Singhai, Sharmeela and Jain, P.C. (2003). Extracellular production of α-amylase by some thermophilic fungi from saw dust and compost. In: Microbiology and Biotechnology for Sustainable Development, Jain, P.C. Ed., CBS publishers and distributors, New Delhi.

Sinha, N. and Sengupta, S. (1995). Simultaneous production of α-arabinofuranosidase and xylanase by *Termitomyces clypeatus. World J. Microbiol. Biotechnol.* 11: 359-360.

Sjostrom, E. (1993). Wood chemistry, Fundamentals and Applications. Academic Press, San Diego, USA.

Smith, J.E. (1996). Substrates for Biotechnology In: Biotechnology. 3rd ed. Cambridge University Press, Cambridge, 19-30.

Srinivasa, B.R., Swaminathan, K.R., Ganapathy, C., Roy, R.P., Murthy, S.K. and Vithayathil, P.J. (1991). The primary structure of xylanase from *Thermoascus aurantiacus. Protein Sequence Data Anal.* 4: 15-20.

Srivastava, R. and Srivastava, A.K. (1993). Characterization of a bacterial xylanase resistant to repression by glucose and xylose. *Biotechnol. Lett.* 15, 847-852.

Steiner, W., Lafferty, R.M., Gomes, I., and Esterbauer, H. (1987). Studies on a wild strain of *Schizophyllum commune*: Cellulase and xylanase production and formation of the extracellular polysaccharide Schizophyllan. *Biotechnol. Bioengg.* 30: 169-178.

Sundberg, M. and Poutanen, K. (1991). Purification and properties of two acetyl xylan esterases of *Trichoderma reesei. Biotechnol. Appl. Biochem.* 13: 1-11.

Sunna, A. and Antranikian, G. (1997). Xylanolytic enzymes from fungi and bacteria. *Crit. Rev. Biotechnol.* 17: 39-67.

Tan, L.U.L., Wong, K.K.Y., Yu, E.K.C. and Saddler, J.N. (1985). Purification and characterization of two D-xylanases from *Trichoderma harzianum. Enz. Microb.Technol.* 7: 425-430.

Taneja, K., Gupta, S. and Kuhad, R.C. (2002). Properties and application of a partially purified alkaline xylanase from an alkalophilic fungus *Aspergillus nidulans* KK-99. *Bioresour. Technol.* 85: 39-42.

Teixeira, S., Leggio, L., Pickersgill, R. and Cardin, C. (2001). Anisotropic refinement of the structure of *Thermoascus aurantiacus* xylanase I. *Biol. Crystallograph.* 54: 385-392.

Tenkanen, H., Puls, J. and Poutanen, K. (1992). Two major xylanases *from Trichoderma reesei.* Enz. Microb.Technol. 14: 566-574.

Timmel, T.E. (1964). Wood hemicelluloses. Part I. *Carbohydr. Chem.* 19: 247-302.

Tolan, J.S. and Canovas, R.V. (1992). The use of enzymes to decrease the Cl_2 requirements in pulp bleaching. *Pulp Paper Can.* 93: 39-42.

Torronen, R.L., Messner, M.R., Gonzalez, R., Kalkkinen, N., Harkki, A. and Kubicek, C.P. (1992). The two major xylanases from *Trichoderma reesei:* characterization of both enzymes and genes. *Biotechnology* 10: 1461-1465.

Tuhoy, M.G., Puls, J., Cleyssens, M., Vranska, M. and Coughlan, M.P. (1993). The xylan degrading system of *Talaromyces emersonii* : novel enzymes with activity against aryl-β- D- xylosides and unsubstituted xylans. *Biochem. J.* 290: 515-523.

Vanparidon, P.A., Boomanm, J.C.P., Selten, G.C.M., Geerse, C., Barug, D., Debot, P.H.M., and Hemke, G. (1992). Application of fungal endoxylanase in poultry diets. In: Xylans and Xylanases, Visser, J., Beldman, G., Someren, M.A.K. and Voragen, A.G.J. Eds., Elsevier, Amsterdam, 371-378.

Viikari, L., Tenkanen, M., Ratto, M., Buchert, J., Kantelinen, A., Bailey, M., Sundquist, J. and Linko, M. (1992). Important properties of xylanases for use. In: Biotechnology in Pulp and Paper Industry, Kuwahara, M. and Shimada, M. Eds., Uni Publishers Co. Ltd. Tokyo. 101-113.

Viikarii, L., Kantelinen, A., Poutanen, K. and Rauna, M. (1990). Characterization of pulps treated with hemicellulolytic enzymes prior to bleaching. In: Biotechnology in Pulp and Paper Manufacture, Kirk, T.K. and Chang, H.M. Eds., Butterworth-Heinemann, Boston, 145-151.

Vollmond, T. (1997). Bleaching and brightness of textiles. (Assignees: Novo Nordisk, Denmark) Patent No. WO 9725469.

Wang, P.Y., Johnson, B.F. and Schneider, H. (1980). Fermentation of D-xylose by yeasts using glucose isomerase in the medium to convert D-xylose to D-xylulose. *Biotechnol. Lett.* 2: 273-278.

Whistler, R.L. and Richards, E.L. (1970). Hemicelluloses. In : The Carbohydrates: Chemistry and Biochemistry, Pigman, W., Histon, D. and Herp, A. Eds., Academic Press, New York, 447-469.

Wilkie, K.C.B. (1979). The hemicelluloses of grasses and cereals. *Adv. Carbohydr. Chem. Biochem.* 36: 771-774.

Wong, K.K.Y. and Saddler, J.N. (1992). Applications of hemicellulases in the food, feed, and pulp and paper industries. In: Hemicellulose and Hemicellulases, Coughlen, P.P. and Hazlewood, G.P., Eds., Portland Press, London, 127-143.

Wong, K.K.Y., Tan, L.U.L. and Saddler, J.N. (1988). Multiplicity of β-1,4-xylanase in microorganisms: Functions and applications. *Microbiol. Rev.* 52: 305-317.

Woodward, J. (1984). Xylanases: Functions, properties and applications. In: Enzyme and Fermentation Biotechnology, Wisemen, A. Ed., Tennesse, 10-30.

Ximenes, F.A., De Sousa, M.V., Puls, J., da Silva, F.G. Jr, and Filho, E.X.F. (1999). Purification and characterization of a low-molecular-weight xylanase produced by *Acrophialophora nainiana. Curr. Microbiol.* 38: 18-21.

Yang, J.Y., Lou, G. and Ericksson, K.E. (1992). The impact of xylanase on bleaching of kraft pulps. *Tappi J.* 75: 95-101.

Yasui, T., Nguyen, B. and Nakanishi, K. (1984). Inducers for xylanase production by *Cryptococcus flavus. J. Ferment. Technol.* 62: 353-359.

Yoshino, S., Oishi, M., Moriyama, R., Kato, M. and Tsukagoshi, N. (1995). Two family G xylanase genes form *Chaetomium gracile* and their expression in *Aspergillus nidulans. Curr. Genet.* 29: 73-80.

Yoshioka, H., Nagato, N., Chavanich, S., Nilubol, N. and Hayashida, S. (1981). Purification and properties of thermostable xylanase from *Talaromyces byssochlamydoides* YH-50. *Agric. Biological Chem.* 45: 2425-2432.

Ziser, L., Setyaveti, I. and Withers, S.G. (1995). Syntheses and testing of substrates and mechnism-based inactivators for xylanases. *Carbohydr. Res.* 274: 137-15.

CHAPTER - 13

VACCINATION AGAINST FUNGAL INFECTIONS : EMERGING STRATEGIES

Sudhir K. Jain*, P.C. Jain, S.C. Agrawal** and S.K. Deshmukh*****

*Department of Applied Microbiology, College of Life Sciences, Cancer Hospital and Research Institute, Gwalior 474 009, M.P.

**Department of Applied Microbiology and Biotechnology, Dr. Hari Singh Gour University, Sagar 470 003, M.P.

***Quest Institute of Life Sciences, Nicholas Piramal India Limited, L.B.S. Marg, Post Box No. 17753, Mulund (W), Mumbai 400 080, Maharashtra

INTRODUCTION

Most fungi are saprophytes playing an important role in the degradation of organic matter in nature while some of these are used in fermentation of food products, production of antibiotics alcohol's, organic acids, etc. Thus, man has always considered them as friendly organisms and has paid comparatively little attention to the potential risk they pose to human health. This may be because only a small number of fungi (about 200 species) are pathogenic to man or animals. However, some of the fungal infections are fatal even though the morbidity and mortality rate due to mycoses is less as compared to bacterial and viral infections. In the recent past, there was a widespread diffusion of mycotic disease of medical interest particularly in the opportunistic fungal infections. Medical surgery, organ transplantation, premature birth, chemotherapy, radiation therapy, frequent use of immunosuppressive drugs, corticosteroids and broad spectrum antibiotics predispose an individual to opportunistic fungal infection. Further, genetically immunocompromised individuals and individuals suffering from AIDS and leukemia are highly susceptible to opportunistic fungal infections. Under such conditions even non-pathogenic fungi cause fatal infections, thereby blurring the dividing line between saprophytic and pathogenic fungi.

According to the degree of parasitic adaptation, pathogenic fungi can be categorized in to following three groups:

(1) Genuine and obligatory parasitic dermatophytes (e.g., *Trichophyton* spp., *Microsporum* spp., *Epidermophyton* spp.).

(2) Exosaprophytic fungi: These are commonly present in the out-side medium whose conidia are almost systemically infectious when inhaled or on entering the organism through other routes (e.g., *Histoplasma, Coccidioides, Blastomyces, Paracoccidioides*)

(3) Opportunistic fungal pathogens : These are endo- or exosaprophytic fungi, which have a pathogenic parasitic effect only on hosts, made susceptible (e.g.,*Candida albicans, Mucor* spp. *Rhizopus* spp., *Cryptococcus neoformans* and *Aspergillus fumigatus*)

INCIDENCE OF FUNGAL INFECTIONS

It is well-established that *Trichophyton, Epidermophyton* and *Microsporum* are genuine parasites of human and animals causing tinea corporis, tinea capitis, tinea pedis, tinea cruris and tinea unguim, etc. Certain endo- and exo-saprophytic fungi like *Candida albicans, Cryptococcus neoformans, Aspergillus fumigatus*, certain species of *Mucor* and *Rhizopus* cause opportunistic infections like candidiasis, cryptococcosis, aspergillosis and zygomycosis. Until recently, the key opportunistic fungal pathogens with which clinicians had to contend were *C. albicans, A. fumigatus* and the members of zygomycetes. *Candida albicans* and other *Aspergillus* infections are also frequent. A wide variety of fungi now isolated from neutropenic patients were not previously recognized as human pathogens. Many of these are soil or plant pathogenic fungi that are posing certain problems to clinicians in diagnosis and treatment. These emerging fungal pathogens belong to the genera, *Fusarium, Paecilomyces, Trichosporon, Exophiala, Bipolaris, Malassezia, Penicillium marneffei* and *Pythiosis insidiosi,* etc. Various studies in different parts of the world have also revealed the occurrence of common exosaprophytic parasitic fungi in soil (Ajello, 1953; Frey, 1965; Ajello and Padhye, 1972; Ajello, 1974; Chmel and Vlacilikova, 1977).

The occurrence of dermatophytes and saprophytic fungi in soils of India and their isolation using hair bait technique have been reported by many workers (Jain and Agrawal, 1977; Deshmukh and Agrawal, 1982 and 1983; Deshmukh, 1985) while recent reports made by Jain *et al.* (2002, 2003 and 2004) revealed the prevalence of onychomycosis, candidiasis and phaeohyphomycosis in normal as well as diseased patients in India.

ANTIFUNGAL TREATMENT

Even after tremendous advances the fungal diseases are still posing a formidable problem to the present day medicine. The efficacy of the treatment of fungal infections is dependent on several factors: (i) Drug factors- such as potency of the drug, tissue penetration and distribution within the body (ii) Organism factors- such as virulence, susceptibility to a given drug and development of resistance (iii) Host factors- such as the underlying immune status including the alteration of the normal mucosal flora, neutropenia, humoral and cell mediated immunity. The obstacles in the management of fungal diseases by most of these agents is due to the eukaryotic nature of both fungal and human cells, their metabolism resemble significantly with each other. Therefore, whenever these drugs are administered for treatment they also pose problem. There is poor penetration of the drug in the tissue because the fungi infect relatively poorly vascularized areas. Moreover, slow growth of fungi and the granulomatous response of the host tissue also make the drug less penetrable to reach the target site. In addition, poor absorption from gastrointestinal tract also demands the use of parenteral route and hence the toxicity of these drugs is very high. Excepting a few, most of the antifungal drugs developed so far are fungistatic, and only a few are clinically effective.

The antifungal agents developed so far are polyene macrolide antibiotics amphotericin B, nystatin, pimaricin, hamycin, etc. Other antibiotics like griseofulvin; (chemical) synthetic chemical preparations like thiocarbamates, alkylamines, azoles, etc.; and some miscellaneous agents like flucytosine, selenium sulfide, potassium iodide, haloprogin, sulfonamides, gentian violet paint, etc. are also used for the control of mycotic infections. Perfumes and essential oils from higher plants have been use in medicines and cosmetics (Jain and Agrawal, 1995; Jain and Agrawal, 2002). The volatile vapors of these are interfering the cellular metabolism of microorganisms and thus inhibiting or stimulating their growth (Jain and Agrawal, 1978; Singh and Deshmukh, 1984). Various essential oils, fixed oils and volatile organic compounds, have been reported to possess strong fungistatic activity. These include essential oils of *Amomum subulatum, Cinnamomum zeylanium, Balsamodendron mukul, Cyperus scariosus*, Euginol, *Zingiber officinalis, Ocimum sanctum, Curcuma longa, Moschus moschiferus, Jasminum officinarum, Mesua ferra, Jasminum sambac*, etc., against otomycotic fungi (Jain and Agrawal, 1976; Deshmukh and Agrawal, 1981; Deshmukh *et al.*, 1986; Jain and Agrawal, 1995; Chauhan *et al.*, 2001; Jain and Agrawal, 2002).

Fatty oils obtained from mustard, coconut, groundnut, amla and soyabean oil mixed with seeds of *Trigonella*, ajwain and garlic bulbs showed excellent antimycotic activity against otomycotic fungi (Jain and Agrawal, 1992; Jain *et al.*, 1993). Some volatile organic substances like ammonia, carbon disulfide, petroleum benzene, carbon dioxide, methanol, acetic acid, hydrogen peroxide were also found to have fungitoxic effect on fungi causing otomycosis (Jain and Agrawal, 1994).

IMMUNITY TO FUNGAL DISEASES

Human beings are continuously exposed to fungi, yet they rarely get fungal diseases. The delicate balance between the host and these otherwise harmless pathogens may turn into a parasitic relationship, resulting in the development of severe infections. Efficient responses to the different forms of fungi require different mechanisms of immunity. Dendritic cells (DC) are uniquely able at decoding the fungus-associated information and translating it in qualitatively different T helper (Th) immune responses. Myeloid DC phagocytosed yeasts and hyphae of *Candida albicans* and conidia and hyphae of *Aspergillus fumigatus*. Phagocytosis occurred through distinct phagocytic morphologies, involving the engagement and cooperativity of distinct recognition receptors. However, receptor engagement and cooperativity were greatly modified by opsonization. The engagement of distinct receptors translated into disparate downstream signaling events, ultimately affecting cytokine production and costimulation. *In vivo* studies confirmed that the choice of receptor and mode of entry of fungi into DC was responsible for the polarization and patterns of susceptibility or resistance to infection. Adoptive transfer of different types of DC activated protective, nonprotective and regulatory T cells, ultimately affecting the outcome of infection. The conclusions are that the selective exploitation of receptors and mode of entry into DC may determine the full range of host's immune relationships with fungi and have important implications in the design of vaccine-based strategies (Claudia *et al.*, 2002).

The natural immunity to fungal diseases is quite high, but the experience about immunity to the fungal diseases is limited as compared to bacterial, viral or parasitic infections. There are important differences in the host defenses that are primarily operative against different groups of mycoses. T- Cell mediated immune mechanism is predominant in containment of most fungal diseases like chronic mucocutaneous candidiasis, dermatophytosis, cryptococcosis, coccidioidomycosis, blastomycosis, paracoccidioidomycosis, sporotrichosis and for a relatively few infections like zygomycosis, invasive aspergillosis, disseminated candidiasis in which the host produces anti-neutrophilic response. Clinical observations and animal models of many mycoses suggest that although cell mediated immunity is important in prevention of fungal infection, protection against disseminated mycoses may depend upon humoral immunity (Matthews *et al.*, 1988; Romani and Howard, 1995).

Owing to our incomplete knowledge of fungal virulence factors and the host's antifungal immune responses, very few attempts have succeeded in obtaining immunization against mycotic infections with safe and reliable vaccines. The present chapter highlights the advances made in development of vaccines against fungal diseases.

APPROACHES FOR VACCINE DEVELOPMENT

Investigators are still in the process of unraveling the determinants of pathogenic fungi. Indeed, they have discovered several immunogenic substances from various fungi that can mediate protection in animal models. Researches have adopted several different kinds of approaches in order to achieve their objective of eliciting a safe and protective immune response like:

(i) live vaccine-using spores or variant non-pathogenic forms.

(ii) killed vaccine - using whole cell or cellular components.

(iii) recombinant live strains as vaccines - using recombinant DNA technology, the "Virulence" genes are knocked out to develop altered strains of pathogenic fungi, which can no longer cause disease but can induces appropriate immune response in the host.

(iv) vaccination by passive immunization - using polyclonal or monoclonal antisera, the host could be passively immunized.

(v) subcellular components as vaccine - using ribosomal fractions or cytoplasmic extracts, the immune system could be stimulated to induce various levels of resistance to infection.

(vi) fungal peptides as vaccines – using some antigenic peptides from fungal pathogens, protective immune response could be generated. Peptides also induce the production of immuno-protective cytokines, antibodies and cytolytic T cells. They can even be edited to remove suppressive epitopes that might prevent a useful immune system response. This is one of the active areas of research, since recent experiments in developing multiple – antigen peptide based vaccine in malaria has given encouraging results. Based on this model a fungal vaccine that consists of linked multiple epitopes may be designed.

(vii) fungal surface polysaccharides - virtually all medically important fungi carry an impressive array of polysaccharides on their cell surfaces, and hence their use in development of vaccines is possible.

(viii) idiotypic vaccination - monoclonal antibodies are used to raise anti-idiotypic antibody, which exhibits antimicrobial activity against pathogenic microorganisms, (a new concept of achieving systemic and secretary immunoprotection). Anti-idiotypic antibodies are those that exert internal imaged antimicrobial toxins because of the occurrence of complementary idiotype - like cell wall receptors. Antibiobodies are raised by immunization of animals with a monoclonal antibody that neutralizes the activity of *Pichia anomala* killer toxin characterized by a wide spectrum of antimicrobial activity. The increasing titers of antibiobodies in response to the monoclonal antibodies administered as vaccines, confers remarkable immunoprotection.

(ix) autoantibody as vaccines - the immune response to microbial infection often focuses on high conserved self-like antigens such as heat shock proteins (hsp). Cohen and Young (1991) suggested that autoantibody against a conserved epitope can be protective (Matthews *et al,* 1991). Thus, monoclonal antibodies against a conserved epitope can be used as vaccines. An immune response to these dominant antigens has already been anticipated by the preformed regulatory network which channel the autoimmune response down pathways which prevent the development of autoimmune diseases.

VACCINATION AGAINST BLASTOMYCOSES

Blastomyces dermatiditis is a fungus, which causes fatal lung infection - blastomycosis, even in people with healthy immune systems. According to experts live vaccine is the best way to achieve adequate immune response for blastomycosis. However, there is always a chance that such variants will spontaneously revert to the disease causing form. Using recombinant DNA technology to irreversibly knock out a virulence gene can eliminate this risk. To do so, one must find and inactivate a "Virulence"gene which is a relatively difficult task in a fungus that has a genome of 25 million base pairs of DNA. A particular virulence gene, WI-1 in this fungus has been found to encode for a surface protein molecule that allows the fungus to stick and stay in the lung (Wuthrich *et al.,* 2000). The WI-1 protein also interferes actively with the immune response by disrupting the balance of defensive molecules called cytokines produced by the immune system. The WI-1 protein is probably the major reason for *B. dermatiditis* infection which is so potent that 10 to 100 cells of this organism are enough to cause fatal infection in healthy mice. When this gene was knocked out with recombinant DNA technology, it was found that the altered *B. dermatiditis* strain could no longer cause disease in laboratory mice.

Understanding of fundamental mechanisms of vaccine immunity will allow proper use and optimization of vaccines. Vaccination with a genetically engineered, live, attenuated strain of *Blastomyces dermatitidis* carrying a targeted deletion at the BAD1 locus confers sterilizing immunity against experimental lethal pulmonary infection. Wuthrich *et al.* (2002) have found in their study that alpha beta T cells are required for durable vaccine immunity, whereas other T and B cells are dispensable. In immune-competent animals, CD4(+) T cell derived cytokines TNF-alpha and IFN-gamma mediate vaccine immunity. These factors have been found to be dispensable in immune-deficient animals, which rely on alternate mechanisms for robust vaccine immunity.

Studies by Wuthrich *et al.* (2002) classified the cellular and molecular bases behind the first genetically engineered fungal vaccine. They also illustrate a sharp difference in vaccine mechanisms between immune-competent and immune-deficient hosts, which underscores the plasticity of residual immune elements in compromised hosts.

VACCINATION AGAINST COCCIDIODOMYCOSES

It has long been established that T-cell-mediated immunity might play a pivotal role in a host's defence against fungal infections (Cox, 1983, Shubitz *et al.*, 2002). In particular, it is considered the principal mechanism of resistance of *Coccidioides immitis*, the causative agent of coccidioidomycosis. High titers of antibodies to *C. immitis* antigens, as detected by complement fixation, are a poor prognostic sign (Pappagianis and Zimmer, 1990). *C. immitis* antigen stimulates the cell-mediated rather than humoral response, might be essential for the subsequent isolation of fungal macromolecules, eliciting immunoprotection, and for the ultimate development of a vaccine against coccidioidomycosis. It has been demonstrated that a soluble conidial wall fraction (SCWF) of *C. immitis* is a potent elicitor of T-cell proliferation in a murine lymph node assay (Cole *et al.*, 1987). The antigenic components of the SCWF in the molecular size range of 43-66 kDa have been reported to be the most immunoreactive. A cDNA expression library of the mycelial form of *C. immitis* was constructed in λgt11 and the library was screened with anti-SCWF by Kirkland *et al.* (1991). Positive clones, that expressed β-gal fusion proteins (FPs), were tested for reactivity by the T-cell immunoblot method (Lee *et al.*, 1989) using a murine T-cell line which showed a specific proliferative response to the presence of the SCWF (Kirkland *et al.*, 1991) Western blot analysis of the SCWF, using anti-FP antibody, identified a single *C. immitis* native protein of approximately 47 kDa (Kirkland *et al.*, 1991). The anti-FP antibody localized the protein in the cell wall of the arthroconidia as well as in the plasma membrane and wall of the spherules of *C. immitis* (Kirkland *et al.*, 1991). A Northern blot of total mycelia RNA with the 200-bp cDNA as the probe identified a single 1.9-kb band. Examination of the GenBank database for amino acid sequence homology with the 7.3 kDa *C. immitis* protein revealed 63% identity with human 4-hydroxyphenylpyruvate dioxygenase (4-HPPD), a liver enzyme with a molecular size of 45 kDa that catalyses the formation of homogenisate from 4-hydroxyphenylpyruvate (Ruetschi *et al.*, 1993). Identification of T-cell reactive component of the SCWF led to the cloning of the full-length gene to examine the significance of its homology to the human enzyme. Alignment of the *C. immitis* and human 4-HPPD amino acid sequence revealed 53% identity and 75% homology, taking into account the conservative substitutions. The partially sequenced *C. immitis* gene predicted a molecular mass of 42 kDa for the protein. Human 4-HPPD has shown 87-90% identity to an F antigen reported in rats and mice (Ruetschi *et al.*, 1993). The mammalian F antigen, which was first defined in mice (Fravi and Lindenmann, 1968), is a liver protein associated with self-tolerance (Oliveira, 1989). It is of possible significance that the tolerance is maintained at the T-cell but not at the B-cell level and that high intravenous doses of the self F protein can produce F protein specific suppressor cells (Lukic and Mttchison, 1984). T-cell assay indicated that the *C. immitis* protein is a candidate molecule for eliciting immunoprotection against coccidioidal infection, its homology to the mammalian F antigen suggests that, this antigen may suppress the T-lymphocyte response (Abuodeh *et al.*, 1999). Animal studies will eventually provide further insight into this immunological conundrum. The complexity, however of the dual interaction between humoral and cellular immune responses calls for innovative approaches for active and passive fungal immunoprophylaxis.

Coccidioides immitis antigens which stimulate a T helper cell 1 (Th1) pathway of host immune response have been considered to be the essential components of a vaccine against coccidioidomycosis. Recombinant urease (rURE) and recombinant heat shock protein 60 (rHSP60) of *C. immitis* were expressed in *Escherichia coli* and tested as vaccine candidates in BALB/c mice. Data obatained by Kun *et al.* (2001) support the need for further evaluation of the *C. immitis* urease as a candidate vaccine against coccidioidomycosis.

VACCINATION AGAINST DERMATOPHYTOSES

Veterinary mycology has laid the basis for experimental, large-scale prevention of mycoses, particularly with the dermatophytoses. *Trichophyton verrucosum*, affecting cattle, calves and other animals, has received most attention due, to compelling economic reasons, since in some countries ringworm- infected animals can not be marketed (Liven and Sternwig, 1985). *T. verrucosum* is the only dermatophyte for which a permanent state of resistance to reinfection may develop. *T. verrucosum* may be transmitted to humans causing severe lesions which can leave permanent scars after healing (Rippon, 1988). Healing and subsequent resistance to reinfection were found to be related to the development of a positive delayed – type cutaneous hypersensitivity reaction. Kielstein (1968) indicated that activation of the cell- mediated component of the immune system is an important mechanism involved in resistance. A state of immunity against infection with *T. verrucosum* develops not only after natural infection but following vaccination as well. A vaccine, based on a living strain of *T. verrucosum* and thus limited in its use in certain countries, has been successfully utilized in Russia (Sarkisov *et al.*, 1971). Nevertheless, attempts to develop alternative vaccines are worthwhile endeavors.

The RF of *Microsporum canis* were also reported to induce such a reaction (Elad and Segal, 1994). The studies of several workers (Georg and Camp, 1957; Bradford, 1976; Spoljar and Tomasic, 2000) on animals using two subcellular fractions i.e., crude ribosomal fraction (CRF) and cytoplasmic fraction (CF) prepared for ammunization indicated that, although both the *T. verrucosum* subcellular fractions may be candidates as a basis for the development of a vaccine against bovine trichophytosis, the CRF is better suited for this purpose.

VACCINATION AGAINST CANDIDIASIS

Superficial and deep-seated candidiasis are important opportunistic diseases in immunocompromised patients. The mucocutaneous forms of this mycosis may also affect a large number of normal subjects. It has been assumed that approximately two-third of fertile women experience at least one episode of candidal vaginitis during their life span (Odds, 1988). Oral and oesophageal candidiasis occur quite frequently in AIDS patients .The systemic candidiasis is an increasing cause of death among immunocompromised AIDS patients. This suggests that humoral immunity might play an important role in preventing candidiasis infection and hence an antibody response to specific or immunodominant antigens may be useful in the prevention and treatment of systemic candidiasis (De-Bernardis *et al.*, 2002).

A number of studies, concerned with the stimulation of the killer toxin-microbial cell wall receptor interaction (Idiotypic-anti-idiotypic interaction), have been carried out to establish conceptually new prophylactic and therapeutic models against candidiasis (Polonelli *et al.*, 1991). Antibiobodies have been defined an anti-idiotypic antibodies that exert an antibiotic activity against pathogenic micro-organisms susceptible to internal imaged antimicrobial toxins because of the occurrence of complementary idiotype – like cell wall receptors (Polonelli *et al.*, 1991a). Polonelli *et al.* (1990) have raised antibiobodies with *in vivo* and *in vitro* activity against *C. albicans* cells by interaction with specific cell wall receptors by immunization of animals (rabbits, rats, mice) with a monoclonal antibody (MAb KT4) that neutralizes the activity of a *Pichia anomala* killer toxin (Polonelli and Morace, 1986). When administered as a vaccine by parenteral (Polonelli *et al.,* 1993) or intravaginal (Polonelli *et al.,* 1994) routes, MAb KT4 has been shown to confer a remarkable systemic or secretory immunoprotection to animals against high concentrations of infecting *C. albicans* cells administered by the intravenous or intravaginal routes. This has been attributed to the production of increasing titers of antibiobodies in serum and in vaginal fluid thus defining the new concept of idiotypic vaccination. Secretory antibiobodies, elicited in the vaginal fluid of rats by intravaginal idiotypic vaccination with MAb KT4 have been able passively to transfer immunoprotection to unvaccinated animals (Polonelli *et al.*, 1994). The internal imaging of the yeast killer toxin by antibiobodies implies the simulation of the yeast killer toxin-cell wall receptors by the idiotype of the neutralizing monoclonal antibody. Thus, theoretically, the immune system should recognize the specific yeast killer toxin-cell wall receptors of sensitive microorganisms as the idiotype of the monoclonal antibody used as an immunogen in idiotypic vaccination.

The occurrence of antibiobodies in animals immunized with *C. albicans* cells suggested the possibility that natural antibiobodies could be detected in the normal antibody response of individuals with candidiasis. Human natural antibiobodies have been effectively characterized mostly in the vaginal fluid of women affected by recurrent vulvovaginal candidiasis as well as in the saliva and serum of AIDS patients in comparison with control groups. The exciting finding of the possible fungal nature of *Pneumocystis carinii* (Edman *et al.*, 1988) has permitted the *in vitro* verification of the sensitivity of the trophozoites of that micro-organism to the *P. anomala* killer toxin and to visualize by immunofluorescence specific yeast killer toxin superficial receptors in their cell walls (Aliouat *et al.*, 1993) using MAb KT4. The hypothesis of a new approach to idiotypic vaccination and anti-idiotypic therapy against *P. carinii* pneumonia by the corresponding engineered antibody-combining site becomes very attractive. Preliminary results have shown that antibiobodies may be present in the bronchoalveolar washings of HIV-infected patients with pneumocystosis. Human recombinant antibiobodies, developed by the polymerase chain reaction to isolate the variable region genes from the peripheral blood lymphocytes of patients recovering from systemic mycoses, may represent conceptually new therapeutants that show no antagonism with conventional antifungal drugs (Barbas *et al.*, 1992).

The epitopes of both candidal and human hsp 90 have been mapped with sera from patients with systemic candidiasis who had antibody to the 47kDa antigen (Matthews *et al.*, 1991a; Al-Dughayam *et al.*, 1994). All patients examined recognized the conserved epitope LKVIRK on candidal hsp 90, and seven out of nine recognized a similar site on human hsp 90, KILKVIRK. Three patients with invasive aspergillosis and one with malaria also recognized the latter, but not by six patients with systemic lupus erythematosus. Sera from two patients with antibody to this epitope, taken before starting antifungal chemotherapy, were given prophylactically to mice challenged intravenously with a lethal dose of *C. albicans* (Matthews *et al.*, 1991). Mortality fell from 100% to 50% or less after 24 h the same effect was observed with a purified immunoglobulin fraction from one patient's serum. Normal human serum and the immunoglobulin fraction of normal serum had no effect on mortality. Prophylactic administration of a monoclonal antibody, raised against the synthetic peptide LKVIRK, was associated with a fall to 66% and, with a less lethal dose; mortality fell by 40% after 24h in all of the experiments. The antibody was administered to the mice 1 h before challenge.

A phage antibody library was used as a source of single-chain Fv (ScFv) against the overlapping peptide LKVIRK NIVKKMIE (A) and NKILK VIRK NIVKK (B). They were assessed in a similar mouse model. This time the agent was administered several hours after a lethal dose of *C. albicans*. Mortality at 24 h was 10% with the ScFv against peptide B, 94% with the ScFv against A, 64% with saline and 76% with the helper phage M13K07. The peptide NKILK VIRK NIVKK is in the middle of the proposed protein-binding site of human hsp 90 (Edman *et al.*, 1988). Human and yeast hsp 90 probably serve a similar chaperon function within the cells, which entails binding to a large variety of proteins. If circulating candidal hsp 90 binds to human serum proteins, causing them to malfunction, this could underlie the mechanism whereby antibody to this epitope is protective.

Candida albicans mannan extracts encapsulated in liposomes were previously used to stimulate mice to produce antibodies protective against candidiasis. Mannan-protein conjugates without liposomes were tested as vaccine candidates (Han *et al.*, 1999). Mannan extracts were coupled to bovine serum albumin, and isolated conjugates consisted of carbohydrate and protein at a ratio of 0.7-1.0. Vaccination of mice with the conjugate and an adjuvant yielded antiserum that contained *Candida* agglutinins. Vaccinated mice challenged with yeast cells had a mean survival time of 56 days, compared with <13 days for control groups. The antiserum protected naive animals against disseminated disease. Naive mice given the antiserum intravaginally developed 79% fewer fungal colony-forming units, compared with control groups. The serum-protective factor was stable at 56°C and was removed by adsorption with yeast cells. It is concluded that the conjugate vaccine can induce protective antibody responses against experimental disseminated candidiasis and vaginal infection caused by *Candida.* In some cases of atopic dermatitis (AD), a possible pathological contribution to disease development by *Candida albicans* has been suggested. AD patients with severe symptoms showing positive capsulated hydrolic carrier polymer radioallergosorbent test (CAP-RAST) against *C. albicans* demonstrated significantly higher levels of serum IgE Abs than did AD patients with mild symptoms. Based on the clinical facts, it can be postulated that elimination of *C. albicans* by mucosal vaccination may lead to the restoration of severe symptoms in AD patients. Mice which were systemically challenged with *C. albicans*-associated antigen, manganese superoxide dismutase (MnSOD) or secreted aspartic proteases 2 (SAP2), together with alum, exhibited hyper IgE Abs. Systemically primed mice were then immunized with MnSOD or SAP2 plus cholera toxin (CT) as mucosal adjuvant through the nasal route. Interestingly, nasally immunized mice showed increased levels of *Candida* Ag-specific IgA Ab in fecal and nasal washes as well as in saliva samples but unchanged levels in Ag-specific IgE responses. Consistent with the Ab levels, high numbers of *Candida* Ag-specific IgA Ab-forming cells were induced in mononuclear cells isolated from intestinal lamina propria, nasal passages and salivary glands of nasally vaccinated mice with Ag plus CT. Furthermore, nasal immunization using MnSOD or SAP2 together with CT resulted in the elimination of colonized *C. albicans* from the intestinal tract. These results also suggest a potential role of mucosal vaccination in the control of *C. albicans* in patients with allergic diseases, including AD, although more research is needed to establish this therapeutic approach for mucosal vaccination (Suenobu *et al.*, 2002)

The gene coding for enolase from *Candida albicans* was recently cloned, sequenced, and expressed in a eukaryotic system. Although the recombinant protein elicits a cellular immune response in animals colonized with this fungus, the protective efficacy of this immunogenic protein has not been evaluated.

VACCINATION AGAINST PENICILLOSIS

Most infections caused by dimorphic fungi are usually acquired through inhalation. Wong *et al* (2002) investigated the usefulness of Mp1p (a secreted cell wall antigen encoded by MP1)-based vaccines for generation of protective immune responses against *Penicillium marneffei* infection using a mouse model, and compared the relative effectiveness of intramuscular MP1 DNA vaccine, oral mucosal MP1 DNA vaccine delivered by live-attenuated *Salmonella typhimurium*, and intraperitoneal recombinant Mp1p protein vaccine. They concluded that the intramuscular MP1 DNA vaccine offers the best protection against *P. marneffei* infection but the mechanism remains to be elucidated. Their observation prompted further clinical trials on the use of MP1 DNA immunization on asymptomatic human immunodeficiency virus carriers in *P. marneffei* endemic areas (Wong *et al.*, 2002).

VACCINATION AGAINST ASPERGILLOSIS

Invasive pulmonary aspergillosis is an emerging devastating infection in the immunocompromised host that is treated with corticosteroids for neoplastic disease or for organ transplantation. By use of a model of invasive pulmonary aspergillosis in corticosteroid-treated CF-1 mice, prior infection and 2 *Aspergillus fumigatus* vaccine preparations (sonicate and filtrate) administered intranasally and subcutaneously were tested for efficacy in protecting against subsequent lethal *A. fumigatus* infection. The sonicate vaccine was more protective, but the subcutaneous route was more effective (Ito and Lyons, 2002).

Aspergillus fumigatus, an opportunistic fungal pathogen, is responsible for multiple airway diseases of an allergic and a nonallergic nature. Cenci *et al.* (2000) assesed the ability of *Aspergillus* crude culture filtrate Ags and the recombinant allergen Asp f 2 to induce protective antifungal responses in mice with invasive pulmonary aspergillosis. Similar to what occurred upon nasal exposure to viable *A. fumigatus* conidia, treatment of immunocompetent mice with *Aspergillus* crude culture filtrate Ags resulted in the development of local and peripheral protective Th1 memory responses, mediated by Ag-specific CD4+ T cells producing IFN-gamma and IL-2 capable of conferring protection upon adoptive transfer to naive recipients. Protective Th1 responses could not be observed in mice deficient of IFN-gamma or IL-12 and did not occur in response to Asp f 2, which, on the contrary, elicited high level production of inhibitory IL-4. The results show that Ags of *Aspergillus* exist with the ability to induce both Th1- and Th2-type reactivity during infection, a finding that suggests a possible mechanism through which potentially protective immune responses are inhibited in mice with the infection. However, the occurrence of Th1-mediated resistance upon vaccination with *Aspergillus* crude culture filtrate Ags, suggests the existence of fungal Ags useful as a candidate vaccine against invasive pulmonary aspergillosis.

In a murine model of invasive pulmonary aspergillosis, dendritic cells (DCs) pulsed with *Aspergillus* antigens induced the activation of CD4(+) Th1 cells capable of conferring resistance to the infection. Bozza *et al.* (2002) showed that the combined, local delivery of unmethylated CpG oligodeoxynucleotides (ODNs) and the Asp f 16 *Aspergillus* allergen resulted in the functional maturation and activation of airway DCs capable of inducing Th1 priming and resistance to the fungus. Therefore, ODNs act as a potent adjuvant for the vaccine-induced protection against the fungus by promoting dominant Th1 response to *Aspergillus* antigens and allergens.

VACCINATION AGAINST *HISTOPLASMA*

Vaccination with heat shock protein 60 (Hsp60) from *Histoplasma capsulatum* induces a protective immune response in mice. Deepe and Gibbons (2002) explored the cellular and molecular requirements for the efficacy of recombinant Hsp60 in mice. Depletion of CD4(+), but not CD8(+), cells during the inductive phase of vaccination abolished protection, as assessed by survival and by the fungal burden in lungs and spleens. In the expressive phase, the elimination of CD4 (+) or CD8 (+) cells after immunization did not significantly alter fungal recovery or survival from a lethal challenge. Depletion of both subpopulations after Hsp60 vaccination resulted in a failure to control a lethal infection and a higher fungal burden in lungs and spleens. Cytokine release by spleen cells from mice vaccinated with Hsp60 produced substantially more gamma interferon and interleukin-10 and -12 than that of cells from mice immunized with either *H. capsulatum* recombinant Hsp70 or bovine serum albumin. The generation of gamma interferon, but not of interleukin-10, was dependent on T cells, in particular CD4 (+) cells. Treatment of Hsp60-immunized mice with monoclonal antibody to gamma interferon or interleukin-10 or -12 in the inductive phase of vaccination was accompanied by increased recovery of yeast cells from lungs and spleens and 100% mortality. Likewise, the neutralization of gamma interferon or interleukin-12 abolished the protective effect of Hsp60 in the expressive phase. These results delineate the complexity of the regulatory elements necessary for vaccination against this fungus. Epitopes of a protective immunogen from *Histoplasma capsulatum,* namely, heat shock protein 60, are being mapped. In addition, considerable interest is focusing on the product of a yeast phase-specific gene from *H. capsulatum, yps3 that* is not detected in the mycelia phase. Although located in either the cell wall or cell membrane of the yeast, little is known about the function of this protein. However, because it stimulates human lymphocytes to proliferate, this protein is a target of interest to investigators interested in stimulating a protective immune response to this fungal pathogen.

HIS-62 is a glycoprotein that has been isolated from the cell wall and cell membrane fraction of the pathogenic fungus *Histoplasma capsulatum*. It is a target of the cellular immune response to this fungus, and it protects mice against a lethal intravenous inoculum of *H. capsulatum* yeast cells. Gomez *et al.* (1995) cloned the gene encoding this antigen to reveal its biological nature and studied the immunological activity of recombinant antigen. The amino acid sequences of the NH_2 terminus and internal peptides were obtained by Edman degradation. Degenarate oligonucleotides were used to isolate a gene fragment of HIS-62 by PCR. One 680-bp segment that corresponded to the known peptide sequence was amplified from *H. capsulatum* DNA. This DNA was used to screen a genomic library, and the full-length gene was isolated and sequenced. The deduced amino acid sequence of the gene demonstrated approximately 70 and approximately 50% identity to heat shock protein 60 (hsp 60) from *Saccharomyces cerevisiae* and hsp 60 from *Escherichia coli*, respectively. A cDNA was synthesized by reverse transcription PCR and was expressed in *E. coli.* Recombinant protein reacted with a monospecific polyclonal rabbit antiserum raised against native HIS-62, with monoclonal HIS-62-reactive T cells, and with splenocytes from mice immunized with viable yeast cells. Moreover, vaccination with the recombinant protein conferred protection in mice against a lethal intranasal inoculation with yeast cells. Thus, HIS-62 is a member of the hsp 60 family, and the recombinant hsp 60 is protective against pulmonary histoplasmosis in mice.

VACCINATION AGAINST *CRYPTOCOCCOSIS*

Vaccination and infection can elicit protective and nonprotective antibodies to the fungus *Cryptococcus neoformans* in mice. The effect of nonprotective antibodies on host defense is unknown. Nussbaum *et al* (1996) used mixtures of protective and nonprotective monoclonal antibodies (mAbs) to determine if nonprotective mAbs blocked the activity of the protective mAbs. Antibody isotype and epitope specificity are important in determining the ability to prolong survival in mice given a lethal *C. neoformans* infection. Three different nonprotective immunoglobulin (Ig) G23 mAbs to cryptococcal capsular polysaccharide were used to study the interaction between the IgG3 isotype and protective IgG1 and IgG2a mAbs in murine cryptococcal infection. One IgG3 mAb reduced the protective efficacy of an IgG1 with identical epitope specificity. A second IgG3 mAb with different epitope specificity also reduced the protection provided by the IgG1 mAb. The protective efficacy of an IgG2a mAb was also dramatically decreased by still another IgG3 mAb. To our knowledge this is the first report of blocking antibodies to a fungal pathogen. The results have important implications for the development of vaccines and passive antibody therapy against *C. neoformans*.

Mice immunized with two different cryptococcal antigen preparations, one a soluble culture filtrate antigen (CneF) in complete

Freund's adjuvant (CFA) and the other heat-killed *Cryptococcus neoformans* cells (HKC), develop two different profiles of activated T cells. CneF-CFA induces CD4+ T cells responsible for delayed-type hypersensitivity (DTH) reactivity and for amplification of the anticryptococcal DTH response, whereas HKC induce CD4+ and CD8+ T cells involved in anticryptococcal DTH reactivity and activated T cells which directly kill *C. neoformans* cells. Murphy *et al.* (1998) assessed the level of protection afforded by each of the two different T-cell profiles against challenge with viable *C. neoformans* cells, thereby identifying which activated T-cell profile provides better protection. CBA/J mice immunized with CneF-CFA had significantly better protective responses, based on better clearance of *C. neoformans* from tissues, on longer survival times, and on fewer and smaller lesions in the brain, than HKC-immunized mice or control mice similarly infected with *C. neoformans*. Both immunization protocols induced an anticryptococcal DTH response, but neither induced serum antibodies to glucuronoxylmannan, so the protection observed in the CneF-CFA immunized mice was due to the activated T-cell profile induced by that protocol. HKC-immunized mice, which displayed no greater protection than controls, did not have the amplifier cells. Murphy *et al.* (1998) proposed that the protective anticryptococcal T cells are the CD4+ T cells which have been shown to be responsible for DTH reactivity and/or the CD4+ T cells which amplify the DTH response and which have been previously shown to produce high levels of gamma interferon and interleukin 2. Our results imply that there are protective and nonprotective cell-mediated immune responses and highlight the complexity of the immune response to *C. neoformans* antigens.

Cell-mediated immune (CMI) responses defined by delayed-type hypersensitivity (DTH) reactivity to cryptococcal culture filtrate antigen (CneF) can be either protective or nonprotective against an infection with *Cryptococcus neoformans* (McGaha and Murphy, 2000). The protective and nonprotective anticryptococcal DTH responses are induced by different immunogens and have differing activated-T-cell profiles. They examined the effects of blockade of the interaction between cytotoxic T lymphocyte antigen 4 (CTLA-4) and its ligands B7-1 (CD80) and B7-2 (CD86) on the anticryptococcal DTH responses and protection. They found that CTLA-4 blockade at the time of immunization with the immunogen that induces the protective response, CneF, in complete Freund's adjuvant (CFA) or the immunogen that induces the nonprotective response, heat-killed cryptococcal cells (HKC), enhanced anticryptococcal DTH reactivity. In contrast, blocking CTLA-4 after the immune response was induced failed to enhance responses. Blockade of CTLA-4 in an infection model resulted in earlier development of the anticryptococcal CMI response than in control mice. Concomitant with increases in DTH reactivity in mice treated with anti-CTLA-4 Fab fragments at the time of immunization; there were decreases in cryptococcal CFU in lungs, spleens, and brains compared to controls. Blockade of CTLA-4 resulted in long-term protection, as measured by significantly increased survival times, only in mice given the protective immunogen, CneF-CFA. Anti-CTLA-4 treatment did not shift the response induced by the nonprotective immunogen, HKC, to a long-term protective one. These results indicate that blockade of CTLA-4 interactions with its ligands may be useful in enhancing host defenses against *C. neoformans.* Although *Cryptococcus neoformans* causes serious infections in AIDS patients, cryptococcosis in immunologically immature infants, as in immunocompetent adults, is rare. Aguirre *et al* (1997), investigated the resistance of neonates to *C. neoformans* and to determine whether they could be efficiently vaccinated as neonates against challenge with the yeast as adults, the course of infection was monitored in the lungs of mice infected intranasally with yeast cells. Neonates were less able than adults to reduce yeast burdens less than 24 h postinoculation and less able to control the progressive growth of yeast over several weeks. However, in both neonates and adults, yeast's were substantially eradicated by 6 to 8 weeks after infection. Numbers of all classes of leukocytes recovered from lung lavages of infected neonates and adults were similar. Significant differences appeared only on day 14, when neonates had more neutrophils and adults had significantly more CD4+ CD45RB cells with low fluorescence intensity. When vaccinated neonates were rechallenged after reaching adulthood, they expressed resistance to *C. neoformans* as effectively as did mice immunized as adults and survived an intravenous challenge that was lethal for unimmunized controls. Thus, exposure of neonatal mice to viable *C. neoformans* yeast, which persists in the lungs for many weeks, does not result in immunological tolerance to a yeast challenge in adult mice, as predicted by immunological dogma, but instead immunizes them. Therefore, even in immunologically immature individuals, the immune system serves to protect against pathogens rather than simply to distinguish self from nonself .

The prevalence and specificity of serum antibodies to *Cryptococcus neoformans* proteins was studied in mice and rats with experimental infection, in individuals with or without a history of potential laboratory exposure to *C. neoformans*, human immunodeficiency virus (HIV)-positive individuals who developed cryptococcosis, in matched samples from HIV-positive individuals who did not develop cryptococcosis, and in HIV-negative individuals. Rodents had little or no serum antibody reactive with *C. neoformans* proteins prior to infection. The intensity and specificity of the rodent antibody response were dependent on the species, the mouse strain and the viability of the inoculum. All humans had serum antibodies reactive with *C. neoformans* proteins regardless of the potential exposure, the HIV infection status, or the subsequent development of cryptococcosis. Our results indicate (i) a high prevalence of antibodies reactive with *C. neoformans* proteins in the sera of rodents after cryptococcal infection and in humans with or without HIV infection; (ii) qualitative and quantitative differences in the antibody profiles of HIV-positive individuals; and (iii) similarities and differences between humans, mice, and rats with respect to the specificity of the antibodies reactive with *C. neoformans* proteins. The results are consistent with the view that *C. neoformans* infections are common in human populations, and the results have implications for the development of vaccination strategies against cryptococcosis (Chen *et al.*, 1999).

CHALLENGES

Research efforts on the lines described above may soon become fruitful bringing vaccines for several fungal diseases for use of common man. However, the major challenge that the researchers face is to find appropriate mode of delivery of the immunogens derived from fungal pathogens. Some vaccine candidates like peptides show poor immunogenecity when they are injected into animals. So they must be inserted into a suitable vector for delivering a wide range of vaccine components. Another reason for vaccine failure is the presence of trace contaminants, which may produce blocking antibodies that nullify the protective effects of the vaccine. The route of administration of vaccines is an important challenge, since most of the pathogenic fungi infect the host via the mucosal surfaces of the gastrointestinal or respiratory tracts for which they have specific tropisms. Attempts to induce protection by using vaccines targeted at the secretary immune system should be considered an important objective.

FUTURE PROSPECTS

The vaccine development efforts aimed at preventing fungal infections are currently being augmented by several new technologies and approaches that are being brought to these challenge. It is becoming increasingly easier to define precisely which fungal epitopes can induce protective responses by the host and then applying this information to the development of candidate vaccines. Combined applications of molecular genetics and sophisticated chemical instrumental analysis have enabled researchers to map fungal components epitopes that may induce protective antibodies. Tools have now become available to study gene manipulation in fungal pathogen to allow molecular approaches to be used. Genetic studies may also be accelerated by the availability of the whole genomes of the medically important pathogens. The task is challenging but there are reasons to expect considerable progress in the coming years.

REFERENCES

Abuodeh, R.O., Shubitz, L.F., Siegel, E., Snyder, S, Peng, T., Orsbom, K.I., Brummer, E., Stevens, D.A. and Galgiani, J.N. (1999) Resistance to *Coccidioides immitis* in mice after immunization with recombinant protein or a DNA vaccine of a proline – rich antigen. *Infection and Immunity* 67(6): 2935 – 2940.

Aguirre, K.M., Garvy, B.A. and Johnson, L.L. (1997). Acquired resistance to *Cryptococcus neoformans* in adult mice vaccinated as newborns. *Infec Immun* 65(5): 1688-94.

Ajello, L. (1953). The Dermatophyte *Microsporum gypseum* as a saprophyte and parasite. *Journal of Investigative Dermatology* 21: 157-171.

Ajello, L. and Padhye, A.A. (1972). Keratinophilic fungi of the Galapagos islands. *Mykosen* 17:239-243.

Ajello, L. (1974). Natural history of the dermatophytes and related fungi *Mycopath Mycol Appli* 53: 93-10.

Al-Dughaym, A.M., Mathews, R.C. and Burnie, J.P. (1994). Epitope mapping human heat shock protein 90 with sera from infected patients. *FEMS Immunology and Medical Microbiology* 8: 43-48.

Aliouat, E.M., Cailliez, J.C. Siguy, N., Deecas, E., Polonelli, L., Gerloni, M., Conti, S. and Camus, D. (1993). Inhibitory effect of a yeast killer toxin to the in vitro *Pneumocystis carinii* attachment. *Serodiagnosis and Immunotherapy of Infections Disease,* 5: 102-106.

Barbas, C.F., Crowe, J.E., Cababa, D., Jones, T.M., Zebedee, S.L., Murphy, B.R., Chanock, R.K. and Burton, D.R. (1992). Human monoclonal Fab fragments derived from a combinatorial library bind to respiratory syneytial virus F glycoprotein and neutralize infectivity. *Proceedings of the National Academy of Sciences, USA,* 89: 10164-10168.

Bozza, S., Gaziano, R., Lipford, G., Montagnoli, C., Bacci, A., Difrancesco, P., Kump, V.P., Wangner, H. and Romani, L. (2002). Vaccination of mice against invasive aspergillosis with recombinant *Aspergillus* proteins and CpG oligodeoxynucleotides as adjuvants. *Microbes Infect* 4(12): 1281-90.

Bradford, M.M. (1976). A rapid and sensitive method for the quantitation of microgram quantities of protein utilizing the principle of protein dye binding. *Analytical Biochemistry,* 72: 248-254.

Cenci, E., Mencacci, A., Bacci, A., Bistoni, F., Kurup, V.P. and Romani, L. (2000). T cell vaccination in mice with invasive pulmonary aspergillosis. *Journal of Immunology* 165 (1): 381-8.

Chauhan, U.K., Soni, P., Singh, V. and Jain, S.K. (2001). Antimicrobial activities of the Rhizome of *Curcuma longa* linn. *Indian Medical Gazette* CXXXV(8): 241-243.

Chen, L.C., Goldman, D.L., Doering, T.L., Pirofski, L. and Casadevall, A. (1999). Antibody response *Cryptococcus neoformans* Proteins in rodents and humans. *Infect Immun* 67(5): 2218-24.

Chmel, L. and Vlacilikova, A. (1977). Keranofilne huby, V Niektorych Podnych Typoch A factory Ovplyvnujuce Ich Zostupenie. *Biologia* (Brastislava) 32 : 53-59.

Claudia, M., Bacci, A., Silvia, B., Gaziano, R., Sperca, A. and Romani, L. (2002). The interaction of fungi with dendritic cells : Implications for the immunity and vaccination. *Curr Mol Med* 2(6):507-24.

Cohen, I.R. and Young, D.B. (1991). Autoimmunity, microbial immunity and the immunological homunculus, *Immunology Today,* 12: 105-110.

Cole, G.T., Kirkland, T.N. and Sun, S.H. (1987). An immunoreactive, water soluble conidial wall fraction of *Coccidioides immitis. Infection and Immunity,* 55: 657-667.

Cox, R.A. (1983). Cell mediated immunity In: D.H. Howard (Ed.*) Fungi Pathogenic for Humans and Animals,* Part B, *Pathogenicity and Detection,* Vol. I. Marcel Dekker, New York, 61 - 98.

De-Bernardis, F., Boccanera, M. Adrioni, D., Girolamo, A. and Cassone, A. (2002). Intravaginal and intranasal immunizations are equally effective in inducing vaginal antibodies and conferring protection against vaginal Candidiasis. *Infection and Immunity* 70(5): 2725– 9.

Deepe, G.S. and Gibbons, R.S. (2002). Cellular and molecular regulation of vaccination with heat shock protein 60 from *Histoplasma capsulatum. Infect Immun* 70(7): 3759-67.

Deshmukh, S.K. and Agrawal, S.C. (1981). Antimycotic activity of some essential oils. *Indian Drugs* 19 (1) : 24.

Deshmukh, S.K. and Agrawal, S.C. (1982). *Malbranchea sulfurea* (Miche) Sigler and Charmichael: A new record from India. *Current Science.* 51(16): 785-786.

Deshmukh, S.K. and Agrawal, S.C. (1983). Prevalence of dermatophytes and other keratinophilic fungi in soils of Madhya Pradesh (India). *Mykosen* 26(11):574-577.

Deshmukh, S.K. (1985). Isolation of dermatophytes and other keratinophilic fungi from soil of mussoorie (India), *Mykosen* 28 (2): 98-101.

Deshmukh, S.K., Jain, P.C. and Agrawal, S.C. (1986). A note on mycotoxicity of some essential oils. *Fitoterapia* LV II: 4.

Edman, J.C., Kovacs, J.A., Masur, H., Sannti, D.V., Elwood, H.J. and Sogin, M.L. (1988). Ribosomal RNA sequence shows *Pneumocystis carinii* to be member of the fungi, *Nature,* 334: 519-522.

Elad, D. and Segal, E. (1994). Immunogenicity in guine pigs of a crude ribosomal fraction from *Microsporum canis, Vaccine,* 12: 134-138.

Fravi, G. and Lindenmann, J. (1968). Induction by allogeneic extracts of liver specific precipitating autoantibodies in the mouse. *Nature,* 218: 134-138.

Fray, D. (1965). The isolation of keratinophilic and other fungi from soils collected in Australia and New Guinea. *Mycologia* 57: 202-215.

George, L.K. and Camp, L.B. (1957). Routine nutritional tests for the identification of dermatophytes. *Journal of Bacteriology,* 74: 113-121.

Gomez, F.G., Allendoerfer, R. and Deepe Jr, G.S. (1995). Vaccination with recombinant heat shock protein 60 from *Histoplasma capsulatum* protects mice against pulmonary histoplasmosis. *Infection and Immunity* 63 (7) : 2587 – 2595.

Han, Y., Ulrich, M.A. and Cutler, J.E. (1999). *Candida albicans* mannan extract – protein conjugates induce a protective immune response against experimental candidiasis. *Journal of Infectious Diseases* 179(6):1477- 84.

Ito, J.I. and Lyons, J.M. (2002). Vaccination of corticosteroid immunosuppressed mice against invasive pulmonary aspergillosis. *Journal of Infectious Diseases* 186(6): 869-71.

Jain, P.C. and Agrawal, S.C. (1983). An addition to Indian mycoflora. *Current Science* 52(12): 612.

Jain, P.C. and Agrawal, S.C. (1976). Activity of plant extracts against some keratinophilic species of *Nannazzia. Indian Drugs* 13 : 25 –26 .

Jain, P.C. and Agrawal, S.C. (1977). Keratinophilic fungi from the soils of Mount Abu (India) *Geobios* 4:136-138.

Jain, P.C. and Agrawal, S.C. (1978). Notes on the activity of some odoriferous organic compounds against some keratinophilic fungi. *Trans. Mycol. Soc. Japan.* 19:197-200.

Jain, S.K. and Agrawal, S.C. (1992). Sporostatic effect of some oils against fungi causing otomycosis. *Indian Journal of Medical Sciences* 46(1):1-6.

Jain, S.K., Agrawal, S.C. and Malaiya, S. (1993). Antimycotic effect of fixed oils treated with herbal seeds on the growth of fungi causing otomycosis. *Ancient Science of Life* XIII (1and2):160-164.

Jain, S.K. and Agrawal, S.C. (1994). Fungitoxic effect of some organic volatile substances against fungi causing otomycosis. *Mycoses* 37:299-301.

Jain, S.K. and Agrawal, S.C. (1995). Antimycotic activity of some essential oils. *Asian Journal of Experimental Science* 9(1and2): 1-3.

Jain, S.K. and Agrawal, S.C. (2002). Fungistatic activity of some perfumes against otomycotic pathogens. *Mycoses* 45 :88-90.

Jain, S.K., Singh, V. and Gupta, S. (2002). Immunological study of onychomycosis caused by *Candida albicans. Journal of Immunology and Immunopathology* 4 (1 and 2):101-103.

Jain, S.K., Agrawal, S.C. and Jain, P.C. (2003) Subcutaneous phaeohyphomycosis on face caused by *Cladophialophora bantiana. Mycoses* 46 (5 and 6) : 137-139.

Jain, S.K., Shrivastava, A. and Agrawal, S.C. (2004). Incidence of Candidiasis in Leukoplakia cases in cancer suspected patients. *In : Microbiology and Biotechnology for sustainable development* (Ed) Jain, P.C.. CBS publishers and distributors, India.

Kielstein, P. (1968). Immunologische Untersuchungen waehrend Der spontanen and experimentellen Rindertrichophytic. *Wiss Z Karl Mark Universitat Leipzig,* 17: 177 – 178.

Kirkland, T.N., Zhu, S., Kruse, D., Hsu, I., Seshan, K.R. and Cole, G.T. (1991). *Coccidioides immitis* fractions which are antigenic for immune T lymphocytes. *Infection and Immunity,* 59: 3952 – 3961.

Kun, L.I., Jieh-Juen, Yu, Chiung-Yu Hung, Paul, Lehman, F. and Garry, T. Cole (2001). Recombinant urease and urease DNA of *Coccidioides immitis* elicit on immunoprotective response against coccidioidomycosis in mice. *Infect Immun* 69(3): 2878-87.

Lee, S.P., Stoker, N.G., Grant, K.A., Handzel, Z.T., Hussain, R., McAdam, K.P.W.J. and Dockrell, H.M. (1989). Cellular Immune responses of leprosy contacts to fractionated *Mycobacterium leprae* antigens. *Infection and Immunity* 57: 2475 – 2480.

Liven, E. and Stenwig, H. (1985). Efficacy of vaccination against ringworm in cattle. *Nordisk veterinaermedicin,* 37: 187.

Lukic, M.L. and Mitchison, N.A. (1984). Self-specific and allo-specific suppressor T cells evoked by intravenous injection of F protein. *European Journal of Immunology*, 14 : 766 – 768.

Matthews, R., Burnie, J., Smith, D., Clark, L., Middley, J., Conolly, M. and Gazzard, B. (1988). *Candida* and AIDS evidence for protective antibody. *Lancet*, ii : 263-266.

Matthews, R.C., Burnie, J.P. and Lee, W. (1991). The application of epitope mapping in the development of a new serological test for systemic candidosis. *Journal of Immunology Methods*, 143: 73-79.

Matthews, R.C., Burnie, J.P. Howai, D., Rowland, E and Walton, F. (1991a). Autoantibody to heat shock protein 90 can mediate protection against systemic candidosis, *Immunology*, 74 : 20-24.

McGaha, T. and Murphy, J.W. (2000). CTCA-4 down regulates the protective anticryptococcal cell mediator immune response. *Infect Immun* 68(8): 4624-30.

Murphy, J.W., Schafer, F., Casadevall, A. and Adesina, A. (1998). Antigen induced protective and nonprotective cell mediated immune components against *Cryptococcus neoformans. Infec. Immun* 66(6):2632-9.

Nussbaum, G., Yuan, R., Casadevall, A. and Scharff, M.D. (1996). Immunoglobulin G3 blocking antibodies to the fungal pathogen *Cryptococcus neoformans. J Exp Med* 1; 183(4): 1905-9.

Odds, F.C. (1988). *Candida and Candidosis*, 2nd edn. Bailliere Tindall, London

Oliveira, D.B.G.(1989) F protein and immune suppression genes. *Immunology, suppl.*2; 26-32.

Pappagianis, D. and Zimmer, B.L. (1990). Serology of coccidioidomycosis, *Clinical Microbiology Reviews*, 3: 247-242.

Polonelli, L. and Morace, G. (1986). Reevaluation of the yeast killer phenomenon. *Journal of Clinical Microbiology*, 24: 866-869.

Polonelli, L. Conti, S., Gerloni, M., Magliani, W., Chezzi, C. and Morace, G. (1991). Interfaces of the yeast killer phenomenon. *Critical Reviews in Microbiology*, 18: 47-87.

Polonelli, L., Conti, S., Gerloni, M., Magliani, W., Castagnola, M., Morace, G and Chezzi, C. (1991a). "Antibiobodies" : antibiotic – like antiidiotypic antibodies. *Journal of Medical and Veterinary Mycology*, 29: 235-242.

Polonelli, L., Fanti, E., Conti. S., Campani, L., Gerloni, M. Castagnola, M., Morace, G. and Chezzi, C. (1990). Detection by immunofluorescent and anti idiotypic antibodies of yeast killer toxin cell wall receptors of *Candida albicans. Journal of Immunological Methods*, 132: 205-209.

Polonelli, L., Lorenzini, R., De Bernardis, F.L., Gerloni, M., Conti, S. Morace, G., Magliani, W. and Chezzi, C. (1993). Idiotypic vaccinations: Immunoprotection mediated by anti-idiotypic antibodies with antibiotic activity. *Scandinavian Journal of Immunology*, 37: 105-110.

Polonelli, L., De Bernards, F.L., Conti. S., Boccanera, M., Gerloni, M., Morace, G., Magliani, W., Chezzi, C. and Cassone, A. (1994). Idiotypic intravaginal vaccination to protect against candidal vaginitis by secretory, yeast killer toxin – like antiidiotypic antibodies. *Journal of Immunology*, 152: 3175-3182.

Rippon, J.W. (1988). Dermatophytosis and dermatomycosis *In : Medical Mycology,* 3rd edn. W.B. Saunders and Co., Philadelphia, P.A. pp. 169-275.

Romani L and Howard, D.H. (1995). Mechanisms of resistance to fungal infections. *Current opinion in Immunology,* 7 : 517 – 523.

Ruetschi, U. Dellsen, A., Sahlin, P. Stenman, G., Rymo, L. and Lindstedt, S. (1993). Human 4-hydroxyphenylpyruvate dioxygenase. Primary structure and chromosomal localization of the gene. *European Jounral of Biochemistry* 213: 1081-1089.

Sarkisov, A.K. Petrovich, S.V., Nikiforov, L.I., Yablochnik, L.M. and Korolev, V.P. (1971). Immunization of cattle against ringworm. *Veterinarya* 2: 54-56.

Sero, G.D., Mencaca G., Cenci, E., Fe dostaini, C., Montagnoli, C., Bacci, A., Mosci, P., Kopf, M. and Romani, L. (1999). Antifungal type 1 responses are upregulated in IL-10-deficient mice. *Microbes and Infection*1(14) :1169-1180.

Shubitz, L., Peng, T., Perrill, R., Simons, J., Orsbom, K. and Galgiani, J.N. (2002). Protection of mice against *Coccidioides immitis* intranasal infection by vaccination with recombinant antigen 2/PRA, *Infection and Immunity* 70(6): 3287 – 9.

Singh, K.V. and Deshmukh, S.K. (1984). Volatile constituents from members of Liliaceae and spore germination of *Microsporum gypseum* complexes *Fitoterapia* LV(5):297-300.

Spoljar, B.H. and Tomasic, J. (2000). A novel ELISA for determination of polysaccharide specific immunoglobulins. *Vaccine* 25; 19(7-8): 924-30.

Suenobu, N., Kweon, M.N., Kiyono, H. (2002). Nasal vaccination induces the ability to eliminate *Candida* colonization without influencing pre- specific IgE. Abs:A possibility for the control of *Candida* related atopic dermatitis. *Vaccine* 26; 20(23-24) : 2972-80.

Wong, L.P., Woo, P.C., Wu, A.Y. and Yuen, K.Y. (2002). DNA immunization using a secreted cell wall antigen Mp1p is protective against *Penicillium marneffei* infection *Vaccine* 26; 20(23-24) : 2878-86.

Wuthrich, M., Filutowicz, H.I. and Klein, B.S. (2000). Mutation of the WI-1 gene yields an attenuated *Blastomyces dermitiditis* strain that induces host resistance. *Journal of Clinical Investigation* 106(11): 1381-89.

CHAPTER - 14

BIOLOGY OF EDIBLE MUSHROOMS

M.P. Thakur

Department of Plant Pathology, Indira Gandhi Agricultural University, Raipur - 492 006, Chhatisgarh

INTRODUCTION

The word 'mushroom' means different things to different people and countries. The part of the organism that we see and call a mushroom is really just the fruiting body. Unseen is the mycelium- tiny threads that grow throughout the substrates collecting nutrients by breaking down the organic material. A clear definition is thus required in view of the specialized studies on the economic value of edible mushroom and their products. In a broad sense, "Mushroom is a macro fungus with a distinctive fruiting body which can be either epigeous or hypogeous and large enough to be seen with naked eyes and to be picked by hand" (Chang and Miles, 1992). Mushroom can be roughly divided into four categories (Chang, 2002b): (I). those which are fleshy and edible, fall into the edible mushroom category, e.g. *Agaricus bisporus* (II). Mushrooms which are considered to have medicinal applications, are referred to as medicinal mushrooms e.g. *Ganoderma lucidum* (III). Those which are proven to be or suspected of being poisonous are named poisonous mushrooms e.g. *Amanita phalloides* (IV). A miscellaneous category, which includes a large number of mushrooms whose properties remain less, well defined and tentatively be grouped as 'other mushrooms'. Chang (2002a) further included one more category of mushroom as 'inedible mushroom' constituting roughly five categories of mushrooms.

The world of mushrooms have always been fascinating and mystic to man owing to their sudden appearance in numbers, groups, rings, bunches and also in isolation as a single attractive and fascinating structures. Their sudden appearance has often led to localized beliefs and notions such as their association with thunder-storm and lightening and their typical growth could not be described better by any other terminology than 'mushroom growth'. Man has been fascinated with mushroom since time immemorial and references about the mushrooms are available in most ancient literature like Vedas and Bible (Verma, 1997). Theofrastus (372-287 BC) the great Greek Philosopher wrote about the food value of mushrooms during middle ages when mushrooms comprised royal dishes for the Greek and Roman Empires.

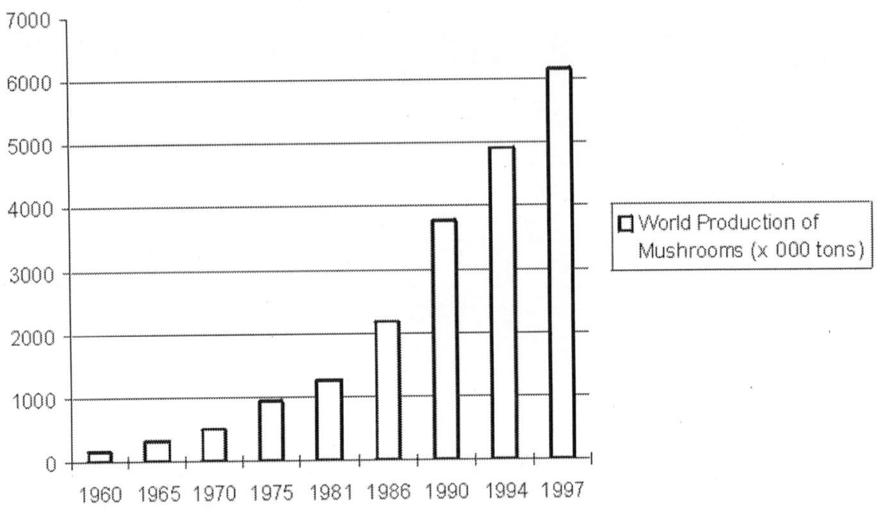

Fig. 1. World Production of mushrooms

Mushroom has become an ideal activity for generation of employment in urban, peri-urban and rural areas. Mushroom fungi has unique capacity to convert a wide variety of agriculture, forest and food processing waste in to a highly nutritious food and feed which have multibeneficial effects on human beings. A huge quantity of lignocellulosic waste (1.09 x 11^{11} tons of dry matter) is generated every year in the world including 357 million tons in India. But, it is not being properly utilized. Mushroom cultivation is one of the ways by which the voluminous waste can be recycled. World production of 7.2 million tons of mushroom could hardly utilize a negligible part of the waste available. However, there is still an ample opportunity to convert the available waste through mushroom cultivation. Mushroom cultivation generates the worth about 25 billion US $ from edible mushrooms including 6.0 billion US$ from medicinal mushroom (Wassar *et al.*, 2000).

Mushrooms are highly rich in nutritive values like proteins, vitamins, minerals, crude fibers, carbohydrates, etc. and medicinal values like -polysaccharides, triterpenoids, linghzi 8, germanium, proteins, RNA's. Edible mushrooms are industrially important in paper industry, production of enzymes, in fermentation technology, etc. (Rai, 1995; Chang and Buswell, 1996; Rai, 1997a, b; Singh and Rai, 1998; Thakur, 1999). They are very well known for their unique taste, texture and aroma (Chang and Miles, 1989).

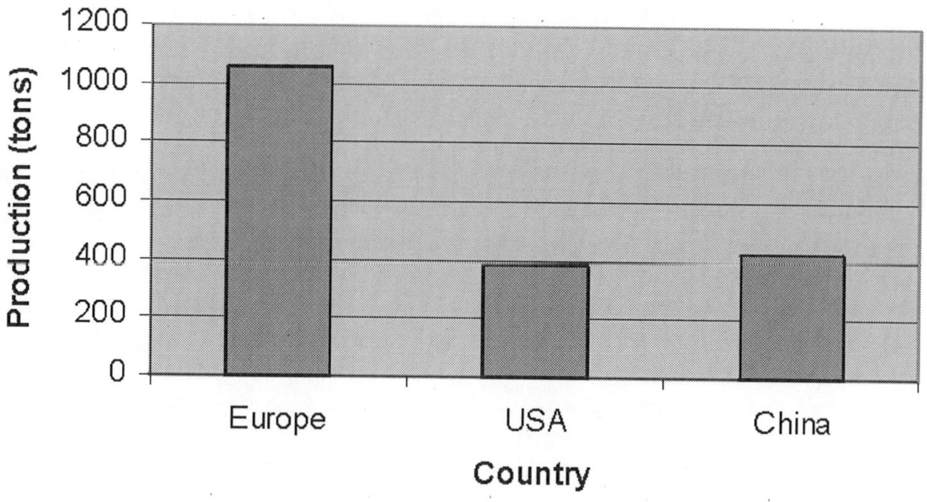

Fig. 2. World production of *Agaricus bisporus*

There are 5000 species of edible mushrooms, of which 20-25 species are cultivated commercially all around the world (Chang, 2002b). India cultivate mainly white button mushroom followed by oyster, paddy straw and white- milky mushroom (Sharma, 1997). A need was, therefore, felt to review the progress being made in the research and development of edible mushrooms all around the world with special reference to Indian context.

PERSPECTIVES ON MUSHROOM CULTIVATION

Mushroom is a horticultural crop grown worldwide for income generation on cottage scale as well as for commerce. Mushroom cultivation is labor intensive and generates new jobs especially in tropical or less developed countries. In India, it has become an ideal income generating activity especially in rural areas for small scale and landless farmers, which besides generating income also supplements nutrition (Kumar and Bindra Ban, 1997) for the socio-economic development of the society (Madan, 1997). Mushrooms are relatively fast growing organisms. Some tropical mushrooms can be harvested and consumed within 10 days after spawning. Mushroom can be cultivated year round using primitive farming technique in rural areas on highly industrialized techniques in urban and peri-urban communities. Mushrooms have some unique colour, taste, aroma and textures, which attract their consumption by humans (Chang, 2002b). Mushrooms like all other fungi lack chlorophyll and are non-green organisms (Chang, 1999). They cannot convert solar energy through the process of photosynthesis to organic matter as green plants do, but they can produce extensive enzymes for their own nutrients for growth and fruiting (Chang, 2001). They can degrade agricultural and forestry lignocellulosic materials into high quality fungal protein (Buswell and Chang, 1994) or mycomeat (Miles and Chang, 1988) consumed directly in the form of fruiting body. Different mushrooms have different lignocellulolytic enzymes profile (Cai *et al.*, 1994, 1998, 1999, Buswell *et al.*, 1995; 1996a,b). It is therefore, represents, the example of most efficient conversion of plant wastes into edible food (Zadrazil, 1985). This also demonstrates the impressive capabilities of mushrooms for "biosynthesis". Mushroom cultivation is one of the viable answers to reduce the pollutants to some extent by utilizing the agro-industrial wastes and its byproducts in a nutritionally and economically beneficial manner (Bahl, 1997). Indoor mushroom cultivation can utilize the vertical space in view of the increasing pressure on agricultural land due to fast expanding civilization and population growth. Mushroom cultivation under controlled conditions permit rich harvest during successive cropping free from the vagaries of weather, to which other crops are exposed. The protein conversion efficiency of edible mushrooms per unit of land and per unit of time is far superior compared to animal sources of protein (Bano and Rajarathnam, 1988). Harvesting and separation of edible biomass from the substrates is the easiest method (Zadrazil and Grabbe, 1983).

Thus, mushroom cultivation offers promising scope of meeting the worldwide food shortage for the rapidly increasing population by almost 80 million people per year.

WORLD PRODUCTION OF PLANT SUBSTRATES

Natural plant wastes are the basic substrates for growth and yield of mushroom. The world has an immense amount of lignocellulosic material resource. It is also called cellulosic biomass, containing 49.1% of cellulose, 21.8% of hemicellulose and 21.8% of lignin as primary component for cultivation of edible mushrooms. Lignocellulosic material is estimated to amount to 1.09 x 11^{11} tons of dry matter on land annually (Chang, 1989). The world's annual yield of cereal straws in 1999 is estimated to be 3570 million tons, which constitutes the principal objects for conversion into useful products by man's activities. The vast quantities of lignocellulosic waste generated annually through the activities of agriculture, forestry and food processing industries can be suitably utilized for cultivation of edible mushroom by solid-state fermentation and submerged culture. Bioconversion of the huge lignocellulosic biomass waste materials using both edible and medicinal mushrooms through a designed package of multidisciplinary technologies has been recently named as the non green revolution (Chang, 1999).

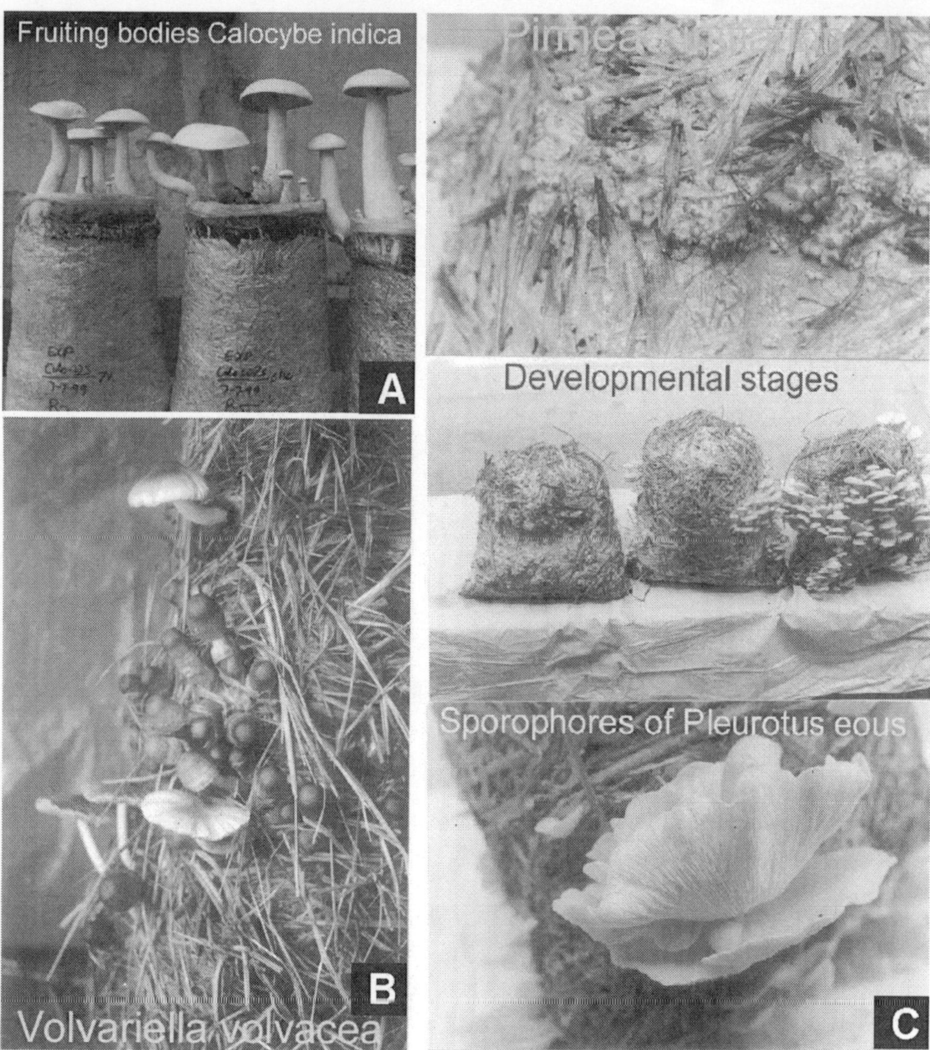

Fig. 3. A Fruiting bodies of *Calocybe indica,* B. *Volvariella volvacea,* C. *Pleurotus* sp.

Of various strategies developed to utilize the waste, one of the most significant in terms of producing a higher value product is mushroom cultivation. Secondly, many of these mushrooms produced a variety of metabolites of immense interest to the pharmaceutical/neutriceutical and food industries. It is estimated that about two third of the biomass production occurs on land and remaining in the ocean. Most terrestrial plant material occurs in forests (65%) with a bit more than 15% generated in grassland and cultivated land. In addition, it is estimated that millions and millions of tons of sawdust, wood chips, coffee pulp, spent coffee, brewery spent grain, cotton seed hull, textile cotton waste are generated every year around the world. These materials are burnt on site, buried and dumped at unplanned and uncontrolled landfills, which ultimately pollute the environment. Time has come for us not to expect the earth to produce more, but to expect man to do more with what the earth produces.

Tiwari and Pandey (2002) estimated that about 357 million tons of agro-waste is generated annually in India, of which 170 million tons are left for burning or cause environmental pollution. As per one of the estimates, the production of rice straw and wheat straw in our country is about 85.99 and 70.77 million tons respectively. If even one percent of the generated waste is properly utilized for mushroom cultivation, India will become a major mushroom growing country in the world.

PRODUCTION OF EDIBLE MUSHROOMS

The artificial cultivation of edible mushroom, *Auricularia auricula-judae* (Bull.) Wettst. was first recorded around 600 A.D. in China (Chang and Miles, 1987). The mass production was then researched and developed only after World War-II. The world production of cultivated edible mushrooms merely was 170,000 tons in 1960 but, it was estimated to be >6 million tons in 1999 (Delcaire, 1978; Sharma, 1997; Chang, 1999). In 1997, Asia contributed 74.4% of the total world mushroom tonnage, Europe, 16.3% and North America, 7.0%. In contrast, both Africa's and Latin America's shares were <1% of the total production. In 1997, China alone contributed 22.62% of the global production and appeared as the largest producer in the world followed by USA, Netherlands and France. In 1978, the production of edible mushroom in China was estimated to be only 400,000 tons but it increased steadily from 0.4 million tons to 1.0 million ton in 1990, 2.6 million tons in 1994, about 4.00 million tons in 1997 and 4.4 million tons in 1998 and 4.5 million tons in 1999 (Chang, 2002b).

Among different edible mushrooms cultivated globally, white button mushroom (*Agaricus bisporus*) was on the top and its share varied from 72.66% in 1975 to 31.8% in 1997 of the world mushroom production. The United States of America was the largest producer of *Agaricus* in the world followed by China and France. The world production of *Agaricus bisporus* was 136,000 tons in 1960 (Delcaire, 1978) but the production reached to approximately 1,971,000 tons in 1979. In 1999, the production of *Agaricus*, in Europe reached 1,059,000 tons, in the United States 386,000 tons (O' Brien, 2000) and in China 426,000 tons (Huang, 2000). In 1999, the production of China crossed to that of USA and became the world leader in the production of *Agaricus* as well as *Lentinus*. Shiitake mushroom (*Lentinus edodes*) is the second most important edible mushroom whose share varied from 14.09% in 1975 to 25.4% in 1997 and likely to increase more than that of white button mushroom as per the recent estimate (Chang, 2002). Japan was by far the worlds largest producers of Shiitake in the 1970s with China second, followed by Taiwan and South Korea. As late as 1983, Japan accounted for more than 82% of the total world production of Shiitake. At the same time, China represented only 9.4%. Only 4 years later in 1997, China took over Japan for the first time as the leading producer. In 1997, China share had risen to 85.1% thus increasing production by 1,060% and Japan, Korea and Taiwan decreased production by 46.5%, 23.2% and 82.2% respectively. Now, China has become the world's largest producer, exporter and consumer of Shiitake. The production *L. edodes* in China in 1998 was 1,338,000 tons and in 2000 was 1,600,000 tons (Chang, 1999, Chang, 2002a).

India like China and some other Asian neighbors was a late starter, making a humble beginning in the early sixties. The production of edible mushrooms in India was only 1000 tons in 1985 but, it quickly reached to 40,000 tons in 1997 (Dhar, 1997). India produced 80-85% of the white button mushroom, 15-19% of *Pleurotus* and the remaining 1% of other mushrooms of the total annual production (Sharma, 1997).

During the current decade, a number of modern mushroom farms with environment-controlled cultivation system have appeared on the scene, which spread throughout the country. These developments have led to almost five times increase in the total production of mushroom in the country. The annual growth rate of 35-40% was recorded as against the global growth rate of 7%. It has been projected that if the present growth rate is sustained, the country's mushroom production may cross 600,000 tons in by 2020. India is dominated by cultivation of only white button mushroom unlike other countries (Japan, China, Korea), which grows varieties of mushrooms. Thus, there is a need to diversify the production of other mushrooms like oyster, white milky, black ear etc., which are suited well to agro climatic conditions of our country.

EXPLOITATION OF EDIBLE MUSHROOMS

Nutritional values: Edible mushrooms are a good source of food protein. They exhibit wide variations in their protein content. Even varietal and strained differences in protein content have been reported. The protein content of the cultivated species ranges from 1.75-5.9% of their fresh weight (Flegg and Maw; 1976, Chang, 1980) and an average of 3.5-4.0% is considered to be more representative. This means that the protein contents of edible mushroom in general, is about twice that of onion (1.4%) and cabbage (1.4%), and four times and 12 times those of oranges (1.0%) and apples (0.3%). In comparison, the protein content of common meats is: pork – 9-16%; beef, 12-20%; chicken 18-20%; fish, 18-20% and milk, 2.9-3.3%. On a dry- weight basis, mushroom normally contain 19-35% protein as compared to 7.3% in rice, 12.7% in wheat, 38.1% in soybean and 9.4% in corn (Crisen and Sands, 1978; Bano and Rajarathnam, 1982; Li and Chang, 1982). In terms of the amount of crude protein, mushrooms rank below animal meats but well above most other foods including milk, which is an animal product (Chang *et al.*, 1981). Furthermore, mushroom proteins contain all the amino acids required by man.

They are considered as a potential substitute of muscle protein on account of their high digestibility (digestibility co-efficient around 89%), good amino acid content and about 1000 times higher production of mushroom protein/unit area. Crisen and Sands (1978) determined the nutritional value of foods based on their content of essential amino acids. The Essential Amino Acid Index (EAA. Index) rates dietary protein in terms of an essential amino acid pattern based on known adult human dietary requirements. They proposed the use of a nutritional index which can be calculated as Nutritional Index = (EAA index x percentage protein)/100.

EAA Indices and amino acid scores of the most nutritive mushrooms (highest values) rank in potential nutritive value with those of meat and milk, and are significantly higher than those for most legumes and vegetables. The least nutritive mushrooms ranks appre-ciably lower but are still comparable to some of our common vegetables. Bisaria *et al.* (1987) also investigated amino acid composition, the protein score, essential amino acid index and biological value of the fruit bodies of *P. sajor-caju* on various agroresidues. Mushroom protein is not only cheaper but is almost as nutritional as muscle protein. It consists high concentration of essential amino acids and can prove a good supplement to those cereal diets, which lack in some essential amino acids (Lysine). Many fungal proteins are similar to those produced by animals, which make them especially useful as supplements to cereals.

Mushrooms are also an excellent source of several vitamins and excel many fruits and vegetables. They are particularly rich in thiamin (B_1), riboflavin (B_2), niacin, biotin, ascorbic acid (Lau *et al.*, 1985) and ergosterine (Huang *et al.*, 1995). Mushrooms are good source of minerals. However, they are particularly rich in phosphorus, potash, and iron but are low in sodium. About 1/3rd of the total iron in mushroom is in the available form. Mushrooms are low in calories, starch and fat (0.6-3.1% of dry weight). At least 72% of the total fatty acids are found to be unsaturated in all these mushrooms. Unsaturated fatty acids are essential and significant in our diet and to our health. Fresh mushrooms contain relatively large amount of fibers, which are also part of healthy diet. On the basis of all these qualities, the nutritional values of mushrooms as compared to other foodstuffs appear to be quite high and mushrooms no doubt, represent one of the world's greatest untapped resources of nutritious and palatable food of the future.

Medicinal values : The second major attribute of mushrooms, their medicinal properties has long been recognized in China, Korea and Japan. Chinese have been using mushrooms as medicines for the last 3000 years, Japanese for the last 2000 years and considered it as the "Mushroom of immortality" while history of Korean using mushroom is also very old. They considered it to be "longevity mushroom". However, there has been a recent upsurge in interest in traditional remedies for the treatment of various physiological disorders through various pharmacologically active compounds of mushrooms which have recently been isolated, characterized, many proprietary products have been developed and are being marketed (Rai, 1997b). Recent upsurge of interest in mushroom is not only as health vegetables (food) but also as a source of biologically active compounds of medicinal values including use as complementary medicine/dietary supplements for various ailments. This new class of compounds termed 'mushroom nutriceuticals' are extractable from either the fungal mycelium or fruiting bodies. These active compounds are polysaccharides (high molecular weight strings of sugars), triterpenoids, immunomodulatory proteins, lentinans (Chihara, 1990, 1993), adenosine, lingzhi-8, polysaccharide-Krestin (Sakagami and Takeda, 1993), PSP-polysaccharide bound peptide from *Trametus versicolor* (Yang *et al.,* 1993), PSPC-protein bound polysaccharide complex from *Tricholoma labayense* (Liu *et al.,* 1995, Wang *et al.,* 1995a, Ng *et al.,* 1999) isolated from several edible and medicinal mushrooms. Some of these active compounds have been developed as top selling anti-cancer pharmacological drugs and are being sold in Japan and USA. These are, PSK (Krestin), OK-432, Lentinan, Tegafur and Carmofur.

Table 1. Some anti-cancer drugs from mushrooms sold in Japan

Name	Krestin	Lentinan	Schizophyllan
Source	*Coriolus versicolor*	*Lentinus edodes*	*Schizophyllum commune*
Active principle	Protein bound polysaccharide	B-1, 3-D- glucans	B-1, 3-D- glucans
Route	Oral	Injection	Injection
Type of cancer	Gastrointestinal, brest, lungs	Stomach	Cervical
Market value	$ 600 million	$ 31 million	—

Source

Between 80-85% of all medicinal mushroom products are derived from the fruiting bodies, which have been either commercially farmed or collected from wild while only about 15% of all products are based on extracts from mycelia, for example, PSK, PSP, PSPC, etc.. However, owing to increased quality control and year round production, mycelial products are the wave of the future.

The market value of medicinal mushrooms and their derivatives dietary supplements worldwide was about US $1.2 billion in 1991, 3.6 billion in 1994 (Chang, 1996) and 6.0 billion US $ in 1999 (Wasser *et al.,* 2000). The market value of *Ganoderma* based nutriceuticals alone in 1995 was estimated at US $ 1,628.4 million (Chang and Buswell, 1999). In India too, market of *Ganoderma-* based neutriceuticals is catching up very fast and is estimated to be about US $ 20.0 million in 2000-2001 (Surinder, 2002). Ninety nine percent of all sale of medicinal mushrooms and their derivatives occurred in Asia and Europe, with < 0.1% in North America. The 1999 US market for dietary supplements based mainly on mushrooms was estimated to be US $ 35.0 million. The North America demand in recent years has been increasing between 20-40% annually depending upon the species.

Cochran (1978) and Chang and Miles (1989) have reviewed the literature on antibiotic activities (antiviral, antifungal and antibacterial), antitumor and hypolipidemic effects of mushroom but the most important and significant researches are the anti-tumor, hypocholesterolomic, hypolipidemic, hypertension and hypoglycemic effects (Rai, 1997b). Ikekawa *et al.* (1969) indicated the antitumor effects of the edible mushrooms. Lentinan, a polysaccharide from *Lentinus edodes* inhibited the growth of mouse sarcoma 180 in mice. Another polysaccharide KS 2 extracted from the culture medium of *Lentinus edodes* suppressed sarcoma 180 as well as Ehrlich ascites carcinoma. Significant hypolipidemic and hypocholesterolemic effects of *Lentinus edodes* have been shown to lower plasma cholesterol level in animals and man (Kaneda and Tokuda, 1966; Suzuki and Oshima, 1976; Rai, 1997b). Tyrosinase is a protein purified from *Agaricus bisporus* and represent a valuable source of biologically active compounds with potential for protecting cellular DNA from oxidative damage (Shi *et al.*, 2002). Bisco *et al.* (2002) recently extracted the exopolysaccharide (glucose) of 180-200 KD and 200-220 KD from the submerged culture of *L. edodes* and *Pleurotus ostreatus* respectively. Besides, polysaccharides, a number of low molecular weight substances, such as triterpenoids, steroids and phenols inhibiting cancer cell growth have been reported (Mizuno, 1996; Wasser and Weis, 1999). Numerous articles and monographs contain detailed information on the pharmacologically medicinal values of these species (Bisco and Dudka, 1987; Solomko *et al.*, 1987; Hobbs, 1995). A novel single chained, ribosome inactivating protein (RIP) was recently isolated from fruiting bodies of *Volvariella volvacea* (Yao *et al.*, 1998). The mushroom RIP, designated volvarin exhibited a potent inhibitory action on protein synthesis in the rabbit reticulocyte lyasate system. The aqueous extracts of *Pleurotus sajor-caju* (Tam *et al.*, 1986) and *V. volvacea* have been reported to produce hypotensive effects in normotensive rats. Feeding powered maitake (*Grifola frondosa*) mushroom to spontaneous hypertensive rats, resulted in a lowering of the blood pressure. Dried powder of *Auricularia polytricha* and *Tremella fuciformis* when fed to the rats, lowered the serum total cholesterol and the low density lipoprotein cholesterol levels (Cheung, 1996).

Biologically active compound extracted from edible and medicinal mushrooms have extraordinary low toxicity even at a high dose (Chang and Bushwell, 1996). Long viewed as tonics, they are now known to be capable of profoundly improving quality human health. It has been estimated that there are about 1800 species of mushrooms with potential medicinal properties (Chang, 2001). Both these mushrooms and their root-like structure (called mycelium) produced several medicinal or nutriceuticals (general immune-enhancing) compounds. Central among these are the polysaccharides (*Ganoderma lucidum*) comprising high molecular weight sugar polymers. It has been suggested that the antitumor and-cancer effects of the polysaccharides are based on the enhancement of the body's immune systems (including activated macrophages, natural killar cells, cytotoxic T-cells and their secretary products such as the tumor necrosis factor, reactive nitrogen and oxygen intermediates) rather than direct cytologial effects (Toth *et al.*, 1983; Mizuno *et al.*, 1995; Liu *et al.*, 1996; Chang and Buswell, 1999). The killing mediated by cytotoxic T-lymphocytes (Soto-Velazco *et al.*, 2002) and natural killer cells, represents an important mechanism in immune defense against tumors, virus infected cells, parasites and other foreign invaders. In addition, it helps in alleviating the side-effects of chemotherapy and radiation treatments (through cell- level regenerative effects). There are many studies in Asia and particularly in Japan and China, documenting life span increases of cancer patients undergoing conventional cancer treatment plus mushroom extract consumption or injection (Mizuno *et al.*, 1995; Liu, 1999; Chang and Mshigeni, 2000, 2001). At the same time, owing to the enhancement of the immune system, the extract can help reduce the possibility of infection by other diseases. Triterpenoids, steroid-like compounds, glucans, proteins, ganoderic acid, ergosterol, polynucleotides, LingZhi-8 and lectins, have been shown to have antitumor and immunomodulatory activities (Wang *et al.*, 1995b, 1996, 1997).

The growing global consciousness regarding the use of natural products have made mushroom based medicinal products, an important place to occupy in the sound health management system. The rapid increase of public interest in the use of medicinal mushrooms has led to first International Conference on Medicinal Mushroom at Shanghai in 1999 and 2nd International Conference on Medicinal Mushroom at Thailand in July, 2003. Also published in "International Journal of Medicinal Mushroom" since 1999. Thus, the development of mushroom products had an important impact on health care and economic growth at National and Regional Levels worldwide.

Use in industries : Most of the basidiomycetous fungi including edible mushrooms have the capacity to grow on a wide spectrum of lignocelluloses materials by secreting a range of degradatory and oxidizing enzymes, which have several industrial applications.

Degradatory enzymes produced by commonly cultivable edible mushrooms are ligninolytic, polysaccharide degrading and protein degrading. Ligninolytic enzymes including phenol oxidases, peroxidases, catalases and laccases, which are responsible for lignin degradation without interfering with cellulose and hemicelluloses. Some *Pleurotus* spp. are being utilized in delignification of lignocellulosic wastes to produce pulp for cardboard or paper manufacture. Several edible mushrooms for enzymes involved in lignin degradation have been screened (Valmaseda *et al.*, 1990, 1991; Pelaezet *et al.*, 1995; Ansu *et al.* 2001). Polysaccharide degrading enzymes like xylanase and polygalactu-ronase of *Pleurotus* spp. offer an alternative to boiling with NaOH for removal of non-cellulosic materials in the retting of flax fibers. Enzyme treated fibers produce high quality yam spun from the untreated fibers. In mushroom food processing, a protein degrading extract of *Pleurotus ostreatus* containing chitinases, chitobiases and gluconase is able to soften the tough stipes of *Agaricus bisporus, Pleurotus* and *Lentinus edodes*.

Oxidizing enzymes produced by white-rot fungi were believed to have first industrial application in decolourization of Kraft liquid. The oxidizing enzymes produced by *Pleurotus ostreatus* and *Coriolus versicolor* can degrade and metabolize not only native lignin but also Kraft lignins and lignisulfonates (Livernoche *et al.,* 1983). Mycelial pellets of *C. versicolor* have been used for the continuous decolorization with glucose as co-substrate, while *Lentinus edodes* can remove colour with no additional carbon source. The phenol oxidizing enzymes of *Pleurotus* or *Auricularia* can decolorize such fermentation wastes. Decolourization of sugarcane molasses is also accomplished in the food industry by immobilized *Coriolus hirsutus* (Tamaki *et al.,* 1989).

As a source of flavourants: Mushrooms are known by unique colour, taste, aroma and texture characteristics which attract their consumption by humans. Due to increasing demand for natural ingredients and flavour in food industries, edible mushrooms are gaining increasing attention as a source of flavour compounds. The flavor based gastronomic appeal is, after, all one of the main reasons for consuming wild and commercially grown edible mushrooms. Certain non-volatile substances, such as L-glutamic acid, short chain fatty acids, carbohydrate, protein, nucleotides may contribute to the characteristic flavours in different mushroom species; although, volatile fraction of edible mushrooms such as aliphatic alcohols, aldehydes, ketones, easters, lactones, and aromatic cinnamyl derivatives are largely responsible for characteristics flavour in the mushrooms. *Lentinus edodes* produces 'lenthionin' which is a cyclic sulfur containing compound. Dried mycelium of edible mushrooms also have a slight flavor due to 1-octen-3-ol and improved food value (Schildler, 1989). Submerged fermentation technique for the production of fungal biomass that is rich in flavour is also developed.

CULTIVATION OF EDIBLE MUSHROOMS

Mushroom science dealing with mushroom cultivation and production, embodies principles of microbiology, environmental technology, and solid-state fermentation involving the conversion of lignocellulosic biomass waste materials, into foods for humans. The technology for mushroom cultivation can be primitive as in rural farming in developing countries. It can also be highly industrialized. The latter requires advanced knowledge of mushroom biology and technology and the use of sophisticated equipments, which are largely a monopoly of the developed countries. Mushroom cultivation is both a science and an art. The science is developed through research, while art is perfected through curiosity and practical experience. Mushroom growth dynamics involve some technological elements, which are in consonance with those exhibited by our common agricultural crop plants, for example, sunflowers, when the plants switch from vegetative growth to reproductive growth, retarded tips for further growth (elongation) in an obvious phenomenon of mature. The same principle applies in mushroom production. After the vegetative phase (mycelial) has reached maturity, what the mushroom farmer needs next is the induction of fruiting. This is the time the mycelial growth tips should be retarded by regulating the environmental factors. These factors are generally called "triggers" or "environmental shocks" (Chang, 2001). Switching on the light providing fresh air and lowering temperatures can trigger fruiting.

It should be noted that the cultivation of mushrooms deals with living organisms: not only the mushroom itself but also other microorganisms, including both harmful and beneficial ones. Therefore, the methods employed in mushroom cultivation require some modification from one region to another, since different environmental conditions and different species of microorganisms may be encountered. Therefore, it should be emphasized that the science, the art and the practice of mushroom cultivation have led to some useful general concepts about various stages of the production process. Indeed, the biological complexity involved in the interaction of various microorganisms, different natural substrates, and a diversity of environmental conditions require multitude of methods. The major practical steps/segments of mushroom, cultivation are (Chang, 1998):

(1) Selection of an acceptable mushroom species.
(2) Requirement of a good quality fruiting culture.
(3) Development of robust spawn.
(4) Preparation of selective substrate/compost.
(5) Care of mycelial (spawn) running.
(6) Management of fruiting and mushroom development.
(7) Harvesting mushrooms carefully.

Before any decision to cultivate a particular mushroom is made, it is important to determine if that species possesses organoleptic qualities acceptable to the indigenous population, or to the international market for export; if suitable substrate for cultivation are plentiful, and if environmental requirements for its growth and fruiting, can be met without excessively costly systems of mechanical control.

Edible mushrooms cultivated commercially around the world

(1) *Agaricus bisporus* - White button mushroom (*A. bitorquis*)
(2) *Lentinus edodes* - Shiitake mushroom
(3) *Pleurotus* spp. - Oyster mushroom
(4) *Auricularia* spp. - Black ear mushroom
(5) *Volvariella volvacea* - Paddy straw mushroom
(6) *Flammulina velutips* - Enokitake, winter mushroom

(7) *Tremella fuciformis* - Silver ear, white jelly fungus
(8) *Hypsizygus marmoreus* - Crab mushroom
(9) *Pholiota nameko* - Nameko, viscid
(10) *Grifola frondosa* - Maitake mushroom

Mushrooms cultivated commercially in India

(1) White button mushroom (*Agaricus bisporus*/bitorquis)
(2) Oyster or Dhingri Mushroom (*Pleurotus* spp.)
(3) Paddy straw mushroom (*Volvariella volvacea*)
(4) White milky mushroom (*Calocybe indica*)

I. White button mushroom

It is synonymously known as white mushroom, button mushroom, champignon or simply the common cultivated mushroom. It has developed over the past 50-85 years from a beginning as a risky venture to a largely predictable and controlled industrial process. The mushroom industry in the U.S. began about 1915, and by 1980 growers produced > 225,000 MT of mushroom. In other countries, the production of button mushroom was started only after World War - II and now there are 200 countries in the world growing button mushroom. It could have been possible only through the background information from research and the well developed technology and management practices, which led to constitute over 70% of the world production of edible mushroom in 1970's. But it has been reduced to about 32% in recent years. In India, the beginning of *Agaricus* cultivation was made in sixty's and seventy's. In 1985, the production was hardly 1000 tons, which reached to 25,000 MT in 1995 and 40,000 in 1997 (Dhar, 1997). The production and productivity was also increased to 18-22 Kg / m^2 in 1990.

1. Requirement of fruiting culture: The first hybrid strains of *A. bisporus* (U_1 and U_3) developed by Dutch scientists in 1981 were obtained by Indian scientists and evaluated. (Bhandal and Mehta, 1968b; Kumar and Chandra, 1983; Patil *et al.,* 1989; Dhar and Kapoor, 1989). Some efforts on isolation were made to improve the strains through multispores, monospores and tissue culture (Kumar and Munjal, 1981; Phutela and Garcha, 1986). Some success was obtained when three single spore isolates of *A. bisporus* as NCS-100, NCS-101 and NCH, 102 was released by NRCM in 1997 (Dhar *et al.,* 2002). Two high yielding strains of *A. bitorquis* NCB-6 and NCB-11 have also been identified in several years of evaluation trials at Solan and multilocational centers were released by the National Research Centre for Mushroom, Solan in 2000 (Dhar, 1995b, Dhar *et al.*, 2002). Some new high yielding strains of *A. bitorquis,* (NCB 6 and NCB-14) resistant to false truffle have been identified in evaluation trials at NRCM, Solan and the work is still in progress. (Verma *et al.,* 2002). No attempt has been made in India to breed a hybrid either by conventional methods or by bio-technological tools (Verma, 1997). But, a holistic approach, combining molecular techniques for the development of high quality hybrids, traditional methods of single spore selection, mutation breeding and somatic hybridization of *A. bisporus* was launched at Solan in 1998. Of 96 *A. bisporus* lines from indigenous wild collection and exotic germplasm maintained at Solan, 67 lines have been evaluated and potential parental lines have been identified. Similarly, seven lines of *A. bitorquis* have also been evaluated to identify suitable parental lines. More than 2500 single spore isolates of *A. bisporus* have been developed and characterized. More than 1000 homokaryotic breeding lines have been isolated to combine yield and false truffle resistance. A number of inter-strain hybrids of *A. bitorquis* have been developed and are currently under evaluation (Verma *et al.*, 2002).

The non-hybrid strains of *A. bisporus* introduced from outside (Western Europe and USA) are still in use by most of the seasonal growers. Some of the popular strains used are S-11, 310, S-76, 791, hybrid strain U 3, U 1 etc. (Dhar *et al.*, 2002).

2. Preparation of compost: Mushrooms are heterotrophic organisms and lack the ability to form organic compounds for their cellular growth from the carbon dioxide in the atmosphere. Thus, their nutritional requirements must be met by materials absorbed from the substrate either before of after the substrate is broken down by the composting process. Composting is a biological process in which the activities of numerous types of microorganisms are involved (Chang and Hudson, 1967). These organisms convert a rich mixture of organic materials into a lignin humus complex suitable for the growth of A. *bisporus* but not so for the growth of competitor microorganisms (Stanek, 1972). Composting takes place in two distinct phase. In phase I composting, the raw materials are mixed and wetted so that the microorganisms that are naturally present may begin the degradative metabolic activities. Phase II composing is a pasteurization process, which eradicates insects and contaminating microorganisms and prepares the compost as a selective substrate favouring the growth of *Agaricus*. This process has been adopted worldwide because of its many advantages. However, this two phase process has some important disadvantages (i) it is time, space and labour consuming (ii) it requires high initial investment and (iii) it is not always environmentally friendly. A technical alternative to this method was introduced in 1962, accordingly to which it was possible to produce *A. bisporus* on a non-composted, sterilized (121^{0}C, two hours) saw dust based mixture. Later, Murphy (1972) reported on his efforts to produce this mushroom without phase I composting and Mee (1978) demonstrated that it was also possible to produce *A. bisporus* as dried, non composted "Cold" manure. Finally, Sanchez and Royse (2001) reported a procedure for preparing mushroom substrates without composting on ingredients used for the cultivation of shiitake. Similarly, Sanchez *et al.* (2002) formulated a pasteurized mixture of non composted raw materials to produce a brown variety of the white mushroom (*A. bisporus*) with range in yield from 17.2

kg/m^2 (42% BE) to 31.4 kg/m^2 (77% BE). The addition of non-composted sheep manure was found to increase the yield.

A variety of crop residues have been tried as base material in India (Rana and Goyal, 1994; Saini, 1994), a number of formulations are today in use, a number of additive were used (Shandilya, 1980; Rao *et al.*, 1994; Garg, 1986) following short (SMC) and long method of composting (LMC). Other methods of substrate preparation were also tried (Dhar and Kapoor, 1988; Dhar, 1993, 1996) but Indian growers are using either by SMC or LMC with bulk pasteurization.

Mushroom compost production is a highly complex ecological process which is not fully understood till date. It is essentially an aerobic process (Laborde, 1994). One of the most significant changes in mushroom compost production in the past decade has been the movement to carry out composting under environmentally controlled and enclosed conditions. Enclosure is a significant change both in process configuration and in the potential for more sophisticated process management. It can be carried out with controlled ventilation than in a traditional method of turned stacks in an outdoor yard. It has several advances like reduction of odors and overall environmental effects, improving the consistency of compost quality (Laborde *et al.,* 1993; Vijay and Gupta, 1995; Millar, 1997), reduction of space requirements, improved materials handling and a reduction in raw materials, losses during composting (Laborde, 1994; Macauley and Perrin, 1994). Enclosed and controlled environmental composting is now being conducted on a commercial scale at small number of facilities in Europe (Laborde, 1993), North America, Australia and on a smaller scale in India (Dhar, 1996; Vijay and Ahlawat, 2000). Indoor compost can be produced by two methods which are known as INRA method and Anglo Dutch method. INRA method is double phase higher temperature process while Anglo Dutch method is a single stage process (Vijay and Ahlawat, 2000).

Finally, prepared compost should have certain distinct parameters, which ultimately governs the quality and yield of *A. bisporus*. These are colour of the compost, texture, pH, moisture content, concentration of ammonia, pests and diseases etc. The need to precisely monitor all these parameters has long been felt but now the researches on application of near infrared spectroscopy (NIRS) in assessing substrate quality had made it possible. The application of NIRS, as a tool for assessing substrate quality has been investigated over the past 5 years in collaboration with major compost producers in Northern Ireland. NIR calibrations for dry matter (moisture content), pH, nitrogen dry matter, carbon, ash, microbial population and fiber fractions using spectra of fresh composts have been developed and validated using a wide range of phase I and phase II samples. In addition, calibration for predicting potential mushroom yield of compost has also been developed from a large number of compost comparative trials (Sharma *et al.,* 2002). Integration of NIR monitoring and tunnel control systems will be advantageous for optimizing compost production.

3. Crop management : In India, low cost seasonal farming and climate controlled round the year cultivation are the two production systems commonly adapted. Some modifications to these methods were suggested with no appreciable change (Dhar *et al.* 1985; Shandilya, 1988a, b; Upadhayay and Vijay, 1988; Dhar, 1991, 1995a). However, some improvements of crop management practices like composting methods, compost ingredients, filling rates and growing systems could help to standardize the package of practices for different situations (Jain and Singh, 1982; Patil and Shinde, 1983 and Bisht *et al.*, 1986). Some attempts were made on identification of casing materials like cattle manure, tree bark, dried moss, forest litter, bagasse, vermi compost, rice husk, ash, etc. with a lot of variation in yield of A. *bisporus* indicating a positive correlation of intrinsic properties of the casing layer (Garcha and Shekhon, 1981; Vijay *et al.*, 1987; Gupta and Dhar, 1993; Bharadwaj, 2000; Sharma and Guleria, 2000; Shandilya and Phutela, 2000; Bhatt and Singh, 2002a). The ability to absorb and release water effectively and quickly, porosity for air exchange and chemical properties of the casing layer are some of the factors affecting yield potential of the casing mixture. In addition, the microbiological activity in the casing layer also plays significant role in the productivity of the system. The mechanism behind sporophore formation following casing is one of the challenging aspects of this phenomenon. Several physiological and environmental factors including involvement of volatiles, a role of CO_2, stimulatory effects of microbial population and nutritional status of the casing layer activity interact to induce the fructification process. In India, casing mixture is highly vulnerable to extreme of variation from batch to batch resulting in partial understanding of the several factors involved in interaction. The studies on harmful and beneficial microflora of casing were also made by various workers (Shandilya and Agarwala, 1983; Sharma, 1991 and Vijay and Gupta, 1992). Studies on beneficial microflora particularly bacteria like *Pseudomonas putida*, *Bacillus* sp., *Azotobacter chroococcum* have received greater attention in recent days as they produce substances which improve the growth of *A. bisporus* mushrooms and also reduce the growth and parasitic activities of mycoparasitic fungi of *A. bisporus* (Jhune *et al.*, 1990; Klopper, 1992 and Ahlawat and Rai, 1997). Similarly, five isolates of bacteria inoculated in casing were reported to reduce the growth of pathogens/competitors, like *Verticillium fungicola, Fusarium moniliforme, Sepedonium chrysosporium, Trichoderma harzianum, Mycogone perniciosa*) and thus favoured an increase in yield of *A. bisporus* (Bhatt and Singh, 2002a).

Agronomic and environmental crop management has received a special attention in technically advanced countries, which requires high tech machines and sophisticated equipments. The bed depth, casing depth, bed moisture, temperature, RH, CO_2, air changes, watering in cropping room and other related factors have more or less been standardized for obtaining healthy crop of mushroom in environmentally controlled cropping rooms. All the parameters for spawn running, case running, pinning initiation, fruit body maturation and others are controlled with precision via computers. A number of other agronomic practices like supplementation, ruffling and CACing have been worked out *for A. bisporus* and *A. bitorquis* but none of the practices worked well in India (Dhar and Kapoor, 1990; Bahl, 1991; Vijay and Gupta, 1991, 1992; Vijay and Gupta, 1995; Dhar and Gupta, 2000).

Garcha *et al.* (1987a) recorded enhanced yield in *A. bisporus* due to addition of minerals like, iron, zinc and copper to casing soil. Similarly, supplementation of casing with soybean meal or cotton seed meal resulted in 20-40% increase in yield after treatment with 4000 ppm of formaldehyde (Dhar and Kapoor, 1990; Gupta and Vijay, 1992; Dhar and Gupta, 2000 and Vijay *et al.*, 2002).

4. Disease/pest management : Diseases and pests have been considered to be important factors responisible for lowering productivity/ production of button mushrooms in India particularly under seasonal growing (Sohi, 1992; Sharma, 1994; Sandhu, 1995; Sharma and Vijay, 1996; Madaan and Singh, 2000; Jarial and Sharma, 2000). Sharma (1997) has given a list of insect pests, nematodes, mites and fungi infesting/infecting during cultivation of *A. bisporus.* Bhatt and Singh (2002b) observed *Hypomyces rosellus* (the cause of cow-web disease) in samples collected from seasonal mushrooms of North Indian States. Its incidence varied from 0-17.85%, Thus, lot of information is available on insect and pest damage. However, not much work has been carried out in India on different aspects of integrated pest management for mushrooms. Work on use of resistant germplasm, use of biological agents, manipulation of growing conditions and use of botanical pesticides needs to be addressed.

II. Oyster Mushroom

The *Pleurotus* mushroom is generally referred to as 'Oyster mushroom' or Dhingri in India because the pileus or cap is shell like, spatulate and tongue-shaped and the stipe eccentric or lateral. The name *Pleurotus* has its origin from Greek word 'pleuro' means formed laterally or lateral position of the stalk or stem. Oyster mushroom has long been recognized as a highly potential converter of cheap cellulosics into valuable protein at a very nominal cost. It has wide adaptability due to number of commercially cultivated species suitable for round year cultivation under varied temperate and tropical conditions like India. It has simple cultivation technology and longer shelf life. After initial emphasis on cultivation methods and their improvements during sixty's and seventy's, researches in this versatile mushroom during eighty's and ninety's have encompassed other aspects also (Falck, 1917; Kaufurt, 1935; Block *et al.*, 1958; Bano and Shrivastava, 1962; Zadrazil, 1974; Chang *et al.*, 1981).

Oyster mushroom is cultivated in about 25 countries of far east Asia, Europe and America. It is the 3rd largest cultivated mushroom in the world and its annual world production is around 797,000 tons

(Chang, 1996). China alone contributes 88% of the total world production. The other major producing countries are South Korea, Japan, Italy, Taiwan, Thailand and Philippines. At present, India produces only small quantities (1500 tons) of oyster mushroom in the states of Orissa, Karnataka, Maharastra, Andhra Pradesh, Madhya Pradesh, West Bengal and in the North-Eastern states of Meghalaya, Manipur, Mizoram, Assam and Chhattisgarh.

1. Procurement of strains/species: An inherent advantage with this mushroom is the occurrence of a large number of species. All the varieties or species of oyster mushroom are edible, except *P. olearius* and *P. nidiformis* which are reported to be poisonous. There are 38 species of the genus recorded throughout the world. In recent years, 25 species cultivated commercially in different parts of the world are: *P. ostreatus, P. flabellatus, P. florida, P. sajor-caju, P. sapidus, P. eryngii, P. fossulatus, P. opuntiae, P. cornucopiae, P. yuccae, P. platypus, P. djamore, P. tuber-regium, P. australis, P. purpureo-olivaccus, P. populinus, P. levis,* etc. (Balakrishnan and Nair, 1995; Verma, 1996; Verma, 1997; Upadhaya *et al.*, 1998). Most of these species have the pH requirement of 5-6, RH of 70-90% and exhibit a wide variation in their temperature requirement (Jandaik and Kapoor, 1974; Garcha, 1984; Ghosh and Chakravarty, 1986; Suharban and Gokulapalan, 1995; Verma *et al.*, 1995; Patil *et al.*, 1995). With so many species exhibiting varied temperature (15-30^{0}C) and substrate preferences, most of the studies revolved round their cultivation, under varying agroclimatic conditions and on various substrates (Harsh *et al.*, 1981; Verma, 1985; Patel and Rafique, 1991; Hariprasad, 1991; Sawalgi and Sawalgi, 1994; Upadhayay *et al.*, 1996; Patra and Pani, 1997).

Attempts were also made to improve the available strains by various modern techniques. Pandey and Tewari (1989) came across a natural sporeless mutant, the much desirable trait to alleviate allergy problem. Other scientists also tried to improve the strains of *Pleurotus* by protoplast fusion, hybridization, etc. (Jandaik, 1987; Singh *et al.*, 1990; Ghosh and Chakravarty, 1991; Pandey and Tewari, 1994). Chinese were able to develop 26 strain of *Pleurotus* including one hybrid 17. These strains were highly adaptable to wide range of agroclimatic conditions ranging from 4-16^{0}C to 10-34^{0}C temperature.

2. Crop Management: *Pleurotus* mushroom can degrade any kind of agricultural, forests and industrial wastes which contains cellulose (49.1%), hemicellulose (21.8%) and lignin (21.8%). A wide variety of substrates like straw of cereals and millets, hulled maize cobs, cotton waste, banana pseudostem, rubber wood waste and its saw dust, oil palm mesocarp waste, arecanut waste, several weeds, water hyacinth, fermented coffee pulp, fodder grasses, cut logs, soybean straw, coir waste, lemon grass waste, used tea leaves, populus leaves, apple pomace have been successfully tested for cultivation of oyster mushroom (Khander *et al.*, 1991; Mathew *et al.*, 1991; Babu and Nair, 1991; Evini *et al.*, 1995; Upadhayay *et al.*, 1996; Jandaik, 1997; Sharma *et al.*, 1999;

Hussain *et al.*, 2000; Upadhayay and Verma, 2000; Thakur *et al.,* 2000; Thakur *et al.*, 2001a; Madhusndhanan and Chandramohanan, 2001; Madrigal. *et al.,* 2002; Nunez and Mendoza 2002). Substrates used for cultivation of *Pleurotus* need not to be fermented. Much research and experience throughout the world, particularly during the last two or three decades has resulted in many methods of treatment for substrates such as soaking in water, thermal sterilization, and semi-sterilization, dipping in hot water, steam pasteurization, aerobic thermophillic fermentation, chemical sterilization, submerged anaerobic or semi anaerobic meso-phillic fermentation etc. to prevent the competitor/parasitic fungi (Lelley, 1985; Bano *et al.*, 1987; Vijay and Sohi, 1987; Jadhav and Jagtap, 1991; Singh and Dwivedi, 1991; Lelley and Jansen, 1993; Stamets, 1993; Muez, 1994; Oei, 1996; Jandaik, 1997; Nunez and Mendoza, 2002). Most of the substrate used for oyster mushroom cultivation is a poor source of nitrogen (0.5 to 0.8%) and at the time of fructification when most of the nitrogen is utilized for mycelial growth, the depleted nitrogen in the substrate becomes inadequate and limits mushroom yield. The supplementation of the substrate is, therefore, required prior to spawning with oilseed meals and cake, dal powder, rice and wheat bran (Bahukhandi and Munjal, 1990; Thakur *et al.,* 2002a) pasteurized chicken manure (Vijay and Upadhayay, 1989); rice bran (Gunasekharan and Graham, 1987; Rao, 1991); wheat bran, soybean flour, brewers grain, cotton seed cake, deoiled soybean meal (Bahram, 1989; Jandaik 1989; Sharma, and Doshi, 2000; NRCM, 2001-02, Upadhayay *et al.*, 2002) and starchy material (Nath, 2001) to enhance the yield and dry matter.

Other agronomical practices like cultivation systems- cut wood log cultivation, bed cultivation system, plastic bag cultivation system, bottle or jar cultivation system, modified bed cultivation system, vertical tray system, type of containers - nylon net, polybag (Sohi, 1983); polythene sheet (Rangad, 1987), selves (Dutta *et al.*, 1991), size of containers (Thakur *et al.*, 2002a), spawning rate (Shanmugam, 1986, Verma *et al.,* 1991), spawn generation (Thakur *et al.*, 2002a); spawn substrates (Thakur *et al.*, 2002a), spawn age, spawning methods, urea treatment of substrates and beds (Nath, 2001), effect of hollowness of mushroom beds (Thakur *et al.*, 2001b,c) were attempted by several workers with varying results. The results of the above studies are still not consistent and needs further verification, before adoption by the growers.

3. Diseases and pests management: Diseases induced by fungi, bacteria and viruses have been reported in oyster mushroom and losses in yields have also been worked out in few diseases (Sharma and Kumar, 2000). Four fungal diseases caused by *Cladobotryum* spp. (Upadhayay *et al.,* 1987); *Gliocladinm* spp. (Bharadwaj *et al.,* 1987; Sharma and Jandaik, 1987) *Arthrobotrys pleuroti* (Ganeshan, 1987) and *Sibirina fungicola* (Sharma and Jandaik, 1983), have been reported, most of which were found to be controlled by Bavistin (50-100 ppm) application. They were also able to suppress most of the common competitor moulds reported on *Pleurotus* beds (Vijay *et al.*, 1985; Das and Suharban, 1991; Thakur *et al.*, 2001b). Thakur *et al* (2001b) reported that paddy straw when treated with Bavistin and formalin (75+500 ppm) could reduce the microbial load (496 $cfug^{-1}$) by three fold compared to untreated straw (1324 $cfug^{-1}$). Apron was reported to be another systemic fungicide with no much reduction on growth of *Pleurotus* spp (Shukla *et al.*, 1998) even up to 100 ppm under *in vitro* conditions. Others have also reported some fungicides to control the competitor molds (Dhar and Kapoor, 1990). Effect of homeopathic drugs has also been found to be effective in promoting mycelial growth of *Pleurotus*, suppressing the competitor moulds and thereby increasing the fresh yield of oyster mushroom (Thakur *et al.*, 2001d). Bacterial and viral diseases of oyster mushroom have received very little attention in India. Three bacterial diseases, viz., rot (*Pseudomonas alcaligens*) by Biswas *et al.* (1983) from West Bengal; brown spot (*P. stutzeri*) by Mellesha and Shetty (1988a, 1991b) and a yellow blotch (*P. agarici*) by Jandaik *et al.* (1993) were reported to be controlled by streptocycline application.

A single viral disease affecting *Pleurotus florida* has also been reported (Krishana Reddy *et al.*, 1993). Among the insect pests, sciarid, phorid, cecid flies and spring tails have been reported from West Bengal (Chakravorty *et al.*, 1987), Tamil Nadu (Krishnamoorthy *et al.*, 1991), Haryana and Chandigarh (Johar and Kaushal, 1991) and from H.P., Rajasthan and Punjab (Thapa and Seth, 1983), Chattisgarh (Soni, 2002). Chemical control and other methods of control of all these pests by chemicals and other methods have been reported but they need to be reexamined as a component of IPM.

Inspite of having all the desirable characters, favourable conditions and much simpler cultivation technology, oyster mushroom has not gained much momentum in our country except in few pockets of Andhra Pradesh and Karnataka, where a few small seasonal farms have come up. A large scale farm solely cultivating this mushroom for commerce is yet to be started. It may perhaps be attributed to unstable yield, spore allergy and breakage of fruiting bodies in transit.

III. Paddy straw mushroom

Paddy straw mushroom is also known as Straw Mushroom, Chinese Straw mushroom, Chinese mushroom, Nanhua mushroom and warm mushroom (*Volvariella volvacea*). It is the third most important mushroom cultivated in the world with an annual production of 287 metric tons (Chang and Miles, 1991). It grows in the tropics and subtropics and has been traditionally cultivated in rice straw for many years in China and in South East Asia (Chang, 1965b, Chang, 1982). Around 1932-35, straw mushroom was introduced in Philippines, Malaysia and other South-East Asian countries by overseas Chinese (Baker, 1934, Benemerito, 1936). In 1971, cotton wastes were first introduced as heating material for growing the straw mushroom (Chang, 1972, Yau and Chang, 1972) and in 1973, cotton wastes had completely replaced the traditional paddy straw to grow the mushroom (Chang, 1974). This was a turning point in the history of straw mushroom cultivation because the cotton waste compost through the pasteurization process, brought the cultivation of mushroom into an industrial scale first in Hong Kong and then in Taiwan, Thailand and China. In India, Su and Seth (1940) have first cultivated *V. diplasia* but the scientific cultivation using spawn was successfully demonstrated by Thomas *et al.*, (1943). Being a tropical mushroom (32-34°C), it is highly suitable for northern plains as well as coastal and plateau areas of South India. It is a fast growing mushroom, the time required from spawning to harvesting is only 8-10 days. No other vegetable or cultivated mushroom can be served as a table dish within such a short time. However, very little research support has been extended to this tasteful mushroom by Indian workers particularly during the last 2 decades.

1. Germplasm and their improvement: There are more than 100 species, sub-species and varieties of *Volvariella* which have been reported from different parts of the world (Rath, 1962; Ghosh *et al.*, 1967). But, only 13 species are known from India, out of which three species viz., *V. volvacea, V. diplasia* and *V. esculenta* are under cultivation (*Roy et al.*, 1978). *Volvariella* is a primary homothallic fungus, monospores being self fertile, hybridization is a difficult proposition. Evaluation of existing cultures have not yielded tangible results (Garcha *et al.*, 1987c). However, 29% increase in yield potential derived from a *V. diplasia* culture was achieved by Kalra and Phutela, (1991) which is too low compared to Graham, (1975) who obtained 125 and 199% higher yield from 2 monosporous isolates of *V. volvacea*. The other possibility of obtaining an improved strain is through mutation, mycelial, fusion, protoplast fusion, etc. and some of the scientists have already attempted these techniques with varying success (Muthukrishnan, 1971; Jaganathan, 1972; Krishnamohan and Jayarajan, 1978; Kapoor, 1987; Khanna *et al.*, 1991). However, use of recombinant DNA technique for strainal improvement of this mushroom is yet to make a beginning in this country.

2. Crop management : Paddy straw mushroom in India was traditionally grown on rice straw, though several workers tried other substrates like oil palm bunch waste (Naidu, 1971), oil palm pericarp (Goh and Graham, 1975), straw of wheat, barley, oat (Garcha and Kalra, 1979), bajra (Gupta *et al.,* 1970), cotton waste (Hu *et al.,* 1973; Chang, 1974; Chang, 1993; Thakur *et al.*, 2002b), dry banana leaves (Wade, 1970), bagasse, flax tow (Tzeng, 1974), sisal baggasse, sugarcane bagasse + cotton (Khan *et al.,* 1991) water hyacinth, water hyacinth + cotton waste in the ratio of 1:1 (Saeed *et al.,* 1994), kallar grass (Abdul *et al.,* 1995), wood waste + pine apple in the ratio of 1:2 (Lim *et al.*, 1977), with varying in yield from 1.2 to 76.12% biological efficiency. Of various substrates studied, cotton waste substrate and partial compost made out of it had excellent performance on yield (Garcha, *et al.*, 1989). Cotton waste (cotton seed hull) as the substrate has been widely used in countries like Thailand, Hong Kong, Indonesia and Vietnam and it has attained a status of an agro industry. When the above substrates were supplemented with red powder, wheat bran, soybean flour, cooked apple extract, ground lime stone, rice bran, soybean meal, groundnut shell at 2.5 to 5%, significant increase in yield of straw mushroom has been reported by several workers (Ramakrishna *et al.*, 1968; Garcha, 1975; Ramaswamy and Kandaswamy, 1976; Kaul, 1978; Chang, 1979; Chang Ho and Ho, 1979; Annual Report, 2001-02).

Mother spawn of *V. volvacea* can be prepared using cereal grains such as sorghum, rye, wheat and millet while planting spawn can be suitably prepared in several substrates like cotton wastes, oil palm wastes, tea wastes, coffee hulls, tobacco midrib, rice straw, banana leaves, water hyacinth leaves, saw dust mixture etc. with supplementation of atleast 10-20% or rice bran, corn meal, wheat bran or legume leaves meal (Matiru and Quimio, 1992; Quimio, 2002). Production of pink chlamydospores by the mycelium is the characteristics of *V. volvacea* cultures and is an indicative of good growth, which served as added inocula on the mushroom beds. It is very important to note that the cultures of *V. volvacea* do not survive at chilling temperatures between 5-10^{0}C. They cannot, therefore, be either refrigerated or kept in mineral oil or distilled water. The only means of long storage without change in quality appears to be being "flash-frogen in liquid nitrogen" (Stamets, 1993).

Several techniques of straw mushroom cultivation have recently been reviewed by Jayarajan and Krishnamohan (1994) and Khanna (1995). The prevalent methods of its outdoor cultivation and some improvements like spraying of spore suspensions, mixing of salt/urea, spent straw mixing have been discussed (Garcha *et al.,* 1987, 1989 Quimio *et at.*, 1990; Thakur *et al.* (2002b). They have also described different methods of indoor cultivation commonly used in plains of North India (Garcha, 1984) and South India (Samajpati, 1979). Helix tyre, type, type beds, square size beds (Pani and Das, 2001), hollo interior beds, were reported to be very good for enhanced yield (Garcha and Kalra, 1979; Garcha *et al.*, 1980; 1987, 1989). Box method (Alicbusan, 1975), bed method, (Annual Report, 1992-93 to 1996-97), cage method (Verma *et al.,* 1989a; Godara, 2001) and spiral method (Godara, 2001; Thakur *et al.*, 2002b) of straw mushroom cultivation have been found to be highly suitable.

3. Disease and pests management : Some fungal diseases like button rot (*Sclerotium sp.*) and a bacterial button rot have been recorded from India in the early seventies (Muthukrishnan, 1971). Later, Bhavani Devi and Nair (1986) have reported *Rhizoctonia solani* as causing malformation and reduction in the number of mushroom primordia. A number of competitor moulds have been recorded from different substrates like *Aspergillus flavus, Sclerotium rolfsii, Coprinus* spp., *Mucor* sp., etc. (Gupta *et al.*, 1970; Bahl, 1984; Purkayastha and Das, 1991, Annual Report, 1999-2000), which can be minimized either by pasteurization of straw or by fungicidal sprays (Sharma and Verma, 1985). However, Kanniayan and Prasad (1978) recommended a combination of malathion (0.25%), dithane Z-78 (0.2%) and tetracycline (0.025%) spray for the prevention and management of pests, diseases and competitors. Some management practices, like pH of > 7.00 to 8.00, pasteurization of substrate for > 2 hours at 60^0 or 70^0C and use of fresh, reasonable N content of substrate have also been advocated to keep the disease problems under check.

From the above review, it appears that very little research support has been extended to this tasteful mushroom. It has not been the focus of attention of the Indian Mushroom Biologists at least for the last 10-15 years to make it more adaptable on a commercial scale.

IV. White Milky Mushroom

White milky mushroom (*Calocybe indica*) was first reported from India in 1994 and first cultivated in 1976 (Purkayastha and Chandra, 1976) on a mixture of soil, sand and maize meal (12:6:1) and in soil jars. It grows in nature on humus soil under the road side trees in forest. It is sold in city and village markets, collected from the forests of West Bengal and liked because of its attractive robust white sporocarps, long shelf-life and taste. This mushroom can be cultivated throughout the year with a temperature ranging from 28-30^0C and relative humidity of > 85%. It can be cultivated on a wide range of cellulosic substrates viz; paddy straw, maize stalks, vetivar grass, palmarosa grass, wheat straw, wheat + paddy straw (1:1), sorghum stalks, sugarcane bagasse, groundnut haulms, soybean hay, black gram hay (Annual Report, 1999-2000, 2000-2001; Theradimani *et al.,* 2001). The substrates used were sterilized by chemical treatment, hot water or steam and supplemented with maize meal, wheat bran, rice husk or lucerne hay at 5 and 7.5% concentrations and added to the substrate at the time of spawning as well as casing respectively (Doshi *et al.,* 1989).

The substrate after sterilization is filled in trays or poly bags and spawned @ 250 g of spawn/2 ft^2 tray or @ 5% (wet weight of substrate) by layering method. Spawn run is completed in 18-20 days at 27-32^0C. After spawn run, casing soil is prepared by mixing dried loam soil or garden soil and sand (1:1) and calcium carbonate @ 12% of soil and sand mixture or 2 years old cow dung patties or FYM + spent compost thoroughly steam sterilized is used to 3.4 cm thickness. After 15-16 days of casing, fruit body is ready for harvest.

For white milky mushroom, improved production technology has been evolved with 40-70% BE at IIHR, Bangalore, RAU, Udaipur, IGAU, Raipur, NCRM, Solan. The production technology was further refined by Theradimani *et al.* (2001) at TNAU, Coimbatore with 145% BE. The Regional Research Station at Aruppukottai introduced APK2 strain of *C. indica* for commercial cultivation during January 1998, which clicked very well and the area under this mushroom in Tamil Nadu has assummed a commercial status.

Krishnamoorthy (1995) has thoroughly studied the production technology of this mushroom. Some work on agronomical practices like effect of light and growth regulators have recently been addressed by Theradimani *et al.* (2001). They have reported that the beds incubated in polythene chambers lined with blue high density polythene sheet (HDP) yielded more mushroom/bed, with increased pileus diameter and reduced stipe length. It was followed by red HDP sheet. Similarly, kinetin growth regulator at 100 ppm recorded maximum number of mushrooms formed/bed followed by mannitol 100 ppm. Stimulation of the fruiting bodies of *C. indica* with Gibberellic acid (100 ppm) and Indol Acetic Acid (100 ppm) had also been reported.

The competitor molds sometimes pose a severe problem in cultivation of *C. indica.* The common molds encountered were the species of *Sclerotium, Trichoderma, Coprinus, Aspergillus,* and *Rhizoctonia* during spawn run. It has been noticed that *Trichoderma harzianum* sporulates in the fruit bodies of *C. indica* specially after first harvest resulting in suppression of growth and development of fruit bodies (Annual Report, 2000-01).

CONCLUSIONS AND FUTURE STRATEGIES

The cultivation of edible mushrooms on lignocellulosic wastes represents the most successful example of solid state fermentation and generation of valid and valued form of biomass. The mushroom cultivation has become one of the most intensive and technically demanding of all vegetable cultivation practiced throughout the world. The biochemical properties of mushrooms have several applications in industries. It can be exploited to get the improved nutritional quality of various lignocellulosic wastes for animal feed stock and to alleviate the problems of malnutrition, energy and environmental degradation.

Mushroom production continues to increase rapidly in both developed and developing countries of the world. India is fast emerging as a major mushroom growing country but it is still not figured in the international mushroom map despite of having abundant raw material, enormous domestic market, cheap manpower, strong technical base, favorable environmental conditions and good government support.

An effort should, therefore, be made to explore several unexplored biological activities of edible mushrooms, to search for new species, develop newer strains through modern biotechnological methods, conserve germplasm of edible mushrooms, maintain biological diversity, evolve cheap production technology, minimize price, develop suitable farm designs, processing technologies and make wider publicity about the merits of the mushrooms.

REFERENCES

Abdul, R., Khan, S. Mohamd, M. and Imam Ul-Haq (1995). Cultivation of Chinese mushroom (*Volvariella volvacea*) in liquid and solid culture on Kallar grass, *Pakistan Journal of Phytopathology* 7 : 5.

Ahlawat, O.P. and Rai, R. D. (1997). Effect of *Azotobacter* and *Phosphotica* fertilizers on the White button mushroom (*Agaricus bisporus*), *Mushroom Research.* 6 (2) : 69-74.

Alicbusan, R.V. (1975). Mushroom growing (*Volvariella volvacea*),*The Philippine Biota* , 9: 120.

Annual Report (1992-93). All India Coordinated Mushroom Improvement Project, Department of Plant Pathology, Indira Gandhi Agricultural University, Raipur Centre.

Annual Report (1993-94). All India Coordinated Mushroom Improvement Project, Raipur Department of Plant Pathology, Indira Gandhi Agricultural University Raipur Centre.

Annual Report (1994-95). All India Coordinated Mushroom Improvement Project, Department of Plant Pathology, Indira Gandhi Agricultural University Raipur Centre.

Annual Report (1995-96). All India Coordinated Mushroom Improvement Project, Department of Plant Pathology, Indira Gandhi Agricultural University Raipur Centre.

Annual Report (1996-97). All India Coordinated Mushroom Improvement Project, Department of Plant Pathology, Indira Gandhi Agricultural University, Raipur Centre.

Annual Report (1999-2000). All India Coordinated Mushroom Improvement Project, Department of Plant Pathology, Indira Gandhi Agricultural University, Raipur Centre.

Annual Report (2000- 2001). All India Coordinated Mushroom Improvement Project, Department of Plant Pathology, Indira Gandhi Agricultural University, Raipur Centre.

Annual Report (2001-2002). All India Coordinated Mushroom Improvement Project, Department of Plant Pathology, Indira Gandhi Agricultural University, Raipur Centre.

Ansu, O., Geetha, D. and Suharban, M. (2001). Lignocellulose degradation by oyster mushroom, *Mushroom Research* 10(1): 36-40.

Babu, M.K. and K.R. Nair. (1991). Mushroom cultivation on oil palm factory wastes 2. Performance of five species of *Pleurotus* (Oyster Mushroom) *Adv. Mushroom Sci.* P. 36 (Abstr.)

Bahl, N. (1984). *Handbook of Mushrooms.* Oxford & IBM, New Delhi. pp. 123.

Bahl, N. (1991). Protein and lipid supplementation for increased yield of white button mushroom. *Indian Phytopath.* 44 (2) : 229-31.

Bahl, N. (1997). Mushroom cultivation in Delhi Metropolis-Training and Development. In: Mushroom Biology and Production, Rai, R.D., Dhar, B.L. and Verma, R.N., eds, MSI, Solan (H.P.), 391-397.

Bahram, H. (1989). Edible oyster mushroom on straw. *Mushrooms* J. 9 : 97-108.

Bahukhandi, D. and R.L. Munjal, (1990). Studies on evolving high yielding strains of *Pleurotus sajor-caju* thorough hybridization. *Indian Phytopath.* 43.7-73.

Baker, J.A. (1934). Mushroom growing in Wellesley and Penanag Provinces, *Malaysian Agric. J.* 22:25-28.

Balakrishnan, B. and Nair, M.C. (1995). Production technology of oyster mushroom (*Pleurotus* spp.). In: Advances in Horticulture, Vol. 13-Mushroom. (Eds. K.L. Chadha and S.R. Sharma), Malhotra Publish. House, New Delhi. pp. 109-116 .

Bano, Z. and Rajarathanam, S. (1982). *Pleurotus* mushroom as a nutritions food. In : Tropical mushrooms – Biological Nature and Cultivation Methods, Chang, S.T. and Quimio, T.H. eds., The Chinese University Press, Hong Kong, 363.

Bano, Z. and Rajarathnam, S. (1988). *Pleurotus* mushrooms. Part II. Chemical composition, nutritional value, post-harvest physiology, preservation and role as human food. *CRC Critical Review. Food Science and Nutrition,* 27 : 87-158.

Bano, Z. and Srivastava, H.C. (1962). Cultivation of *Pleurotus* species on paddy straw. *Food Science,* 12 : 363-65.

Bano, Z., Rajarathnam S. and Nagaraja N. (1987). Some important studies on *Pleurotus* mushroom technology. *Indian Mush. Sci.* 12 (2) : 67-71.

Bassham, J.A (1975). Biotechnol Bioeng. Symp. No.2, 9.

Benemerito, A.N, (1936). Mushroom culture in Canton, *Philip. Agric.*, 24: 624-34.

Bhandal, M.S. and K.B. Mehta. (1968b). Strategies of breeding *Agaricus bisporus.* Intl. Symp. Sci. Technol. Cult. Edible Fungi, Pennsylvania, p. 26 (Abstr.).

Bharadwaj, G. (2000). Vermicompost – A new casing media for white button mushroom *Agaricus bisporus* under Indian conditions, *Journal of mycology and Plant Pathology,* 30 (2) : 289.

Bharadwaj, S.C., Jandaik C.L. and Beig G.M. (1987). *Gliocladium virens* - a new pathogen of *Pleurotus* species. *Mush. J. Tropics* 7: 97-100.

Bhatt, N. and Singh R.P. (2000). Chemical control of mycoparasite of button mushroom (*Agarious bisporus*), *Journal of mycology and Plant Pathology,* 30 (2) : 291.

Bhatt, N. and Singh, R.P. (2002a). Casing soil bacteria as biocontrol agents against the mycoparasitic fungi of *Agaricus bisporus.* In : Mushroom Biology and Mushroom Products, Sanchez, J.E., Huerta. G. and Montiel, E., Eds, Universidad Autonoma Del Estado De Morelos.

Bhatt, N. and Singh, R. P. (2002b). Cow web disease of *Agaricus bisporus.* Incidence. losses and effective management. In : Mushroom Biology and Mushroom Products, Sanchez, J.E., Huerta. G. and Montiel, E., Eds, Universidad Autonoma Del Estado De Morelos, Mexico, 161-170.

Bhavani Devi, S. and Nair M.C. (1986). Popularisation of *Volvariella* spp. in the tropics.. In : Beneficial Fungi and their Utilization Eds. M.C. Nair and S. Balakrishnan, Scientific Publishers, Jodhpur, pp. 25-33.

Bisaria, R., Madan, M., Bisaria, V.S. and Mukhopadhayay, A.N. (1987). Amino acid composition of the mushroom, P*leurotus sajor-caju*, cultivated on different agro-residues. Biological Wastes 20:251-259.

Bisco, N.A., Babitskay, V.G., Bilay, V.T. and Mitropolskaya, N. Yu. (2002). Biologically active substances from *Lentinula edodes* and *Pleurotus ostreatus* In : Mushroom Biology and Mushroom Products, Sanchez, J.E., Huerta. G.

and Montiel, E., Eds, Universidad Autonoma Del Estado De Morelos, Mexico. 383-386.

Bisco, N.A. and Dudka, I.A. (1987). Biology and cultivation of edible mushroom of the genus *Pleurotus*. Naukuva Dumka.

Bisht, I.S., Singh, R.N and. Kanaujia, J.P (1986). Mycoflora of synthetic compost and their relative cellulolytic ability. *Indian J. Mycol. Pl. Pathol.* 16: 232-234.

Biswas, P., Sarkar, B.B Chakravarty, D.K. and Mukherjee. M.L. (1983). Evaluation of two locally available millets for spawn production of *Agaricus bisporus. Indian Phytopathology.* 36 : 777-778.

Block, S.S., Tsao, G. and Han, L. (1958). Production of mushrooms from saw dust. *J. Agri Food Chem.* 6 : 923-27.

Buswell J.A., Cai Y.J. Chang S.T., Peberdy J.F., Fu S.Y., and Yu H.S. (1996a). Lignocellulolytic enzyme profiles of edible mushroom fungi. *World J Microbiol Biotechnol,* 12, 537-542.

Buswell, J.A., Cai, Y.J., and Chang, S.T. (1995). Effect of nutrient nitrogen and manganese on manganese peroxidase and laccase production by *Lentinula edodes. FEMS Microbiol Lett,* 128, 81-88.

Buswell, J.A., Cai, Y.J., and Chang, S.T. (1996b). Ligninolytic enzyme production and secretion in edible mushroom fungi. In: Mushroom biology and mushroom production, Proceedings of the second international conference of mushroom biology and mushroom products. Royse D.J., ed. Pennsylvania State University, PA, pp. 113-122.

Buswell, J.A. and Chang, S.T. (1994). Biomass and extracellular hydrolytic enzyme production by six mushroom species grown on soybean waste. *Biotechnol Lett.* 16, 1317-1322.

Cai, Y.J., Buswell, J.A., and Chang, S.T. (1994). Production of cellulases and hemicellulases by the straw mushroom, *Volvariella volvacea, Mycol Res*, 98, 1019-1024.

Cai, Y.J., Buswell, J.A., and Chang, S.T. (1998). B- Glucosidase components of cellulolytic system of the edible straw mushroom, *Volvariella volvacea. Enzyme Microb Technol,* 22, 12-129.

Cai, Y.J., Chapman, S.J., Buswell, J.A. and Chang, S.T. (1999). Production and distribution of endoglucanase, cellobiohydrolase and B-glucosidase components of the cellulolytic system of *Volvariella volvacea*, the edible straw mushroom. *Appl Environ Microbiol,* 65, 533-559.

Chang, S.T. (1965). Cultivation of the straw mushroom in S. E. China. *World Crops.* 17, 47-49.

Chang, S.T (1972). The Chinese mushroom (*Volvariella volvacea*): morphology, cytology, genetics, nutrition, and cultivation. The Chinese University Press, Hong Kong, p.99.

Chang, S.T. (1974). Production of the straw mushroom (*Volvariella volvacea*) from cotton wastes. *Mush. J.* 21, 358-353.

Chang, S.T. (1980). Mushroom as human food. *Bio Science,*30, 399-401.

Chang, ST (1982). Cultivation of *Volvariella* mushroom in South East Asia. In : Tropical mushrooms - Biological nature and cultivation methods, Chang,

S.T. and Quimio, S.W., Eds, The Chinese University Press, Hong Kong, p.182.

Chang, S.T. (1989). Recycling of solid wastes with emphasis on organic wastes. Green production, 16-23.

Chang, S.T. (1993). Mushroom biology : the impact on mushroom production and mushroom products. In: Mushroom biology and mushroom products. Chang S.T., Buswell J. A., and Chiu S.W., eds. The Chinese University Press, Hong Kong, pp. 3-20.

Chang, S.T.(1996). Mushroom research and development-equality and mutual benefit. In: International Conference on Mushroom Biology and Mushroom Products. Royse D. J., ed.; Pennsylvania State University, PA, pp. 1-10.

Chang, S.T. (1998). Development of novel agroscience industries on bioconversion technology. In: Frontiers in biology: the challenges of biodiversity. Chou C. H. and Shao K. T. eds. Academia Sinica, Taipei, pp. 217-222.

Chang S.T. (1999). Global impact of edible and medicinal mushrooms on human welfare in the 21st century: non-green revolution. *International. J. Medicines. Mushroom,* 1, 1-7.

Chang S.T. (2001). A 40-Year journey through bioconversion of lignocellulosic wastes to mushrooms and dietary supplements, *International Journal of Medicinal Mushrooms.* 3, 299-310.

Chang, S.T. (2002a). Past and present trend in the production of *Lentinula edodes* in Asia. In : Mushroom Biology and Mushroom Production, Sanchez, J.E., Huerta, G. and Montiel, E. eds, Universided Autonoma Del Estada De Merelos, Mexico. 1-8.

Chang, S.T. (2002b). Mushroom and mushroom biology. 2nd International Training Workshop on Mushroom Cultivation at Edible Fungi Institute Sanghai, China Lecture, 19-37.

Chang, S.T. and Buswell, J.A. (1996). Mushroom nutriceuticals. *World J Microbiol* Biotechnol, 12, 473-476.

Chang, S.T. and Buswell, J.A. (1999). *Ganoderma lucidum* (Curt.: Fr.) P. Karst. (Aphyllophoromycetideae): a mushrooming medicinal mushroom. *Int. J. Med. Mushr.* 1, 139-146.

Chang, Y. and Hudson, H.J. (1967). The fungi of wheat straw compost, I. Ecological studies, *Tranaction and British Mycological Society*, 50:649-666.

Chang, S.T. and Miles P. G.(1987). Historical record of the early cultivation of *Lentinus* in China. *Mushr. J. Tropics,* 7, 31-37.

Chang, S.T. and Miles, P.G. (1989). Edible mushrooms and their cultivation, CRC Press, Bacakoton, Florida, USA.

Chang, S.T. and Miles, P.G. (1991). Recent trends in world production of cultivated edible mushrooms, *Mush. J.* 504: 15-17.

Chang, S.T. and Miles, P. G. (1992). Mushroom biology – a new discipline. *Mycologist,* 6, 64-65.

Chang, S.T. and Mshigeni, K.E. (2000). *Ganoderma lucidum* – paramount among medicinal mushrooms. *Discov Innovat*, 12, 97-101.

Chang, S.T. and Mshigeni, K.E. (2001). Mushrooms and human health: with special reference to their growing significance as potent dietary supplements. University of Namibia, Windhoek, 52 pp.

Chang, S.T. Lau, O.W. and Cho, K.Y. (1981). The cultivation and nutritive value of *Pleurotus sajor-caju, European Journal of Applied Microbiology and Biotechnology*, 12 : 58,

Chang, Ho, Y. and Ho, T.M. (1979). Effect of nitrogen amendment on the growth of *Volvariella volvacea, Mushroom Science*, 10:619.

Chakravorty, D.K., Sarkar, B.B.. Dutta, S. and Chatterjee. M.L. (1987). Bionomics and control of sciarid fly *Lycoriella auripila* Winn. in sub-tropical mushroom *Pleurotus sajor-caju* (Fr.) Sign. *Indian Mush. Sci.* 2 : 146-150.

Chihara, G. (1990). Lentinan and its related polysaccharides as host defence potentiator: their application to infectious diseases and cancer. In *'Immunotherapeutic prospects of infectious diseases'* (K. Noel Masihi and W Lange ed.) Springer Verlag, Heidelberg. 1990. pp. 9-18.

Chihara, G. (1993). Medical aspects of Lentinan isolated from *Lentinus edodes* (Berk) Sing. In Mushroom Biology and Mushroom Products. S. Chang, JA Buswell and SW. Chiu eds). The Chinese University Press, Hong Kong. pp. 221-226.

Cochran, K.W. (1978). Medical effects. In : The Biology and cultivation of Edible mushrooms (Chang S.T. and Hayes W.A., eds.) Academic Press, New York. PP. 169-187.

Crisan, E.V. and Sands, A. (1978). Nutritional value of edible mushrooms. In: The Biology and Cultivation of Edible Mushrooms. (eds. Chang, S.T., Hayes, W.A.) Academic Press, New York, pp. 137-178.

Das, L and M. Suharban. (1991). Fungal parasites of oyster mushroom in Kerala. Advances in Mushroom Science, p. 91 (Abstr.).

Delcaire J.R. (1978). Economics of cultivated mushrooms. In.: The biology and cultivation of edible mushrooms. Chang, S.T. and Hayes H.A., eds., Academic Press, New York, pp. 726-793.

Dhar, B.L. (1991). Mushroom farm design suiting Indian condition. *Indian Mushrooms*. pp. 33-39.

Dhar, B.L. (1993). Mushroom composting for *Agaricus bisporus/bitorquis. In* Advances in Mushroom Biotechnology ed. M.C. Nair, *et al.* Scientific Publish on Jodhpur. pp. 84-90.

Dhar, B.L. (1995a). Mushroom farm design. In : *Advances in Horticulture*, Vol. 13 - Mushrooms EDS. K.L. Chadha and S.R. Sharma, Malhotra Publish. House, New Delhi, pp 597-619.

Dhar, B.L. (1995b). Production Technology of *Agaricus bitorquis. In Advances in Horticulture*, Vol. 13 - Mushrooms. Chadha K.L. and Sharma S.R., Malhotra Publish. House, New Delhi, pp. 99-107.

Dhar, B.L. (1996). Composting of *Agaricus* - Recent Developments and present position in India. In : Mushroom Biology and Mushroom Products, DJ Royse (ed.) Penn. State Univ. pp. 251-257.

Dhar, B.L. (1997). Mushroom industry in India. *In* : Advances in Mushroom Biology and Production, Rai, R.D., Dhar, B.L. and Verma, R.N., eds, MSI, Solan (H.P.), 369-378.

Dhar, B.L. and Gupta, Y. (2000). Improved production technology of white button mushroom, *Journal of Mycology and Plant Pathology*, 30 (2) : 281.

Dhar, B.L. Yadav, M.C. and Verma R.N. (2002). Evaluation of white button mushroom strains and their availability in India. Book of Abstract of IV Conference on Mushroom Biology and Mushroom Products, held at Mexico, 9.

Dhar, B.L. and Kapoor, J.N. (1988). Different formulation for composting. *Indian Journal of Mycology Plant pathology* 18 : 93 (Abstr.)

Dhar, B.L. and Kapoor, J.N. (1989). Strain evaluation of Agaricus bisporus. *Mushroom Science* 12 (1) : 55-61.

Dhar, B.L. and Kapoor, J.N. (1990). Post-composting nutrient supplement for increased mushroom yield. *Indian. Phytopath.* 43 : 74-76.

Dhar, B.L., Vijay, B., Upadhyay, R.C. and Sohi, H.S. (1985). Effect of compost depth in polythene bags on yield of *A. bisporus. Indian J. Mycol. Pl. Pathol.* 15 : XI (Abstr.).

Doshi, A., Sidana, N. and Chakravarti, B.P. (1989). Cultivation of summer mushroom, *Calocybe indica*, (Purkayastha and Chandra) in Rajasthan. *Mush. Sci.* 12 (2) : 395-400.

Dutta, M., Deb, G. and Chakravarti, D.K. (1991). Cultivation technology of oyster mushroom in shelf under W. Bengal condition. *Advances in Mushroom Science* p. 30 (Abstr.)

Evini, M., Prema, P. and Jaya Kumar, J. (1995). Cultivation trials of *Pleurotus ostreatus* Kummer on lime water pretreated coir waste and paddy straw. *Mush. Res.* 4 (2) : 77-80.

Falck, R. (1917). Uber die waldkultur des Austernpilzes (*Agaricus ostreatus*) auf Laubholzstubben. *Zeitschrift fur Forstund Jagdwesen.* 49 : 159-165.

Flegg, P.B. and Maw, G.A. (1976). Mushrooms and their possible contribution to world protein needs, *Mushroom Journal*, 48: 396.

Ganeshan, G. (1987). A new disease of oyster mushroom by *Arthrobotrys pleuroti* nov.- a first report. *Mushroom inform.*4 (10): 12-13.

Garcha, H.S. (1975). Mushroom culture: Communication Centre, Punjab Agricultural University, Ludhiana, Punjab.

Garcha, H.S. (1984). Manual of Mushroom Growing. Punjab Agricultural University, Ludhiana., India.

Garcha, H.S., Chang, S.T. , Phutela, R.P., Sodhi, H.S., Dhanda, S., Khanna, P.K. and Mok. S.N. (1989). Studies on substrate manipulation for growing *Volvariella* in India. *Mush. J. Tropic* 9: 147-154.

Garcha, H.S. and Kalra, K.L. (1979). Paddy straw mushroom in North India. *Mush. Sci.* 10: 645-652.

Garcha, H.S., Sodhi, H.S., Phutela, R.P. and Khanna, P.K. (1987). *Proc. Int. Mushroom Symp. On Edible Fungi Pennsylvania.* pp. 533-544.

Garcha, H.S. and Sekhon, A. (1981). Evaluating casing materials for mushroom culture in Punjab (India). *Mushroom Science,* 11 (1) : 411-417.

Garcha, H.S., Khanna, P.K., Sodhi, H.S., Dhanda, S., Sidhu, A. and Phutela, R.P. (1987a). Nutrient supplementation for *Agaricus bisporus* (Lange) sing. Cultivation. In : Cultivating Edible Fungi. Eds. P.J. Wuest, D.J. Royse and R.B. Beelman, Elsevier Science Publishers, The Netherlands, pp. 101-108.

Garcha, H.S., Sekhon, A.K. and Phutela, R.P. (1980). Studies on spawn substrate relationship for cultivation of tropical mushroom. Proceeding of symposium, PAU, Ludhiana. Pp. 351-354.

Garcha, H.S., Sodhi, H.S. Phutela, R.P. and Khanna, P.K. (1987c). Evaluating strains of paddy straw mushroom (*Volvariella spp.*) in India. In : Cultivation Edible Fungi. Eds. P.J. Wuest, D.J. Royse and R.B. Beelman, Elsevier Science Publishers, The Netherlands, pp. 533-554.

Garg, P. (1986). Use of sunflower meal (De-oiled cake) during composting for lower cost mushroom production. In : Mushroom Biology & Mushroom Products, DJ Royse (Ed.) Penn. State Univ. pp. 133-136.

Ghosh, N. and Chakravarty D.K. (1986). Introduction of a new Indian edible mushroom. (*Pleurotus citrinopileatus* for cultivation. *Indian Agric.* 30 : 125.

Ghosh, N. and Chakravarty, D.K.. (1991). Studies on evolving new strains of *Pleurotus sajor-caju* by selective dikaryotistion. *Adv. Mushroom Sci.* p. 28 (Abstr.).

Ghosh, R.N., Pathak, N.C. and Tewari, I. (1967). Studies on Indian *Agaricales..Indian Phytopathol.* 20: 237-242.

Godara, D.R. (2001). Studies on production and preservation techniques of paddy straw mushroom (*Volvariella volvacea).* M.Sc. Thesis submitted to the Department of Plant Pathology, Indira Gandhi Agricultural university, Raipur. pp. 1-79.

Goh, S.C. and Graham, K.M. (1975). Studies on the paddy straw mushroom (*Volvariella volvacea*).VI Supplementation of oil palm pericarp compost with waste water, *Malany. Agric. Res.* 4:109-116.

Graham, K.M. (1975). Studies on the paddy mushroom (*Volvariella volvacea*). V. Performance of single-basidiospore isolates, *Malasian Agriculture Research,* 5 :31.

Gunasegaran, K. and Graham, M. (1987). Effect of organic additives on yield of the phoenix mushroom grown on cellulosic waste. *Mush. J. Tropics*, 8: 23-28.

Gupta, G.K., Bajaj, B.S. and. Surayanarayana, D. (1970). Studies on the cultivation of paddy straw mushroom (*Volvariella volvacea* and *V. diplasia*). *Indian Phytopath.* 23: 615-620.

Gupta, Y. and Vijay. B. (1992). Post-composting supplementation in *Agaricus bisporus* under seasonal growing conditions. *Mush Res.* 1 : 115-117.

Hu, K.M., Song, S.F. and Liu, P. (1973). Experiments on Chinese mushroom cultivation. I. The comparison of cultivating materials, *Journal of Taiwan Agriculture Research*, 22 : 145-154.

Hariprasad. (1991). Experiment studies on *Pleurotus* cultivation. *Adv. Mushroom Sci.* p. 101 (Abstr.)

Harsh, N.S., Bisht, K. and Upreti, J.C. (1981). Utilization of waste paper and tea leaves to cultivate *Pleurotus ostratus. Intl. Biodet. Bull.* 17 : 77-78.

Hobbs, Ch. (1995). Medicinal mushrooms: An exploration of tradition, healing and culture, Botanica press.

Huang N. L. (2000). Present situation and prospects of edible fungi industry in China. Edible fungi China, 19, 3-5.

Huang, J.W., Hu, C.K. and Tzeng, D.D.S. (1995). Experiment of growing media for straw mushroom (*Volvariella volvacea*), *Pl. Pathol. Bull.*, 4 (2) : 76-82.

Hussain, R., Bhan, M.K. and Kaul, B.L. (2000). Lemon-grass-waste, a potent substrate for *Pleurotus* cultivation and spawn production, *Journal of mycology and Plant Pathology* 30 (2): 286.

Ikekawa, J; Uehara, N; Meada Y; Nakamishi, M and Fukuoka, F. (1969). Antitumor activity of aqueous extracts of some edible mushrooms. *Cancer Res.* 29 : 734-735.

Jadhav, S.W. and Jagtap, V.D. (1991). Chemical sterilization of wheat straw used for growing oyster mushroom. Advances Mushroom Science p. 41 (Abstr.)

Jaganathan, G. (1972). Studies on certain mutants of *Volvariella diplasia* (Berk. & Br.) Sacc. M.Sc. Thesis, Annamalai University.

Jain, V.B. and Singh, S. P. (1982). Effect of spawning method on the yield of *A. brunnescens*. Peck. *Prog. Hort.* 14 (4): 246-248.

Jandaik, C.L. (1987). Breeding potentials of *Pleurotus* species in India with special reference to *Pleurotus sajor-caju. Indian Mush. Sci.* 2 : 355-360.

Jandaik, C.L. (1989). Response of *Pleurotus sajor-caju* to supplementation prior to and after pasteurization of straw. *Indian Phytopath.* 42 : 284-85.

Jandaik, C.L. (1997). History and development of *Pleurotus* cultivation in the world and future prospects. In: Advances in Mushroom Biology and Production. Rai, R.D, Dhar, B.L. and Verma, R.N. Eds., Mushroom Society of India, Solan, India. 181-192.

Jandaik, C.L. and Kapoor, J.N. (1974). Studies on cultivation of *Pleurotus sajor-caju. Mushroom Science* 9 (1) : 667 72.

Jandaik, C.L., Sharma, V.P. and Raina, R. (1993). Yellow blotch of *Pleurotus sajor-caju* (Fr.). Singer- bacterial disease new to India. *Mush Res.* 2 (1) : 45-48.

Jarial, R.S. and Sharma,V.P. (2000). Survey, incidence and losses caused by false truffle, *Journal of mycology and Plant Pathology,* 30 (2) : 291.

Jayarajan, R. and Krishnamohan G. (1994). Paddy straw mushroom (*Volvariella diplasia*) - Current status and future prospectus. In: Advances in Mushroom Biotechnology Eds. M.C. Nair, C. Gokulapalan and Lulu Das. Scientific Publishers, Jodhpur. pp. 147-150

Jhune, C.S., Cha, D.Y. and Yon, C.H. (1990). The antifungal microorganisms to *Mycogone perniciosa* Magn. In cultivated mushrooms, *Agaricus* (Lange) Sing. *Korean Journal of Mycology,* 18 : 2, 96-101.

Johar, K. and Kaushal, S.C. (1991). New report of cecid as pest of *Pleurotus* species. Advance Mushroom Science pp. 83-84 (Abstr.)

Kalra, R. and Phutela, R.P. (1991). Strain selection and development in *Volvariella.* Advance Mushroom Science p. 49 (Abstr.)

Kanedo, T and Tokuda, S. (1966). Effect of various mushroom, preparations on cholesterol levels in rats. *J. Nutri* 90 : 371-376.

Kannaiyan, S. and Prasad N.N. (1978). Production of paddy straw mushroom in India- Review. *Indian Mushroom Science* 1: 287-93,

Kapoor, J.N. (1987). Cytology, sexuality and breeding of *Volvariella volvacea. Indian Mush, Sci.* 2: 361-363.

Kaufert, F. (1935). The production of asexual spore by *Pleurotus corticatius. Mycologia,* 27 : 333-40.

Kaul, K.N. (1978). Prospects of large scale cultivation of paddy straw mushroom in Uttar Pradesh. In Indian Mushroom Science, edited by Atal, CK, Bhan, BK and Kaul, TN,pp. 305-10. Indo-American Literature House, New Delhi.

Khan, S.M., Haq, R., Dongar, M.A. and Maher, M.J. (1991). .Mushroom Science XIII.2 Proceeding of the 13th International Congress on the Science and Cultivation of Edible Fungi held at Dublin, -584.

Khandar, R.R., Vaishav, M.V., Akbari, L.F. and Andharia, J. H. (1991). Effect of various plant substrates on sporophore production of *Pleurotus sajor-caju.* Adv. Mushroom Sci. pp. 38-39. (Abstr.)

Khanna, P.K. (1995). Production Technology of paddy straw mushroom (*Volvariella* spp.). In : Advances in Horticulture Vol. 13. – Mushrooms, Chadha, K.L. and Sharma, S.R., Malhotra Publishing House, New Delhi, pp. 117-134.

Khanna, P.K., Kaur, J., Garcha, H.S. and Khanna, V. (1991). Isolation, culture and reversion of protoplasts of *Volvariella diplasia,* Indian Mushroom, pp. 145-153.

Klopper, J. W. (1992). Plant growth promoting Rhizobacteria as biological control agents. In: Soil Environmental Management, Melling, F.B. Jr. ed., Marcel Dekker, New York, 255-274.

Krishna Reddy, M., Pandey, M., and Tewari, R.P. (1993). Immunodiagnosis of mycovirus infecting oyster mushroom (*Pleurotus florida*). Golden Jubilee Symp. Hort. Research, Changing Scenario, 24-28 May, Bangalore, P. 248 (Abstr.).

Krishnamohan, G. and Jeyarajan, R. (1978). Studies on X-ray mutants of paddy straw mushroom. *Indian Mush Sci,* 1 : 343-347.

Krishnamoorthy A.S. (1995). Studies on the cultivation of milky mushroom (*C. indica*) Ph.D. Thesis Tamil Nadu Agricultural University, Coimbatore, India.

Krishnamoorthy, A.S., Marimuthu, T., Sivaprakasam, K. and Jeyarajan, R. (1991). Occurrence and damage caused by phorid fly on oyster mushroom. *Mush. J. Tropics* 11: 23-27.

Kumar, S. and Brinda Ban, V. (1997). Mushroom cultivation and upliftman of women in India. In : Advances in Mushroom Biology and Production, Rai, R.D., Dhar, B.L. and Verma, R.N., eds, MSI, Solan (H.P.), 399-406.

Kumar, S. and Chandra, K.D. (1983). Evaluation of different strains of *Agaricus brunnescens* in India. *Indian. J. Mush.* 9 :15-21.

Kumar, S. and Munjal, R.L. (1981). Studies on evolving new strains of the cultivated mushroom (*Agaricus brunnescens* Peck) by single spore isolation method. *Indian J, Mush.* 7 : 37-42.

Laborde, J. (1993). Required qualities of indoor composting *Mushroom Information*, 9 : 5-27.

Laborde, J., Lanji, G., Francesscutti, R. and Giordani, E. (1993). Indoor composting : General principles and large scale development in Italy. In: Mushroom Biology and Mushroom Products, S.T. Chang, J.A. Buswell and S.W. Chiu (eds.), The Chinese University Press, Hong Kong. pp. 93-114.

Laborde, J. (1994). Management of some physical factors involved in composting. *Mushroom Informarion.*12: 5-15.

Lau, O.W., Shiu, K.K. and Chang, S.T. (1985). Determination of ascorbic acid in vegetables and fruits by different pulse polarography, Journal of Food Science and Agriculture, 36-733.

Lellcy, J. (1985). Pilze aus dem eigenem Garten, BLU Verlagsgellschaft, Munchen.

Lellcy, J. and Jenssen, A. (1993). Productivity improvement of oyster mushroom substrate with a controlled release nutrient, *Mushroom News*, 41 : 6-13.

Li, G.S.F. and Chang, S.T. (1982). Nutritive value of *Volvariella volvacea.* In : Tropical Mushrooms-Biological Nature and Cultivation Methods, Chang, S.T. and Quimio, T.H., Eds., The Chinese university Press, Hong Kong, 199.

Lim, K., Tee, L.L. and Avadhana, P.N. (1977). In Second Int. Mycol. Cong. Tampa. pp 386 (Abs.).

Livernoche, D., Jurasek, L., Desrochers, K., Dorica, J. and Valiky, I.A. (1983). Removal of colour from Kraft mill wastewaters with cultures of white-rot fungi. *Biotechnol. Bioeng.*, 25 : 2055-2065.

Macauley, B.J. and P.S. Perrin (1994). Environmentally controlled composting. In: Agaricus Compost, N.G. Tan (ed.), Australian Mushroom Growers Association, Windsor, NSW, Australia. pp. 177-185.

Madan, M. (1997). Role of Mushroom cultivation in rural development in India. In : Advances in Mushroom Biology and Production, Rai, R.D., Dhar, B.L. and Verma, R.N., eds, MSI, Solan (H.P.), 407-414.

Madaan, R.L. and Singh S. (2000). Prevalence of false truffle in *Agaricus bisporus* Cultivation, *Journal of Mycology and Plant Pathology,* 30 (2) : 289.

Madhusudhanan, K. and Chandramohanan (2001). Production technology of oyster mushroom, In: Frontiers in Fungal Biotechnology and Plant Pathogen Relations,

Manoharachary, C. Bagyanarayan, G., Bhadraiah, B., Satyaprasad, K, Reddy, B.N. and Nagmani, A. Eds. Allied Publishers Ltd., New Delhi, pp. 324-330.

Madrigal, J., Motolinia, C. and Arias, A. (2002). Biomass production of Pleurotus ostreatus and Lentinula edodes on tequila vinasses. In: Mushroom biology and Mushroom Products, Sanchez, J.E., Huerta, G. and Montiel, E., Eds., Universities Autonoma Del Estado De Morelos, Mexico, 331-336.

Mallesha, B.C. and Shetty, K.S. (1988a). A new brown-spot disease of oyster mushroom caused by *Pseudomonas stutzeri. Curr. Sci.* 57 : 1190-1192.

Mallesha, B.C. and Shetty, K.S. (1991b). Characteristics of *Pseudomonas stutzeri* causing brown spot disease of oyster mushroom. *J. Fd. Sci. Technol.* 28 : 187-188.

Mathew, J., Kothandaraman, R. and Joseph, K. (1991). Cultivation of oyster mushroom in rubber processing waste - a possible solid waste utilization method. *Adv. Mushroom Sci.* p. 35 (Abstr.).

Matiru, V. and Qnimio, T.H. (1992) Enhancement of growths of *Volvariella volvacea* (Bull ex. Fr.) Sing on saw dust. *Mushroom Research,* 1 : 65 –141.

Mee, H.M. (1978). Mushroom composting U.S. patent 4 127 964.

Miles P. G. and Chang S.T. (1988). "Mycomeat" – a food produced from soybean slurry by fungal mycelium. In: Recent Advances in Biotechnology and Applied Biology. Chang S.T., Chan K.Y. and Woo N.Y.S., eds., The Chinese University Press, Hong Kong, pp. 577-586.

Millar, F.C. (1997). Enclosed phase-I mushroom composting system considerations of the underlying technology and methods of implementations. In: Advances in Mushroom Biology and Production, Rai, R.D., Dhar, B.L. and Verma R.N. eds, MSI, Solan, 129-138.

Mizuno T., Sakai T. and Chihara G. (1995). Health foods and medicinal usage of mushrooms. *Food Rev Int,* 11, 69-81.

Mizuno, T. (1996). A development of anti tumor polysaccharides from mushroom fungi *Foods Ingred Journal* (Japan), 167: 69-85.

Muez, M.A. (1994). Bases para el cultivo de pleurotus. In : International Jornadas Tecnicas del Champinony otros hongos comestibles. Patronato de promocion Economica. Diputacion Provincial de Cuenca. 129-141.

Murphy, W. S. (1972). Development of a mushroom production medium without phase I composting.*Mushroom News,* 20 : 4 - 22.

Muthukrishnan, P. (1971). Studies on *Volvariella diplasia* (Berk, & Br.) Sacc. M.Sc. Thesis. Annamalai University. 199 pp.

Naidu, N. R. (1971). Cultivation of paddy straw mushroom, *Volvariella volvacea* (Fr.) Sing. using oil palm bunch waste as medium. *The planter* 47 : 190-193.

Nath, P.K. (2001). Improvement in the production of *Pleurotus sajor-caju* (oyster mushroom) with nutrient supplementation. In: Frontiers in Fungal Biotechnology and Plant Pathogens Relations, Manoharachary, C., Bhagya Narayan, G., Bhadraiah, B., Satyaprasad, K., Reddy, B.N. and Nagamani, A. Eds., Allied Publishers Limited, New Delhi, 453-458.

National Research Centre for mushroom (2000-01). Annual Report, All India coordinated Mushroom Improvement Project, Solan (H.P.). 80 p.

Ng, T.B., Wang H.X., Liu W.K., Ooi V.E.C., and Chang S.T. (1999). Immunostimulatory effects of intramycelial and extramycelial polysaccharo-peptide complexes prepared from mushroom (*Coriolus versicolor*) mycelia and mycelial culture medium. In: Advance research in PSP. Yang Q.Y., eds. The Hong Kong Association for Health Care, 236-239.

Nunez, P.J. and Mendoza, C.G. (2002). Submerged fermentation of lignocellulosic waste under moderate temperature conditions for oyster mushroom growing substrates. In : Mushroom Biology and Mushroom Products,

Sanchez, J.E, Huerta G. and Montiel, E. Eds., Universsidad Autonoma Del Estado De Morelos, Mexico, 271-278.

O'Brien A. (2000). Update on the European mushroom industry November 2000. *Mushr J,* 611, 17-20.

Oei, P. (1996). Mushroom cultivation. Tool publications, Leiden.

Pandey, M. and Tewari, R.P. (1989). A natural sporeless variant of *Pleurotus. Indian J. Mush.* 15 : 25-28.

Pandey, M. and Tewari, R.P. (1994). Strategies for selection and breeding of edible mushrooms. Adv. Mushroom Biotechnology. Eds. M.C. Nair, C. Gokulapalan & Lalu Das. Scientific Publishers, Jodhpur. pp. 61-69.

Pani, B.K. and Das, S.R. (2001) Seasonal productivity of paddy straw mushroom (*Volvariella spp.*) in Orissa, *Mushroom research,* 10 (1) : 41-42.

Patel, R.B. and Rafique, A.N. (1991). Cultivation of *Pleurotus sajor-caju* in Gujarat. Adv. *Mushroom Sci.* p. 98 (Abstr.).

Patil, B.D., Jadhav, S.W. and Sathe, A.V. (1995). Mushroom Flora of Maharashtra. In: Advances in Horticulture, Vol. 13-Mushroom. Eds. K.L. Chadha and S.R. Sharma, Malhotra Publish. House, New Delhi, pp. 317-328.

Patil, B.D. and Shinde, P.A. (1983). Effect of spawning rate and method of spawning on yield of white button mushroom (*Agaricus bisporus*). *J. Maharashtra Agric. Sci.* 8 (1): 82.

Patil, B.D., Jadhav, S.W. and Perane, R.R. (1989). Evaluation of different strains of *Agaricus bisporus* for yield potential. *J. Maharashtra Agric. Univ.* 14 (1) : 109-110.

Patra, A.K. and Pani, B.K. (1997). Utilization of cotton waste and waste paper for production of oyster mushroom (*Pleurotus*). In: Advances in Mushroom Biology and Production, Rai, R.D., Dhar, B.L. and Verma, R.N. eds.), Mushroom society of India, National Research Centre for Mushroom, Solan (H.P.) pp. 205.207.

Phutela, R.P. and Garcha, H.S. (1986). Improving yield potential of the cultivated mushroom *Agaricus bisporus* (Lange) Sing through monospore cultures. *Punjab Hort. J.* 26 : (1-4) : 117-124.

Purkayastha, R.P. and Chandra, A. (1976). A new techniques for *in vitro* production of *Calocybe indica*- an edible mushroom of India. *Mushroom Journal,* 40: 112 -113.

Purkayastha, R.P. and Das, A.K. (1991). Regulation of sporophore production of paddy straw mushroom and its competitor *Coprinus* by cultural practices. *Indian Mush. Sci.* pp. 153.59.

Quimio, T.H., Chang, S.T. and Royse, D..J. (1990). Technical guidelines for mushroom growing in the Tropics. Fao,Rome.*Taiwan Mushroom* 5:10-16.

Quimio, T. H. (2002). Updates on spawn production of *Volvariella volvacea,* the tropical straw mushroom. In: Mushroom biology and Mushroom Products, Sanchez, J.E., Huerta, G. and Montiel, E., Eds., Universities Autonoma Del Estado De Morelos, Mexico, 337-343.

Rai, R.D. (1995). Nutritional and medicinal values of mushrooms. In : Advances in Horticulturel Vol. 13, Sharma, S.R. and K.L. Chadha, eds, Malhotra Publishing House, New Delhi.

Rai, R.D. (1997a). Food value of Mushrooms. In : Booklet on nutritive and medicinal values of edible mushrooms. Vigyan Samiti, Udaipur. Pp 9-13.

Rai, R.D. (1997b). Medicinal mushrooms. In : Advances in Mushroom Biology and Production, Rai, R.D. Dhar, B.L. and Verma, R.N. eds, Mushroom Society of India, NRCM, Solan, pp. 355-368.

Ramakrishnan, K, Lalithakumari, D, Shanmugan, N and Krishnamurthyi (1968). A simple technique for increasing yield of straw mushroom (*Volvariella volvacea*), *Madras J.* 55:194-198.

Ramaswamy, K. and Kandaswamy, T.K. (1976). Certain observations on the effect of cooked apple extract on paddy straw mushroom, *Indian J. Mushroom.* 2:4-7.

Rana, R.S. and Goyal, S.P. (1994). Utilization of Brassica straw as substrate for mushroom cultivation. Abstr. Natl. Symp. Mushrooms. NCMRT, Solan. April 8-10., 1994, p. 34.

Rangad, C.O. (1987). Mushroom cultivation in Meghalaya, Indian Mush. Sci. 2 : 46-48.

Rao, M.S., Pandey, M., Reddy, P.P. and Tiwari, R.P. (1994). Effect of certain botanicals on the growth of *Agaricus bisporus.* Abstr. Natl. Symp. Mushrooms. NCMRT, Solan. April 8-10, 1994, p. 34.

Rao, S.C. (1991). Evaluation of substrates and supplements for the oyster mushroom cultivation in the state of Goa. Adv. Mushroom Sci. p. 34 (Abstr.).

Roy, A., Sur, C. K. and Samajpati, N. (1978). Edible mushrooms of West Bengal III *Volvariella, Indian Agric.* 22: 61-62.

Saeed, M., Khan S.M. and Imam-Ul-Haq, M. (1994). Studies on the cultivation of Chinese mushroom (*Volvariella volvacea*) on water hyacinth (*Eichhornia crassipes*). *Pakistan J. Phytopathol.* 6:130-134.

Saini, L.C. (1994). Low cost technology of *Agaricus bisporus* with mustard straw. Abstr. Natl. Symp. Mushrooms, NCMRT, Solan. April 8-10, 1994, p. 51.

Sakagami, H. and Takeda, M. (1993). Diverse biological activity of PSK (Krestia) a protein bound polysaccharide from *Coriolus versicolor* (Fr.) Quel. In : Mushroom Biology and Mushroom Products, Chang, S. T., Buswell, J.A. and Chiu, S.W., Eds., The Chinese University Press, Hong Kong, 237-245.

Samajpati, N. (1979). Nutritive value of some Indian edible mushrooms, *Mushroom Sci.* 10:629-633.

Sanchez, J.E. and Royse, D.J. (2001). Adapting substrate formulas used for shiitake for production of brown *Agaricus bisporus*, *Bioresource Technology*, 77 : 65-69.

Sanchez, J.E. and Royse, D.J. and Hernandez, G. (2002). Development of non-composted substrate for production of *Agaricus bisporus.* In: Mushroom Biology and Mushroom Products, Sanchez, J.E., Huerta, G. and Montiel, E., Eds., Universities Autonoma Del Estado De Morelos, Mexico, 265-270.

Sandhu, G.S. (1995). Management of mushroom. insect pests. In: Advances in Horticulture Vol.13, Mushroom (eds. K.L. Chanda and S.R. Sharma) MPH, New, Delhi. pp. 261-75.

Sawalgi, V. and Sawalgi, V. (1994). Evaluation of substrates and supplements for there *Pleurotus* spp. *Indian Journal of Mycology* 24 (3) : 190-191.

Schindler. F. (1989). Method of producing a mushroom aroma in mushroom cell masses. US Patent 4, 810, 504.

Shandilya, T.R. (1980). Wheat straw and chicken manure mixture as high yielding substrate for *A. bisporus. Taiwan Mushroom.* 4 : 6-12.

Shandilya, T.R. (1988a). Low cost technology for mushroom cultivation. *Indian J. Mush.* 14 : 20-31.

Shandilya, T.R. (1988b). Polythene sacks or wooden trays for growing white button mushrooms. *Indian Hort.* 33 (2) : 4-5.

Shandilya, T.R. and Agrawala, T.K. (1983). Evaluating farm yard manure and spent compost as casing substrate in the cultivation of *Agaricus bisporus. Indian. J. Plant Path.* 1 : 34-37.

Shandilyua, T.R. and Phutela, R.P. (2000). Future needs for casing research in India, *Journal of Mycology and Plant Pathology,* 30 (2) : 281.

Shanmugam, N. (1986). Biology and cultivation of Pleurotus in India. In : Beneficial Fungi and their Utilization. Eds. M.C. Nair and S. Balakrishnan, Scientific Publishers, Jodhpur, pp. 17-24.

Sharma, A.D. and Jandaik, C.L. (1983). Some preliminary observations in the occurrence of Sibirina rot of cultivated mushrooms in India and its control. *Taiwan Mush.* 6 (1, 2) : 38-42.

Sharma, A.D. and Jandaik, C.L. (1987). Occurrence of brown rot – a new disease in *Pleurotus* beds. Indian. *Mush. Sci.* 2 : 123-125.

Sharma, S. G., Jyoti and Singh, V.K. (1999). Biological efficiency and cellulase activities of early and late fruiting *Pleurotus* spp., *Mushroom Research,* 8(1): 23-26.

Sharma, H.S.S., Kilpatric, M., Lyons, G. and Murray, J. (2002), Near infrared spectroscopy- an adaptive process control system for *Agaricus bisporus* compost production. In: Mushroom Biology and Mushroom Products, Sanchez, J.E., Huerta, G. and Montiel, E., Eds., Universities Autonoma Del Estado De Morelos, Mexico, 254-264.

Sharma, J.P. and Verma, R.N. (1985). Effect of various concentrations of fungicides in vitro on *Sclerotium* state of *Corticium rolfsii, Indian Phytopath.* 38 : 358-359.

Sharma, M. and Guleria, D.S. (2000). Effect of different casting materials on the mycelial characters and productivity of *Agaricus bitorquis* (Qnel.) Sqcc. *Journal of Mycology and Plant Pathology,* 30 (2) : 288.

Sharma, S.R. (1991). Mycoflora of casing soils. Indian Mushrooms (Nair, M.C., Ed.). KAU, Vellanikara, pp. 56-58.

Sharma, S.R. and Kumar, S. (2000). Trends in management of diseases and pests of mushrooms, *Journal of mycology and Plant Pathology,* 30 (2) : 284.

Sharma, S.R. and Vijay, B. (1996). Prevalence and interaction of competitor and parasitic moulds and *Agaricus bisporus. Mush. Res.* 5 : 13-18.

Sharma, S.S. and Doshi, A. (2000). Effect of organic amendments on the yield of *Pleurotus citrinopileatus* and *Pleurotus* sp-503, *Journal of Mycology and Plant Pathology*, 30 (2) 290.

Sharma. S.R. (1994). Diseases of Mushrooms and their Management. Tech. Bull No. 7. NCMRT, Publication. 52pp.

Shi,Y., James, A.E., Benzie, I.F.F. and Buswell, J.A. (2002). Genoprotective effects of Selected mushroom species. In: Mushroom Biology and Mushroom Products, Sanchez, J.E., Huerta, G. and Montiel, E., Eds., Universities Autonoma Del Estado De Morelos, Mexico, 357-366.

Shukla, C.S.; Thakur M.P. and Agrawal, K.C. (1998). Effect of systemic fungicides on growth and inhibition of, different species of *Pleurotus, Advances in Plant Science* 11 (1) : 45- 50.

Singh, A.N. and Bharat Rai (1998). Exploitation of edible mushrooms : an overview. In: Trends in microbial exploitation, Bharat Rai, Upadhayay, R.S. and Dubey, N.K. Eds., International Society for Conservation of Natural. Resources, Department of Botany, BHU, Varanasi. 150-158.

Singh, R.P. and Dwivedi, R.R. (1991). Standardization of substrate for production of *Pleurotus sajor-caju, Adv. Mushroom Sci.* p. 36 (Abstr.).

Singh, Ravinder P., Garcha, H.S. and Khanna, P.K. (1990). Protoplast production, regeneration and their use for isolation of hyper ligninolytic mutants of *Pleurotus florida.* 30th Annu. Conf. Assoc. Microbiol. India, Annamalia University, Annamalai Nagar, 9-11 January, (Abstr.).

Sivaprakasam, K., Rajan, F.S. and Jeyarajan, R. (1986a). *Pleurotus citrinopileatus*, new record: a new Indian edible mushroom *Curr. Sci.* 55: 1203.

Sohi, H.S. (1983). Role of edible mushrooms in recycling of agricultural waste and as an alternative protein source: Present status of mushroom cultivation in India. In: *Frontiers of Research in Agriculture.* Ed. S.K. Roy. Indian Statistical Institute, Calcutta, 565-579.

Sohi, H.S. (1992). Important of edible mushrooms in Indian diet and factors responsible for low production of cultivated mushrooms. *Indian Phytopath.* 45: 147-57.

Solomko, E.F., Eliseeva, G.S., Rjabchuk, V.A. and Pchelinceva, R.K. (1987). The composition of fruit bodies and mycelium of *Pleurotus ostreatus* (Fr.) kumm. *Applied Biochemistry and Microbiology,* 2 : 230-236.

Soni, M.K. (2002). Seasonal incidence of major insect pests of mushroom with management studies on phorid and Sciarid flies. Submitted to Department of Entomology, Indira Gandhi Agricultural University Raipur, 85 p.

Soto- Velazco, C., Lopez, M.C., Vazquez, E., and Alvarez, I. (2002). Cultivation of Ganoderma lucidum and its effect on the production of lymphocytes. In : Mushroom Biology and Mushroom Products, Sanchez, J.E., Huerta, G. and Montiel, E., Eds., Universities Autonoma Del Estado De Morelos, Mexico, 379- 382.

Stamets, P. (1993). Growing Gourmet and Medicinal Mushroom. Ten Speed Press.

Stanek, M. (1972). Microorganisms inhibiting the mushroom compost during fermentation, *Mushroom Science*, 8 : 797-811.

Su, U.T. and Seth, L.N. (1940). Cultivation of straw mushroom, *Indian Farm.* 1: 332-33.

Suharban, M. and Gokulapalan, C. (1995). Mushroom cultivation in Kerala. In : Advances in Horticulture, Vol. 13 – Mushrooms. Eds. K.L. Chadha and S.R. Sharma, Malhotra Publish. House, New Delhi, pp. 449-454.

Surinder, T. (2002). The magic mushroom, *Business World*, p. 52. (11th Nov. 02).

Suziki, S. and Oshima, S. (1976). Influence of Shiitake (*Lentinus edodes*) on human serum cholesterol. *Mushroom Sci.* 9 (1) : 43-469.

Tam, S.C., Yip, K.P., Fung, K.P. and Chang, S.T. (1986). Hypotensive and renal effects of an extract of the edible mushroom, *Pleurotus sajor-caju. Life Sci* 38, 1155-1161.

Thakur, M.P. (1999). Food and medicinal values of mushrooms. In: Health care and development of medicinal plants, Puri, S. and William, A. eds, Baba printer, Geeta Nagar, Raipur (C.G.), pp. 107-119.

Thakur, M.P., Chaurasia, V.K. and Shukla, C.S. (2000). Cultivation of *Pleurotus columbinus* (oyster mushroom) on different substrates and its correlation with prevailing weather conditions, *Journal of Mycology and Plant Pathology,* 30 (3) : 393-397.

Thakur, M.P., Godara, D.R., Shukla, C.S. and Sharma, R.L. (2002b). Recent advances in production technology of paddy straw mushroom. In: Advances in mushroom biology and production NRCM, Solan (in press).

Thakur, M.P., Gupta, A. and Shukla, C.S. (2001c). Effect of hollowness of mushroom beds on growth and yield of oyster mushroom, *Journal of Mycology and Plant Pathology,* 31(1): 117 (Abs.)

Thakur, M.P., Gupta, A. and Shukla, C.S. (2001d). Effect of homeopathic drugs on the growth and fruiting of *Pleurotus florida*, oyster mushroom, *Journal of Mycology and Plant Pathology*, 31(1) : 117 (Abs.).

Thakur, M.P., Lakpale, N., Shukla, C.S. and Kumar, S.M. (2002a) Chhattisgarh mein oyster mushroom: Anusandhan awam uplabdhiyan, Technical Bulletin-2, Directorate of Extension, Indira Gandhi Agricultural University, Raipur. 34 p.

Thakur, M.P., Ram, R. N. and Shukla, C.S. (2001a). Effect of environmental conditions and substrates on vegetative and fruiting stages of *Pleurotus florida* (oyster mushrooms) In: Frontiers in Fungal Biotechnology and Plant Pathogens Relations, Manoharachary, C., Bhagya Narayan, G., Bhadraiah, B., Satyaprasad, K., Reddy, B.N. and Nagamani, A. eds, Allied publishers Limited, New Delhi. 275-281.

Thakur, M.P., Ram, R.N. and Shukla, C.S. (2001b). Mycoflora associated with paddy straw substrate during different stages and cropping months of oyster mushroom (*Pleurotus florida*), *Journal of Mycology and Plant Pathology*, 31 (1) : 59-63.

Thapa, C.D. and Seth, P.K. (1983). Spring-tails (*Lepidocyrtus spp.*) – The tiny pests of mushrooms and their control. *Indian J. Mush.* 9 : 40-43.

Theradimani, M., Meena, B. and Krishnamoorty, M. (2001). Innovative techniques for the improvement of sporophore size and yield of milky mushroom (*Calocybe indica*), *Mushroom Research*, 10 (1) : 23-26.

Thomas, K.M., Ramakrishnan, T.S. and Narasimhan, I.L. (1943). Paddy straw mushroom. *Madras Agric. J.* 31 : 57-59.

Tiwari, R.P. and Pandey, M. (2002). Sizeable income generating venture, The Hindu Survey of Indian Agriculture, 6 : 165-167.

Toth, JO, Luu, B: and Ourisson, G. (1983). Ganoderic acid T-Z, cytotoxic triterpenes from *Ganoderma lucidum* (Polyporaceae). *Tetrahedron Lett.* 24: 1081-1084.

Tzeng, D.S. (1974). Studies on nutritional requirements and improved of techniques in cultivation of straw mushroom *Volvariella volvacea* sing. M.Sc. Thesis, Natl Chung.Hsinh. Univ., Taiching, Taiwan.

Upadhyay, R.C. and Verma, R.N. (2000). Non-conventional substrates for growing oyster mushrooms. *Mushroom Research,* 9 (1) : 35:38.

Upadhyay, R.C. and Vijay, B. (1988). Optimum compost depth in polythene bags for *Agaricus bisporus* cultivation. *Indian J. Agaric. Sci.* 58 : 778-779.

Upadhyay, R.C. Vijay, B. and Verma, R.N. (1998). Cultivation technology of oyster mushroom- recent advances. In: Recent Advances in the Cultivation Technology of Edible Mushrooms, National Research Centre for mushroom, Solan (H.P.), pp. 205-214.

Upadhyay, R.C., B. Vijay and R.N. Verma. (1996). Use of industrial tealeaf waste for cultivation of oyster mushrooms. In: *Mushroom Biology and Mushroom Products Royse* (ed.). Penn. State Univ., pp. 423-428.

Upadhyay, R.C., Sohi, H.S. and Vijay, B. (1987). *Cladobotryum apiculatum*, a new mycoparasite of *Pleurotus* beds. *Indian Phytopath.* 40 : 294 (Abstr.).

Upadhyay, R.C; Verma, R.N., Singh, S.K. and Yadav, M.C. (2002). Effect of organic nitrogen supplementation in *Pleurotus* species. In : Mushroom Biology and Mushroom Products, Sanchez, J.E., Huerta, G. and Montiel, E., Eds., Universities Autonoma Del Estado De Morelos, Mexico 225-232.

Valmaseda, M.G., Almendres and Martinez, A.T. (1990). Substrate dependent degradation patterns in the decay of wheat straw and beech wood by ligninolytic fungi. *Appl. Microbid. Biotechno.* 33:481-484.

Valmaseda, M.G., Almendes and Martinez, A.T. (1991). Chemical transformation of wheat straw constituents after solid -state fermentation with selected lignocellulose degrading fungi *Biomass Bioenergy* 1:261-266.

Verma, R.N. (1985). A new technology of Pleurotus cultivation. *N.E.H. Complex Newslett.* 8 (2) : 9-11.

Verma, R.N. (1989a). Improved cultivation methods for paddy straw mushroom. *ICAR Newsletter,* 8(4) : 1-2.

Verma, R.N. (1996). Mushrooms. IN: *"Fifty Years of Agricultural Research in India"*. Eds. R.S. Paroda and K.L. Chadha, ICAR, New Delhi.

Verma, R.N. (1997). Recent Advances in mushroom research in India. In : Advances in Mushroom biology and Production, Rai, R.D., Dhar, B.L. and Verma, R.N. Eds, MSI, Solan, 1-29.

Verma, R.N., Singh, G.B. and Singh, S.M. (1991). Spawn-cycling – A new approach to cultivation of oyster mushroom. *Mushroom Information* 6 : 6-10.

Verma, R.N., Singh, G.B. and Mukta Singh, S. (1995). Mushroom Flora of North Eastern Hills. In : Advances in Horticulture, Vol. 13-Mushroom. Eds. K.L. Chadha and S.R. Sharma. Malhotra Publish. House, New Delhi, pp. 329-349.

Verma, R.N., Yadav, M.C., Upadhyay, R.C., Dhar, B.L., Singh, S.K. and Kumar, S. (2002). Genetic improvement of mushrooms : A holistic approach, Book of Abstract of IV Conference on Mushroom Biology and Mushroom Products, held at Mexico, 9-10.

Vijay, B. and Ahlawat, O.P. (2000). Advances in substrate preparation of white button mushroom, *Journal of Mycology and Plant Pathology,* 30 (2) : 280.

Vijay, B. and Gupta, Y. (1991). Microflora of *Agaricus bisporus* compost – a review. Adv. *Mushroom Sci.* p. 20 (Abstr.).

Vijay, B. and Gupta, Y. (1992). Studies on manipulation of casing microflora on the yield of A. bisporus (Lange) Sing. *Mush. Res.* 1 (1) : 61-63.

Vijay, B. and Gupta, Y. (1995). Production Technology of *Agaricus bisporus.* In : Advances in Horticulture, Vol. 13- Mushroom. Eds. K.L. Chadha and S.R. Sharma, Malhotra Publish. House, New Delhi, pp. 63-98.

Vijay, B. and Sohi, H.S. (1987). Cultivation of oyster mushroom *Pleurotus sajor-caju* (Fr.) Singer on chemically sterilized wheat straw. *Mush. J. Tropics* 7 : 67-75.

Vijay, B. and Upadhyay, R.C. (1989). Chicken manure as a new nitrogen supplement in oyster mushroom cultivation. *Indian J. Mycol. Pl. Pathol.* 19 : 297-298.

Vijay, B., Saxena, S. and Sohi, H.S. (1987). Studies on new casing media for *Agaricus bisporus.* (Lange) Sing. *Mushroom J.* 178 : 313-315.

Vijay, B., Sohi, H.S. and Upadhyay, R.C. (1985). Control of green mold *Trichoderma viridae* by use of Bavistin in the cultivation of *Pleurotus sajor caju* (Fr.) Sing., *Indian Phytopathology,* 39;159.

Vijay, B., Sharma, S.R. and Lakhanpal, T.N. (2002). Effect of treating post-composting supplements with different concentrations of formaldehyde on the yield of Agaricus bisporus In: Mushroom biology and mushroom products, Sanchez, J.E. Huerta, G. and Montiel, E., eds, Universidad autonoma del estado de Morels, Mexico, 239-242.

Wade, (1970). Top banana-mushroom. *Singapore Trade Ind.,* Apr. pp.70-71.

Wang, H.X., Liu, W.K., Ng, T.B., Ooi, V.E.C. and Chang, S.T. (1995a). Immunomodulatory and antitumor activities of a polysaccharide-protein complex from a mycelial culture of Tricholoma sp. A local edible mushroom, *Life Science,* 57 : 260-281.

Wang, H.X., Ng, T.B., Liu, W.K., Ooi, V.E.C., and Chang, S.T. (1995b). Isolation and characterization of two distinct lectins with anti-proliferative activity from the cultured mycelium of the edible mushroom *Tricholoma mongolicum Int. J. Peptide Protein Res.* 46, 508-513.

Wang, H.X., Liu, W.K., Ng, T.B., Ooi, V.E.C. and Chang, S.T. (1996). The immunomodulatory and antitumour activities of lectins from the mushroom *Tricholoma mongolicum. Immunopharmacology* 31, 205-211.

Wang, H.X., Ng, T.B., Ooi, V.E.C., Liu, W.K., and Chang, S.T. (1997). Actions of lectins from the mushroom *Tricholoma mongolicum* on macrophages, splenocytes and life-span in sarcoma-bearing mice. *Anticancer Res.* 17, 416-420.

Wasser, S.P. and Weis, A.L. (1999). Medicinal properties of substances occurring in higher Basidiomycetes mushrooms : Current Perspectives (Review). *International Journal of Medicinal Mushrooms,* 1:31-62.

Wasser, S.P., Nevo, E., Sokolov, D., Reshetnikov, S., and Timor-Tismenetsky, M. (2000). Dietary supplements from medicinal mushrooms: diversity of types and variety. *Int. J. Med Mushr.* 2, 1-19.

Yao, Q.Z., Yu, M.M., Ooi, L.S.M., Ng, T.B., Chang, S.T., Sun, S.S.M. and Ooi, V.E.C. (1998). Isolation and characterization of type I ribosome-inactivation protein from fruiting bodies of the edible mushroom (*Volvariella volvacea*). *J. Agric. Food Chem. 46,* 788-792.

Yang, Q.Y., Hu, Y.J., Li, X.Y., Yang, S.X. Liu, X.J. Liu, T.F., Xu, G.M. and Liao, M.L. (1993). A new biological response modifier-PSP. In : Mushroom Biology and Mushroom Products, Chang, S.T. *et al.*, eds. The Chinese University, Press, Hong Kong. 247-259.

Yau, C.K. and Chang, S.T. (1972). Cotton waste for indoor cultivation of straw mushroom. *World Crops.* 24, 302-303.

Zadrazil, F. (1974). The ecology and industrial production of *Pleurotus ostreatus, P.florida, P. cornucopiae* and *P. eryngii. Mushroom Sci.* IX (2) : 621-52.

Zadrazil, F. (1985). In: Degradation of Lignocellulosics in Ruminants and Industrial Processes. (eds. Van Der Meer, J.M., Rijkens, M.P., Ferrante, B.A.) Elsevier Publisher, London, pp. 55-62.

Zadrazil, F. and Grabbe, K. (1983). In: Biotechnology Vol. 3. (ed. Dellweg, H.) Verleg, Florida, 145-165.

CHAPTER – 15

PHANSOMBA, A FOLK MEDICINAL MUSHROOM FROM THE WESTERN GHATS OF MAHARASHTRA : FUTURE PERSPECTIVES

Jitendra Vaidya, Prasad Lamrood and Shekhar Bhosle

Department of Botany, University of Pune, Pune – 411 007. M.S.

INTRODUCTION

Since last few years, mushrooms have gained a wider popularity not only because of their taste but also because of their medicinal properties. Several mushrooms have been screened for their different bioactivities and many compounds like polysaccharides (mainly includes β-D-glucan, heteroglycans, activated hexose corellated compounds, pachymanans, proteioglycans etc.), Terpenoids (like ganoderic acid, ganoderol, ganodermic acid etc.), germanium, ergosterol have been isolated and commercialized (Hobbs, 1996). These compounds or products obtained from the mushroom are now categorized as neutraceutical or Biological Response Modifiers (BRMs) and many of these compounds or products are approved as an authentic medicine to strengthen the immune system or cure a particular or an array of ailment(s) (Hobbs,1996; Wasser and Weis, 1999; Stamets, 2000; Wasser, 2002).

Amongst these several mushrooms, like *Ganoderma*, *Grifola*, *Coriolus*, *Cordyceps*, *Lentinus*, *Hericium*; *Phellinus* is yet another potent mushrooms which, sadly has not been extensively studied for its medicinal uses. Amongst several species of *Phellinus*, only *Phellinus linteus* has received much attention in Korea and under a National Scheme, several companies came forward to work on the various aspects of this mushroom through the research institutes, universities, and collages (Mizuno, 2000). The mushroom is popularly known as Sang-Hwang in Korea; Song Gen in China and Meshimakobu in Japan. In recent years, from the cultured mycelium product (PL2 and PL5 strains that are similar to ATCC 26710) "meshima capsules" have been developed as a medicine in Korea and it was approved as a medicine manufactured by Korean New Pharmaceutical Co., in 1997. (Mizuno, 2000). This Korean mushroom is described as Sangmogi (mulberry ear) at Decoction chapter of Oriental Medical Thesaurus, a Korea's old medical book, and as Sangshin, Sanghwanggo or Sang-i in Chinese medical books including Botanical List of Medical Herbs by Ming dynasty's Lee Shi-jin, Encyclopedia on Chinese Medicines, Encyclopedia on Oriental Medicine, etc. The medicinal literature in China such as *Shin-nou-honzokyo* (Imperial Medicine Handbook), *Honzo-komoku* (Medicine handbook), *Chuuyaku-daijitenn* (Great Dictionary of Chinese Medicines), *Chuugoku-yakuyo-shinnkinn* (Chinese Medicinal Fungi), and *Toyoigaku-daijitenn* (Great Dictionary published in 1967, cited in Mizuno, 2000) describes the efficacy of this mushroom as - it is used for the treatment of palsy, gonorrhea and abdomian pain, and effective for helping urination disorders, stomach problems, hematuria, gripes, penis ache, prolapsus ani, melena, over work, menses disorder, vitiated blood, lymphatic tumor, metrorrhagia, nose bleed, facial and diarrhea. While, in Korea the mushroom is used mainly to treat various cancers e.g. cancers in the digestive system such as stomach, gullet, duodenum, colon and rectum and as an immune modulating/boosting medicine. (Mizuno, 2000; Kwon, 2000, 2001; http://www.nonkong.co.kr/english/sang.html)

Similarly, in India, especially in Western Ghats many species of *Phellinus* are routinely used as a folk medicine to cure teeth, tongue, and throat related ailment, to stop excessive salivation in case of children, against diarrhea. (Andhalkar, 1988; Vaidya and Bhor, 1991; Vaidya, 1995; Vaidya and Lamrood, 2000). The fungus is well known by the name Phansomba, Phanas-alambi and Phanas-banda. Few indian ayurvedic books such as: *Indian Polyporaceae* (Bakshi, 1971, 1976), *Glossary of Indian medicinal plants* (Chopra *et al.*, 1956), *Aushadhi sangrah - medicinal plants from India: Identity, characters and uses* (Desai, 1927), *Pharmacographia Indica - a history of the principal drugs of vegetable origin Part III* (Dymock *et al.*, 1890), *Dravyagunvigyan ayurvedic materia medica* (Gogate, 1972), *Bombay materia medica and their therapeutics* (Khory, 1887), *Indian medicinal plants. Part II* (Kirtikar, 1918), *Indian Materia Medica, Vol. I* (Nadkarni, 1954), *Sarthbhavpraksh (Marathi and Sanskrit), 1st ed.* (Nanal, 1929), *Gharguti Aushadhe* (Sathe, 1982), *Medicinal Plants of Maharashtra* (Sawant, 1974), *Unani dravyaguna-darsh. Vol II* (Singh, 1974), *Nighantu adarsh. Part II. A treatise on vegetable materia medica* (Shah, 1928). Describes the medicinal properties of this mushroom. However, unfortunately there is no previous study carried out to investigate the biochemical composition of the mushroom.

Many a times it is also found during our routine survey from the market, field pharmaceutical industry as well as from the local learned vaidue who supplies this mushroom, that samples of *Ganoderma*, knowingly or unknowingly are being adulterated with the species of

Phellinus and sold as samples of Phansomba. *Ganoderma* is mainly used by the tribal to cure tumerous growth in throat, inflammation of knee, cataract, asthma etc. Surprisingly their claims were proved to be positive using samples of *Ganoderma.* (Vaidya and Bhor, 1991; Vaidya and Rabba, 1991 and 1993; Vaidya and Lamrood, 2000)

As we know that the species of *Ganoderma* are sold as an ancient medicine under the trade name Reishi or Ling Zhi in China, Japan, Korea, Thailand, Malyasia and other Asian countries (Willard, 1990; Hobbs, 1996; Hseu *et al.,* 1996; Chang and Buswell, 1999; Chiu *et al.,* 2000). The Reishi is even sold in Indian market with trade name Ganotherapy (therapy used with the help of powder of *Ganoderma* in the form of capsules or tea bags or ointment) in some parts of Western India. The use of ganotherapy is becoming more and more popular for curing the chronic diseases, which are not easily curable by allopathic treatment (since allopathic treatment has several side effects if used for a long period).

About four years back, there were only few hundred patients using Ganotherapy and now record suggests that the number has increased in Pune it self to few thousand. With this view in mind the samples of *Ganoderma* adulterated as Phansomba, the investigations were clubbed as adulterated Phansomba (not authentic Phansomba) have shown to be administered from many parts of Western Ghats, Saputara and Himalaya under different names few are: Jarh Phorh, Kulu Chati etc. (Harsh *et. al.,* 1993, 1996)

Our investigation have shown that many of the specimens collected as adulterated Phansomba those belong to species of *Ganoderma* have reported same active principles like terpenoids, ergosterol, as well as germanium, lectins, polysaccharides with that of Reishi.

The present chapter is a compilation of our group work carried out over the years (1989 – till date) on various aspects like survey, collection, identification, pharmacognostic studies, biochemical studies with respect to specific bioactive molecule, anticancer, antimicrobial activity, toxicity studies, isolation of novel compound(s), Phase I clinical trial etc. and the results gathered from these various studies strongly revealed that the species that are used in Western Ghats are equally competant with that of Korean Sang-Hwang or Reishi and has a much wider perspective to work on different aspects.

Thus, the future scenario in medicine research will concern only obtaining bioactive compounds from folk medicinal fungi from India. Not only Phansomba but even from other mushrooms as well. A new field of science will emerge where instead of antibiotics, many such nutraceuticals will become popular for maintaining general health. It is also likely that, similar to Befungin (which is an accepted Russian Pharmacopoeia drug obtained from a Chaga fungus) (Hobbs, 1996; http://www3.sk.sympatico.ca/david079/chaga.htm; http:// www. mushworld. com/tech/view.asp?cline=11&cata_id=1290&vid=514), there will be similar extracts from species of *Phellinus* or *Ganoderma* for curing incurable cancer or AIDS or Hepatitis like ailments.

Our main goal is to get such extractives like Befungin from commonly available species of *Phellinus* and *Ganoderma* causing diseases to Phanas (*Artocarpus integrifolia* L.), Tamarind (*Tamarindus indica* L.), Avla (*Phyllanthus emblica* L.), Behda (*Terminalia bellerica* Roxb.) by a standard simple technique which are generally prescribed in Ayurvedic preparations or even preparing the tablet of the sample or their mycelium from such medicinal mushrooms. It is also possible to mix the powder along with other drugs to cure diseases like "Mukhpaka" and trade such powder as "Mukhsuddhi". Another possibilities may be using extracts of this fungus along with herbal gel to be used for wound dressing or the smoothening or softening of the skin. One can enlist the out come of this work in many fields like preparing ointment, lotions, toothpaste, decoction as Asava or in some homoeopathic drug preparations.

In the present investigation the samples were surveyed by different workers routinely for last several years from different regions of the south west Maharashtra (Table 1)

Table 1. Sample Survey

Surveyor	Year	Locality	Sample
Vaidya, Bhor, Rabba	1989 – 1992	Adali, Sawantwadi, Alibaugh, Pune, Mumbai	VS, SS, VFS
Vaidya, Rabba	1992 – 1995	Bhimashankar, Pune and Mumbai	VFS
Vaidya, Lamrood, Deshpande, Sawant	1998 – 2000 2001 – 2003	Kolhapur, Sawantwadi, banda, Nardave, Kankavali, Deogarh, Coorg	VS, SS, PIS, VFS

VS – Sample Collected from Vaidu; SS – Sample Collected From Shop; PIS – Sample Collected from Pharmaceutical Industry; VFS – Sample collected with Vaidu from the Forest.

Collection of specimens were carried out in sexual stages from above mentioned localities. Since, some of the fresh collections of *Ganoderma* specimens showed occurrence of hyper-parasite, the specimens were sun dried and were preserved at 4-10°C in refrigerator. All dried samples were systematically arranged and maintained in brown envelops and then stored in cardboard boxes. Spore prints were obtained from fresh basidiocarps on cellophane paper and maintained in air tight plastic bags.

For detailed microscopic examinations the specimens of *Phellinus* temporary slides were made by cutting free hand sections by chopping method. The sections were treated with 10% KOH solution and stained with 1% Phloxine and Cotton Blue after washing in distilled water. Lactoglycerine was used as a mounting medium for semipermanent slide. Permanent slides were prepared in P.V.A. (Polyvinyl Alcohol) medium as per Omar *et al.* (1979) method. All specimens were deposited in the

Herbarium Poonensis at the University of Pune, Department of Botany, under the accession number MF- PH. Identification was done by using the key of (Gilbertson and Ryvarden, 1986; Larsen and Cobb-Poulle, 1990).

For the morphological details of *Ganoderma* specimens, thin sections were taken from the cutis layer, context layer and from the tube layer. Spores were isolated from a block of tube layers, technique described by Steyaert (1972). To loosen the hyphae the sectioned material was treated with 10% KOH washed with water and stained with 1% phloxine, which was again washed with water and finally stained with cotton blue. Lactoglycerine was used as mountant. The slides were observed under Bausch and Lomb compound microscope having a combination of 10X eye piece and 10X, 45X oil immersion (i.e.100X), objectives. Color scheme followed in this paper is as per (Jordan and David, 1995)

The spores were observed under Olympus BX-40 at phase contrast (100X objective) and the dermis sections at 40X objective of the same. Photographs were taken at the mentioned magnifications using Olympus BX-40 attached with photomicrography unit. The camera lucida sketches of spores were made by using Metzer mirror type camera lucida, fitted to Bausch and Lomb compound microscope. Materials were numbered as MF-GA-01 onwards (i.e. Medicinal Fungi - *Ganoderma* Sample 1).

The table 2 shows comparative account of species identified by different workers over these years.

Table 2. Phansomba sample survey for last 18 years

BHOR (1984)*	RABBA (1994)	LAMROOD & DESHPANDE (2002)	LAMROOD & SAWANT (2002-03)
P. badius	*P. fastuosus*	*P. llovdii*	*P. merilli*
P. chinchonensis	*P. adamantinus*	*P. merrillii*	*P. fastuosus*
P. durrissimus	*P. calcitratus*		*P. lloydii*
P. gilvus	*P. carteri*	Host other than ARTOCARPUS	*P. sublinteaus*
P. linteus	*P. conchatus*	*P. adamantinus*	*Bark and*
P.merrillii	*P. merrillii*	*P. aureobrunneus*	*wood of*
pectinatus		*P. badius*	*Jack fruit tree*
P. pachyphioeus	Host other than	*P. coffeatoporus*	
P. robiniae	*ARTOCARPUS*	*P. crocatus*	
P. senex	*P. mcgregori*	*P. griseoporus*	
P. sublinteus	*P. sanctigeorgii*	*P. lamaensis*	
Bark and	*P. melanoderms*	*P. linteus*	
wood of	*P. sublintus*	*P. microcystideus*	
Jack fruit tree	*P. torulosus*	*P. microcystideus.*	
	P. caryophyllii	*P. minutiporus*	
	P. swieteniae	*P. orientalis*	
	P. pectinatus	*P. pappianus*	
	P. callimorphus	*P. pectinatus*	
	P. tropicalis	*P. poonensis sp. nov.*	

P. umbrinellus	*P. rickii*
P. callimorphus	*P. sublinteus*
P. chryseus	
P. shaferi	
P. glaucescens	
P. membranaceus	
P. johnsonianus	
P. ribis	
P. sanjianii	

* The year in the parenthesis indicates the species of *Phellinus* collected in that year and not the year of publication.

Similarly, taxonomy of *Ganoderma* was also done by using the keys of (Steyaert, 1972 and 1980; Ryvarden and Johansen, 1980; Gilbertson and Ryvarden, 1986; Ryvarden, 2000).

Table 3. Possible adultrated samples of *Ganoderma* as Phansomba

BHOR (1984)*	BHOSLE (2002)
G. applanatum	*G. lucidum*
G. lucidum	*G. orbiformum*
	G. applanatum
	G. capense
	G. curtisii
	G. sessiliformae
	G. testaceum

*The year in the parenthesis indicates the species of *Ganoderma* collected in that year and not the year of publication

Total *Phellinus* spp. = 48, Total *Ganoderma* spp. = 8

After the taxonomic work was over the Phansomba samples were compared with the species that are sold as Sang-Hwang in Korea and based on the field collection experience and the information collected from the ayurvedic doctors, vaidus, our studies reveled that the samples of phansomba that were collected from the market as well as from the field are from different hosts and can be categorized in to three types as:

1. Applanate to imbricate up to 2 cm thick, occuring abundantly, rusty brown, azonate and glancing brown to coffee brown on the pore surface, light in weight.
2. Thick more than 5 cm, ungulate to applanate, upper surface zonate, velvety, sulcate, reddish-brown, mostly light in weight, pore surface rusty brown, glancing, margin prominent.
3. Bracket, woody to hard, not light in weight, blackinsh brown, sulcate, zonate, margin thin, brown glancing pore surface.

Any of such conk occurring on the jack fruit (*Artocarpus integrifolia* L.), Mango (*Mangifera indica* L.), Jambhul (*Syzygium cuminni* Roxb.), Babhul (*Accacia* spp.), Tamarind (*Tamarindus indicus* L), Giripushpa

(*Glyricidia sepium* Roxb.) tree are generally collected and sold in the market irrespective of their names.

Most of the samples are generally collected from wild old plantation tree from natural habitat causing heart rot disease in live standing tree of age more that 40 years. Maximum conks were seen on oldest large, broad trunk trees (Deshpande, 2002). Usually the conk appear to the height of loft which is generally collected with the help of a special cutting knife which is also used to collect the fruits of mango. One old tree (over 20-25 years old) may yield around 2-3 kg of conk. However, the cost of this mushroom not more that 1-2$ per kilogram which is very meager as compared to the cost of Sang-Hwang mushroom which is sold for about $ 2000. Table 4 gives comparative account of the *Phellinus* species:

Table 4. Morphological comparison of medicinal species of *Phellinus*

Characters	*Phellinus linteus* (Sang-Hawang)	*Phellinus baumii* (Sang-Hawang)	*Phellinus sublinteus* (Sang-Hawang & Phansomba)	*Phellinus merrillii* (Phansomba)	*Phellinus fastuosus* (Phansomba)	*Phellinus lloydii* (Phansomba)
Basidiocarp	Perennial, Hard, Woody, dimidiate, applanate to ungulate.	Same rarely effuso reflexed	Same	*Same*	Same	Same
Upper Surface	Brown-black shades, con-centrically sulcate, tom-entose at first then glabrous	Same	Same	Same	Same	Same
Margin	Acute to obtuse, sterile below	Same	Same	Same	Same	Blunt, rounded
Pore Surface	Brown-red shades, falvus tawny	Same	Same	Same	Same	Bay to Ferrugium
Pores / mm	5 – 8	9 – 11	6 – 8	6 – 8	5 – 7	4 – 5
Tubes	Concolorous, (6-15 mm)	Same (10-20mm)	Same (10-15mm)	Upto 5mm	Same (6-15mm)	Upto 6mm
Hymenial Setae	Present	Present	Present	Absent	Absent	Absent
Basidio-spores	Brown, Globose, 4 – 5.5μm	Pale Yellow 3-3.5μm	Brown, 4-5μm, Globose	Pale Yellow to Rusty brown, Globose to subglobose 4.6μm	Golden brown, globose – ellipsoid, 4 – 4.5μm	Yellow Brown, Globose 4 – 5.5μm

After taxonomic part the comparison of Phansomba and Sang-Hwang is given for their medicinal properties (see table 5). Similar comparison is given for *Ganoderma* (adulterant of Phansomba) and Reishi (a Chinese medicinal mushroom) (see table 6).

Table 5. Comparisons of Medicinal Mushroom "Indian" (Phansomba) And "Korean"(Sang-Hawang)

Medicinal Properties	Phansomba	Sang – Hawang
Known	Tongue, teeth, throat related diseases, Excessive salivation, dysentery, Diarrhea	Treatment of palsy, gonorrhea, abdominal pains, urination disorders stomach problems, hematuria, nose bleed, facial and Diarrhea
Literature	Same as above, anticancer, anti- inflammatory, anti-oxidant	Anti-tumor, anti-cancer, anti-diabetic.
Bioactivity/ Clinical Trials	"Mukhapak" diseases, 60 patients treated successfully as well as for edema and ulcerations on mucus membrane of buccal cavity, anticancer activity using SiHa cell lines, toxicity using *Allium* root tip assay	Trials on mouse and rats for anti-tumor, anti-cancer, anti diabetic activity, Cell toxicity using A549, SR-OV-3, SK-OV-3, Sk-MEL-2, HCT-15 and XF-498 cell lines derived from humans
Pharmacognostic studies	Ash, moisture, minerals, metals, crude fibers, total carbohydrates, total proteins, amino acids, organic acids, phenols	Moisture, crude protein, dietary fibers, carbohydrates, crude ash, K_{cal}, sodium, arsenic and lead
Biochemical studies	Lectins, Terpenoids, Ergosterol, Alkaloids, germanium, Polysaccharides.	Polysaccharides, carbohydrate analysis and polysaccharide fraction (KP).

Table 6. Comparisons of samples of "Phansomba" belonging to the species of *Ganoderma* with Chinese medicine "Reishi"

Medicinal Properties	*Ganoderma* (adulterant of Phansomba)	Reishi
Known	Anti-tumor, Anti-inflammatory	Anti-cancer, Anti-tumor, Anti-diabetic, Anti-inflammatory, Anti-thrombotic, Immuno-stimulating, antiviral, anti-microbial, anti-oxidant, Anti-HIV-1, Anti-HIV-1 protease, Nerve stimulant, Hepato-protective, anti-nociceptive, cosmetic applications
Literature	Anti-tumor, anti-cancer	Same as above
Clinical Trials	Insomnia	Several human trials and on test animals
Pharmacognostic studies	Same as Phansomba	Same as Phansomba

Biochemical studies	Same as Phansomba	Mainly concentrated on Polysaccharide and Triterpenes (more than 100 different triterpenes reported), GMP, AMP Nucleosides etc.

Besides taxonomic work we have also carried out the biochemical investigation of the Phannsomba samples and Table 7, 8, 9 and 10 summarizes the work carried by on Vaidya and Bhor (1991) on *Phellinus* and *Ganoderma*. After comparison of these results with that of Sang-Hwang and Reishi it was strongly revealed that the results are in accordance with that of Korean and Chinese medicinal mushrooms.

Table 7. Chemical composition of eleven samples of phansomba powder extract in percentage as per Dymock *et. al.* (1890)

Name of the extract	*Phellinus fastuosus* (5 samples)	*Phellinus merrillii* (6 samples)
Ether extract	1.8 ± 0.02*	1.2 ± 0.03
Alcohol extract	2.5 ± 0.034	2.4 ± 0.034
Aqueous extract	3.4 ± 0.06	2.7 ± 0.06
Alkaline extract	25.36 ± 1.34	25.57 ± 1.34
Crude fibers	58.14 ± 2.8	58.93 ± 2.8
Ash	4.2 ± 0.06	3.6 ± 0.06
Moisture content	4.6 ± 0.075	5.6 ± 0.075

* Mean ± SD

Table 8. Metal content in the ash (ppm).

Name of the metal	*Phellinus fastuosus* (5 samples)	*Phellinus merrillii* (6 samples)
Copper	2.7 ppm	1.78 ppm
Ferrous	12.2 ppm	11.6 ppm
Magnesium	24.6 ppm	23.8 ppm
Mangenese	0.5 ppm	0.6 ppm

(Source: Vaidya and Bhor, 1991)

Table 9. Quantitative estimation of phansomba samples (percentage).

Estimation of	*Phellinus fastuosus* (5 samples)	*Phellinus merrillii* (6 samples)
Ash	4.2 ± 0.16*	3.6 ± 0.14
Polyphelols	0.42 ± 0.1	0.43 ± 0.07
Total Carbohydrates	79.4 ± 0.66	81.5 ± 0.85
Proteins	0.45 ± 0.13	0.33 ± 0.12
Sugars	0.42 ± 0.03	0.21 ± 0.04

* Mean ± SD, (Source: Vaidya and Bhor ,1991)

Table 10. Qualitative estimation of powder of phansomba samples.

Estimation of	*Phellinus fastuosus* (5 samples)	*Phellinus merrillii* (6 samples)
	B- alanine	—
	Aspartic acid	Aspartic acid
Free amino acids	Arginine	Arginine
	L-cystein	L-cystein
	Methionine	Methionine
	L-cystein	Glutamic acid
Bound amino acids	L-cystine	L-cystine
	Leucine	Leucine
Lipids	ND	ND
	Erythritol	Atranorin
	Lecanoric acid	Lecanoric acid
Organic acids	Salazinic acid	Salazinic acid
	Socellic acid	Socellic acid
	—	Erythrin
	Catechol	Catechol
	Gallic acid	Gallic acid
	P-hydroxybenzoic acid	P-hydroxybenzoic acid
Phenols	4-methylresorcinol	2-methylresorcinol
	Salicylic acid	Orcinol
	Vanillic acid	Phloroglucinol
	—	Salicylic acid
	Arabinose	Fructose
Sugars	Galactose	Raffinose
	Maltose	Maltose
Terpenoids	Present	Present

ND = Not detected; (Source: Vaidya and Bhor, 1991)

Table 11. Comparative account of analysis of "Phansomba" samples

Name of the Test	*Phellinus*	*Ganoderma*
Ash	WP	3.2%
Moisture Content	5 – 7%	WP
Minerals/Metals	WP	K: 2 – 6ppm, Na: 2 – 20ppm
Crude Fibers	5.5 - 33%	WP
Total Carbohydrates	40 – 50%	32 – 45%
Total Proteins	11-14%	9 – 14%
Amino acids	About 4%	2 – 6%
Organic Acids	WP	WP
Lectins	*Phellinus badius*, *P. membranaceous*	Mild in *G. stipitatum*
Alkaloids	ND	WP
Terpenoids	Variety of Triterpenes	Variety of Triterpenes
Ergosterol	0.2%	0.3%
Germanium	WP	WP
Polysaccharide	0.2%	WP

WP = Work in progress; ND = Not Detected;

(Source: Current study of Lamrood on *Phellinus* and Bhosle on *Ganoderma*)

Apart for the taxonomic and biochemical work of Phansomba (*Phellinus*) and *Ganoderma* (adulteration of Phansomba), following work also has been carried out.

Antimicrobial Studies

Fifteen basidiocarps of *Phellinus* spp. that are sold as samples of phansomba were collected and tested for their antimicrobial activity against human pathogenic microorganisms. *Phellinus adamantinus; P. coffeatoporus; P. merrillii; P. sublinteus* and *P. minutiporus* showed strong activity against *Escherichia coli* (MTCC 724 and MTCC 739); *Klebsiells pneumoniae* (MTCC 432); *Proteus mirabilis* (MTCC 1429) and *Candida ablicans* (MTCC 3017 and MTCC1637), *Phellinus linteus* and *P. lloydii* exhibited strong activity against *Escherichia coli* (MTCC 724 and MTCC 739); *Klebsiells pneumoniae*; *Proteus mirabilis* and *Candida ablicans*, *Phellinus rickii* (PH-14)- strong activity against *Escherichia coli*; *Klebsiells pneumoniae*; *Proteus mirabilis; Candida ablicans, Phellinus fastuosus* - strong activity against *Escherichia coli; Candida ablicans, Phellinus pappianus* - strong activity against *Escherichia coli; Klebsiells pneumoniae, Phellinus merrillii* - strong activity against *Escherichia coli* , *Phellinus badius* - moderate activity against *Escherichia coli; Klebsiells pneumoniae* and *Candida ablicans, Phellinus griseoporus* and *P. crocatus* - moderate activity against *Escherichia coli* and *Candida ablicans, Phellinus aureobruneus* - moderate activity against only *Escherichia coli.* On the basis of the experimental evidence in this investigation, it appears that *Phellinus* could be useful source for exploiting 'antimicrobial compounds'. The work is in progress for the activity test against other microorganisms as well as the partial purification of the active compound in this laboratory.

Recently, crude terpenoid and polysaccharide extracts were also tested for their antimicrobial action and the results are encouraging.

Novel Compound Study

A methanolic extract of *Phellinus merrillii* was fractionated and three compounds were purified. The compounds are not named yet and further work on structural determination is going on but it was revealed that two compounds are a straight chain molecule while one have aromatic rings attached to a sugar base (correct identification of these compounds is yet to be completed)

Lectin Studies

A total of 75 herbarium specimens belonging to the species of *Phellinus*, *Ganoderma* and *Daedaleopsis* including samples collected from ayurvedic shops from Pune, Mumbai and Konkan as well as from field in the plantation trees were screened for their lectin activity by using hemagglutination assay (Lis and Sharon, 1972). Out of 75 samples; 68 samples were of *Phellinus*, 5 samples were of *Ganoderma*, and 2 samples were of *Daedaleopsis*. The activity was screened by using human (of blood type A Rh +ve, B Rh -ve and O Rh +ve) and rabbit erythrocytes. Out of these 75 samples, only 3 samples showed hemagglutination against human erythrocytes and 15 samples showed hemagglutination against rabbit erythrocytes.

Toxicity Studies

Different concentrations of cold-water extract of medicinal mushroom *Phellinus merrillii* were tested for their toxicity by using *Allium* assay. Treated *Allium cepa* root tips did not show any chromosomal aberrations at a dose of 0.029 gm to 2.000 gm. However, decrease in mitotic index; root length and percent germination was observed with increasing concentration (from lower concentration to higher concentrations) whereas cell size was observed to be increasing. Thus, the results observed in *Allium* indicate no serious toxicity issues, indicating that the compounds in question may be safe in human therapy, and that the encouraging results indicate the great value of further study. Recently different concentrations of hot water extract were tested for their toxicity and EC_{50} was found to be 16.3 gms which clearly indicates that the sample may be safe in human therapy.

Ergosterol Analysis

A new modified method was employed for ergosterol analysis (unpublished data) for 21 species of *Ganoderma*, the percentage ergosterol was found to be in range of 0.065 to 0.349, a range reported for *Ganoderma lucidum* (Reishi) is 0.03 – 0.04% (Mizuno, 2000).

Anti Cancer Study

Preliminary experiments to check anticarcinogenic activity using human cervical carcinoma cells (SiHa cells) were carried out with two species of Phansomba, *Phellinus merrillii* (Murr.) Ryv. and *Ph. fastuosus* (Lev.) Ryv. The study suggests that the hot and cold water extracts as well as hot and cold 50% ethanol extracts of these species show anticarcinogenic activity. Our studies revealed that cold and hot water extracts of *Ph. merrillii* and cold and hot ethanol extracts of both species have high cytotoxic effect on SiHa cells. *Ph. merrillii* cold and hot water extracts demonstrated higher cytotoxic effect than those from *Ph. fastuosus*, while cold and hot ethanol extracts from both species entirely inhibited SiHa cell growth at a concentration of 500 μg (Bhonde *et. al.*, 2002).

FUTURE SCENARIO

Our investigation have shown that many of the specimens collected as adulterated Phansomba those belong to species of *Ganoderma* have reported same active principles like Terpenoids, Ergosterol, as well as germanium, lectins, polysaccharides with that of Reishi (*G. lucidum*).

The present report is a compilation of our group work carried out over the years on various aspects like survey, collection, identification, pharmacognostic studies, biochemical studies with respect to specific bioactive molecule, anticancer, antimicrobial activity, toxicity studies, isolation of novel compound(s), Phase I clinical trial etc. and the results gathered from these various studies strongly revealed that the species that are used in Western Ghats are equally competent with that of Korean Sang-Hwang or Reishi and has a much wider perspective to work on different aspects. Thus, the future scenario in medicine research will concern only obtaining bioactive compounds from folk medicinal fungi from India. Not only Phansomba but even from other mushrooms as well. A new field of science will emerge where instead of antibiotics, many such nutraceuticals will become popular for maintaining general health. It is also likely that, like Befungin (which is an accepted Russian Pharmacopoeia drug obtained from a Chaga fungus), there will be similar extracts from species of *Phellinus* or *Ganoderma* for curing incurable cancer or AIDS or Hepatitis like ailments. Our main goal is to get such extractives like Befungin from commonly available species of *Phellinus* and *Ganoderma* causing diseases to Phanas (*Artocarpus integrifolia* L.), Tamarind (*Tamarindus indica* L.), Avla (*Phyllanthus emblica* L.), Behda (*Terminalia bellerica* Roxb.) by a standard simple technique which are generally prescribed in ayurvedic preparations or even preparing the tablet of the sample or their mycelium from such medicinal mushrooms. It is also possible to mix the powder along with other drugs to cure diseases like “Mukhpaka” and trade such powder as “Mukhsuddhi”. Another possibilities may be using extracts of this fungus along with herbal gel to be used for wound dressing or the smoothening or softening of the skin. One can enlist the out come of this work in many fields like preparing ointment, lotions, toothpaste, decoction as asava or in some homoeopathic drug preparations.

REFERENCES

Andhalkar, R.V. (1988). The uses of Phansomba in management of Mukhpaka (Sarvasara). M.D. (Ayu.). Thesis, University of Pune.

Bakshi, B. K. (1976). *Forest Pathology – Principles and Practices in Forestry.* FRI & Colleges, Dehradun, pp. 400.

Bakshi, B.K. (1971). Indian Polyporaceae. Published by, Indian council of Agriculture Research, New Delhi, India.

Bhonde, R.R., Lamrood, P.Y. and Vaidya, J.G. (2002). Anricarcinogenic activity of two samples of Phansomba, *Phellinus merrillii* (Murr.) Ryv. and *Ph. fastuosus* (Lev.) Ryv., on SiHa cell lines. *Int J Med Mushr,* 4 : 133-138.

Chang, S.T. and Buswell, J.A. (1999). *Ganoderma lucidum* (Curt.: Fr.) P.Karst. (Aphyllophoromycetideae)- A Mushrooming Medicinal Mushroom. *International Journal of Medicinal Mushrooms.* 1: 139-146.

Chiu. *et al.* (2000). Developmental Plasticity of Hong Kong Lingzhi as a Response to the Environment. Abstract from ISMS Congress 2000, May 15-19,

Maastricht, The Netherlands. (http://www.agro.nl/pc/isms/abstracts/ses8-30.htm)

Chopra, R.N.; Nayar, S.L. and Chopra, L.C. (1956). Glossary of Indian medicinal plants. ICAR Pub., New Delhi, pp 9 and 27.

Deahpande, G.G. (2002). Some live standing tree diseases caused by Aphyllophorales from Campus of Pune University and some other places from Pune City. M. Phil. Thesis, University of Pune.

Desai, V.G. (1927). Aushadhi sangrah - medicinal plants from India: Identity, characters and uses. Shree Gajanan Book Depot, Dadar, (Bombay) Mumbai, India, pp 285-289.

Dymock, W.; Wardern, C.J. and Hooper, D. (1890). Pharmachographia Indica - a history of the principal drugs of vegetable origin. Part III. Education Society Press, Byculla, (Bombay) Mumbai, pp 629-635.

Gerard, J. (1597). Herbal of General history of plants. Norton, London. (Dover Publ. New York Reprint, 1974).

Gilbertson, R.L. and Ryvarden, L. (1986). North American Polypores. Published by Fungiflora, Oslo, Norway.

Gogate, V.M. (1972). Dravyagunvigyan Ayurvedic materia medica. Maharashtra Vidyapeeth Granth Nirmiti Mandal. Continental Prakashan, Pune, India.

Gottlieb, A.M. and Wrigth, J.E. (1999a). Taxonomy of *Ganoderma* from the southern South America: subgenus *Ganoderma. Mycol. Res.* 103(6): 661-673.

Gottlieb, A.M. and Wrigth, J.E. (1999b). Taxonomy of *Ganoderma* from the southern South America: subgenus *Elvingia. Mycol. Res.* 103(10) : 1289-1298.

Harsh, N.S.K., Rai, B.K. and Tiwari, D.P. (1993). Uses of *Ganoderma lucidum* in Folk Medicine. *Indian J. Trop. Biod.* 1 : 324-326.

Harsh, N.S.K., Tiwari, C.K. and Rai, B.K. (1996). Forest Fungi in the Aid of Tribal Women of Madhya Pradesh. *Sustainable Forestry*, 1 : 10-15.

Hobbs, C. (1996). Medicinal mushrooms: An exploration of tradition, healing and culture. Botanica Press, Santa Cruz, CA

Hseu, R. S., Monocalvo, J.M. and Wang, H.F. (1996). Molecular methods for *Ganoderma lucidum* complex systematic. Taipei International *Ganoderma* Research Conference p. 11-12. Taipei, Taiwan, R.O.C.

http:// www.nonkong.co.kr/english/sang.html

http://www.mushworld.com/tech/view.asp?cline=11&cata_id=1290&vid=514

http://www3.sk.sympatico.ca/david079/chaga.htm

Jordan, M. and David C. (1995). Biology of Higher Fungi in *Encyclopedia of Fungi of Britain and Europe.*

Khory, R.N. (1887). Bombay materia medica and their therapeutics. Pri Raninas Union Press, pp 563-565.

Kirtikar, K.R. (1918). Indian medicinal plants. Part II. Pub. Sudhir Nath Basu, Bhuwaneshwar Ashram, Bhandurgun, Allahabad, pp 1394 - 1396.

Kwon, H.J. (2000). Medicinal Effects of *Phellinus linteus.* http://www.mushworld.com

Kwon, H.J. (2001). Rare, Unknown but Marvelous Medicinal Mushroom: Sang-Hwang, *Phellinus linteus*; Mulberry Yellow) http://www.mushworld.com

Larsen, M.J., and Cobb-Poulle, L. A. (1990). *Phellinus* (Hymenochaetaceae). A survey of the world texa. Synopsis Fungorum 3, Fungiflora, Oslo, Norway.

Lis, H., and Sharon, N. (1972) Soybean aggulutinin. *Methods of Enzymology,* Ginsberg V Ed., New York 38: 360-368.

Mizuno, T. (2000). Development of an antitumor response modifier from *Phellinus linteus* (Berk. Et Curt.) Teng (Aphyllophoromycetideae) (Review). *Int J Med Mushr,* 2 : 21-34.

Nadkarni, K.M. (1954). Indian Materia Medica, Vol. I 3rd edit. Popular Book Depot, (Bombay) Mumbai, pp. 51-52 and 202.

Nanal, P.G. (1929). Sarthbhavpraksh (Marathi and Sanskrit), 1st edit. Pune, India, 206.

Omar M.B., Bolland L. and Heather W.A. (1979). A permanent mounting medium for fungi. *Bulletin of the British Mycological Society.* 13(1): 231-232.

Ryvarden L. (2000). Studies in neotropical polypores 2: a preliminary key to neotropical species of *Ganoderma* with a laccate pileus. *Mycologia.* 92(1): 180-191.

Ryvarden, L. and Johansen, I. (1980). A preliminary Polypore Flora of East Africa. Published by Fungiflora, Oslo, Norway.

Sathe, K.N. (1982). Gharguti Aushadhe. Ayurved Bhawan, Kakawati, (Bombay) Mumbai, India, 225 pp.

Sawant, S.Y. (1974). Medicinal Plants of Maharashtra. Continental Prakashan, Pune, India, pp 1-23.

Shah, B.C. (1928). Nighantu adarsh. Part II. A treatise on vegetable materia medica. Ahmedabad. pp. 622-625.

Singh, D. (1974). Unani dravyagunadarsh. Vol II. Unani Pharmacopoeia, Ayurved and Tibbi Academy, Lucknow, U.P. India. pp. 204, 920.

Stamets, P. (2000). Growing gourmet and medicinal mushrooms. 3d ed. Ten Speed Press, Berkeley, CA.

Steyaert, R.L. (1972). Species of *Ganoderma* and related Genera, mainly of the Bogor and Lieden Herbaria. *Personia* 7(1) : 55-118.

Steyaert, R.L. (1980). Study of some *Ganoderma* species. *Bull. Jard. Bot. Nat. Belg.* 50 : 135 - 186.

Vaidya, J. G. and Bhor, G.L. (1991). Medicinal important wood rotting fungi with special emphasis on phansomba. *Deerghayu Int,* 7 : 16-19.

Vaidya, J.G. and Lamrood, P.Y. (2000). Traditional medicinal mushrooms and Fungi of India. *Int. J. med. Mushr,* 2(3) : 209-214.

Vaidya J.G. (1991). Phanas alombe - atrue mushroom or phansomba. In: Proceedings of national symposium on mushrooms. Nair, M.C. ed., pp. 216-219.

Vaidya J.G. (1995). Biology of the Fungi. 1st ed. Satyajeet Prakashan. Kothrud, Pune – 38.

Vaidya, J.G. and Rabba, A.S. (1993). Fungi in folk medicine. *Mycologist.* 7 : 131-133.

Wasser, S.P. and Weis, A.L. (1999). Medicinal properties of substances occurring in higher Basidiomycetes mushrooms: Current perspectives (review). *Int J med mushr*, 1 : 31-62.

Wasser S.P. (2002). http://www.mushworld.com.

Willard, T. Research by Ken Jones. (1990). Reishi Mushroom: Herb of spiritual Potency and Medicinal Wonder. Sylvan Press, 1420 Gilman Blvd., #2265, Issaquah, WA 98207.

CHAPTER - 16

EFFECT OF STORAGE FUNGI ASSOCIATED WITH LOBIA (*Vigna sinensis* (Linn) Savi ex Hassk) SEEDS

Yoganand Y. Barve and R. P. Thakre

Post Graduate Dept. of Botany, Nagpur University Campus, Amravati Road, Nagpur - 440 010 (Mah.)

INTRODUCTION

The leguminous crops constitute the most important source of plant proteins in human diet. In some parts of the world, such as India, Africa where people cannot afford animal proteins are largely dependant on plant sources. *Vigna sinensis* (Linn) Savi ex Hassk is one of the principle pulses in common use in India. Cowpea is valued primarily for its pods and seeds used as food and the plants left after harvesting are used as valuable feed for livestock. Cowpea is considered antibilious and is prescribed in liver complaints with jaundice. Two varieties of cowpea were taken in to consideration during present study viz. Pusa Barsati and Asparagus Black.

THE TECHNIQUES

Collection of experimental seeds: The seed samples were collected from different cultivators from different parts of Vidarbha region. Each type of seed collected from different parts were bulked together and drawn randomly for further investigations.

Monthly checkup of germinability and seed health under storage: The seed samples were stored in small cotton bags under normal room temperature condition for one year i.e. Dec. 1990 to Nov.

1991. After every month four hundred seeds were taken out randomly and percent germination as well as percent fungal incidence were recorded. Percent germination was done by roll towel technique as suggested by Musket (1948). Externally as well as internally borne fungi were isolated by Blotter test and agar plate method (ISTA, 1966). Total percentage of fungal incidence (Sahai *et al.*, 1982) was calculated using the formula

$$\text{Percentage frequency} = \frac{\text{Total no. of seeds infected by particular fungus}}{\text{Total no. of seeds studied}} \times 100$$

Changes in seed N % and crude protein % due to fungal flora: To study the effect of fungi on changes in seed protein, the seeds were stored in separate cotton bags in glass desiccators maintained at 50% RH, prepared by saturated salt solution of sulfuric acid (Eds. Jonston *et al.*, 1983). The control seeds were surface sterilized (Cristensen *et al*, 1965), dried and stored in separate desiccators at same RH. At the interval of every 30 days, the seed samples were drawn randomly and powdered separately in grinder and estimated quantitatively for Nitrogen and crude protein percentage by semi-microkjeldahl method suggested by Davy's *et al.* (1969) and Sadasivam *et al.* (1992).

Aflatoxin elaboration: At the end of year five seed samples from each variety were drawn randomly and screening was done by BGY test under long UV (360 nm) light. 5g BGY positive samples were extracted by chloroform on wrist action shaker for 10 min. After centrifugation, the supernatant dried over water bath and the residue was dissolved in 1 ml chloroform and stored in small vials at 4^{0}C in dark. Detection was done by TLC using chloroform : acetone (9:1 v/v) solvent system. The spots were visualised under long wavelength UV (360 nm) light (Stahl, 1969).

RESULTS

Germinability: *V. sinensis* var. *asparagus* black seeds showed 92.5% germination with 78.5% normal germination and 7.5% dead seeds and var. Pusa Barsati seeds showed 87% total average germination with 13% nonviable seeds. At the end of storage period of one year it declined to 59% and 48% respectively. The mortality of seeds increased to 41% and 52% respectively (Table 1 and Table 2). In Pusa Barsati, it reduced in the summer months but increased again. Whereas such observations were not found in Asparagus Black.

Seed health testing: Based on their frequency of incidence in different periods throughout the year, the fungal organisms are categorised in different groups.

Category 'a': Occurring throughout the year. Category 'b': Occurring during winter only (Sep.-Feb.). Category 'c': Occurring during summer only (Mar.-Aug.). Category 'd': Occurring rarely without showing any specificity to the time of occurrence.

Table 1. Monthly germination percentage of *Vigna sinensis* var. *asparagus* black seeds

S.N.	Month/year	Total Avg. germination %	Normal germination %	Abnormal germination %	Dead seeds %
1	Dec. '90	92.50	78.50	14.00	7.50
2	Jan. '91	94.50	78.50	16.00	5.50
3	Feb. '91	87.50	62.00	25.50	13.50
4	Mar. '91	80.00	60.00	20.00	20.00
5	Apr. '91	71.00	51.00	20.00	29.00
6	May '91	68.00	63.50	4.50	32.00
7	Jun. '91	65.00	45.00	20.00	35.00
8	Jul. '91	65.00	48.00	17.00	35.00
9	Aug. '91	67.00	56.00	11.50	32.50
10	Sep. '91	65.00	47.50	17.50	35.00
11	Oct. '91	60.00	38.00	22.00	40.00
12	Nov. '91	60.00	30.00	29.00	41.00

Table 2. Monthly germination percentage of *Vigna sinensis* var. *pusa barsati* seeds

S.N.	Month/year	Total Avg. germination %	Normal germination %	Abnormal germination %	Dead seeds %
1	Dec. '90	87.00	80.00	7.00	13.00
2	Jan. '91	73.50	55.50	18.00	26.50
3	Feb. '91	70.00	54.00	16.00	30.00
4	Mar. '91	71.00	59.00	12.00	29.00
5	Apr. '91	59.50	41.00	18.50	40.50
6	May '91	59.50	36.00	13.50	40.50
7	Jun. '91	61.50	40.00	21.50	38.50
8	Jul. '91	62.00	39.00	23.00	38.00
9	Aug. '91	60.00	38.00	22.00	40.00
10	Sep. '91	58.00	29.00	29.00	42.00
11	Oct. '91	53.00	20.00	33.00	47.00
12	Nov. '91	48.00	21.00	27.00	52.00

Table 3 depicts the percentage frequency of incidence of fungal organisms associated with seeds of *V. sinensis* var. *asparagus* Black. Total 26 fungi were isolated out of which *Alternaria alternata, Aspergillus candidus. A. effusus, A. flavus, A. fumigatus, A. niger, Curvularia lunata, C. tetramera, Fusarium moniliforme, F. oxysporum, Mucor heimalis, Penicillium multicolor, P. notatum, P. oxalicum, Rhizopus nigricans* were present throughout the year and represented category 'a'. *Alternaria humicola, Cunninghamella* sp. and *Penicillium chrysogenum* represented category 'b'. Fungi representing category 'c' were *Aspergillus glaucus, Aureobasidium* sp., *Mucor plumbeus* and *Syncephalastrum recemosum.*

Rare occurrence of *Neurospora* sp., *Penicillium purpurogenum, P. stekii* and *Phytophthora undulata* was seen. These fungi were included in category 'd'.

Table 3. Monthly incidence of fungi during storage on seeds of *V. sinensis* var. *asparagus* Black

Name of Organism		Incubation Period											
		Dec. '90	Jan. '91	Feb. '91	Mar. '91	Apr. '91	May '91	Jun. '91	Jul. '91	Aug. '91	Sep. '91	Oct. '91	Nov. '91
Alternaria alternata	a*	13.8	15.9	10.0	5.0	-	-	9.1	21.0	36.0	25.0	10.0	12.0
(Fr.) Keissler	b*	5.6	7.0	4.0	-	-	-	-	8.0	13.0	7.0	2.0	-
A. humicola		-	22.7	-	-	-	-	-	-	-	-	-	-
Oudemans		-	-	38.9	-	-	-	-	-	-	-	-	-
Aspergillus candidus		10.0	7.6	-	-	5.7	-	3.0	-	-	-	-	-
Link.		-	-	-	7.7	10.0	-	-	-	6.0	-	-	-
A. effusus Tiraboschi		6.7	-	-	-	-	13.7	-	8.8	10.0	-	14.3	-
		-	4.8	-	-	-	5.0	-	-	-	9.5	9.5	-
A. flavus Link.		8.8	9.1	33.3	42.0	30.2	42.9	23.8	57.1	33.3	33.3	28.5	9.5
		-	-	-	40.0	-	-	-	10.0	14.3	61.9	38.1	-
A. fumigatus Fres.		13.3	-	37.5	47.0	20.2	47.6	9.5	19.0	19.0	12.1	-	42.9
		-	7.7	-	10.0	-	52.4	76.2	33.3	38.1	-	-	13.3
A. glaucus Link.		-	-	8.3	5.0	10.0	9.0	-	-	-	-	-	-
		-	-	-	-	-	-	-	-	-	-	-	-
A. niger Van Tieghem		13.6	4.5	-	-	47.6	100.	40.9	81.0	66.7	61.9	9.5	-
		5.0	-	-	-	-	-	-	-	14.3	-	-	-
Aureobasidium sp		-	-	9.0	-	10.6	12.0	-	-	-	-	-	-
Viola & Boyer		-	9.5	-	-	-	-	5.8	-	-	-	-	-
Cunninghamella sp		10.0	-	-	-	-	-	-	-	-	-	48.6	20.0
Matruchot		-	-	-	-	-	-	-	-	-	-	-	-
Curvularia lunata		9.5	22.7	30.2	15.4	-	-	-	4.8	15.2	-	-	21.3
(Walker) Boedijn		4.8	-	-	-	9.1	-	3.0	-	-	-	-	4.8
C. tetramera		15.2	30.0	41.6	13.2	-	10.0	71.5	15.6	9.1	4.8	-	-
(McKinney) Boedijn		5.0	10.0	-	-	-	-	15.6	-	-	-	-	-
Fusarium moniliforme		-	4.5	25.0	31.0	-	4.7	40.1	-	-	15.0	27.0	9.1
Schldon		-	-	4.8	5.0	-	-	-	-	14.3	-	-	-
F. oxysporum		-	18.2	45.6	61.4	65.1	20.0	18.2	30.2	-	23.8	-	39.5
Schlecht		4.8	5.0	-	9.1	12.0	-	-	-	-	25.0	11.0	21.1
Mucor heimalis		-	9.1	-	19.0	-	-	-	33.3	40.1	9.1	-	-
Wehmer		-	-	-	-	-	-	-	4.8	-	-	-	-
M. plumbeus		-	-	4.2	23.0	-	-	16.7	-	-	-	-	30.0
Bonorden		-	-	-	-	4.8	-	-	-	-	-	-	-
Neurospora sp Pers.		-	22.7	-	-	-	30.2	-	-	-	-	-	-
		-	-	-	-	-	-	-	-	28.6	-	-	-
Penicillium		19.1	40.9	45.0	-	-	-	-	-	-	-	19.0	9.5
chrysogenum Thom		-	-	-	-	-	-	-	-	-	-	-	-

Continue....

Name of Organism	Incubation Period											
	Dec. '90	Jan. '91	Feb. '91	Mar. '91	Apr. '91	May '91	Jun. '91	Jul. '91	Aug. '91	Sep. '91	Oct. '91	Nov. '91
P. multicolor Gri.-Man.	20.9	4.8	10.0	9.0	5.0	4.8	-	21.6	-	-	33.3	40.1
& Por	4.8	-	-	-	-	-	-	-	-	-	5.0	-
P. notatum Westling	-	18.2	20.0	51.5	19.1	-	4.8	26.3	33.3	9.0	-	-
	-	-	-	-	-	-	-	-	-	-	-	-
P. oxalicum Thom	9.1	21.5	-	-	-	-	-	47.6	18.2	23.0	30.3	23.8
	-	-	-	-	-	-	-	-	-	4.8	-	-
P. perpurogenum Stoll	-	-	-	-	-	-	-	-	9.5	-	-	-
	-	-	-	-	-	-	-	-	-	-	-	-
P. steckii Zaleski	-	-	-	-	-	-	-	-	-	-	-	-
	-	-	-	-	-	-	-	-	4.8	-	-	-
Phytophthora undulata	-	4.5	-	-	-	-	-	-	-	-	-	-
	-	-	-	-	-	-	-	-	-	-	-	-
Rhizopus nigricans	-	-	-	20.0	-	10.0	-	-	-	-	61.9	43.3
Ehrenberg	-	-	-	-	-	-	-	-	-	-	-	-
Syncephallastrum recemosum	-	-	-	-	9.5	12.0	-	-	-	-	-	-
(Cohn) Schroeter	-	-	-	-	-	-	-	-	-	-	-	-
Total Fungal Frequency	179.7	300.9	367.	414.	258.8	374.	338.0	422.	423.7	325.1	348.1	340.0

a* - Blotter test; b* - Agar plate test

V. sinensis var. *pusa barsati* seeds were found to be associated with 26 fungi (Table 4). Category 'a' was constituted of *Alternaria alternata, Aspergillus flavus, A, fumigatus, A. niger, Fusarium moniliforme, F. oxysporum, Penicillium chrysogenum* and *P. multicolor*. Fungi representing category 'b' were *Aspergillus effusus, Chaetomium bostrychodes, Penicillium frequentans, Phytophthora* sp. and *Pythium* sp. Category 'c' was represented by *Aspergillus nidulans-alba, Mucor plumbeus* and *Papulospora* sp. *Acremonium vitis, Aspergillus candidus, A. flavipus, A. terricola, A. versicolor, Curvularia subarrium, Helminthosporium anomatum, myceliophthora* sp and *Penicillium tardum* were sporadic in their appearance and constituted category 'd'.

Table 4. Monthly incidence of fungi during storage on seeds of *V. sinensis* var *pusa barsati* using blotter and agar plate method

Name of Organism		Incubation Period											
		Dec. '90	Jan. '91	Feb. '91	Mar. '91	Apr. '91	May '91	Jun. '91	Jul. '91	Aug. '91	Sep. '91	Oct. '91	Nov. '91
Acremonium vitis	a*	-	-	-	-	-	-	9.52	-	-	-	-	-
Cattaneo	b*	-	4.54	-	-	-	-	-	-	-	-	-	-
Alternaria alternata (Fr.)		21.9	33.3	19.1	4.6	-	-	9.5	-	-	28.9	41.6	30.0
Keissler		5.0	-	-	-	9.1	-	4.8	-	19.1	5.0	20.0	-
Aspergillus candidus		10.0	-	-	-	-	-	9.1	-	-	4.8	-	-
Link.		-	-	-	-	-	-	-	-	-	-	-	-

Continue....

Name of Organism	Incubation Period											
	Dec. '90	Jan. '91	Feb. '91	Mar. '91	Apr. '91	May '91	Jun. '91	Jul. '91	Aug. '91	Sep. '91	Oct. '91	Nov. '91
A. effusus Tiraboschi	-	-	-	-	-	-	-	-	14.3	-	-	-
	-	18.2	-	-	-	-	-	-	-	19.0	19.0	9.5
A. flavipes (Bai.& Sar)	-	-	-	-	-	-	-	-	-	-	-	-
Thom & Church	-	-	-	-	-	4.8	-	-	-	-	-	-
A. flavus Link.	-	8.7	58.3	38.1	9.5	4.8	-	-	33.3	57.1	28.6	12.6
	-	9.1	-	28.6	14.0	-	9.5	28.6	61.9	38.1	76.2	-
A. fumigatus Fres.	9.5	-	12.5	19.0	4.8	42.9	-	4.8	23.8	4.8	14.3	-
	5.0	4.5	-	-	90.0	47.6	100.0	4.8	33.3	91.9	14.3	5.0
A. nidulans-alba E.	-	-	-	-	-	-	-	14.3	-	-	-	-
Yuilll	-	-	-	-	-	4.8	-	-	-	-	-	-
A. niger Van Tieghem	42.9	13.0	9.5	4.8	19.0	-	-	9.5	33.3	19.0	4.8	10.0
	-	-	-	-	-	-	-	-	33.3	23.8	28.6	-
A. terricola Marchal	-	-	-	-	-	-	4.8	-	-	-	-	-
	-	-	-	-	-	-	-	-	-	-	-	-
A. versicolor (Yuill)	4.8	-	-	-	-	-	-	9.5	-	19.1	-	-
Tiraboschi	-	-	-	-	-	-	-	-	-	-	-	-
Chaetomium	-	-	12.1	-	-	-	-	14.3	-	-	-	-
bostrychoides Zoef	-	4.5	25.0	-	-	-	-	-	-	-	-	-
C. subarium (Nees)	-	-	-	-	-	-	-	4.8	-	-	-	-
Boedijn	-	-	-	-	-	-	-	-	-	-	-	-
Fusarium chlamydo-	-	-	-	-	-	-	-	-	-	-	-	49.0
sporum Wolle. & Rein	-	-	-	-	-	-	-	-	-	-	-	-
Fusarium moniliforme	12.0	17.4	24.5	30.9	-	39.1	17.1	-	10.0	-	9.5	66.7
Schldon	5.0	4.8	-	-	-	-	-	-	4.8	-	-	-
F. oxysporum Schlecht	19.0	26.1	42.8	25.5	10.0	4.8	5.0	21.7	-	9.5	-	45.0
	14.3	22.7	12.0	-	-	-	-	10.0	5.7	-	-	30.0
Helminthosporium	-	-	9.5	4.8	-	-	-	-	-	-	-	-
anomatum Gil. & Abbot	-	-	-	-	-	-	-	-	-	-	-	-
Mucor plumbeus	-	-	4.2	30.0	24.4	-	12.0	40.8	-	-	-	19.0
Bonorden	-	-	-	-	-	-	-	-	-	-	-	-
Myceliophthora sp	-	-	-	-	-	-	-	-	-	-	5.0	-
Goddard.	-	-	-	-	-	-	-	-	-	-	-	-
Papulospora sp Preuss	-	-	-	-	-	-	9.5	12.0	10.0	-	-	-
	-	-	-	-	-	-	-	-	-	-	-	-
Penicillium chrysogenum	9.5	-	28.6	12.0	-	-	40.4	38.5	-	-	28.6	19.1
Thom	5.0	4.8	4.8	-	10.0	-	-	-	-	-	4.8	-
P. frequentans Westling	-	-	-	-	-	-	-	-	-	-	16.6	-
	-	-	-	-	-	-	-	-	-	-	-	-
P. multicolor Gri.-Man.	28.1	41.6	30.0	-	12.7	-	4.8	19.1	11.8	-	13.5	-
& Por	-	-	4.8	-	3.0	-	-	10.0	-	-	-	-

Continue...

Name of Organism	Incubation Period											
	Dec. '90	Jan. '91	Feb. '91	Mar. '91	Apr. '91	May '91	Jun. '91	Jul. '91	Aug. '91	Sep. '91	Oct. '91	Nov. '91
P. tardum Thom	9.5	-	-	-	4.8	-	-	-	-	-	-	-
	-	-	-	-	-	-	-	-	-	-	-	-
Phytophthora sp de Berry	9.5	8.7	-	-	-	-	-	-	-	-	-	-
	-	-	-	-	-	-	-	-	-	-	-	-
Pythium sp *pringsheim*	14.3	4.3	-	-	-	-	-	-	-	52.4	-	-
	-	9.1	-	-	-	-	-	-	-	-	-	-
Total Fungal Frequency	225.3	230.8	297.5	198.2	211.2	148.6	226.3	242.4	294.6	373.4	325.1	295.9

a* - Blotter test, b* - Agar plate test

Changes in seed N % and crude protein % due to fungal flora: Control seeds of *V. sinensis* var. *asparagus* Black were assayed at zero month storage which showed 27.63% protein content. In case of infested seeds, for first two months the decline was minimum but in third month there was sharp reduction in protein content (25.13%). It showed 9.05% reduction over control. At the end of storage period the total reduction was 38.69% over the control. In case of surface sterilised seeds the total reduction was only 2.5% as compared to initial protein level (Table 5).

Table 5. Changes in protein contents in *Vigna sinensis* var. *asparagus* black seeds due to fungal association during storage

S.N.	Incubation period (Months)	Nitrogen percentage		Protein percentage		% change in protein over control	
		a*	b*				
1	0	4.42	4.42	27.63	27.63	0.00	0.00
2	1	4.42	4.35	27.63	27.19	0.00	-1.62
3	2	4.40	4.30	27.50	26.88	-0.47	-2.71
4	3	4.38	4.02	27.38	25.13	-0.90	-9.05
5	4	4.38	4.00	27.38	25.00	-0.90	-9.52
6	5	4.37	3.76	27.31	23.50	-1.16	-14.95
7	6	4.37	3.60	27.31	22.50	-1.16	-18.57
8	7	4.36	3.41	27.25	21.31	-1.38	-22.87
9	8	4.37	3.08	27.31	19.25	-1.16	-30.32
10	9	4.35	2.85	27.19	17.81	-1.59	-35.54
11	10	4.35	2.80	27.19	17.50	-1.59	-36.67
12	11	4.31	2.71	26.94	16.94	-2.50	-38.69

a* - Control seeds; b* : Infested seeds

When surface sterilised as well as infested seeds of *V. sinensis* var. *pusa barsati* were assayed at 0 month of storage, crude protein content in both the samples were found to be 24.18%. During the early stages of storage i.e. 0-5 months there was gradual decrease in the protein content of infested seeds but from 6th month onwards there was acute quantitative reduction throughout the storage period. At the end of storage period 47.27% reduction was recorded in infested seeds as compared to control seeds (Table 6).

Table 6. Changes in protein contents in *Vigna sinensis* var. *pusa barsati* seeds due to fungal association during storage

S.N.	Incubation period (Months)	Nitrogen percentage		Protein percentage		% change in protein over control	
		a*	b*				
1	0	3.87	3.87	24.19	24.19	0.00	0.00
2	1	3.87	3.81	24.19	23.81	0.00	-1.53
3	2	3.87	3.72	24.19	23.25	0.00	-3.85
4	3	3.85	3.70	24.06	23.13	-0.50	-4.38
5	4	3.85	3.60	24.06	22.50	-0.50	-6.95
6	5	3.84	3.59	24.00	22.44	-0.74	-7.24
7	6	3.81	3.18	23.81	19.88	-1.53	-17.78
8	7	3.80	3.00	23.75	18.75	-1.78	-22.46
9	8	3.80	2.82	23.75	17.63	-1.78	-27.08
10	9	3.79	2.30	23.69	14.38	-2.02	-40.57
11	10	3.79	2.17	23.69	13.56	-2.02	-43.92
12	11	3.79	2.04	23.69	12.75	-2.02	-47.27

a* - Control seeds; b* : Infested seeds

Detection of aflatoxin from seed samples : In present investigation *A. flavus* was one of the most dominant fungus associated with seeds. Thus, screening of seed lots was undertaken. The data on chemical analyses of aflatoxin from both varieties under study is presented in Table 7. From each variety, five samples were drawn randomly and tested for presence of aflatoxin by BYG test. Only two samples of var. *asparagus* Black, VSA4 and VSA_5 gave BYG positive test showing high concentration of B_1 aflatoxin. Only VSA_4 showed low concentration of aflatoxin B_2 where as all the five samples of var. *pusa barsati* seeds gave BYG positive test showing presence of high concentration of B_1 in VSP_1 and VSP_2. VSP_3 and VSP_4 showed presence of moderate amount of aflatoxin B_1 while VSP_5 contained low concentration. VSP_1 and VSP_5 gave florescence of aflatoxin B_2 with low intensity.

Table 7. Detection of aflotoxin in seed samples

S.N.	Name of crop	Sample No.	BYG test	Aflatoxin B1	Aflatoxin B2
1.	*Vigna sinensis* var. *asparagus* black	VSA1	N	—	—
		VSA2	N	—	—
		VSA3	N	—	—
		VSA4	P	++	+
		VSA5	P	++	—
2.	*Vigna sinensis* var. *pusa barsari*	VSA1	P	+++	+
		VSA2	P	+++	—
		VSA3	P	++	—
		VSA4	P	++	—
		VSA5	P	+	+

DISCUSSION

Monthly checkup of germinability and seed health under storage: The results presented in Table 1 and Table 2 shows clearly that; in general, germination decreases with the increase in the storage period. The germination was found low in summer season, which indicates the particular temperature requirement for seeds to germinate. Otherwise, with the extension of storage period regular loss of germinability was visible. This finding was confirmed by earlier reports (Christensen, 1973). In present investigation no positive correlation was found between germination percentage and fungal incidence as recorded by blotter as well as agar plate methods. Neither number of fungi detected from seeds in particular incidence nor total infection percentage of seed could be correlated to germination percentage. Like in variety *asparagus* Black, though the fungal frequency was highest in the month of March 1990 the germination percentage was not lowest. On the contrary, fungal frequency was not highest in the month of Nov. 1991 but the germination percentage was least. Therefore, one may presume that percent incidence of fungi as such has no direct bearing on percent germination and survival of seedlings. Mathur *et al.* (1972) and Mohan *et al.* (1994) also made such type of observations. The data presented in Table 3 and 4 shows that the frequencies of fungi changed according to the changes in environmental conditions, particularly temperature, which is responsible for changing in their existence in/on seeds. It shows that in variety *asparagus* Black, in the month of May, the total fungal frequency was not least which is attributed to the fact that *A. fumigatus* was present in high percentage. Earlier Thakre (1975) showed that *A. fumigatus* was thermotolerant. The fungi occurring throughout the year showed varying percentage of incidence. These fungi are obviously more versatile in their food requirement and capacity of tolerance for varying environmental conditions.

Detection of aflatoxin from seed samples: Presence of different levels of aflatoxin B1 and B2 can be attributed to the individual capacities of isolates to produce different components of aflatoxin or varietal resistance shown by seeds against its production. Nagarajan *et al.* (1972) identified the inhibitory factor as a protein of low molecular weight, which was present in high concentrations in resistant variety of maize. Earlier detection of aflatoxin contamination was chiefly restricted to oily commodities. Reports regarding their presence in proteinaceous seeds are comparatively meager. However, the presence of aflatoxins in leguminous seeds reflects its wide distribution in agricultural commodities and calls for immediate attention and careful screening to avoid disastrous consequences.

REFERENCES

Bilgrami, K.S. and R.N. Verma (1978). 'Physiology of fungi' Vikas Publishing House Pvt. Ltd. New Delhi. p. 505.

Christensen, C.M. and H.H. Kaufmann (1965). Deterioration of stored grains by fungi. *Ann. Rev. of Phytopathology* 3: 69-84.

Christensen, C.M. (1973). Loss of viability in storage: Mycoflora. *Seed Science and Technology.* 1: 547-562.

Davy's M.N.G. and N.W. Pirie (1969). International Biological Programme (IBP) *Biotech. Bioengr.* 11 : 528.

International Seed Testing Association (1966). International rules for seed testing. *Int. Seed Testing Assn.* 31: 1-151.

Johnston, A. and C. Booth (eds.) (1983). 'Plant pathologist's pocket book' The Cambrain News Ltd. Abrystwyth, Wales.

Mathur, S.B., J.I. Mallaya and P. Neergaard (1972). *Proc. Int. Seed Testing Asso.* 35: 157-163.

Mohan, C. and K.M. Leelavathi (1994). Seed borne fungi of rice in Kerala and the probable role of some seed borne pathogens on germination and survival of seedlings. In: 'Vistas in Seed Biology' T. Singh and P.C. Trivedi. Eds., Printwell Publishers, New Delhi. 229-233.

Musket, A.E. (1948). Technique for the examination of seeds for the presence of seed-borne fungi. *Trans. Br. Myco. Soc.* 30 : 74-83.

Nagarajan, V. and R.V. Bhat (1972). Factor responsible for varietal differences in aflatoxin production in maize. *J. Agric. Food Chem.* 38: 568-571.

Sadasivam, S. and A. Manikam (1992). 'Biochemical Methods for Agricultural Science'. Wiley Eastern Ltd. New Delhi. p. 246.

Sahai, A. and B. S. Mahrotra (1982). Mycoflora associated with the seeds of forest trees and their effect on germination. *Proc. Ind. Natn.. Sci. Acad.* Part B 48 (5): 706-713.

Stahl, E. (1969). 'Thin layer chromatography. A laboratory handbook.' Springer International Student Edition, Springer-Verlag, Berlin. P. 1041.

Thakre, R.P. (1975). 'Ecological and physiological studies of soil fungi (Thermophilic) from coal mine localities in M.P. Ph. D. Thesis, Sagar University, Sagar.

CHAPTER – 17

MYCOPESTICIDES : THEIR POTENTIAL AND CHALLENGES

M.V. Deshpande

Biochemical Sciences Division, National Chemical Laboratory, Pune-411 008

INTRODUCTION

The agriculture pests known so far include more than 10,000 species of insects and fungi. With every major outbreak of plant pathogen, the concept of plant protection has been revolutionized. Disaster in Ireland caused by fungal attack on potatoes in 1840s necessitated the development of control agents as one of the means for sustainable agriculture. While in the late 1960s, the insect pest on cotton crop in Mexico when became tolerant to organophosphate insecticides triggered the process of rethinking about use of chemicals for the fungal and insect control (Deshpande, 1998a). The total market for pesticides in 2005 will be US $ 35 Billion (30 for chemicals and 5 for Biopesticides). Under biopesticides though BT, *Bacillus thuringiensis* takes major share fungi, nematodes, viruses, etc. are also contributing a lot. Control of insect pests, particularly by their natural enemies comprising of parasites, predators and microbes including fungi in agro-ecosystem is a continuous process. Although viruses are being used in fields for the pest control, their disadvantages are, rapid deactivation by environment, requirement of living host for the production, their effectiveness only when ingested by the host and time from ingestion of a lethal dose until death under field conditions is long (3-15 days). Similarly, bacterial pesticides are effective after ingestion and they have wide host range, therefore must be used with some care. Unlike bacteria and viruses, fungal agents act by contact and do not need ingestion.

Furthermore, fungi can be mass-produced very easily and are host-specific. Fungi intended for use as bio-control agents in agriculture are known as myco-pesticides. The characteristic properties of mycopesticides in use are as mentioned in Table 1. Hereafter, this term will be used to imply fungi and fungal products (mycolytic and cuticle degrading enzymes, inhibitors, antibiotics, etc) which are used either singly or in combination to control plant-pathogenic fungi and insects (Milner, 1987; Deshpande, 1998b). The fungus- fungus and –insect interaction, large scale production of live propagules for their use in mycopesticide formulation, as well as production of mycolytic and cuticle degrading enzymes will be discussed in the subsequent sections.

Table 1. Characteristic properties of mycopesticides

Advantages	Distinctive characterisitics
No accumulation in food chain	Limited shelf life
Absence of toxic residues in soil and in groundwater or drinking water	Effective within limited temperature and humidity range
Safe for mammals, in general	Contact effect, death delayed in accordance with disease course
Can be re-isolated on selective medium	*Allergy risk*
Application is simple and easy	
Mass production is possible	
Relatively narrow host spectrum	

FUNGUS-FUNGUS INTERACTION

A wide spectrum of fungi is known to show antagonism with each other (Table 2). Weindling (1932) for the first time noted the potential of *Trichoderma lignorum* in controlling plant-pathogenic fungi. Mycoparasitism, *i.e.* a direct attack of a fungus on to the other fungal mycelium followed by nutrient utilization by the parasitic fungus is of two types. In the necrotrophic (destructive) parasitism the relationship results in to the death and destruction of one or more components of the host while in biotrophic (balanced) parasitism the development of the parasite is favoured by a living rather than dead host structure. The latter one has a restricted host range and the specialized structures are produced by the parasite to take up the nutrients from the host. Initially parasitic fungus locates the host mycelium or spores, using chemical signals originating from the host fungus. Attachment and lectin, sugar-binding protein mediated recognition between a parasite and its host appear to constitute an essential step (Inbar and Chet, 1997). *Trichoderma*, for instance, attaches to the host hyphae by coiling, hooks or appressorium-like bodies and penetrates the host by secreting cell wall lytic enzymes (Lorito *et al*, 1993; Carsolio *et al*, 1999; Elad and Kapat, 1999). The production of cell wall lytic enzymes, such as chitinase and β–1,3-glucanse, by *Gliocladium roseum* in the interaction with *Botrytis allii* was reported (Li *et al.*, 2002). Li *et al.* (2002) further reported the penetration of *G. roseum* through *B. cinerea* cell walls by producing germ tubes without the formation of appresoria. The fungus like *Sporidesmium sclerotivorum* produces haustoria in the host *Sclerotinia sclerotiorum* and triggers the production of cell wall lytic enzyme activities by the host itself (Schroth and Hancock, 1981). For *Colletotrichum muscae,* a pathogen of banana, it has been shown that even broad host-range mycoparasites which attack two or more pathogens discriminated significantly between *C. muscae* strains due to their differential susceptibility to one or more mechanisms of mycoparasitism (Krauss *et al.*, 2001). Therefore, they suggested the use of the mixture of different mycoparasitic strains. Krauss and Soberanis (2001) have also reported the use of the mixture of mycoparasitic strains to control cocoa pod diseases. The cocoa pod diseases, moniliasis, witches broom and black pod are caused by *Moniliophthora roreri, Crinipellis pernicosa* and *Phytophhora palmivora,* respectively. Five strains of *Clonostachys rosea* and three strains of *Trichoderma* sp. in different combinations were reported to be effective in controlling cocoa pod diseases.

Table 2. Some Common mycoparasitic fungi

Fungus	Host	Crop	Specific examples and comments
Fusarium	*R. solani, Fusarium species*	Wilt of Radish, and banana, seedling blights of wheat and maize.	Within same genus mycoparasitism is observed. *F. oxysporum, F. semitectum, F. udum*
Gliocladium	*Pythium ultimum, R. solani, Botrytis cinerea*	Damping-off disease of cotton	*G. virens* with commercial potential. *G. roseum* parasitize on *Botrytis cinerea*
Peniophora	*Fomes annosus*	Pine stumps	*e. g. P. gigantiea*
Piptocephalis	*Choanephora Cucurbitarum*	Fruit rot, leaf blight of okra, squash, pumpkin, beans, cucumber, etc	*e.g. P. virginiana* further work is necessary
Pythium	*F.solani, Pythium myriotylum, P. ultimum, P. ultimum. Sclerotinia sclerotiorum*	Browning root rot of cereals, rhizome rot of ginger, damping off of cucumber, sugar beet and cress	*P. nunn, P. oligandrum, P. acanthophoron* can parasitize a range of fungi
Rhizopus	*F.oxysporum* var. *lycopersici*	Tomato wilt	*R. nigricans*
Tremella	*Peniophora laeta Peniophora incarnata Phlebia radiata*	Occurs on hardwoods with basidiomycetes	*T. mesenterica T. encephala T. mycophaga*
Trichoderma	*R. solani, Sclerotium rolfsii, F. solani, F. culmorum, P. ultimum, S. sclerotiorum*	Potato wilt, damping off of cucumber	*T. harzianum, T. viride,* one of the well studied system for fungus-fungus interaction

Compiled from Schroth and Hancock, 1981; Deshpande, 1999; Krauss and Soberqnis, 2001; Li *et al.,* 2002

FUNGUS-INSECT INTERACTION

Recognition that fungi have potential for biological control dates back to Louis Pasteur (Milner, 1987). During 1880s *Metarhizium* species was used to control wheat chafer, *Anisoplia austriaca* and the sugar beet curculio, *Cleonis punctiventris*. Fungi constitute a largest group of more than 500 species which can parasitize insects. The common insect pathogenic fungi are listed in Table 3. The insect-pathogenic fungi are present in most of the taxonomic groups. The genera such as *Metarhizium, Beauveria, Verticillium, Nomuraea, Entomophthora, Neozygites,* etc. are commonly encountered in nature. In most of the cases asexual conidia/spores are the means for the spreading of infection in an insect population. Similar to mycoparasitism, pre-infection requirements during entomopathogenesis are adhesion and germination of spores to form either germ tubes or appresoria especially for the penetration through the cuticle. Invasion through the cuticle is facilitated with these structures and cuticle degrading enzymes. The proliferation of hyphal bodies (irregular growth) or unicellular yeast-like cells in hemolymph and the production of enzymatic and non-enzymatic components are all involved in the insect killing process. Indeed, fungus has to combat with the host defense responses such as melanization, cellular reactions like phagocytosis, nodulation or encapsulation and hemocytic defense reaction such as polyphenol oxidase activation leading to the production of reactive molecules (St. Leger *et al*, 1987; Charnley, 1989; Gillespie and Moorhouse, 1989). The fungus *Hirsutella thompsonii* has been proved to be pathogenic to various mites such as *Phyllocoptruta oleivora* on citrus, *Acalitus vaccini* on blueberry, *A. guerreronis* on coconut, to name a few (Sreerama Kumar and Singh, 2000). The interaction of *M. anisopliae, Nomuraea rileyi* and *Beauveria bassiana* with noctuid larvae (larvae which hide either in soil or inside leaf clusters in daytime and feed on leaves at night) of *Helicoverpa armigera* and *Spodoptera litura* has been reported extensively in the literature (Tang and Hou, 1998; Deshpande and Tuor, 2001). Adults of the common or European cockchafer (white grub, *Melolontha melolontha* L.) feed on forest and fruit-tree leaves such as oak, beech, maple, chestnut, plum, etc. A host specific strain of *Beauveria brongniartii* has been tested in large-scale field tests in Switzerland over last two decades (Keller *et al*, 1997; Inglis *et al.*, 2001).

DUAL INTERACTION

The intriguing feature that has been reported so far in few of the organisms is the dual pathogenecity towards both insects and fungi, mediated through the production of specific hydrolytic enzymes. For example, *Trichoderma harzianum,* a mycoparasitic fungus, can parasitize the elm bark beetle, *Scolytus,* while *Verticillium lecanii,* pathogen of aphids, can also attack rust fungi (Deshpande, 1999). The specific hydrolytic enzymes, that are assumed to be involved are chitinases, proteinases, glucanases and lipases produced by the above which facilitate the entry of the pathogen in the respective hosts (Jun *et al.*, 1991).

Table 3. Some common entomogenous fungi

Fungus	Insect	Crop	Comments
Aschersonia	Scale insect, white fly	Citrus, sugarcane, etc.	
Beauveria	Spider, beetles, brown plant hoppers, European cockchafer	Sugarcane, rice, cardamom, cabbage, etc.	*B. bassiana*- already in commercial use; *B. brongniartii* commercially used to control European cockchafer
Cephalosporium	Coffee green bug, brinjal leaf beetle	Coffee, Brinjal, vegetables etc.	Not very common.
Conidiobolus	Aphids and diverse insects		*C. coronatus,* weak pathogen
Entomophthora	Aphids (alfa alfa, wheat, crucifer, etc), chest-nut beetle, coffee green bug, house fly, etc.	Alfa alfa, chest-nut, cruciferous vegetables, wheat, coffee, etc.	*E.muscae, E. grylli,* Studied for commercialization
Hirsutella	*Pyrilla* (sugarcane leaf hopper), mango leaf hopper, coconut mite	Sugarcane, coffee, mango, coconut, etc.	*H.thompsonii* an important alternative for the biocontrol of coconut mite.
Isaria	*Pyrilla*	Sugarcane	Distinct morphology of erect, rigid sporophores
Lagenidium	Mosquito	-	Mosquito control
Metarhizium	*Pyrilla*, coconut beetle, brinjal mealy bug	Sugarcane, coconut, vegetables, etc	*M. anisopliae,* and *M. flavoviride* in commercial use
Neozygites	Mite, thrips, mealybug	Vegetables	
Nomuraea	*Heliothis , Spodoptera*	Pulses	Important for commercial exploitation
Paecilomyces	White fly	Mango	
Verticillium	Leaf hopper, banana aphid, sugarcane shoot borer	Banana, sugarcane, etc.	*V. lecanii* commercially important, exhibits dual pathogenicity.

Compiled from Keller *et al,* 1997; Deshpande, 1999; Inglis *et al*, 2001.

The insect cuticle is composed of wax, lipids, protein and chitin (Fig. 1). The main component is being chitin, its share is 25-50% of the dry weight of the cuticle. Fungal cell wall also contain chitin (20-60%) as the main structural component (Deshpande, 1998a) (Fig. 1). Thus, insects and fungi share a key structural polymer, chitin, a β-1,4-linked *N*-acetylglucosamine polymer and its production is essential to them (Deshpande, 1998a). Therefore, it may be advantageous to develop a common agent against both the groups. Furthermore, chitin is absent in plants and mammals and thus its metabolism too presents an attractive target for the control of plant pathogenic fungi and insects. Indeed, severity of the infection, competition between targeted and non-targeted species for the survival, the selection of the time of application, are the important factors with respect to the effect of the control agent on other useful species of the environment.

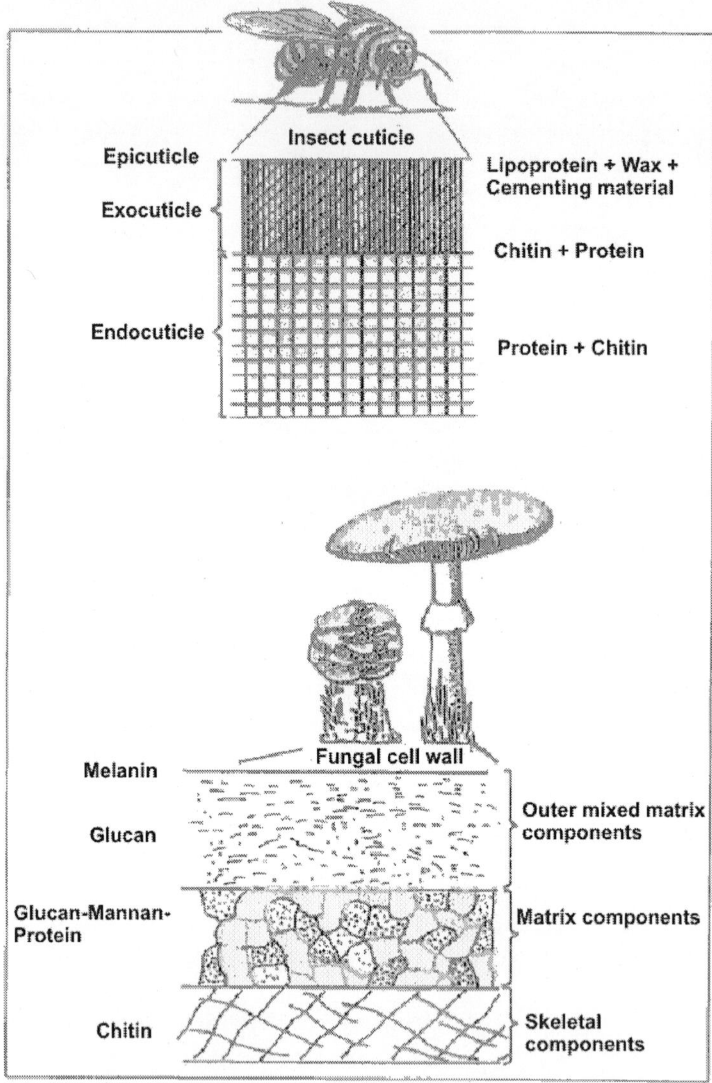

Fig. 1. Structure of insect cuticle and fungal cell wall.

SELECTION OF BIOCONTROL STRAINS

Many past efforts to select biocontrol organisms were empirical. The random selection of organisms from surfaces of healthy plants and their rhizosphere in otherwise heavily infested agricultural sites was the method of choice (Martin and Hancock, 1986). Furthermore, to isolate the insect-pathogenic fungi, *Galleria* bait method (Zimmermann, 1986) can be used (Chart 1). *Galleria* sp., commonly known as greater wax moth, is a severe pest on honey bee. It is highly susceptible to the insect-pathogenic fungi and has tendency to move downwards in the soil.

These characteristics are useful for the isolation of entomo-pathogenic fungi. However, similar type of selective bait method has not been designed successfully so far to isolate potential mycoparasitic strains. For the identification of mycoparasitic strains, the co-culture technique on the solid nutrient media can be used.

Biological control with fungi often gives inconsistent results, and the slow speed of killing as compared to chemical pesticides is hampering their commercial development too. The researchers have identified genes for strain improvement that are involved in adhesion of the spores to the protective cover, enzymatic hydrolysis of fungal cell wall or insect cuticle, regulation of hyphal bodies formation, inactivation of host defence and production of toxic metabolites (Heale *et al.*, 1989; Goldman *et al.*, 1994; Smithson *et al.*, 1995). Though the attempts are being made to improve the performance and consistency of fungal agents by genetic manipulations (St Leger and Roberts, 1997), registration requirements to date have been for natural isolates.

LARGE-SCALE PRODUCTION OF LIVE PROPAGULES

Production of fungi to use as biocontrol agents aims at the maximum possible number of living propagules that can infect pathogenic fungi and insects (Deshpande, 1999). This is a distinguishing feature between production of fungi to be used as biopesticides and fungi meant for other products. The biomass produced needs safe storage, while in latter most of the processes use fresh, and actively growing organism. It is expected that fungal propagules in mycopesticides must be alive at the time of their use. The expected minimum shelf life is 18 months at 20°C.

Both liquid and solid fermentation systems can be used to obtain desirable quantity of the biomass, and the former is a preferred approach. Liquid fermentation has obvious advantages of controlled processing, and both organism itself and its products exhibiting biocontrol activity can be obtained easily. Using a wide range of controls for pH, agitator speed, aeration rate, temperature, etc. the timing for harvesting spores only can be precisely envisaged to obtain suitable biomass for storage. Fungi can be successfully grown by solid state fermentation too. However, the problems faced in solid state fermentation are substrate sterilization, gas exchange, temperature control, maintenance of purity of the culture, long time of the process and product recovery. But it is advantageous because of easy process, cheap raw material and most important is that spores are more tolerant to desiccation as compared to spores produced in submerged fermentation (Jackson *et al.*, 1991). To minimize the fermentation time, high inoculum can be used. However, it is likely that an organism may use all the nutrients for the proliferation of mycelium and the differentiation in spore form may not occur. In fact, different nutritional and environmental factors such as temperature, light and pH affect vegetative growth, asexual and sexual sporulation and spore germination. The understanding of their effect will provide useful information for improving both the efficacy and propagule production.

Another most commonly used technique for the large scale production of infective propagules is a two-stage cultivation in which the fungus is first grown in submerged fermentation and then allowed to sporulate as a surface culture in either semisolid medium or in still culture. These aerial conidia, as discussed before, produced usually possess desired properties of virulence, and resistance to adverse environmental conditions.

FUNGAL MORPHOLOGIES IN MYCOPESTICIDE FORMULATION

The fungal kingdom is a diverse assemblage of organisms with a great variety of structural types (Deshpande, 2000). The vegetative growth forms are mycelia and unicellular yeast, while asexual spores such as conidia, and sporangiospores are produced for the dispersal. The lower fungi sexually produce resistant oospores or zygospores which are commonly used in the biocontrol formulations. Other resistant structures like blastospores and chlamydospores are also used as infective propagules. *Paecilomyces fumosoroseus,* a biocontrol agent against sweet potato whiteflies, *Bemisia argentifolii* and *B. tabaci* can be grown in form of blastospores, which are resistant to simple drying (Inglis *et al.*, 2001). *P. fumosoroseus* like many other entomopathogenic fungi is dimorphic in nature, *i.e.* it can exhibit two forms, mycelial and yeast-form cells, in liquid culture. The yeast-form cells eventually lead to the production of blastospores or hyphal bodies. Since the blastospores are produced from yeast-form cells, media which contained high concentration of nitrogen and carbon supported maximum blastospore production (Jackson *et al.*, 1997). As blastospore formation is not a response of organism to the unfavourable conditions, in the true sense it is not a sporulation. As many entomopathogenic fungi exhibit dimorphic growth, it will be worthwhile to grow organism for blastospore production in a liquid culture. The short fermentation time could be one of the major economic advantages.

In submerged fermentation, *P. fumosoroseus* isolates produce high concentration of blastospores, which are not amenable to simple drying and tend to perish more rapidly during storage. Jackson and co-workers (1997) have reported a medium containing glucose and casamino acids to produce blastospores which survive air drying. Unlike most of the fungi, desiccation tolerance of *P. fumosoroseus* blastospores was not found to be dependent on reduced water activity that enhances accumulation of trehalose and polyhydroxy alcohols. However, these authors have attributed this property to the nutrients provided by casamino acids. While the conidia formation in *M. flavoviride* can be improved by adjusting the proportion of yeast nitrogen base to sucrose in the medium. The depletion in yeast nitrogen increases conidia production (Jenkins *et al.*, 1998).

Different mycoparasitic *T. harzianum* strains grow as mycelia and later on produce chlamydospores in submerged fermentation while on solid substrates the conidia formation is common. Chlamydospores of another mycoparasitic fungus *G. virens* are also produced in stirred tank fermentor and then are being used to make the effective biocontrol formulation (Eyal *et al.*, 1997). In case of *Pythium oligandrum,* an oomycetous parasite of another *Pythium* species, oospores produced by fermentation are used for the seed treatment of sugar beet, cress and carrot to control damping-off diseases (McQuilken *et al.*, 1990).

Myrothecium verrucaria, a potent chitinase producer, was successfully used for the control of *Drechslera teres* infection of barley leaves by the seed treatment (Mostafa, 1993). It exhibits unicellular yeast formation in the liquid culture. A mass of spherical yeast cells can be obtained in the liquid medium containing ammonium salts and low pH, that grow either by budding or transform in to normal mycelium in the fresh medium at higher pH. These cells also produce same levels of quantities of mycolytic enzymes as mycelial cells and are equally efficient in the biocontrol of root infecting fungus like, *S. rolfsii.* Furthermore, because of unicellular nature, the quantitative estimation of the biomass in the mycopesticide formulation is easy.

Entomophthora virulenta is pathogenic to aphids which can produce zygospores in surface and also in submerged fermentation. In the presence of dextrose and yeast extract high number of zygospores was obtained within 9 days in submerged fermentation. As these zygospores showed high frequency of germination (> 70%) they were used to prepare formulation to control aphids (Latge and Soper, 1977). Another pathogen most commonly occurring on aphids is *Erynia neoaphidis*. Similar to other entomophthorales, the primary conidia of *E. neoaphidis* eject actively and start another cycle if they land on the host. Though mycelial phase can be produced *in vitro,* it can not be used in biocontrol formulation. Therefore, algination, i.e. entrapment of mycelia in the alginate beads, for conidia production was used to apply *E. neoaphidis* for the biocontrol of aphids (Shah *et al.*, 1998).

M. anisopliae, Beauveria tenella and *Aschersonia aleyrodes* produce aerial conidia when cultured in mixtures of organic bases and porous materials at 25°C within 4 weeks (Katakura, 1998). These mixtures usually maintain 30% water that supports sporulation. The organic materials used were husk, rice bran, rape seed meal, fish meal, feather meal, or crab shells and porous materials were charcoal, zeolite, etc. One of the useful raw materials available for the development of biocontrol organisms is crab shell waste. The contributing element is the presence of chitin, which can increase the potential for chitinase production, a key enzyme in the biocontrol of pathogenic fungi and insects (Deshpande, 1999; Patil *et al,* 2000).

FUNGAL PRODUCTS

Production of enzymic biocontrol components

The entomopathogenic fungi in submerged fermentation produce extracellularly cuticle degrading enzyme (CDE) complex consisting of chitinase, lipase and proteases mainly, when locust cuticle is used as a sole carbon source. Furthermore, extracellular enzymes appear sequentially in *Metarhizium* and *Beauveria* cultures (St Leger *et al.*, 1986, 1987). The proteolytic enzymes such as esterase, endopeptidase, aminopeptidase and carboxypeptidase are produced within first 24 h of growth. The *N*-acetylglucosaminidase appeared next and endo-chitinase, which attacks chitin polymer randomly made the significant appearance after 4 days (St Leger *et al.*, 1986, 1987). While lipases were undetected up to 5 days. This order of extracellular production of the CDEs is parallel to the sequence of cuticle constitutents. Significant degree of strain variability was noted in the production profiles of *B.bassiana* and *M. anisopliae.* Variability was not only in the quantities of enzymes but also to their secretion pattern on different growth media. It was reported that *B. bassiana* GK2016 strain for protease production required gelatin and on glucose medium it did not produce any proteolytic activity (Bidochka and Khachatourians, 1988). However, some strains of *B. bassiana* produced proteases when grown on glucose medium. Glucose medium in combination with nitrate did not support the chitinase production in *B. bassiana.* Similar to glucose, *N*-acetylglucosamine also regulated the chitinase and protease production. These observations can certainly be useful to tailor biological control agent(s) against specific pests. Natural variation in different strains could be used to rationally develop improved mycoinsectides. However, it is sometimes difficult to correlate activities of these enzymes alone with pathogenicity of the biocontrol organism.

The cuticle degrading enzyme complex from *M. verrucaria* also shows comparable effectiveness to the chemical insecticide in controlling infestation of the pod borer, *Helicoverpa armigera* and a common pest of cotton, pulses and vegetables, *Spodeptera litura* in the field. The III instar larvae, crop damaging stage of the pest, sprayed with CDE complex of *M. verrucaria* showed the morphological abnormalities and eventually death within 6 days. With the increased CDE concentration killing can be achieved faster.

The insect-pathogenic fungi, *B. bassiana, B. brongniartii* and *V. lecanii* produced cuticle degrading enzymes when grown on chitin containing medium. To control insect pest the enzymic treatment before or simultaneously along with the entomopathogenic fungus, was also reported to be suitable (Charnley, 1989).

Fusarium chlamydosporum, a mycoparasite of groundnut rust, *Puccinia arachidis* produces an endo-chitinase which inhibits germination of uredospores of the rust fungus (Mathivanan *et al.*, 1998). Similarly, Patil and co workers (2001) reported the use of chitinoytic enzymes in the biocontrol of the root infecting fungus, *S. rolfsii* on ground nut. Chitinolytic enzymes of *T. harzianum,* one of the most studied mycoparasitic fungus, were found to be inhibitory to a wider range of pathogenic fungi than similar enzymes from other sources (Lorito *et al.*, 1993). Several species of fungi are very potent biocontrol agents of plant pathogenic fungi and insects. *M. verrucaria* produced high levels of extracellular mycolytic enzymes, viz. chitinase, β-1,3-glucanase and proteinase (Vyas and Deshpande, 1989; Deshpande, 1998b). The crude mycolytic preparation was found to be effective against *S. rolfsii,* a root-infecting fungus of groundnut (Patil *et al.*, 2001). Govindsamy *et al.* (1998) also reported use of purified preparation of chitinase from *M. verrucaria* to control a groundnut rust, *P. arachidis.*

Non-enzymic metabolites of fungi as biocontrol components

Fungi secrete a wide array of compounds that exhibit biological activity against pests and sometimes against other useful organisms too (Vey *et al.*, 2001). Though these metabolites serve different functions such as protection of mycopesticide itself from antagonistic microorganisms, inhibition of saprophytic growth on the dead pest, determination of pathogenicity of the bicontrol fungus, antifeedant/repellent, etc., the discussion is restricted to their role in the control of plant pathogenic fungi and insects.

The mycoparasitic fungus *Gliocladium* produces a broad-spectrum antibiotic, gliotoxin and argifin, a chitinase inhibitor that are involved in the killing process of root infecting fungus such as *R. solani* (Vey *et al.*, 2001). While Thines *et al.* (1998) reported the production of glisoprenins (C, D and E), inhibitors of appresorium formation in *Magnaporthe grisea,* a causative agent of rice blast from the cultures of *G. roseum.* As chitinase is one of the enzymes involved in the cell wall synthesis such type of inhibitor should enhance mycoparasitic potential of the biocontrol organism (Patil *et al.*, 2001). Interestingly argifin retards the moulting process in cockroach larvae that again can be attributed to the inhibition of insect chitinase activity (Vey *et al.*, 2001).

T. harzianum usually secretes a family of short-chain polypeptides, potential antifungal agents, which include trichorzianins, harzianins, trichokindins, etc. However, the capacity to produce these antifungal substances is not sufficient to explain its mycoparasitic activity. The synergism between peptaibols and fungal cell wall lytic enzymes in the biocontrol of *Botrytis cinerea* has been reported (Vey *et al.*, 2001). Both the mycolytic enzymes and antifungal substances are produced in parallel under same cultural conditions using cell walls of *B. cinerea* in the growth medium.

The entomopathogenic fungi, *M. anisopliae, B. bassiana, B. brongniartii, P. fumosoroseus, V. lecanii, Tolypocladium* sp. and *H. thompsonii* produce different toxins such as, destruxins, swainsinone, cytochalasin C, bassianin, beuvericin, oosporein, dipcolonic acid, enfrapeptins, hirsutellin A, B, etc (Amiri-besheli *et al.*, 2000; Vey *et al.*, 2001). Different insects vary in their susceptibility to these toxins. For example, lepidopterean insects are highly susceptible to destruxins produced by *M. anisopliae*. Efrapeptins of *Tolypocladium* show insecticidal and miticidal effects against potato beetle, mites, bud-worm and diamondback moth and others. Oosporein produced by *B. brongniartii* affects enzyme malfunctioning by redox reactions. It is effective against cockchafer larvae, however, its concentration in the environment does not pose any risk to humans and mammals. The hyphomycete, *H. thompsonii* produces an extracellular insecticidal protein hirsutellin A. This toxin is effective against citrus rust mite, mosquito larvae, *Spodoptera frugiperda* and others. (Vey *et al.,* 2001).

FINAL REMARKS

Mycopesticides are found to be suitable in the present biocontrol scenario because their natural interactions with pathogenic pests can be utilized (Deshpande and Tuor, 2001). However, inherent limitations are their slower action as compared to the chemicals and the lack of long term stability of infective propagules in the biocontrol preparation. As chitin is the main structural component of insect and fungal protective coats, to obtain desirable effect with the biocontrol organism, it is essential to understand the process of chitin modification such as linkages with other polymers, deacetylation, and its degradation. Sometimes, chitinolytic enzymes cannot overcome the natural defences like melanization of pathogenic fungi and insects which defends host from the attack of parasitic fungus. Interestingly, the insect pathogen like *M. anisopliae* constitutively produces chitin deacetylase and chitosanase that initiate degradation of cuticle in the presence of melanin too (Nahar *et al.*, 2001). Nevertheless, in integrated pest management, cuticle degrading and mycolytic enzymes can be used in enhancing the inhibitory effect of different conventional pesticides and antibiotics.

ACKNOWLEDGEMENT

I am thankful to Department of Biotechnology, New Delhi, and Swiss Agency for Development and Cooperation, Bern, Switzerland for the financial support for the programmes on mycopesticides.

REFERENCES

Amiri-Besheli B., Khambay B., Cameron S., Deadman M.I. and Butt T.M. (2000). Inter- and intra-specific variation in destruxin production by insect pathogenic *Metarhizium* spp., and its significance to pathogenesis. *Myco.Res.* 104: 447-452.

Bidochka, M.J. and Khachatourians, G.G. (1988). *N*-Acetyl-D-glucosamine-mediated regulation of extracellular protease in the entomopathogenic fungus *Beauveria bassiana, Appl. Environ. Microbiol.* 54: 2699-

Carsolio C., Benhamou N., Haran S., Cortes C., Gutierrez A., Chet A. and Herrera-Estrella A. (1999). Role of *Trichoderma harzianum* endochitinase gene, ech42, in mycoparasitism. *Appl. Environ. Microbiol.* 65: 929-935.

Charnley A.K. (1989). Mechanisms of fungal pathogenesis in insects. In: *Biotechnology of Fungi for improving Plant Growth,* Whips J.M. and Lumsden, B. Eds., Cambridge University Press, Cambridge, 85-117

Deshpande M.V. (1998a). *Novel Biopesticides,* CSIR Golden Jubilee Series, CSIR, New Delhi.

Deshpande M.V. (1998b). Cuticle degrading and mycolytic enzymes with special reference to chitinases as biopesticides: status, potential and challenges. In: *Biotechnology-A Challenge for 2005, Proceedings of the Symposium,* All India Biotech Association, New Delhi, 1-4.

Deshpande M.V. (1999). Mycopesticide production by fermentation: Potential and challenges. *Crit. Rev. Microbiol.* 25: 229-243.

Deshpande M.V. (2000). Mycopesticide production: Contribution of fungal morphologies in the biopesticide formulations. *Nat. Bot. Soc.* 55:11-17.

Deshpande M.V. and Tuor U. (2001) Microbial control of pests: Entomopathogenic fungi as mycoinsecticides. Proceedings of the III Asia Pacific Crop Protection Conference-2001. Pesticides Manufacturers & Formulators Association of India, Mumbai, 56-59.

Elad Y. and Kapat A. (1999). The role of *Trichoderma harzianum* protease in the biocontrol of *Botrytis cinerea. Eur. J. Plant pathol.* 105: 177-189.

Eyal, J., Baker, C.P., Reeder, J.D., Devane, W.E. and Lumsden, R.D. (1997). Large-scale production of chlamydospores of *Gliocladium virens* strain GL-21 in submerged culture-biological control agent production by fermentation, *J. Ind. Microbiol. Biotechnol.* 19:163-

Gillespie A.T. and Moorhouse E.R. (1989). The use of fungi to control pests of agricultural and horticultural importance. In: *Biotechnology of Fungi for improving Plant Growth,* Whips J.M. and Lumsden, B. Eds., Cambridge University Press, Cambridge, 55-84

Goldman, G.H., Hayes, C. and Harman, G.E. (1994). Molecular and cellular biology of biocontrol by *Trichoderma* spp., *Trends Biotechnol.* 12: 478-482.

Govindsamy V., Gunaratna K.R. and Balasubramanian R. (1998). Properties of extracellular chitinase from *Myrothecium verrucaria,* an antagonist to the groundnut rust *Puccinia arachidis. Can J. Plant Pathol.* 20: 62-68.

Heale, J.B., Isaac, J.E. and Chandler, D. (1989). Prospects for strain improvement in entomopathogenic fungi. *Pesticide Sci.* 26: 79-92

Inbar J. and Chet I. (1997). Lectins and biocontrol. *Crit. Rev. Biotechnol.* 17:1-20.

Inglis G.D., Goettel M.S., Butt T.M. and Strasser H. (2001). Use of hyphomycetous fungi for managing insect pests. In: *Fungi as biocontrol agents: progress, problems and potential.* T.M. Butt, C.W. Jackson and N. Magan eds. CAB International pp. 23-69.

Jackson M.A., Mcguire M.R., Lacey L.A. and Wright, S.P. (1997). Liquid culture production of desiccation tolerant blastospores of the bioinsecticidal fungus *Paecilomyces fumosoroseus. Mycol. Res.* 101: 35-41.

Jackson M. A., Whipps J.M. and Lynch J.M. (1991). Nutritional studies of four fungi with disease biocontrol potential, Enz. *Microb. Technol.* 13: 456-461.

Jenkins N.E., Heviefo, G., Langewald J., Cherry A.J. and Lomer C. (1998). Development of mass production technology for aerial conidia for use as mycopesticides. *Biocontrol News Inform.* 19: 21N-31N

Jun Y., Bridge P.D. and Evans H.C. (1991). An integrated approach to the taxonomy of the genus *Verticillium. Journal of General Microbiology* 137: 1437.

Katakura (1998). An insect pest controlling material-composed of moulds e.g. *Metarhizium anisopliae,* an organic base and a porous material, *Biotechnol. Abstr.* 98-10909 E1 A2.

Keller S., Schweizer C., Keller E. and Brenner H. (1997). Control of white grubs (*Melolontha melolontha* L.) by treating adults with the fungus *Beauveria brongniartii. Biocontrol Sci. Technol.* 7: 105-116.

Krauss U., Matthews P., Bidwell R., Hocart M. and Anthony F. (2001). Strain discrimination by fungal antagonists of Colletotrichum musae. Implications for biocontrol of crown rot of banana. *Mycol. Res.* 99:997-1003.

Krauss U. and Soberanis W. (2001). Biocontrol of cocoa pod diseases with mycoparasite mixtures. *Biol. Control* 22: 149-158.

Latge, J.P., and Soper, R.S. (1977). Media suitable for industrial production of *Entomphthora virulenta* zygospores, *Biotech. Bioengg.*, XIX: 1269-

Li, G.Q., Huang, H.C., Kokko, E.G. and Acharya, S.N. (2002). Ultrastructral study of mycoparasitism of *Gliocladium roseum* on *Botrytis cinerea. Bot.Bull. Acad. Sin* 43: 211-218

Lorito, M., Harman, G.E., Hayes, C.K., Broadway, R.M. Tronsmo, A., Woo, S.L. and Di, Pietro A. (1993). Chitinolytic enzymes produced by *Trichoderma harzianum*: antifungal activity of purified endochitinase and chitobiosidase, *Phytopathology* 83: 302- 307.

Mathivanan, N., Kabilan, V. and Murugesan, K. (1998). Purification, characterization, and antifungal activity of chitinase from *Fusarium chlamydosporum,* a mycoparasite to groundnut rust, *Puccinia arachidis. Can. J. Microbiol.* 44: 646-651.

McQuilken, M.P., Whipps, J.M., and Cooke, R.C. (1990). Oospores of the biocontrol agent *Pythium oligandrum* bulk-produced in liquid culture, *Mycol. Res.,* 94: 613.

Martin, F.N., and Hancock, J.G. (1986). Association of chemical and biological factors in soils suppressive to *Pythium ultimum. Phytopathology* 76: 1221-1231.

Milner, R.J. (1987). From chinese medicine to mycoinsecticides: Man's exploration of entomogenous fungi, *Mycologist* (Oct.)147-151.

Mostafa, M.M. (1993). Biological control of *Drechslera teres*:ability of antagonists to reduce conidia formatio, coleoptile infection and leaf infection in barley (*Hordeum vulgare*). *Cryptogamie, Mycologie,* 14, 287, 1993 (From CAB Abstracts, Vol. 6, Brunet, J.A. ed, 1995).

Nahar, P., Patil, A.S., Gooday, G.W., Ghormade, V. and Deshpande, M.V. (2001). The constitutive production of chitin deacetylase and chitosanase activities in *Metarhizium anisopliae*: Possible role in the biocontrol of insect pest. In: *Proc. 3rd International Symposium on Chitin Enzymology,* R.A.A. Muzzarelli, ed. European Chitin Society, Ancona, Italy, 511-519

Patil, R.S., Deshpande, A.M., Natu, A.A, Nahar, P., Chitnis, M., Ghormade, V., Laxman, R.S., Rokade, S. and Deshpande, M.V. (2001). Biocontrol of root infecting plant pathogenic fungus, *Sclerotium rolfsii* using mycolytic enzymes and chitin metabolism inhibitors singly and in combination. *Journal of Biological Control* 15:157-164

Patil, R.S., Ghormade, V. and Deshpande, M.V. (2000). Chitinolytic enzymes: An exploration. *Enz. Microb. Technol.* 26: 473-483.

Schroth, M.N. and Hancock, J.G. (1981). Selected topics in biological control. *Ann.Rev. Microbiol.* 35: 453-476.

Shah, P.A., Aebi, M., and Tuor, U. (1998). Method to immobilize the aphid-pathogenic fungus *Erynia neoaphidis* in an alginate matrix for biocontrol, *Appl.Environ.Microbiol.*, 64: 4260-

Smithson, S.L., Paterson, I.C., Bailey, A.M., Screen, S.E., Hunt, B.A., Cobb, B.D., Cooper, R.M., Charnley, A.K., and Clarkson, J.M. (1995). Cloning and characterization of a gene encoding a cuticle-degrading protease from the insect pathogenic fungus *Metarhizium anisopliae, Gene,* 166: 161-165.

Sreerama Kumar P. and Singh, S.P. (2000). *Hirsutella thompsonii*: The best biological control option for the management of the coconut mite in India. *Indian Coconut J.* (Sept. issue) pp.11-17.

St.Leger R.J., Cooper, R.M. and Charnley, A.K. (1986). Cuticle degrading enzymes of entomopathogenic fungi: Regulation of production of chitinolytic enzymes, *Journal of General Microbiology.* 132: 1509-1517 .

St.Leger R.J., Cooper, R.M. and Charnley, A.K (1987). Production of cuticle degrading enzymes by the entomopathogen *Metarhizium anisopliae* during infection of cuticles from *Calliphora vomitoria* and *Manduca sexta, Journal of General Microbiology.* 133: 1371-1382

St.Leger R.J. and Roberts, D.W. (1997). Engineering improved mycoinsecticides, *Trends in Biotechnology* 15: 83- 85.

Tang, L-C and Hou, R.F. (1998). Potential application of the entomopathogenic fungus, *Nomuraea rileyi*, for control of the corn earworm, *Helicoverpa armigera. Entomol. Expt. Appl.* 88: 25-30.

Thines, E., Eilbert, F., Anke, H. and Sterner, O. (1998). Glisoprenins C, D and E, new inhibitors of appressorium formation in *Magnaporthe grisea* from cultures of *Gliocladium roseum.* 1. Production and biological activities. *J. Antibiotics* (Tokyo) 51: 117-122.

Vey, A., Hoagland, R.E. and Butt, T.M. (2001). Toxic metabolites of fungal biocontrol agents. In: *Fungi as biocontrol agents: progress, problems and potential.* T.M. Butt, C.W. Jackson and N. Magan eds. CAB International. 311-346.

Vyas, P.R. and Deshpande, M.V. (1989). Chitinase production by *Myrothecium verrucaria* and its significance for fungal mycelia degradation. *J. Gen. Appl. Microbiol.* 35: 343-350.

Weindling, R. (1932). *Trichoderma lignorum* as a parasite of other fungi. *Phytopathol.* 22: 837-845.

Zimmermann, G. (1986). The *Galleria* bait method for detection of entomopathogenic fungi in soil. *Journal of Applied Entomology* 102: 213-215.

CHAPTER - 18

ENTOMOPATHOGENIC FUNGI FOR THE MANAGEMENT OF MOSQUITO

P. Vikrant and S. Sandhu

Fungal Biotechnology and Invertebrate Pathology Lab., Department of Biological Sciences, R.D. University, Jabalpur, Madhya Pradesh

INTRODUCTION

Insecticides have been the only control measures for the medically important vectors of malaria, dengue and filaria, but recent concern over the continued use of insecticides, insect resistance and particularly, their environmental impact has necessitated the developing of alternative means of biological control. Development of resistance to chemical insecticides and their deleterious effects on the ecology has led to a concept of replacing or supplementing them with biological control agents. Invertebrate pathology, as a vocation, is a recently organized discipline. Its root, however, can be traced to ancient history with reference to solutions for preventing disease in honeybees and silkworms (Steinhaus, 1956; Lacey *et al.*, 2001). Notable early pioneers in invertebrate pathology, such as, Agostino Bassi, Louis Pasteur and Elie Metchnikoff proposed the application of microorganisms for control of insect pests (Steinhaus, 1956; Lacey *et al.*, 2001). These naturally occurring entomopathogens are important regulatory factors in insect populations (Lacey *et al.*, 2001). Entomopathogenic organisms used for microbial control include bacteria, viruses, fungi, protozoa and nematodes. Many species are employed as biological control agents of vector insects of veterinary and medical importance (Burges, 1981; Tanada and Kaya, 1993; Lacey and Kaya, 2000; Lacey *et al.*, 2001). The comparison of entomopathogens with conventional chemical pesticides is solely from the perspective of their efficacy and their cost. The use of microbial control agents render safety for humans and other non target organisms, reduction of pesticide residues in food, preservation of other natural enemies and increased biodiversity in managed ecosystems (Lacey *et al.*, 2001).

A large number of viruses, bacteria, fungi and protozoa can kill or incapacitate mosquitoes. Some of these have the potential for a significant role in the management and regulation of mosquito species, either as naturally occurring entomopathogens or as applied or introduced mosquitocidal agents. The usefulness of these mosquito pathogens, especially the feasibility of development as microbial insecticides is examined by consideration of their effectiveness, safety and specificity, production and propagation and marketability and profitability. Effectiveness, a major consideration in assessing potential, is influenced by several factors including efficacy, dissemination, persistence, compatibility with other regulatory agents and establishment in the host population and by factors such as humidity, temperature, chemicals and formulation, that directly influence the activity of the pathogen and among the promising biological control agents are the entomopathogenic fungi.

The use of fungus as microbial control agents was experimented by several researches from the late 19th century (Lacey *et al.*, 2001). There are 700 species of entomopathogenic fungi have been reported, but only 10 of these have been or are currently being developed for insect control (Hajek and St. Leger, 1994, Lacey *et al.*, 2001) and fungi infecting mosquitoes have been known since the eighteenth century (Chapman, 1974). Jenkins reported 47 fungal species as pathogens of mosquitoes (Chapman, 1974). A broad range of antibiotic from obligate parasites to opportunistic pathogens that can survive saprophytically in the absence of living hosts; have been documented for the entomopathogenic fungi (Lacey *et al.*, 2001).

Genus *Beauveria* and *Metarrhizium*

Among the Deuteromycetes (Anonymous, 1982; Sandhu *et al.*, 1993), *Beauveria bassiana* and *Metarrhizium anisopliae* have been found to be effective pathogens of mosquito larvae under laboratory conditions (Campbell *et al.*, 1978; Daoust, 1982; Daoust, 1983; Sandhu *et al.*, 1993). Balaraman *et al.* (1979) have isolated the fungal agents from sick/dead larvae of *C. fatigans* collected from cesspits and identified them as *Metarrhizium anisopliae* (Metch.) Sor. and *Beauveria tenella* Vuil. No mortality was observed in controls and they have reported that both the agents caused 100% mortality of larvae of *C. fatigans* and *A. stephensi*. But *A. aegypti* larvae were completely refractory. They found that *Metarrhizium* was most virulent against both mosquito species, in comparison to the other agent. *A. stephensi* was more susceptible to infection with *Beauveria*. LC_{50} dosages were varied for the different agents. When double the LC_{50} dosage was used against different larval instars of *C. fatiqans* and *A. stephensi*, it was found that late instars of *C. fatigans* were more susceptible to infection with both the agents, the earlier larval instars of *A. stephensi* were more susceptible. The mortality induced in the different instars of the two species of mosquitoes by the two fungal pathogens. Mortality was observed in different days, even after metamorphosis of the early instars, and carrying the infection to pupal and adult stages also, indicating trans-stadial transmission of the fungal pathogen. It was possible to culture the fungal agents from the few pupae and adults emerging after the larval instars were exposed to infection. Although most fungi proliferate in habitats with acidic to neutral pH range (Balaraman *et al.*, 1979), the isolation of these fungal pathogens from habitats (cesspits) where the pH ranges between 8 and 10, indicates its potential for use in habitats such as cesspits which breed *C. fatigans* (Balaraman *et al.*, 1979). Larvae of most mosquito species are normally found on water surface with tracheal system open to the atmosphere (e.g., *C. pipiens fatigans*) but some species make extensive use of cutaneous respiration, for example *A. aegypti* (MacFie, 1917; Balaraman *et al.*, 1979). The mode of action of fungi on mosquito larvae was found to involve the blocking the respiratory siphon. The higher susceptibility of late instar *C. furtigans* larvae to fungal infection corroborates earlier findings (Roberts, 1974; Balaraman *et al.*, 1979). There is evidence that in cases where conidia are the infective agents germination is stimulated by unidentified compounds of the exocuticle; conversely some insect cuticles inhibit conidial germination (McInnis, 1975; Balaraman *et al.*, 1979). The variations observed in the susceptibility of different instars of *C. fatigans* to fungal infection in their study may be due to the interaction of the above mentioned stimulatory or inhibitory factors of the cuticle with the process of fungal infection. Also the observed lower susceptibility of late instar *A. stephensi* larvae may be due to quantitative differences in the melanin content of the cuticle between the young and old which had been observed to defend the organism against hyphal penetration at the site of attack (McInnis, 1975; Balaraman *et al.*, 1979). The lower susceptibility of *A. stephensi,* when compared to that of *C. fatigans*, may be due to difference in the time taken for the passage of food through the gut. It has been reported that while it takes 60 min for the food to pass to the posterior end in the case of *Anopheles* spp., it takes only 20-30 min in the case of *C. fatigans* (Schildmacher *et al.*, 1950; Balaraman *et al.*, 1979). When the food stopped moving through the gut, certain protozoa and algae were seen to be digested (Howland, 1956; Balaraman *et al.*, 1979) and therefore it is possible that the fungal spores could have been digested also. All the fungal agents have been found to cause mortality in two vector species, in varying degrees. The infection is carried trans-stadially and the few adults which emerged also contained the pathogens. The few reports of B.bassiana infecting mosquitoes in the field and laboratory are also being summarized. They found that B. *bassiana* was effective against larvae of *Anopheles* and *Culex* but not against larvae of *Aedes* when the agent was applied as a conidial dust to the surface of the water. B.*tenella* was noted in the tree holes in California where from 26-91% of the larvae of *Aedes sierrensis* were infected (Chapman, 1974). First instar larvae were more susceptible, while fourth instar larvae were less susceptible. Mosquitoes are not natural hosts of *M. anisopliae* though infects more than 200 species of insects in nature (Chapman, 1974). Both the spores and destruxins A&B of this pathogen were quite lethal to larvae of *Aedes, Anopheles* and *Culex* and can be easily cultured (Chapman, 1974). These agents are also amenable to mass culture and Mycelia of fungi, after disintegration of dead larvae are capable of living on dead organic matter in the breeding habitat as these agents are saprophytes primarily and can parasitize mosquito larvae, when present. Thus, these agents, once introduced, are self perpetuating in the ecosystem and therefore are potential candidates for biological control of mosquitoes. However, further observations are necessary under field conditions (Balaraman *et al.*, 1979; Sandhu *et al.*, 1993).

Genus *Entomophthora*

It has also been known for many years that the fungus *Entomophthora coronata* is pathogenic to insects (Chapman, 1974). The first report that *E. coronata* is pathogenic to higher animals was made by The fungus causes a chronic granulomatous disease of the nasal mucosa and the lips of horses. Since, the disease has been found in man, it is particularly interesting and it was isolated *E. coronata* from adult mosquitoes, *Culex pipiens quinquefasciatus*. This record is apparently the first of this species occurring in mosquitoes and of this species of mosquito being affected by any species of *Entomophthora*. The infection was first noted in adult mosquitoes maintained in an 18X24X12-ft outdoor screened cage where about 2000 insects were being maintained as a control population for other studies. One of the first signs of abnormality was a drastic decrease in egg production. *Entomophthora culicis* has been reported killing hibernating mosquito adults and *Entomophthora aqutica* was noted in the larvae of *Aedes canadensis* and *Culiseta morsitans*. Also *E. coronata* infected *C. p. quinquefasciatus* and *Aedes taeniorhynchus* in the laboratory and many species of *Entomophthora* can be cultured.

Genus *Pythium* and *Aspergillus*

Pythium sp. causes high mortality in *A. sierrensis* from a tree hole, though in the laboratory motile zoospores were apparently incapable of penetrating undamaged cuticle (Chapman, 1974). *Pythium* has been reported as a wound pathogen of mosquito larvae. The susceptibility of mosquito larvae to entomopathogenic fungi *Aspergillus parasiticus* and two isolates of *Pythium* sp. are presented by Nnakumusana (1985). First instars were most susceptible while fourth instars were least susceptible. Temperatures above 25^{0}C reduced susceptibility of mosquito larvae to infection. No infection was found at 35^{0}C. *Aspergillus* is a ubiquitous and easily recognized genus, which is readily isolated from nearly any organic substrate; it is hardly surprising that many of its species belonging to several sections of the genus should affect insects. The pathogenecity of *Aspergillus* species for insects was discussed (Raper and Fennell, 1985;

Nnakumusana, 1985). Culture filtrates of *Aspergillus flavus* isolated from moribund mosquito larvae contained two chloroform extractable compounds toxic to *Culex* larvae (Roberts, 1980; Nnakumusana, 1985). Younger larvae suffered higher mortality rates while 4th instar larvae suffered least infection. Fresh pupae when exposed to infection did not die, probably because it takes only three days for the pupae to emerge and pupal cuticle is much harder than that of 1st instar larvae. A few pupae resulting from infected 4th instar were found infected and died of obvious mycosis. The others that acquired the infection did not die but carried the infection into adult female mosquitoes. Examination showed that dead larvae turned black when the hyphae invaded the larval haemocoel. Hypha was encapsulated to haemocytes and melanin formation occurred. This would account for the blackening of the dead larvae. Melanization was taken to be evidence that foreign bodies continued to live following encapsulation. Since the hyphal invasion of the haemocoel was not extensive at death, it was concluded that the toxin largely accounted for the death of the larvae. The toxin was found to be temperature resistant. The optimum temperatures required for the germination and growth in culture of Uganda isolate of *Aspergillus parasiticus* were between 20^0 and 25^{0}C. At 30^{0}C the fungus germinated but did not grow. The most favorable temperature for growth and germination of some other insect pathogenic fungi has been found to lie between 20^0 and 30^{0}C (Mc Innis, 1982; Nnakumusana, 1985). Although there is no doubt that *Aspergillus* species can truly be entomopathogenic, many reports of this genus from insects may be due to secondary infections by *Aspergillus* or its rapid establishment after the insect's death. But it has been proved beyond doubt that their isolate of *Aspergillus parasiticus* was truly toxic to mosquito larvae. Infection by *Aspergillus parasiticus* markedly reduced the reproductive capacity of *A. gambiae, C. fatigans* and *A. aegypti* regardless of the larval instars that become infected (Nnakumusana, 1985). *Anopheles* larvae were found highly susceptible to *Aspergillus* sp.; a high incidence of *Aspergillus* sp. was reported in *Culex tritaeniorhynchus summorosus* larvae. *Aspergillus parasiticus* was reported causing high mortality in *Culex gelidus*. A sudden decline in oviposition and high adult mortality of *A. gambiae*, which had been exposed to *Aspergillus* sp. in larval stages, was observed. The pathogenecity of *Aspergillus* species for insects has been discussed. Invasion of the insect is usually through the cuticle, but penetration through the gut and spiracles also occurs. Insect death may result from any of the several factors, including histolysis, mechanical interference with the function of the insect's tracheoles and spiracles, or the production of potent mycotoxins. Therefore it can be concluded that successful introduction of *Aspergillus parasiticus* into natural breeding areas of *A. gambiae, C. fatigans* and *A. aegypti* could reduce their vectorial capacity and substantially reduce malaria transmission rates and transmission rates of arboviruses (Nnakumusana, 1985).

Genus *Lagenidium*

Lagenidium giganteum was reported as not being especially pathogenic in a *Culex*. However, *L. culicidum*, readily infected larvae of all species of *Culex* and *Aedes* exposed to it in the laboratory, and in later tests infected *Anopheline* larvae when the infective stage was maintained at the water surface. Studies to date have shown that *Lagenidium giganteum* is a promising biological control candidate (Chapman, 1974). *Lagenidium* infections have been reported from eggs of *Anopheles* and *Armigeres dentatus* and *Polyscytalum* infections have been seen in the eggs of *Psorophora howardii* (Willoughby, 1969; Jaronski and Axtell, 1983; Nnakumusana, 1985). Laboratory studies with *Lagenidium* sp. showed that this pathogen was very infective to *Culex restuans* in the laboratory and a limited field release indicated that larvae of *C. restuans*, *Anopheles* sp. and *Psorophora* sp. developed infections.

Genus *Fusarium*

Fusarium oxysporum isolated from *Aedes detritus* in the field, produced 80% mortality in the larvae of *A. detritus* and *Culex pipiens pipiens* in the laboratory. *Fusarium* was relatively more efficient than *Beauveria* against *C. fatigans*. (Balaraman *et al.*, 1979)

Genus *Paecilomyces*

Paecilomyces farionosus significantly reduces oviposition in female *Leptinolarsa decemlineata* (Chapman, 1974)

Genus *Coelomomyces*

Coelomomyces is a complex genus of chytridio mycetes-blasto-cladiales, which basically infects mosquito larvae (Nnakumusana, 1985). Certainly the greatest emphasis in research on fungi that attack mosquito larvae has been on species of *Coelomomyces* (Laird, 1967; Chapman, 1974; Balaraman *et al.*, 1979). *Coelomomyces* was first described from the mosquito and occurs principally in mosquitoes; however, it is not restricted to this family since infections have been reported in notonectids, sand flies, black flies and chironomid midges (McCrae, 1972; Taylor *et al.*, 1980). In 1964, Jenkins listed 51 mosquito species (*Aedes, Aedomyia, Anopheles, Armigeres, Culex, Culiseta, Psorophora* and *Uranotaenia*) as hosts of 24 species of *Coelomomyces*. *Coelomomyces* is most common in species of *Anopheles* and then in *Aedes* and *Culex*. Although the fungus is reported from all continents except South America, many records for large areas are based on only a few infected specimens. *Coelomomyces indicus* was found extensively parasitizing A*nopheline* larvae in the rice fields of Pondicherry. The experiment was done to assess the essentiality of paddy field mud for inducing *Coelomomyces* infection. In the present study *Cyclops* had been found to act as the intermediate host of *Coelomomyces indicus*. The infection of *A. subpictus* larvae by *Coelomomyces indicus* only in the presence of paddy field mud may be due to specific microenvironment provided by the mud for the infection process. On the basis of the occurrence of infection in mosquito larvae, the presence of meiospores on the exterior of the *Cyclops* and subsequent development of mycelia and oval bodies (which may be gametes of the fungus) it has been concluded that this *Cyclops* is acting as the intermediate host of *Coelomomyces indicus* in the paddy field ecosystem. Though this fungus has the most ideal characteristics of a biocontrol agent, its fastidious nature for specific microenvironment necessary to cause infection *in vivo* makes it extremely difficult to pursue further studies. Considerable amount of work has been done on *Coelomomyces*, but mosquito control using this fungus was not achieved in field trials. The main disadvantages with this fungus are that they are host-specific and that in some species, an obligate intermediate host for completing its life cycle was found necessary. Also in the habitat where this organism is found naturally parasitizing mosquito larvae, a certain balance between the hosts and parasites was noticed. Considerable difficulty was experienced in mass culturing this genus, rendering it unsuitable for use in a large-scale control program.

Genus *Culicinomyces*

Culicinomyces clavosporus Couch, Romney and Rao, a member of the subdivision Deuteromycotina, class Hyphomycetes, was named and described by Couch *et al.* (1974) as a parasite of larvae of the mosquito *Anopheles quadrimaculatus* Say collected from University lake, Chapel Hill, NC, USA. At approximately the same time, Sweeney *et al.* (1973) working in Australia, discovered and also described a conspecific fungus (Sweeney *et al.*, 1982) from larvae in a laboratory colony of *Anopheles amictus hilli* Woodhill and lee. In this situation, Sweeney *et al.* (1973) felt that the fungus was introduced in water from McCarr's Creek (French Forest, Sidney) that was used in culture of the mosquito larvae. In comparing descriptions of the American and the Australian strain of *C. clavosporus*, the two are similar morphologically in having hyaline, branched hyphae; flask-shaped, one to several times septate primary conidiogenous cells; and hyaline, obclavate, nonseptate conidia. The isolate of *C. clavosporus* (American strain) used in their study was (ATCC No. 38490) which was cultured at 23°C on Emerson's YPSS agar (Difco). In *Culicinomyces clavosporus*, formation of the conidium initial from the conidiogenous cell is similar to that reported to occur in *Trichoderma saturnisporum* (Hammil, 1972) and *Metarrhizium anisopliae* (Hammil and Wang, 1971; Hammil, 1972). Conidia of *Culicinomyces clavosporus* bear some morphological similarity to single-celled conidia of *Metarrhizium anisopliae* (Barnett and Hunter, 1972). Whole mounts of the latter reveal a basal appendage not unlike, but more exaggerated than, the basal protrusion (sub basal peg) of *Culicinomyces clavosporus*. Further, the bilaminar cell wall of *Culicinomyces clavosporus* is structurally similar to the wall described for ascospores of *Ascodesmis* (Moore, 1963) and conidia of *Metarrhizium anisopliae* (Zacharuk, 1970). Mycelium developing in living mosquito and chironomid larvae, filling the body cavity with hyaline, septate, branched hyphae. When the larvae die, hyphae penetrate the cuticle in vast numbers and form conidiophores, which may be straight and unbranched or much elongated up to 0.5 mm and branched; conidia arising from jug-shaped cells each terminating in a sterigma on which the conidium is formed. First conidium apical on end of conidiophore or in regularly spaced whorls and thus verticillate. With aging, conidiophores may form vast numbers of jug-shaped mother cells in penicilloid clusters, which may be terminal or lateral on a conidiophore (Couch *et al.*, 1974). From preliminary studies, the following is a list of mosquito species in addition to *An. quadrimaculatus* to which experimental infections with *Culicinomyces clavosporus* have been transmitted in the laboratory at Chapel Hill, North Carolina. *Aedes atropalpus epactius* Dyar, *Anopheles punctipennis* (Say), *An. stephensi* Liston, *Culex erraticus* (Dyar and Knab), *Culex pipiens quinquefasciatus* Say, *Culex restuans* Theobald, *Culex territans* Walker, *Culiseta melanura* (Coquillett), *Psorophora confinnis* (Lynch arribalzaga), *Uranotaenia sapphirina* (Osten Sacken). Also occurring in mosquitoes in water from a lake near Sydney, Australia (Dr. Lee and associates). Chironomid larvae have been infected with the fungus at Chapel Hill. Our preliminary studies indicate that the host range of *Culicinomyces clavosporus* may be registered to certain families of Nematocera only. Culturable on a wide variety of artificial culture media. In cultures of larvae in which lake water is used, certain Trichomycetes as *Smittium* are frequently found in the hindgut of these larvae. The conidia of *Smittium* are broadest in the middle while in *Culicinomyces* the conidia are broadest at the proximal end, i.e., the end of attachment to the conidiophore. Also, the conidia of *Culicinomyces* lack the appendages found on the proximal end of the conidia of *Smittium*. Furthermore, the conidiophores of the two genera are entirely different in structure and in *Smittium* are formed in the hindgut while in *Culicinomyces* the conidiophores are formed almost entirely on the outside of the larva. Finally, *Smittium* is a commensal with the larva (Lichwardt, 1964; Couch *et al.*, 1974) while *Culicinomyces* is a fatal parasite of its host. Species of *Beauveria* attack the larval stages of a wide range of land-inhabiting insect larva. In fact about 14 species of the genus have been described and named. *Beauveria bassiana* and *Beauveria tenella* (Delacr.) Siem., the former with spherical conidia, the later with oval to ellipsoidal conidia. In all of these the conidia are borne on zigzag conidiogenous axes. In *Culicinomyces* the conidiophores are never zigzag, the mother cells are flask shaped and the conidia are obovate to obclavate. Another important difference is that *Culicinomyces* appears to be confined to aquatic larvae while *Beauveria* attacks only terrestrial larvae. Conidiophore production begins at approximately 80 hr after first exposure to the inoculum at 21°C. Penetration of the host cadaver by conidiophores begins first in the thin-walled anal papillae and progresses anteriorly so that in a few hours (2-6) the entire surface of the larva except for the head is covered with a dense pile of conidiophores.

Nevertheless, entomopathogenic fungi, which are causing debilitating diseases in mosquitoes, can be cultured *in vitro* on organic substrates and can be preserved in the living state so that they could be used for experiments, teaching and use in agriculture, forestry and human health. The intervention of biotechnology has enlightened the germplasm of these fungi as a rich source of biologically active novel compounds. During the last spectacular phase of growth in the demand for collection of entomopathogenic fungal germplasm, it is being witnessed as a direct result of increased interest in biotechnological processes and they are being recognized as a major microbial resource and by this these entomopathogenic fungi can be developed into a potential mycoinsecticide for the management of mosquitoes. The genomic analysis and PCR of the virulent and less virulent strain of the entomopathogenic fungi can be done for the identification, characterization of field isolates of fungi with respect to vegetative or sexual compatibility groups; host specificity and pathogenecity, sub species formation, disease diagnosis, specific metabolic capacities of the strains and for the improvement of the strain. Attempts are being made to develop a hybrid through protoplast fusion of entomopathogenic fungi *Beauveria bassiana* and *Metarrhizium anisopliae*. This hybrid was later found to be very potential and efficient in producing enzyme protease than the wild type strains of *Beauveria bassiana* and *Metarrhizium anisopliae*. Shimizu (1987), Bernier *et al.* (1989), Hasija *et al.* (1990), Gadau and Lingg (1992) and Sandhu and Venkateswerulu (1997) have also done a similar sort of work regarding the strain improvement of these entomopathogenic fungi.

CONCLUSIONS

It is hard to say that the knowledge of the fungal pathogenesis for the microbial control of mosquitoes is still fragmentary. However, from the available data it can be concluded that three genera *Culicinomyces, Coelomomyces* and *Lagenidium* out of other entomopathogenic fungi, have demonstrated potential as mycocontrol agents of mosquitoes. They have the ability to recycle and are efficacious under field conditions. *Lagenidium* and *Culicinomyces* offer the additional advantage of being mass-produced on artificial media. Other genera, like *Metarrhizium, Beauveria* and *Tolypocladium* have also been isolated and tested. *Lagenidium giganteum* is a facultative pathogen of mosquitoes, with both sexual and asexual cycles. Isolations have been made in the United States, Fiji, India, Africa, U.K. and Antarctica. It is also reported that a number of mosquito species are susceptible to infection with *Lagenidium giganteum*. As a facultative parasite it can grow vegetatively either as a parasite of mosquito larvae or as a saprophyte in the aquatic environment where it apparently prefers a littoral habitat. The life cycle is typical of the genus *Lagenidium*, with both asexual and sexual reproduction. The parasitic life is initiated by a laterally biflagellate zoospores, which on contacting a mosquito larva encysts on the cuticle and reach to the haemocoel, apparently aided by the trypsin-like protease. Also the development of molecular biological techniques for entomopathogenic fungi such as *Beauveria bassiana* and *Metarrhizium anisopliae*, which are coupled with cloning of genes encoding putative pathogenecity determinants, will create more potential candidates to manage mosquitoes. Strategies for the use of entomopathogenic organisms for insect control are basically the same as that for the biological control agents. They may be used to augment naturally occurring pathogens, conserved or activated in nature, introduced into pest populations as classical biological control agents to become established and exert long-term regulation of the pest or used inundatively for rapid short-term control.

REFERENCES

Anonymous (1982). 6th report of W.H.O. expert committee on vector biology and control tech. Rep. Services 679, W.H.O., Geneva, pp. 39.

Balaraman K., Bheema Rao U.S. and Rajagopalan P.K., (1979). Isolation of *Metarrhizium anisopliae, Beauveria tenella* and *Fusarium oxysporum* (Deuteromycetes) and their pathogenecity to *Culex fatigans* and *Anopheles stephensi. Indian J. Med. Res.* 70: 718-722.

Barnett H.L. and Hunter B.B. (1972). Illustrated genera of imperfect fungi. 3rd ed. Burgess Publishing Co., Minneapolis.

Bernier L., Cooper R.M, Charnley A.K. and Clarkson J.M. (1989). Transformation of the entomopathogenic fungus *Metarrhizium anisopliae* to benomyl resistance. FEMS Microbiol. Lett. 60: 261-266.

Burges H.D. (Ed.) (1981). Microbial control of pests and plant diseases 1970-1980. Academic Press, London.

Campbell R.K., Perring T.M., Barnes G.L., Eikenbary R.D. and Gentry C.R. 1 (978). Growth and sporulation of *Beauveria bassiana* and *Metarrhizium anisopliae* on media containing various amino acids. *J. Invert. Pathol.* 31: 289-295.

Chapman H.C. (1974). Biological control of mosquito larvae. *Annu. Rev. Entomol.* 19: 33-59.

Couch J.N., Romney S.V. and Rao B., 1974. Anew fungus which attacks mosquitoes and related *Diptera. Mycologia* 66: 374-379.

Daoust R.A. and Roberts D.W. (1982). Virulence of natural and insect-passages strains of *Metarrhizium anisopliae* to mosquito larvae. *J. Invert. Pathol.* 39: 107-117.

Daoust R.A. and Roberts D.W. (1983). Studies on the prolonged storage of *Metarrhizium anisopliae* conidia; effect of temperature and relative humidity on conidial viability and virulence against mosquitoes. *J. Invert. Pathol.* 41: 143-150.

Gadau M.E. and Lingg A.J. (1992). Protoplast fusion of auxotrophic strains of *Beauveria bassiana*. In, *Handbook of Appl. Mycol. Fungal. Biotechnol* 4: 101-128.

Hajek A.E. and St. Leger R.J. (1994). Interactions between fungal pathogens and insect hosts. *Annu. Rev. Entomol.* 39: 293-322.

Hammil T.M. and Wang C.J.K. (1971). Fine structure of phialoconidiation in *Metarrhizium anisopliae* (Hyphomycetes). *Am. J. Bot.* 58: 474.

Hammil T.M. (1972). Electron microscopy of phialoconidiogenesis in *Metarrhizium anisopliae. Am. J. Bot.* 59: 317-326.

Hasija S.K., Yadu V., Gupta V. and Rajak R.C. (1990). Production of protease by *Aspergillus niger* isolated from butchery effluent. *Proc. Nat. Acad. Sci.* India. 60: 417-422.

Howland L.J. (1956). The nutrition of mosquito larvae with special reference to their algal food. *Bull Roy Soc. Nz.* 6: 213.

Jaronski S.T. and Axtell R.C. (1983). effects on organic water pollution on the infectivity of the fungus *Lagenidium giganteum* (Oomycetes: Lagenidiales) for larvae of *Culex quinquefasciatus* (Diptera: Culicidae): Field and laboratory evaluation. *J. Med. Entomol.* 19: 255-262.

Lacey L.A., Frutos R., kaya H.K. and Vails P. (2001). Insect pathogens as biological control agents: Do they have a future? *Biological Control* 21: 230-248.

Lacey Y. and Kaya H.K. (Eds.) (2000). Field manual of techniques in invertebrate pathology: Application and evaluation of pathogens for control of insects and other invertebrate pests. Kluwer Academic, Dordrecht.

Laird M. (1959b). *Ecology.* 40: 206.

Lichwardt R.W. (1964). Axenix culture of two new species of branched Trichomycetes. *Amer. J. Bot.* 51: 836-842.

MacFie J.W.S. (1917). The limitations of kerosene as a larvicide with some observations on the cutaneous respiration of mosquito larvae. *Bull. Ent. Res.* 7: 295.

McCrae A.W.R. (1972). *J. Med. Entomol.* 2: 545.

Mc Innis T. (Jr.) (1975). Biochemical basis of pathogenesis of fungal diseases of vectors, In: *Biological Regulation of Vectors* (Ushdhew Publication No. (NIH) 77-1180), pp. 95.

Moore R.T. (1963). Fine structure of mycota. Electron microscopy of the Discomycete *Ascodesmis. Nova Hedwigia* 5: 263-278.

Nnakumusana E.S. (1985). Laboratory infection of mosquito larvae by entomopathogenic fungi with particular reference to *Aspergillus parasiticus* and its effect on fecundity and longevity of mosquitoes exposed to conidial infections in larval stages. *Current Science* 54: 1221-1228.

Raper K.B. and Fennell D.I. (1965). The genus *Aspergillus*, Williams and Wilkins Co., Baltimore.

Roberts D.W. (1980). Pest management with microbial agents. Recent achievements, deficiencies and innovations. Boyce Thompson Institute, Ithaca, New York, pp. 71.

Roberts D.W. (1974). Fungal infections of mosquitoes. Mosquito Control. (Less Gresser De L' Universite Due Qubect), pp. 144.

Sandhu S.S. and Venkateswerulu G. (1997). A rapid method for the isolation of stable niaD and crnA mutants of entomopathogenic fungi *Beauveria bassiana* and *Metarrhizium anisopliae. Fungal. Genetic Newsletter* USA. 44: 50-51.

Sandhu S.S., Rajak R.C. and Sharma Meeta (1993). Bioactivity of *Beauveria bassiana* and *Metarrhizium anisopliae* as pathogens of *Culex tritaeniorhynchus* and *Aedes aegypti*: effect of instar, dosages and time. *Indian Journal of Microbiology* 33(3): 191-194.

Schildmacher H., Darmkanal and Verdauung (1950). bei S'teckmucken larven. *Biol. Zbl.* 69: 438.

Shimizu S. (1987). Protoplast fusion of insect pathogenic fungi. In: *Biotech of Invert. Pathol and* Cell culture. (K. Maramorosch, Ed.). Academic Press, San Diego, pp. 401-414.

Steinhaus E.A. (1956). Microbial control: The emergence of an idea. *Hilgardia* 26: 107-160.

Sweeney A.W., Couch J.N. and Panter C. (1982). The identity of an Australian isolate of *Culicinomyces. Mycologia* 74: 162-165.

Sweeney A.W., Lee K.J., Panter C. and Burgess L.W. (1973). A fungal pathogen for mosquito larvae with potential as a microbial insecticide. *Search* 4: 344-345.

Tanada Y. and Kaya H.K. (1993). Insect pathology. Academic Press, New York.

Taylor B.W., Harlos J.A. and Braust A.R. 1980. *Can. J. Zool.* 58: 1215.

Willoughby L.G. 1969. *Trans. Brit. Mycol. Soc.* 52: 393.

Zacharuk R.Y. 1970. Fine structure of the fungus *Metarrhizium anisopliae* infecting three species of larval Elateridae (Coleoptera). I. Dormant and germinating conidia. *J. Invertebr. Pathol.* 15: 63-80.

CHAPTER - 19

ROLE OF PLANT PRODUCTS IN ECO-FRIENDLY MANAGEMENT OF PLANT DISEASES

T.S. Thind, H.S. Dhaliwal, J.K. Arora and Chander Mohan
Department of Plant Pathology, Punjab Agricultural University,
Ludhiana-141 004.

INTRODUCTION

Plant diseases caused by various pathogens like fungi, bacteria, viruses, mycoplasmas, nematodes, etc. are responsible for causing significant reduction in normal yield levels of different crops. It is estimated that out of the total loss in yield caused by various pests, diseases alone can cause about 20% reduction in crop yields (Anonymous, 1971). The losses could be much more in case of certain devastating diseases attacking susceptible varieties. Timely and effective management of plant diseases is must for achieving desirable production levels. Among various methods employed for controlling diseases, resistant varieties play an important part in disease management. However, breakdown of resistance due to evolution of new races is a major problem in using these varieties for a longer period and hence breeding for disease resistance is a continouos process. In the absence of suitable resistant varieties chemicals play a significant role in plant disease control notably in those crops, where host resistance is not easily available. However, the continuos and irrational use of these chemicals has resulted into several problems like residues in edible plant parts, resistant strains and environmental pollution etc. In the present day agriculture, a lot of emphasis is being given on using eco-friendly means for controlling diseases and pest problems affecting various crops.

Several plant species serve as store houses for different chemical compounds which have found varied uses in pharmaceutical industry. These compou which are generally secondary metabolites of plant metabolism represent different chemistry depending upon plant species.

Among different kinds of plants, those yielding essential oils are considered to possess strong activity against various pathogens (Picman and Towers, 1984; Saller *et al.*, 1998). These oils generally contain compounds like aliphatic and aromatic aldehydes, terpenoids, alkaloids, flavonoids, phenols, ethers/esters and steroids etc. Some of the plants produce glycosides and protein compounds, aromatic amines, aromatic esters etc. which are also reported to possess considerable level of antimicrobial activity.

Table 1. Major chemical groups and their components produced as secondary metabolites in plants

Main Groups	Components
Aliphatic aldehydes	Cinnamaldehyde, perillaldehyde, citral, citronellal, octanal, nonal, decanal, D-camphor etc.
Aromatic aldehydes	Chalcones, benzaldehyde, p-methylbenzaldehyde, cumina-ldehyde, p-anisaldehyde, vanillin, furfural etc.
Phenols	Carvacrol, eugenol, thymol, p-cresol, p-ethylphenol, p-n-propylphenol, guaiacol, cresol, isoeugenol, salicylaldehyde, tannins, coumaric acid, ferulic acid, oleuropien, caffiec acid, protocatechuic acid, cinnamic acid etc.
Ethers / Esters	Anethole, methyl eugenol, methyl isoeugenol, 1,8-cineole, safrole, L-bornyl acetate, isopinocampheol, pinocamphone etc.
Terpenoids	Cedrelone, azadiradione, limonin, limonol, sesquiterpene lactones, sesquiterpene hydroxyesters, germacranolide, elemanolide, linalool, terpinen-4-ol, alpha-terpineol, gamma-terpinene, alpha-cadinol, geraniol, nerol, alpha-terpinene, spatulenol, T-cadinol, arteannuin B, caryophyllene, gamma-thujaplicin, beta-thujaplicin, nepetalactones, citronellol, menthol, alpha-thujone, beta-thujone, curucmin, limonene, solavetivone, lubimin, lubiminoic acid, aethione, lubiminol, eudesamanolides, atticin, camphor, alpha-pinene, beta-pinene, lyratyl acetate, geranial, myrthenol, myrthenal, neral, caryophyllene oxide, beta-caryophyllene, beta-myrene, farnesol, spathulenol, bornane, camazulene, curcumene etc.
Flavonoids	Flavones, isoflavonoids
Alkaloids	Nemokil, nemokil-S, SDS & SDC, indole alkaloid etc.
Steroids	Pentacosane, taraxerol, beta-sitosterol, yayoisaponins A-C, dioscin, aginoside, sitosterol, eicosanol, friedetin, scopoletin, rutin, quercetin, spermidine etc.
Other hydrocarbons	Nyasol, phenylpropanoids, limonoids, lapachol, beta-lapachone, allivin, arylpropanoids etc.
Glycoside	O-beta-D-glycoside
Protein + compound	Trypsin-chymotrypsin inhibitor, phaesin A, phasein B etc.
Aromatic amines	Pyrrolidine amide, pyrrolidine etc.
Aromatic ester	Benzyl benzoate

The potential of various plant products as anti-microbial agents is determined by routine laboratory methods such as colony growth inhibition on product amended media, spore germination assay, well method, filter paper disc assay etc. For the pathogens which can not be cultured and grown on artificial media, the activity of plant products and extracts is studied on plants or plant parts using techniques such as detached leaf assay, leaf disc assay and on potted plants (Nene and Thapliyal, 1993).

The major compounds identified as inherent sources of antimicrobial substances in different plants are summarised in Table 1.

Several of these plant based products have been found to possess antifungal, antibacterial and antiviral properties and have been tried for controlling various diseases affecting crop plants.

This chapter discusses the work done by various investigators on the potential of these phytochemicals/plant oils/phytoextracts in controlling different patho-gens/diseases of agricultural importance.

ALIPHATIC ALDEHYDES

Cinnamaldehyde is a volatile consituent of many essential oils. It has been reported to be the highest in its antifungal activity against organisms like *Aspergillus nidulans, Penicillium frequentans* and *P. cyclopium,* among different aliphatic aldehydes evaluated by Kurita *et al.* (1981). The antifungal effects of perillaldehyde and citral were somewhat weaker than that of cinnamaldehyde but still were found fairly potent. In contrast, the antifungal effects of citronellal, octanal, nonal and decanal were very poor. The citral is present in orange and other essential oils and is formed by oxidation of geraniol. Hussian *et al.* (2000) recorded that lemon-grass (*Cymbopogon jwarancusa*) citral rich oil at 0.1 per cent oil conc. (v/v) completely inhibited the radial growth of *Fusarium solani, Rhizoctonia solani, Pythium spinosum, Curvularia lunata, Phoma sorghina* and *Sclerotium rolfsii* while 86.6 per cent inhibition was recorded in *Alternaria alternata* at the same concentration. On the other hand, lemon grass piperitone rich oil at 0.1 per cent concentration completely inhibited radial growth of *C. lunata* and showed high level of efficacy against *F. solani, P. sorghina* and *S. rolfsii.* It exhibited poor growth inhibition against *R. solani* and *P. spinosum.*

AROMATIC ALDEHYDES

Aromatic aldehyde on oxidation give rise to various acids e.g. benzoic acid, cinnamic acid, salicylic acid, coumaric acid, caffeic acid etc. Popova *et al.* (2001) isolated two chalcones in propolis from El Salvador which showed antifungal activity and moderate toxity to *Artemia salina nauplii.* Garg and Siddiqui (1992) observed that the pure isolate and 1:100 and 1:200 dilutions of cuminaldehyde, isolated from seed essential oil of *Cuminum cyminum,* showed very good activity against *Aspergillus nidulans, Aspergillus niger, Rhizopus nodosus, Colletotrichum capsici,*

Alternaria alternata, Pestalotia psidii, Fusarium moniliformae and *Penicillium expansum.* 1,8-cineole isolated from fruit oil of of *Luvunga scandens* showed very good activity against *A. alternata* and *F. moniliformae.*

Benzaldehyde is an oily substance, which is a component of the glucoside, amygdalin. Kurita *et al.* (1981) observed that the antifungal activity of benzaldehyde was very weak or nil except against *Blastomyces dermatitidis.* Cuminaldehyde (p-isopropylbenzaldehyde) was the highest in antifungal activity among the aromatic aldehydes examined (benzaldehyde, p-methylbenzaldehyde, cuminaldehyde, p-anisaldehyde, vanillin and furfural) and its antifungal effect was fairly potent. Furfural or hydroxymethyl furfural is formed when amino acids react with reducing sugar and Vanillin, another aromatic aldehyde is a component of various glycosides. Present in large amount in vanilla fruits. It is formed during enzymatic oxidation of coniferol in the wood of oak-barrels.

PHENOLS

Phenols and aromatic alcohols are oxidized by enzymes to form aldehydes and acids, some of which are constituents of glycosides, essential oils and tannins. These tannins are compounds which are esters of hydroxyaromatic carboxylic acids or they are condensed tannins which are not esters and whose nuclei are linked by carbon atoms. Guaiacol is also one of the phenols having antifungal activity and occurs in plants both as free and as methyl ether guaiacol. It is readily oxidized to give dark coloured materials. Thymol and phenol present in *Lantana camara* and *Ocimum basilicum* are reported to inhibit the growth of many fungi and bacteria (Anonymous, 1975, 1976). Eugenol isolated from leaves of *Ocimum sanctum* exhibited very strong activity against *Absidia glauca, Aspergillus nidulans, Aspergillus niger, Colletotrichum capsici, Fusarium moniliformae, Pestalotia psidii* and *Rhizopus nodosus* (Garg and Siddiqui, 1992). Hitokoto *et al.* (1980) demonstrated that thymol inhibited toxigenesis by *Aspergillus versicolor, A. ochraceus* and *A. flavus.* Thompson (1989) observed that the overall degree of fungitoxic efficiency of the essential oil components for *Rhizopus, Mucor* and *Aspergillus* was 91 for carvacrol, 25 for eugenol and 0 for anethole, cinnamaldehyde and safrole, respectively.

Nor-Azah *et al.* (2002) observed that leaf oil of *Cinnamomum aureofulvum* was predominantly made of benzyl benzoate, benzyl alcohol and linalool whereas, bark oil contained cinnamaldehyde, linalool and eugenol as major constituents. The bark oil was more effective than leaf oil (and their activity was mainly due to their components) against *Candida* species and *Staphylococcus* species. Mehmood *et al.* (1997) observed that cinnamon oil was the most effective against *Aspergillus.* Likewise, palmerosa was found most effective against *Fusarium* species and lemon grass against *Cladosporium* species. The most active constituents were eugenol, citral and geraniol. Tsao and Zhou (2000) observed that carvacrol and thymol completely prevented spore germination and mycelial growth of *Botrytis cinerea* and *Monilinia fructicola* at 100 μg/ml. Other constituents (eugenol, citronellol, geraniol, citral, citronellal and menthol) exhibited higher inhibition against *M. fructicola.*

Khan and Zubairy (2001) recorded that thymol had strong antifungal activity against *Candida albicans* while camphor exhibited weak fungitoxity. Misra *et al.* (2000) found eugenol and benzaldehyde to be the constituents of the essential oil obtained from the bark of *Cinnamomum zeylanicum.* While tested against *Stachybotrys chartarum* the minimum inhibitory concentration of the oil was observed to be 400 ppm. Al-Mughrabi *et al.* (2001) obtained six phenolic compounds namely coumaric acid, ferulic acid, oleuropein, caffeic acid, protocatechuic acid and cinnamic acid from *Olea europaea* (olive). They reported that these compounds showed strongest activity against *Fusarium oxysporum* and *Verticillium* sp. No effect was observed against *Alternaria* sp. This cinnamic acid is a constituent of the gummy exudations known as balsams and protocatechuic acid occurs in the free state in grape leaves and in some other fruits and has both phenolic and acidic groups which can react together to form ester like compounds known as depsides. When tested against *Aspergillus parasiticus,* cinnamic aldehyde and eugenol caused inhibitory effect at 150ppm and 125 ppm respectively. Since cinnamic aldehyde and eugenol are the respective major components of cinnamon and clove oils, it was concluded that these are the major active antifungal ingredients of these oils (Bullerman *et al.*, 1977).

Khumbar and Dewang (2001) observed that synthetic alkyl derivatives of thymol showed significantly improved antifungal activity over the acyl derivatives against the test fungi namely *Aspergillus niger, A. oryzae, Fusarium oxysporum* and *Alternaria alternata.* The enhanced antifungal activity of synthetic derivatives of thymol indicates that the fungicidal potency of these compounds is related to the presence of certain functional groups *i.e.* an aromatic ring, an olefinic bond or a carboxylic group. Inouye *et al.* (2001) observed that *Thymus linearis* contained thymol and carvacrol as major constituents of its essential oil which inhibited the fungal growth of *Candida ablicans, Aspergillus fumigatus* and *Trichphyton mentagrophytes* by vapour action. Vaughn and Spencer (1994) recorded that salicylaldehyde was most inhibitory compound against *Fusarium sambucinum* when applied as a volatile form. Cinnamaldehyde, salicylaldehyde and thymol prevented complete fungal growth when incorporated in media at 0.1% (v/v) level. Bhargava *et al.* (1998) observed that polyphenolic complex (phenols, flavones, tannins and glycosides) extracted from *Acacia nilotica* bark exhibited considerable antifungal activity.

ETHERS / ESTERS

Rihakova *et al.* (2001) recorded that monolauroylglycerol at concentration level between 0.2 and 1.8 m mol l^{-1} caused the highest inhibition of spore outgrowths whereas 6-o-lauroysucrose at concentration between 0.05 and 1.8 m mol l^{-1} did not exhibit inhibition of spore germination of *Aspergillus niger*.

Letessier *et al.* (2001) recorded that various components of hyssop oil, obtained from *Hyssopus officinalis*, (L-bornyl acetate, isopinocampheol and pinocamphone) reduced growth of *Pyrenophora avenae* and any mixture containing isopinocampheol completely inhibited fungal growth. Hyssop oil reduced germination of *Botrytis fabae* conidia and uredospores of *Uromyces viciae fabae*. Its effect against barley powdery mildew and apple powdery mildew were variable.

TERPENOIDS

Terpenoids and their oxygenated derivatives are major components in essential oils produced by several plant species and account for their antimicrobial activity. They are aliphatic or cyclic hydrocarbons which usually have general formula $C_{10}H_{16}$. The basic structural unit of aliphatic terpenes is isoprene. There are also present some sesquiterpenes having antimicrobial activity in many plant species. These sesquiterpenes have empirical formula $C_{15}H_{24}$. There are both aliphatic and cyclic sesquiterpenes. Nerolidol, an aliphatic sesquite-rpene, occurs in essential oil of orange. Many higher plants contain substances toxic to micro-organisms. These compounds are called phytonicides. Both onion and garlic bulbs contain such compounds which are active against bacteria. The active substance allicin, isolated from garlic, is an oily liquid in pure state and decomposes on storing. It is soluble in alcohol and ether but is not readily soluble in water. Some plants have compounds which impart them resistance against different pathogens e.g. pea plants (*Pisum sativum*) contain pisatin which is responsible for their resistance to fungal diseases. Tellez *et al.* (2001) found that hexane fraction of tarbrush (*Flourensia cernua*) contained mostly monoterpenoids, while ethanol fraction volatiles were primarily sesquiterpenoids. Application of as little as 1.0 μg of the essential oil from the hexane fraction was sufficient to provide visible antifungal activity.

Some other workers have also recorded antifungal activity due to terpenoids of various plants species which are listed in Table 2.

FLAVONOIDS

Bhadauria and Kumar (1999) observed that the aqueous and methanolic extracts and free flavonoids obtained from *Catharanthus roseus* and *Lawsonia alba* were not effective against *Candida albicans*. However, bound flavonoids from *L. alba* showed high fungal growth inhibition. Flavonoids and isoflavonoids present in the extract of heart wood of *Platymiscium yucatanum* were observed to exhibit good antifungal against *Lenzites trabea* and *Coriolus versicolor* by Reyes-Chilpa *et al.* (1998). Maximo *et al.* (2002) tested three new isoflavonoids, ulexin C(1), ulexin D(2) and 7-O-methylisolupalbigenin (from dichloromethane extract of *Ulex airensis*) against *Cladosporium cucumerinum* and reported their antifungal activity.

Table 2. Plant species having terpenoids with antimicrobial activity.

Plant species	Source (plant part)	Type of terpenoids	Target pathogen	References
Solanum aethiopicum	Roots	Solavetivone and lubimin	*Fusarium oxysporum* f.sp. *melongenae*, *Verticillium dahliae*, *Ralstonia solanacearum*	Watanabe *et al.*, 2001
Solanum aethiopicum	Roots	Solavetivone, lubimin, lubiminoic acid, aethione and lubiminol	*Fusarium oxysporum* and *Verticillium dahliae*	Nagaoka *et al.*, 2001
Citrus sinensis	Exocarpic part	Limonene	*Ceratocystis paradoxa, Fusarium moniliformae* and *Curvularia lunata*	Rao *et al.*, 2000
Centaurea thessala sub sp. *drakiensis, C. attica* sub sp. *attica, C. achaia*	Aerial parts	Eudesamanolides, eudesmane (atticin), sesquiterpene lactone	Antifungal activity	Skaltsa *et al.*, 2000 a, b
Chrysanthemum coronarium	Bicoloured flowerheads	Camphor, α-pinene, β-pinene, lyratyl acetate	*Alternaria* sp., *Aspergillus flavus, Pythium ultimum*	Alvarez-Castellans *et al.*, 2001
Lippia alba	Leaf	Geranial, myrthenol, myrthenal, neral, geraniol, caryophyllene oxide, β-caryophyllene, citronellol, linalool, β-myrene, farnesol and spathulenol	*Ustilago scitaminea, Colletotrichum falcatum* and *Curvularia lunata*	Singh *et al.*, 2000a
Artemisia princeps	Leaves	Bornane, camazulene	*Fusarium gramin earum, Gaeumannomyces graminis, Gerlachia nivalis, Helminthosporium sativum, Rhizoctonia cerealis, Phytophthora capsici*	Liu *et al.*, 2001
Cinnamomum camphora	Seeds	Linalool		
Zingiber officinale	Rhizomes	Curcumene	*Rhizoctonia solani*	Agarwal *et al.*, 2001
Toona ciliata	Leaves	Cedrelone	Groundnut rust pathogen	Govindachari *et al.*, 2000
Azadirachta indica	Leaves and seeds	Azadiradione		
Citrus medica		Limonin, limonol, nomilinic acid		

Continue....

Plant species	Source (plant part)	Type of terpenoids	Target pathogen	References
Homolomena aromatica	Rhizome	Linalool, terpinin-4-ol, α-terpineol, gamma-terpinene, α-cadinol, geraniol, nerol, α-terpinene, spatulenol, T-cadinol	*Curvularia pallescens, Aspergillus niger, Fusarium graminearum*	Singh *et al.*, 2000b
Artemisia annua	Leaves	Arteannuin B	*Gaeumannomyces graminis, Rhizoctonia cerealis, Gerlachia nivalis, Verticillium dahliae*	Tang *et al.*, 2000
Lantana indica	Roots	Terpenoids	Antifungal activity	Verma *et al.*, 1998
Litsea cubeba	Fruits	Citral	*Fusarium moniliformae, F. solani, Alternaria alternata, Aspergillus niger*	Gogoi *et al.*, 1997
Salvia pomifera	Aerial parts	α and β-thujone	*Rhizoctonia solani, Sclerotinia sclerotiorum*	Pitarokili *et al.*, 2002
Curcuma aromatica	Rhizome	Curucumin	*Staphylococcus aureus, Saccharomyces cerevisiae*	Singh *et al.*, 2000C
Orange	Peel	D-limonene	*Zygosaccharomyces major*	Murdock and Allen, 1960

Flavonoids present in the extracts of *Pongamia glabra* were found responsible for the inhibition of spore germination and mycelial growth of *Alternaria solani, Macrophomina phaseolina* and *Fusarium udum* (Pan *et al.*, 1985). Flavonoids and protoanemonin compounds in the leaf extracts from *Clemantis goyrisna* inhibited the growth of *Alternaria tenuissima* and *Fusarium nivale* (Misra and Dixit, 1977).

ALKALOIDS

Alkaloids are nitrogenous heterocyclic compounds and are usually present in plants as salts of malic, tartaric, citric or other acids and are water soluble. Some alkaloids like indole alkaloid are found in ergot. Ergot is wintering form of the fungus *Claviceps purpurea* which develops on rye. Lysergic or iso-lysergic acid, are indole derivatives. Kariba *et al.* (2001) observed that alkaloids obtained from *Schizozygia coffaeoides* were fungitoxic to *Trichophyton mentagrophytes, Microsporum gypseum, Cladosporium cucumerinum* and *Candida albicans*. Inamori *et al.* (1999) reported that gamma-thujaplicin isolated from *Thujopsis dolabrata* showed a strong antifungal activity against *Penicillium expansum*,

Aspergillus niger and *Fusarium oxysporum*. Essential oils of *Nepeta grandiflora, N. nuda* and *N. sibirica* were found to contain nepetalactones and showed a high activity against *Bacillus subtilis* and *Escherichia coli* (Tropnikova *et al.*, 1999).

Grohs *et al.* (1999) observed that red cedar (*Thuja plicata*) heart wood extract and hiba (*Thujopsis dolabrata* var. *hondae*) oil contain beta-thujaplicin but the effect of the hiba oil against the wood inhabiting moulds was much more pronounced. Four neem based products, namely Nemokil, Nemokil-S, SDS and SDC were observed to possess antifungal activity against the guava wilt pathogen *Fusarium solani* and the anthracnose pathogen *Pestalotia psydii* by Khan *et al.* (1998). Similarly, Singh *et al.* (2000d) observed that an indole alkaloid isolated from *Alstonia venenata*, in aqueous acetic acid solution, inhibited spore germination of *Fusarium udum, Alternaria brassicicola, Ustilago cynodontis, Aspergillus flavus* and *Erysiphe pisi*.

STEROIDS

Steroids are fatty acid esters of sterols, which are high molecular weight cyclic alcohol derivatives of cyclo-pentano-perhydro-phenanthrene. Sitosterols ($C_{29}H_{49}OH$) are one of the commonest groups of plant sterols and differ from ergosterol in containing only one double bond and having a methyl group in the side chain replaced by an ethyl group. Saponins, are steroids which are amorphous, water soluble, poisonous substances and form soapy, opalescent solutions with a strong foamy action. Poisonous properties of corncockle seed is due to saponins. On hydrolysis saponins generate glucose, galactose, arabinose and methyl pentoses in addition to aglycone. Rutin is widely distributed in plants and has quercetin as its aglycone and is found in many plants e.g. tobacco, buck wheat. This quercetin is colouring material of onion skin and also occurs in pollen of plants such as maize, hops and tea. Walters *et al.* (2001) reported that out of three spermidines isolated from pollens of *Quercus alba*, di-p-carmaroyl-caffeoylspermidine and tri-p-coumaroyl-spermidine reduced mycelial growth of the oat leaf stripe pathogen, *Pyrenophora avenae* and reduced powdery mildew (*Erysiphe graminis* f.sp. *hordei*) infection of barley seedlings when applied as post-inoculation treatment. When used as a pre-inoculation treatment, only di-p-coumaroyl-caffeoylspermidine reduced powdery mildew infection significantly. Torrenegra *et al.* (2000) isolated sitosterol, eicosenol, friedetin and one prenylated coumarin from the petrol extract of *Pentacalia corymbosa* leaves. Scopoletin and rutin were isolated from the chloroform fraction of the ethanol extract and quercetin was isolated from the ethyl acetate fraction of the same extract. The coumarins scopoletin and 7-O-geranylscopoletin were responsible for the antifungal activity against *Fusarium oxysporum, Botrytis cinerea* and *Trichoderma viride*.

Jain *et al.* (1998) observed that ethanolic extract of *Kirganelia reticulata* exhibited antiviral effect against potato virus X. Phytochemical investigation of its stem bark revealed presence of pentacosane, 21-

alpha-hydroxyfriedelan-3-one, taraxerol, lup-20(29)-ene-3beta,24-diol, beta-sitosterol and beta-sitosterol-beta-D-glucoside. Three new steroidal saponins namely yayoisaponins A-C along with dioscin and aginoside were obtained from bulbs of a mutant *Allium ampeloprasum* (Sata *et al.*, 1998). They were found to possess antifungal activity against *Mortierella ramanniana* applied at 10μg/disk.

OTHER HYDROCARBONS

Several other hydrocarbons present in various plant species have also been reported to possess antimicrobial activity by different workers and they are listed in Table 3.

Table 3. Plant species having hydrocarbons with antimicrobial activity

Plant species	Source (plant part)	Type of hydrocarbon	Active ingredient	Target pathogen	References
Garlic (round cloved)	bulbs	—	Allivin	*Botrytis cinerea, Mycosphaerella arachidicola Physalospora piricola*	Wang and Ng, 2001
Styrax ferrugineus	leaves	Arypropanoids	—	Antibacterial and antifungal activity	Pauletti *et al.*, 2000
Gordonia dassaneyakei	Stem bark	—	3-formyl-2,4-dihydroxy-6-methylbenzoic acid, 3-hydroxy-4-(methylcar bonyl)-2,5-dime thyl phenyl ester	*Curvularia* spp., *Colletotrichum gloeosporioides, Rhizoctonia solani, Corynespora cassiicola, Fusarium* sp.	Athuko-ralage *et al.*, 2001
Khaya senegalensis	Seeds	Limonoids	Mexicanolide	Groundnut rust pathogen.	Govinda-chari *et al.*, 1999 a
Swietenia mahagoni		Limonoids	Swietenolide	Groundnut rust pathogen	
Azadirachta indica	Uncrushed leaves	Limonoids		Groundnut rust pathogen	Suresh *et al.*, 1997
Anemarrhena asphodelordes	Rhizomes	—	Nyasol	37 strains of fungi and 4 strains of bacteria	Tida *et al.*, 1999
Malmea depressa	Stem bark	Phenylpropa-noids	2,3,4,5-tetram-ethoxybenzaldeh yde, 2,3,4,5-tetramethoxycin namyl alcohol	*Trichophyton mentagrophytes, Fusarium oxysporum*	Arellanes *et al.*, 1996
Tabebula chrysantha, Oroxylum indicum, Fernandoa adenophylla, Jacaranda filicifolia	—	Dichlorometha ne extracts	Lapachol, β-lapachone	Mycelium and conidia in filamentous fungi	Ali *et al.*, 1998

GLYCOSIDES, PROTEIN COMPOUNDS AND AROMATIC AMINES

When a monosaccharide is combined through its glycosidic hydroxyl with an alcohol, a glycoside is formed. Glycosides are responsible for the taste and aroma of some food products from plants and are also used for medicinal purposes. Chouksey and Srivastava (2001) recorded that O-beta-D-glycoside isolated from *Terminalia arjuna* root extracts had good antifungal activity against *Aspergillus niger.* Pyrrolidine and benzyl benzoate both are aromatic amines. Pyrrolidine is oxidized to give rise to corresponding aldehyde. Benzyl and cinnamyl alcohols occur as acetates, benzoates and other esters in essential oils from jasmine and in some balsams e.g. white Peruvian balsam. Salicin occurs in the bark of poplar and willow.

13- Kda protein, designated as ginkbilobin, was isolated by Wang and Ng (2000) from *Ginkgo biloba* seeds and found that this protein exhibited potent antifungal activity against *Botrytis cinerea, Mycosphaerella arachidicola, Fusarium oxysporum, Rhizoctonia solani* and *Coprinus comatus* with EC_{50} values of 0.25, 6.5, 3.6, 8.7 and 3.4 μM, respectively. Ye *et al.* (2001) observed that broad bean trypsin-chymotrypsin inhibitor exerted a prominent suppressive action on mycelial growth in *Fusarium oxysporum* and *Mycospharella arachidicola* and a less notable inhibitory effect on *Botrytis cinerea.* Ye and Ng (2002) isolated two antifungal proteins phasein A and phasein B from the seeds of the pinto bean (*Phaseolus vulgaris* cv. *pinto*) and observed that phasein A was potent antifungal against *Fusarium oxysporum* and *Physalospora piricola.* Phasein B was more potent than phasein A against *P. piricola* but less effective than phasein A toward *F. oxysporum* and *Rhizoctonia solani*

Alecio *et al.* (1998) extracted pyrrolidine amide from leaves of *Piper hispidum* which showed good antifungal activity against *Cladosporium sphaerospermum.*

PLANT OILS

Plants are known to produce several types of oil with various constituents. Essential oils are mostly insoluble in water but soluble in organic solvents. The essential oils and resins are formed and secreted by special plant organs: oils mostly by glandular hairs and resins in resin ducts. They have definite aromas which are responsible for the smell of many plants. Several of these have been reported to possess good level of antimicrobial activity. *Cinnamomum zeylanicum* oil was reported as a potent fungitoxicant against the storage fungi of paddy and rice at a concentration of 0.4×10^3 μl/l (Tiwari *et al.*, 1994). Tiwari (1997) observed volatile fungitoxicity of the oil of *Cinnamomum zeylanicum* against *Aspergillus niger* and *Aspergillus flavus* (storage fungi of paddy and maize grains) at a very low concentration of 0.4×10^3 μl/l. Fungitoxicity of essential oil from *Tagetes erecta* leaves has been reported against *Pythium aphanidermatum* at a concentration of 2000 µg/ml (Kishore and Dwivedi, 1991). Kobaisy *et al.* (2001) observed that the essential oil of kenaf (*Hibiscus cannabinus*) cv. Everglades proved to be antifungal against *Colletotrichum fragariae, C. gloeosporioides* and *C. acutatum*. Behura *et al.* (2000) observed that leaf oils isolated from *Curcuma longa* showed the maximum inhibition against *Rhizoctonia solani* (86.66%), followed by *Helminthosporium oryzae* (68.00%) and *Fusarium moniliforme* (66.66%). An equal degree of inhibition (57.89%) was observed against *Alternaria padwickii* and *Curvularia lunata*. The rhizome oil showed maximum inhibition against *R. solani* (81.48%), followed by *H. oryzae* (68.00%), *A. padwickii* (55.50%) and *C. lunata* (55.50%). The lowest degree of inhibition was observed against *F. moniliforme* (2.43%). Farrukh *et al.* (2001) recorded that clove oil showed maximum antifungal activity against *Aspergillus niger, Alternaria alternata* and *Fusarium chlamydosporum* followed by oil of peppermint and eucalyptus.

Oils of *Cymbopogon citratus* and *Caesulia axillaris* were observed to be fungistatic in nature when tested against *Aspergillus flavus*, the dominant storage fungus (Dubey *et al.*, 2000). Fumigation of wheat samples with *Caesulia* and *Cymbopogon* oils at MICs indicated their *in vivo* applicability as herbal fumigants. Rajappan *et al.* (1999) reported that oil based dust formulation of neem and pungam (*Pongamia pinnata*) oil were more effective than their cake based dust formulations. Neem oil based EC formulation and oil based dust formulations of neem and pungam can be used to control powdery mildew of green gram. Singh and Handique (1997) observed that essential oil from the leaves of *Hyptis suaveolens* showed significant antifungal activity against *Rhizoctonia solani, Sclerotium rolfsii* and *Sclerotinia sclerotiorum*. The oil appeared to inhibit mycelial growth but did not destroy the viability of the tested fungi. Essential oils of oregano, thyme and sassafras were found effective against test bacterium (*Vibrio parahaemolyticus*) at 100µg/ml by Beuchat (1976). Benjilali *et al.* (1984) tested six essential oils against 13 species from the genus *Penicillium* and nine from *Aspergillus* and observed that the essential oil from thyme was the most effective, followed by those from mugwort, rosemary and eucalyptus. Fungitoxic effects of essential oils from *Callistemon lanceolatus* were studied against 6 fungal pathogens of rice and minimum inhibitory concentration of spore germination was reported to be 300 µg/ml against *Magnaporthe grisea* (Misra *et al.,* 1997). The essential oil of lemon grass (*Cymbopogon citratus*) has been reported to inhibit growth of moulds like *Aspergillus flavus, A. fumigatus, Macrophomina phaseolina* and *Penicillium chrysogenum* by Adegoke and Odesola (1996) in Nigeria. Shahi *et al.* (1997) studied fungitoxic activity of essential oils extracted from four *Eucalyptus* spp. They reported that oil of *E. citriodora* was most effective and completely inhibited the spore germination of *Penicillium italicum* at 1000 µg/ml conc. The oils of *E. dalrympleana* and *E. pauciflora* were less effective and showed fungistatic toxicity against the test fungus at 3000 μg/ml but the oil of *E. laveopinea* showed only partial inhibition at this concentration. Considerable efficacy of essential oils extracted from the leaves of *Chenopodium ambrosioides, Ocimum canum* and from seeds of *Anethum graveolens* and *Pimpinella anisum* has been observed against mycelial growth of sugarcane pathogens *Colletotrichum falcatum, Fusarium moniliformae, Ceratocystis paradoxa* and *Rhizoctonia solani* (Chaudhary *et al.,* 1995). Likewise, olive oil and rapeseed oil were found to significantly reduce the incidence of powdery mildew (*Sphaerotheca fuliginea*) on squash in New Zealand (Cheah *et al.*, 1995).

Mahapatra and Tewari (1994) observed that essential oil extract of *Ocimum sanctum* gave best control of collar rot and yellow root diseases of groundnut caused by *Aspergillus niger* and *A. flavus,* respectively. Using *in vitro* tests, Antonov *et al.* (1997) assessed 21 oil extracts for their effects on conidial germination, germ tube elongation and mycelial growth in *Botrytis cinerea.* Maximum reduction in conidial germination and germ tube length was achieved at 0.1 per cent concentration with oil extracts of *Thymus capitatus* (thyme) and *Syzygium aromaticum* (clove). Fungicidal activity of *Thymus capitatus* essential oil has also been demonstrated against *Penicillium italicum* causing fruit rot in orange in Italy (Arras *et al.,* 1997). Montes and Carvajal (1998) used 11 plant essential oils for maize kernel protection against *Aspergillus flavus* and found that essential oils of *Cinnamomum zeylanicum, Mentha piperita, Ocimum basilicum, Origanum vulgare, Teloxys ambrosioides, Syzygium aromaticum* and *Thymus vulgaris* caused a total inhibition of fungal development on maize kernels. No phytotoxic effect on germination and corn growth was detected with any of these oils. Effects of some formulations of *Salvia officinalis* essential oil have been studied on *Botrytis cinerea* infection in tomato plants. Protection by the essential oil against *B. cinerea* infection was comparable to that obtained with the fungicide dichlofluanid (Moretti *et al.,* 1998). Rao and Narayanasamy (1990) reported neem oil to be significantly superior over the other oils tested viz. pungam and illuppai at all intervals of inoculations (1, 24, 48 and 72 hr). The per cent reduction in infection was 83.3, 75 and 6.6 in neem, pugam and illippai oils, respectively, at 1 hr after inoculation and the per cent reduction in infcction slowly decreased to 45.4, 27.3 and 27.3 in neem, pingam and illuppai oils, respectively, at 72 hr after inoculation.

Khan and Zubairy (2001) recorded that eucalyptus oil had strong antifungal activity against *Candida albicans.* Al-Mughrabi *et al.* (2001) reported that when tested against *Aspergillus parasiticus*, cinnamon and clove oils were inhibitory at 200-250ppm.

Kishore *et al* (1996) reported that clocimum oil and lemon grass oil gave complete inhibition of conidial germination of powdery mildew of *Papaver somniferum* at lowest concentration 250 μg/ml. Singh and Dwivedi (1987) observed that all the oils namely *Eucalyptus globulus, Ocimum canum, Citrus medica* var. *limonum, Citrus sinensis* and *Pinus roxburghii* were toxic against *Sclerotium rolfsii* at higher concentration

(3000 and 4000 ppm). Oils from *E. globulus* and *O. canum* were most effective. They reduced colony diameter, dry weight of hyphal mat and sclerotial number even at 2000 ppm and checked the pathogen completely at 4000 ppm. Sclerotial viability decreased significantly at lower concentration (1000 and 2000 ppm) and at 4000 ppm no sclerotia were viable in the treatment with oils from *E. globulus* and *O. canum*. Reduction in sclerotial germination of 85, 80 and 75% was observed with oils from *C. medica* variety *limonum, C. sinensis* and *P. roxburghii*.

The essential oil of *Cyperus scariousus* proved to be effective against *Rhizoctonia solani, Alternaria brassicae, Penicillium digitatum* and *Phytophthora infestans*. When *Cymbopogon citratus* oil was tested, it proved effective against *R. solani* and *P. digitatum*. The oil of *Mentha piperita* observed to possess good fungitoxic effect against *R. solani, A. brassicae* and *P. digitatum* and the essential oil of *Allium sativum* proved effective against *R. solani* and *P. infestans* (Dhaliwal, 2001; Dhaliwal *et al.*, 2003a; Dhaliwal *et al.*, 2003b). Likewise, in another study Dhaliwal *et al.* (2002) reported that generally all the ten essential oils tested against *Uncinula necator* proved to be fungitoxic but the oils of *Allium sativum* was more effective of the all followed by oils of *Cyperus scariousus* and *Cuminum cyminum*. Oils of *Cymbopogon citratus* and *Eucalyptus camaldulensis* were next in order of efficacy.

PLANT EXTRACTS

Organic solvent based or aqueous extracts of various plant parts of different species have been reported to exhibit antimicrobial activity by various workers both under laboratory and field conditions. A good amount of information is available pertaining to antibacterial, antifungal and antiviral properties of these plant extracts. Chitra and Kannabiran (2001) observed antifungal activity of fruit and flower extracts of *Datura innoxia* against *Colletotrichum capsici* under *in vitro* conditions. Eksteen *et al.* (2001) recorded that the crude extract of *Eucomis autumnalis* showed significant antifungal activity against seven fungal pathogens namely *Botrytis cinerea, Fusarium oxysporum, Sclerotium rolfsii, Rhizoctonia solani, Verticillium dahliae, Botryosphaeria dothidea* and *Pythium ultimum*. The crude extract of *Schrebera alata* came out second. It was reported by Karavaev *et al.* (2001) that extracts from the leaves of medicinal comfrey (*Symphytum officinale*) and cowparsnip (*Heracleum sibiricum*) strongly inhibited the germination of *Erysiphe graminis* conidia and uredospores of *Puccinia graminis*. Erdemoglu and Sener (2001) observed that ethanol extract of *Taxus baccata* heart wood @ 2mg/ml showed significant antimicrobial activity against gram negative bacteria (*Pseudomonas pseudomalli, Salmonella typhi* and *Enterobacter cloacca*) but no activity against gram positive bacteria (*Staphylococcus aureus* and *Streptococcus pyogenes*). The extract used @ 400μg/ml exhibited good antifungal activity against *Nigrospora oryzae, Microsporum canis, Epidermophyton floccosum, Curvularia lunata* and *Pleuralus astreatus*. Out of all the plant extracts tested by Lal *et al.* (1998) against *Alternaria* leaf blight of pigeon pea, the highest antifungal activity was given by leaf extract of *Achyranthes* sp. (61.19%) closely followed by *Azadirachta* sp. (55.58%). However, inhibition of mycelial growth in mentha, lawsonia, codiaeum and ocimum ranged from 39.47 to 48.67%. Under field conditions, a foliar spray of *Achyranthes* gave 40.1% protection against the disease followed by spray of *Azadirachta* leaf extract (37.97%). Hidalgo and Fernandez (1996) recorded that *Swietenia mahagoni* @ 1 and 3% was effective against *Alternaria solani* and at 3% against *Fusarium oxysporum* and *Gibberella fujikuroi* var. *subglutinans*. Similar response was observed with *Annona glabra* @ 3%. *Citrus reticulata* @ 1 and 3%, *Parthenium hysterophorus* and *Annona muricata* @ 3% were effective against *G. fujikuroi* var. *subglutinans. Ricinus communis* and *Leucaena leucocephala* inhibited growth of *Alternaria solani* and *F. oxysporum* f. sp. *lycopersici* by more than 40%.

Antifungal activity of *Azadirachta indica* leaf extract on wheat seeds was recorded by Khan and Shah (1992). Treatment caused a marked reduction in seed mycoflora as well as an enhancement of seed germination. Ganguly (1994) observed that extract of *Vinca rosea* showed maximum inhibition of mycelial growth and spore germination of both the test fungi, namely *Pyricularia oryzae* and *Helminthosporium oryzae,* followed by *Polyalthia longifolia* and *Azadirachta indica.* Extracts of *Datura metel* had lowest antifungal activity. Maximum inhibition of brinjal damping-off pathogen, *Pythium aphanidermatum,* was obtained with cold water extracts of *Polyalthia longifolia* (56.6%). *Desmodium purpureum* and *Magnifera indica* did not show any inhibitory effects (Bhat *et al.,* 1994). Among the hot water extracts, *Parthenium hysterophorus* gave maximum mean growth inhibition of 48.3% with *D. purpureum* being the least effective (2.7%). Cold water extract of *Caesalpinia pauciflora* was highly inhibitory (85.5%) followed by *Manilkara kauki* (78.8%). Hot water extract of *Eucalyptus microtheca* showed 90% inhibition. The aqueous extracts of *Psidium guajava, Strychnos potatorum, Thuja orientalis* and *Lagerstroemia parviflora* exhibited good fungitoxic activity against six pathogens of rice namely *Pyricularia oryzae, Rhizoctonia solani, Fusarium moniliformae, Curvularia lunata, Aspergillus niger* and *A. flavus* (Mishra *et al.,* 1992). Garcia and Garcia (1988) recorded that crude extracts of garlic (*Allium sativum*) was found to be effective against *Aspergillus flavus* and *A. parasiticus.* The remaining plant extracts which showed antifungal activity in decreasing order were *Plumeria acuminata* > *Ageratum conyzoides* > *Phyllanthus niruri* >*Terminalia catappa* > *Gliricidia sepium.*

Yamada *et al.* (1990) observed that aqueous extracts of barley and wheat seeds gave no inhibition of any of several phytopathogenic fungi and bacteria tested *in vitro*, but induced resistance in primary leaves of barley and wheat treated with their respective seed extracts and challenged by a compatible race of *Erysiphe graminis.* Methanol extracts of pretreated leaves also showed good antifungal activity. Awuah (1999) observed that clove powder proved to be most effective in reducing fungal colonization of stored peanuts. Lime fruit peel oil and ocimum leaf powder proved to be moderately effective. Yin and Tsao (1999) recorded that the combination of acetic acid plus *Allium* plant extracts was an effective way to inhibit fungal growth of *Aspergillus niger, A. flavus* and *A. fumigatus*. Govindachari *et al.* (1999 b) observed that hexane extract of neem leaves and its fractions completely inhibited the growth of *Fusarium oxysporum* and *Colletotrichum lindemuthianum* when used @ 400μg/cm^2. Extracts of neem leaf powder were found to possess antifungal properties against *Cochliobolus miyabeanus* and *Glomerella tucumanensis* by Pan and Deb (1997). At the highest conc. tested (1000 ppm), turmeric rhizome powder gave encouraging results against *Fusarium udum* and *C. miyabeanus*. *Acacia nilotica* bark powder had strong antifungal activity against *C. miyabeanus* at 1000 ppm. *Moringa oleifera* leaf, seed and root powder had some antifungal properties against *C. miyabeanus*. *In vitro* studies of extracts of *Terminalia chebula* (leaf, bark and dry fruits), *Punica granatum* (leaf, bark and fruit rind), *Delonix regia* (bark), *Psidium guajava* (leaf and bark), *Emblica officinalis* (bark), *Syzygium cumini* (leaf and bark), *Eucalyptus* sp. (leaf) and *Caesalpinia digyna* (leaf) demonstrated antifungal activity against *Trichophyton tonsurans, T. rubrum, T. simii, Trichosporon beigelii, Microsporum fulvum* and *M. gypseum* as reported by Dutta *et al.* (1998).

Deraniyagala *et al.* (1998) observed that the methylene chloride : hexane fractions (from the extract of leaves of *Ficus racemosa*) particularly 1:1 fraction, exhibited antifungal activity against *Cladosporium cladosporioides*. The 1:1 fraction inhibited the growth of *Curvularia* sp., *Colletotrichum gloeosporioides, Alternaria* sp., *Corynespora cassiicola* and *Fusarium* sp. Kumar *et al.* (1997) observed that *Naragamia alata* extract showed antifungal activity against *Aspergillus niger*. Antibacterial activity was high in *Ixora coccinea* and *Tiliacora acuminata*. The growth of *Fusarium, Verticillium, Rhizoctonia* and *Phytophthora* spp. was significantly inhibited by extracts of *Allium globosum* (Li and Ma, 2001). *A. oreoprasum* also significantly inhibited growth of *Fusarium* and *Verticillium* species. Lakshmanan and Mohan (1989) recorded that water extracts of *Allium sativum, Bougainvillea spectabilis* and *Azadirachta indica* significantly inhibited mycelial growth and sclerotial germination of *Thanatephorus cucumeris in vitro*. Under green house conditions, all the plant extracts significantly reduced the seedling mortality caused by this pathogen. *Azadirachta indica* leaf powder was observed to exhibit antifungal activity against *Helminthosporium oryzae* and *Colletotrichum falcatum* by Pan and Deb (1997). *Curcuma longa* rhizome powder gave encouraging results against *Fusarium udum* and *H. oryzae*. *Acacia nilotica* bark powder and *Moringa oleifera* leaf, seed and root powder were also found to have some antifungal properties against *H.oryzae*. Pandey *et al.* (1993) recorded that whole plant or leaf extract of *Trigonella foenum graceum* completely inhibited the mycelial growth of *Colletotrichum falcatum, C. palloscens* and *Fusarium moniliformae* at 4000 ppm. Aqueous leaf extracts of *Azadirachta indica, Ocimum sanctum* and *Riccinus communis* were found to be most effective in controlling several root diseases of banana (Singh *et al.*, 1993). Kolte and Shinde (1973) observed that the leaf, stem and root extracts of urid and mung adversely affected the growth and sclerotial production by *Macrophomina phaseoli*.

In another study, Shekhawat and Prasada (1971) and Misra and Dixit (1976) reported efficacy of crude extract of *Allium sativum* against the mycelial growth of a number of phytopathogenic fungi. Its components when separated by ethyl acetate were found to inhibit the growth of aflatoxin producing fungi, *Aspergillus flavus* and *A. parasiticus* (Kshemkalyani *et al.* 1990). Bisht and Khulbe (1995) recorded that maximum inhibition of brown leaf spot pathogen of rice was caused by extracts from *Juglans regia* followed by *Allium sativum, Origanum vulgare* and *Allium wallichii.* Efficacy of *Allium sativum* extract and of others viz. *Prosopis juliflora, Gynandropsis pentaphylla, Leucos aspera* and *Vitex negundo* have also been reported against *Rhizoctonia solani* (Sundarraj *et al.* 1996). Daya and Ram (1997) studied *in vitro* fungitoxicity of leaf extracts of *Cassia tora, Azadirachta indica, Anisomeles ovata, Aegle marmelos, Adhatoda vasica, Mentha arvensis, Dalbergia sissoo, Tinospora cordifolia, Pongamia pinnata, Cyperus rotundus, Ocimum adscendens, O. sanctum,* resin extract of *Ferula foetida* and bulb extracts of *Allium sativum* and *A. cepa.* They reported that the leaf extract of *O. sanctum* was found to be most effective and completely inhibited spore germination of *Alternaria brassicae* at 10000 μg/ml. It was followed by *A. sativum* and *A. cepa* in order of efficacy. Sharma and Nanda (2000) recorded that water extract of garlic was effective in rendering the teliospore of *Nevossia indica* unviable but it could not penetrate the pericarp of seed. Bhore *et al.* (1995) used 4 natural plant extracts against two pathogens and recorded that the crude extract of *Callistemon lanceolatus* was effective against *Aspergillus flavus* and its acetone extract was effective against *Curvularia lunata.* Hidalgo and Fernandez (1996) tested extracts from different parts of 7 plant species against *Alternaria solani, Fusarium oxysporum* f.sp. *lycopersici, F. subglutinans* and *Colletotrichum gloeosporioides.* They reported that extracts of *Parthenium hysterophorus* and *Annona muricata* @ 3 per cent were effective against *F. subglutinans. Ricinus communis* and *Leucaena leucocephala* inhibited *A. solani* and *F. oxysporum* f.sp. *lycopersici* by more than 40 per cent. Sarvamangala *et al.* (1993) reported that leaf extract of *Eucalyptus* sp. inhibited conidial germination of *Cercospora moricola* by 91.5 per cent.

Nine plant extracts namely *Azadirachta indica, Bougainvillea spectabilis, Cataranthus roseus, Eucalyptus globulus, Nerium odorum, Ocimum sanctum, Parthenium hysterophorus, Prosopis juliflora* and *Vitex negundo* were tested by Abraham and Prakasam (2000) against *Fusarium solani, Cladosporium oxysporum* and *Geotrichum candidum* from carrot roots. *Azadirachta indica, O. sanctum* and *V. negundo* were found to be more inhibitory to the growth of all the three pathogens. Bansal and Gupta (2000) tested 7 leaf extracts namely *Azadirachta indica, Atropa belladona, Calotropis procera, Ocimum bacillicum, Eucalyptus amygdalina,*

Ailanthus excelsa, Lantana camara against *Fusarium oxysporum* and recorded that *Azadirachta indica* completely inhibited the mycelial growth and spore germination followed by *O. bacillicum* and *L. camara. Atropa belladona* and *E. amygdalina* were least effective. Jagadeeswari and Prakasam (2000) used 7 plant extracts namely *Azadirachta indica, Bougainvillea spectabilis, Catharanthus roseus, Eucalyptus globulus, Ocimum sanctum, Prosopis juliflora* and *Vitex negundo* against *Alternaria alternata* and *Fusarium semitectum.* They reported that extracts of *P. juliflora, O. sanctum* and *V. negundo* were effective against *Alternaria alternata* and inhibited the mycelial growth by 70.5, 73.7 and 69.8 per cent, respectively, while that of *C. roseus* and *V. negundo* @ 10 per cent concentration inhibited the growth of *F. semitectum* by 92.3 and 92 per cent, respectively. Verma *et al.* (1997) recorded that in laboratory tests, plant extracts from *Clerodendron fragrans* were shown to have inhibitory effect on *Fusarium oxysporum* f.sp. *udum.* Leaf extracts of *Clerodendron viscosum, Lantana camara, Vitex negundo, Citrus aurantiifolia* and fenugreek were tested for their fungicidal effect on *Rhizoctonia solani* by Shylaja *et al.* (1997) who found that growth and sclerotial production of *R. solani* were significantly inhibited by leaf extracts of *C. viscosum, L. camara* and *V. negundo. Citrus aurantiifolia* and fenugreek extracts were not effective. Pandey and Pant (1997) used seed extracts of *Coriandrum sativum, Cumin cyminum, Foeniculum vulgare* and *Trachysperumum ammi* for screening against *Pythium aphanidermatum, Macrophomina phaseolina* and *Rhizoctonia solani.* The essential oils from these seeds exhibited strong fungitoxicity and completely controlled mycelial growth of all the test fungi, even at very low concentrations. These oils were not phytotoxic to seed germination, seedling growth, general health and morphology of *Abelmoschus esculentus.* Plant extract of *Cyperus rotundus* has been reported to cause inhibition of *Rhizoctonia solani* (Kotasthane and Lakpale, 1994). Antifungal activities of extracts of 16 plants were determined by Damayanti *et al.* (1996) against *Ceratocystis paradoxa.* They found that *Xanthium strumarium* extract was the most effective followed by garlic extract, *Meriandra bengalensis, Mentha piperita, Curcuma longa, Phlogacanthus thyrsiflorus, Toona ciliata, Vitex negundo, Azadirachta indica, Eupatorium birmanicum, Ocimum sanctum* and *Leucas aspera.* Kumbhar *et al.* (2000) used extracts of four plants namely *Ipomea cornea, Capsicum annum, Allium sativum and Azadirachta indica against Aspergillus niger* and recorded that organic solvent extract of *C. annum* and water extract of *I. cornea* were having higher antifungal activities.

Leaf extracts (20 per cent) of *Datura stramonium, Azadirachta indica, Lantana camara, Ocimum sanctum* and fruit extracts (20 per cent) of *Diospyros cordifolia* were found to be highly fungitoxic to *Fusarium solani* inhibiting mycelial growth in petriplates by 75.04, 73.80, 72.71, 72.64 and 65.91 per cent respectively (Rawal and Thakore, 2000). Sharma and Jarial (2000) recorded that growth of *Agaricus bisporus* and *A. bitorquis* was almost equal or slightly better in medium amended with various botanicals as compared to control except water hyacinth which slightly inhibited the growth of *A. bitorquis*. In case of *Diehliomyces microsporus* minimum growth was in case of water hyacinth (28.0 mm) followed by neem cake (30.8 mm), neem leaves (35.0 mm) and *Eucalyptus* leaves (72.8 mm) as compared to 90.0 mm growth in control. Sarvamangala *et al.* (1993) reported that under field conditions leaf extract of Eucalyptus reduced leaf spot incidence, caused by *Cercospora moricola*, by 63.6 per cent. *In vivo* use of stem distillates from *Ocimum gratissimum* and *Cymbopogon citratus* were observed to inhibit *Phytophthora palmivora* and checked black pod lesion development on 75 per cent of the infection courts in detached cocoa pods as reported by Awuah (1994). In other study Daya *et al.* (1994) found that green leaves of *Aegle marmelos*, *Azadirachta indica* and *Cymbopogon flexuosus* at 2 and 5 per cent concentrations reduced both pre and post emergence collar rot of chickpea caused by *Sclerotium rolfsii*. Saniewska (1998) tested garlic extract successfully for control of rust caused by *Puccinia antirrhini* on snapdragon (*Antirrhinum majus*). Chauhan and Singh (1991) observed that volatiles from extracts of garlic and onion exhibited marked effect on germination of zoospore of *Phytophthora drechsleri* f.sp. *cajani*. However, garlic extract was more effective than onion. Both extracts completely inhibited the germination at 5000 and 10000 ppm respectively. Complete inhibition of zoospore germination was not observed in other extracts even at 50000 ppm. However, the germination percentages were reduced to 16.01, 61.51 and 69.28 in ginger, basil and tulsi, respectively, as compared to 76.44% in control. Thiribhuvanamala and Narsimhan (1998) took aqueous leaf extracts (10%) from 22 plant species belonging to 17 families and screened against three seed borne pathogens of sunflower viz., *Alternaria helianthi*, *Macrophomina phaseolina* and *Fusarium solani*. The results revealed that the leaf extracts from *Delonix regia*, *Pongamia glabra* and *Acacia nilotica* could significantly inhibit the spore germination, mycelial growth and spore production of all the three seed borne pathogens. Vozenilkova (1992) reported that growing brassicas in pots inoculated with *Plasmodiophora brassicae*, in which *Mentha piperita* had been grown for the previous 2 years, provided good control of club root.

The extracts from both matured and young neem leaves were found to be effective in reducing the rice yellow dwarf infection by 33.12 per cent over control when inoculated at 1 hour after application and the reduction was significantly greater than in the rest of the treatments (Rao and Narayanasamy, 1990). The per cent reduction in infection decreased from 50 to 18 in neem leaf extracts as the interval between application of extracts and inoculations increased. Meena and Mariappan (1993) recorded that among neem (*Azadirachta indica*), *Mentha arvensis*, *Aegle marmelos*, *Catheranthus roseus*, *Lantana camera*, *Pongamia pinnata*, *Vitex negundo*, and *Nerium odorum*, leaf extracts and flower extracts of *C. roseus* were found to inhibit the mycelial growth and spore germination of seed borne mycoflora of sorghum seeds viz., *Alternaria tenius*,

Aspergillus flavus, Curvularia lunata, Fusarium moniliformae and *Rhizopus stolonifer.* The extracts of neem, *C. roseus* and *L.camara* were more effective than the rest of the plant extracts. Eswaramurthy *et al.* (1989) reported that leaf extracts of neem, *Agave americana, Ipomoea cornea, Peltophorum ferrugineum, Prosopis uliflora, Acacia arabica* and *Delonix regia* inhibited the mycelial growth of *Sarocladium oryzae* and *Fusarium oxysporum* f.sp. *cepae.* Parimelazhagan and Francis (1999) observed that leaf extract of *Clerodendrum viscosum* completely checked the radial growth of the test fungus *Curvularia lunata.* A gradual decrease in growth was observed in *C. lunata* grown in increasing concentrations of the leaf extract. Mycelial dry weight of the test pathogen was reduced in varying proportions after treatment with leaf extracts. 1:10 dilution of leaf extract was the most effective for the inhibition of the test fungus in terms of its growth.

COMMERCIAL FORMULATIONS

In the recent years some commercial formulations of neem and other plants have been introduced for use against plant diseases under field conditions. Various workers have reported encouraging results with these products.

Neem-based

Bhattacharya and Pramanik (1998) observed that nimin at 2ml/100g urea reduced the disease to a great extent (48.76%) as compared to control. The trial indicated that both neem cake and nimin were effective in reducing the disease of clubroot pathogen (*Plasmodiophora brassicae*) under field conditions. They also observed that very few and small clubs were formed with 2% neem oil suspension treatment. It has also been observed by these scientists that nimin when mixed with urea and applied in field helps in releasing the nitrogen slowly as well as bio active agent against club root pathogen. Azadirachtin from neem caused lysis of cell walls of resting spores present in sick soil and stimulation of fungal antagonists in soil may be an indirect effect. Hulloli *et al.* (1998). used 5 neem based products namely Agricare (Agricare, Bangalore, containing neem extract in fortified base, amino nitrogen, protien moiety, polysaccharide, Mg, Zn and Mo), Plantolyte (Agricare, Bangalore, containing neem extract, amino acids, Zn, Mn, Fe, Cu, Mo, B, Mg, Ca and S), Neemark (West-Coast Rasayan International Pvt., Bombay), RD-9 (azadirachtin, ITC Ltd., Hyderabad), and an alcoholic extract of neem kernal/seed (Division of Agricultural Chemicals, IARI, New Delhi). They observed that both Plantolyte and Agricare inhibited the growth of *Xanthomonas axonopodis* pv. *malvacearum* at 1.0 µg/ml. While in case of Neemark and alcohol extract of neem seeds the MIC was 20 and 10 µg/ml respectively. RD-9 was ineffective even at 3000 µg/ml. The efficacy of Plantolyte and Agricare (MIC of both was 1.0 µg/ml) was compared by disc diffusion and cup method and in both cases Plantolyte was more effective. They also observed that the MIC of certain mixtures was less at lower concentrations than the MIC of individual products indicating that neem formulations (plantolyte and agricare) have synergistic activity when combined with antibiotics and fungicides. Rajappan *et al.* (2000) tested two neem based EC formulations, namely, NO 60 EC (A), NO 60 EC (C) and one neem (*Azadirachta indica*) oil + pungam (*Pongamia glabra*) oil based EC formulation, NO + PO 60 EC (C) developed by TNAU, Coimbatore. They observed that foliar spray with NO 60 EC (A) resulted in 47.4 and 38.8 per cent bacterial blight of french bean incidence in 1998 and 1999 trials, respectively, and were at par with NO 60 EC (C) spray. However, among the three formulations NO 60 EC (C) recorded the highest yield of 3915 and 3810 kg/ha in 1998 and 1999 trials, respectively. A series of neem based products viz. Triact 70 and Trilogy 70 were developed by Thermo Trilogy Corp. in United States which were found effective as preventive treatments to protect against powdery mildew and rusts diseases (Locke, 2002). Similarly, according to Quarles (2002) a neem oil formulation called Trilogy has been approved by the EPA for use on foods, while Rose Defence and Triact (for control of powdery mildew, rust, black spot, *Botrytis*, downy mildew and other common diseases) are designed for use on ornamentals.

Giant Knotweed-based

Daayf *et al.* (1995) observed that Milsana (an aqueous formulation from leaves of *Reynoutria sachalinensis* when applied weekly at a concentration of 2%, provided control of powdery mildew on long english cucumber that was as effective as benomyl. Fruit yield was not affected by the treatment, even though repeated applications of Milsana induced a greener and glossier coloration of the leaves, which became brittle to the touch. A rapid and distinct accumulation of fungitoxic phenolic compounds occurred in leaves treated with Milsana, especially in infected leaves. Glycosidically bound phenolics, hydrolyzed into their free form and displayed antifungal activity in different forms and intensity among the treatments. Milsana appeared to induce a rapid and more abundant accumulation of these compounds.

Mint-based

According to Quarles (2002) a commercial formulation of mint oil (Funga-Stop) is available to help control soil borne pathogens.

FUTURE OUTLOOK

The use of synthetic chemicals in agriculture for the control of plant diseases has posed serious problems of environmental pollution residues and non target side effects on other microbes added by the substantial increase in their developmental costs and time consuming process. There is drastic decline in the introduction of new fungicides in the recent years. Natural products of plant or microbial origin are a vast source of bio-active substances which have been exploited only to a limited extent as alternatives to chemicals for disease control in practise. The advent of powerful analytical tools such as MS, NMR, HPLC, etc. has facilitated identification of important bio-active principles isolated from natural plant products. Based on their structure activity relation many bio-active analogues have been synthesised which has demonstrated good activity, better in certain cases against different plant pathogens. Bio-degradable trait of these compounds makes their use preferred over synthesised compounds. Several commercial plant based products have been introduced in the recent past, which have demonstrated good performance in the field. Although the levels of disease control achieved by these products are not comparable with the synthetic chemicals yet their use has several advantages keeping into view their eco-friendly traits. It is warranted to search for more moieties in the plant derivatives to supplement the resting arsenal of disease controlling agents. These products can serve as good tools in the integrated management of plant diseases.

REFERENCES

Abraham S. and Prakasam V. (2000). Efficacy of botanicals on post harvest pathogens of carrot. *J.Mycol. Pl. Pathol.* 30: 257.

Adegoke G.O. and Odesola B.A. (1996). Storage of maize and cowpea and inhibition of microbial agents of iodeterioration using the powder and essential oil of lemon grass (*Cymbopogon citratus*). *Int. Journ. Biodeterioration and Biodegradation.*37: 81-84.

Agarwal M., Walia S., Dhingra S. and Khambay B.P.S. (2001). Insect growth inhibition antifeedant and antifungal activity of compounds isolated/ derived from *Zingiber officinale* Roscoe (ginger) rhizomes. *Pest Management Sci.* 57: 289-300.

Alecio A.C., Silva B.V.da., Young M.C.M., Kato M.J. and Furlan M. (1998). Antifungal amide from leaves of *Piper hispidum. J.Natural Products* 61: 637-639.

Ali R.M., Houghton P.J. and Hoo T.S. (1998). Antifungal activity of some Bignoniaceae found in Malaysia. *Phytotherapy res.* 12: 331-334.

Al-Mughrabi K.I., Aburjai T.A., Anfoka G.H. and Shahrour W. (2001). Antifungal activity of olive cake extracts. *Phytopathologia Mediterranea* 40: 240-244.

Alvarez- Castellans P.P., Bishop C.D. and Pascual-villalobos M.J. (2001). Antifungal activity of the essential oil of flower heads of garland chrysanthemum (*Chrysanthemum coronarium*) against agricultural pathogens. *Phytochem.* 57: 99-102.

Anonymous (1971). Economics of pesticides use. In: *Pesticides Market Studies* Vol III. pp 87. Pesticides Association of India New Delhi.

Anonymous (1976). The Wealth of India. Raw Materials Vol X. Publication and Information Directorate CSIR New Delhi 591 pp.

Antonov A, Stewart A, Walter M and Callaghan M (1997) Inhibition of conidium germination and mycelial growth of *Botrytis cinerea* by natural products.

Proc 50th New Zealand Plant Protection Confer pp. 159-64, Lincoln University, Canterbury.

Arellanes, A.J., Mata, R., Lotina, H.B., Lang, A.L.A. and Ibarra, L.V. (1996). Phytogrowth inhibitory compounds from *Malmea depressa. J. Natural Products* 59: 202-204.

Arras, G., Arru, S. and Marceddu, S. (1997). *Thymus capitatus* essential oil and carvacrol in post harvest control of *Penicillium italicum* on orange fruits under vacuum conditions. *Agricoltura Mediterranea* 127: 92-96.

Athukoralage, P.S., Herath, H.M.T.B., Deraniyagala, S.A., Wijesundera, R.L.C. and Weerasinghe, P.A. (2001). Antifungal constituent from *Gordonia dassanayakei. Fitoterapia* 72: 565-567.

Awuah, R.T. (1994). *In vivo* use of extracts from *Ocimum gratissium* and *Cymbopogon citratus* against *Phytophthora palmivora* causing black pod disease of cocoa. *Ann. Appl. Biol.* 124: 173-178.

Awuah, R.T. (1999). Inhibition of fungal colonization of stored peanut kernels with products from some medicinal/culinary plants. *Peanut science* 26: 13-17.

Bansal, R.K. and Gupta, R.K. (2000). Evaluation of plant extracts against *Fusarium oxysporum* wilt pathogen of fenugreek. *Indian Phytopath.* 53: 107-108.

Behura, C., Ray, P., Ratu, C.C., Mishra, R.K., Ramachandraiah, O.S. and Charyulu, J.K. (2000). Antifungal activity of essential oils of *Curcuma longa* against five rice pathogens *in vitro. J. essential oil bearing plants* 3: 79-84.

Benjilali, B., Elaraki, A.T., Ayadi, A. and Mohamad, I. (1984). Method to study antimicrobial effects of essential oils : Application to the antifungal activity of six Moroccan essences. *J. fd. protection* 47: 748-752.

Beuchat, L.R. (1976). Sensitivity of *Vibrio parahaemolyticus* to spices and organic acids. *J. fd. Sci* 41: 899-902.

Bhadauria, S. and Kumar, P. (1999). Effect of plant extracts of medicinal plants against *Candida albicans*. Flora and Fauna Jhansi 5: 95-96.

Bhargava, A., Srivastava, A. and Kumbhare, V. (1998). Antifungal activity of polyphenolic complex of *Acacia nilotica* bark. *Indian forester* 124: 292-298.

Bhat, N., Sivaprakakasam, M.K. and Jeyarajan, R. (1994). Antifungal activity of some plant extracts. *Indian J. forestry* 17: 10-14.

Bhattacharya, I and Pramanik, M. (1998). Effect of different antagonist rhizobacteria and Neem products on club root of crucifers. *Indian Phytopath.* 51: 87-90.

Bhore, S.T., Thakore, C.S. and Kadam, V.C. (1995). Efficacy of some plant extracts and commercial plant products against test fungi. *J.Maharashtra Agric. Univ.* 20: 251-254.

Bisht, G.S. and Khulbe, R.D. (1995). *In vitro* efficacy of leaf extracts of certain indigenous medicinal plants against brown leaf spot pathogen of rice. *Indian Phytopath.* 48: 480-482.

Bullerman, L.B., Lieu, F.Y. and Seier, A.S. (1977). Inhibition of growth and aflatoxin production by cinnamon and clove oils cinnamic aldehyde and eugenol. *J. fd. Sci.* 42: 1107-1109.

Chaudhary R.K., Rao G.P. and Pandey A.K. (1995). Fungitoxicity of some essential oils against sugarcane pathogens. *J. Living World* 2: 30-37.

Chauhan V.B. and Singh U.P. (1991). Effect of volatiles of some plant extracts on germination of zoospores of *Phytophthora drechsleri* f.sp. *cajani. Indian phytopath.* 44: 197-200.

Cheah L.H., Cox J.K. and Popay A.J. (1995). Screening of plant extracts for control of powdery mildew in squash. Proc. 48th New Zealand Plant Protection Confer. Pp 340-342 Angus Inn Hastings New Zealand.

Chitra H. and Kannabiran B. (2001). Antifungal effect of *Datura innoxia* on the anthracnose fungus *Colletotrichum capsici in vitro. Adv. Plant Sci.* 14: 317-320.

Chouksey B.K. and Srivastava S.K. (2001). New constituent from the roots of *Terminalia arjuna* : antifungal agent. *Indian J. Chem.- Section-B-organic-including-Medicinal* 40:354-356.

Daayf F., Schmitt A. and Belanger R.R. (1995). The effects of plant extracts of *Reynoutria sachalinensis* on powdery mildew development and leaf physiology of long english cucumber. *Plant Disease* 79:577-580.

Damayanti M., Susheela K. and Sharma G.J. (1996). Effect of plant extracts and systemic fungicide on the pineapple fruit - rotting fungus *Ceratocystis paradoxa. Cytobios* 86: 155-165.

Daya R. and Ram D. (1997). Fungitoxicity of some plant extracts against *Alternaria brassicae. Ann. Agri. Bio. Res.* 2: 25-26.

Daya R., Tewari V.P. and Ram D. (1994). Control of chickpea collar rot (*Sclerotium rolfsii*) with soil application of plant products. *J. App. Biol.* 4: 38-40.

Deraniyagala S.A., Wijesundera R.L.C. and Weerasena O.V.D.S.J. (1998). Antifungal activity of *Ficus racemosa* leaf extract and isolation of the active compound. *J. National Science* Council of Sri Lanka 26: 19-26.

Dhaliwal H.S. (2001). Effectiveness of essential oils against some phytopathogenic fungi. M.Sc. Thesis Punjab Agricultural University India.

Dhaliwal H.J.S., Thind T.S. and Mohan C. (2003 a). Activity of some essential oils against *Rhizoctonia solani,* the cause of black scurf of potato. *J. Mycol. Pl. Patho.* 33(3): 399-402.

Dhaliwal H.J.S., Thind T.S. and Mohan C. (2003 b). Fungitoxic potential of ten essential oils against *Alternaria brassicae* (Berk.) Sacc. *Pesticide Research Journal* 15(2): 142-144.

Dhaliwal H.S., Thind T.S., Mohan C. and Chhabra B.R. (2002). Activity of some essential oils against *Uncinula necator* causing powdery mildew of grapevine. *Indian Phytopath* 55: 76-78.

Dubey N.K., Tripathi P., Singh H.B., Tripathi P. and Kumar S. (ed.) Hasan S.A. (ed.) Dwivedi S. (ed.) Kukreja A.K. (ed.) and Singh A.K. (2000). Prospects of some essential oils as antifungal agents. Proceedings of the national seminar on the research and development in aromatic plants: current trends in biology uses production and marketing of essential oils Lucknow India 30-31 july 1999. Journal of Medicinal and aromatic plant sciences 22: 350-354.

Dutta B.K., Rahman I. and Das T.K. (1998). Antifungal activity of Indian plant extracts. *Mycoses* 41: 11-12.

Easwaramurthy S., Muthusamy S. and Mariappan V. (1989). Inhibitory effect of neem *Acacia Prosopis* and *Ipomoea* extracts on *Sarocladium oryzae* and *Fusarium oxysporum* f.sp. *cepae. Neem News letter* 6: 4-5.

Eksteen D., Pretorius J.C., Nieuwondt T.D. and Zietsman P.C. (2001). Mycelial growth inhibition of plant pathogenic fungi by extracts of South African plant species. *Ann. Applied Biology* 139:243-249.

Erdemoglu N. and Sener B. (2001). Antimicrobial activity of the heart wood of *Taxus baccata. Fitoterapia* 72: 59-61.

Farrukh A., Beg A.Z., Iqbal A., Kumar S. (ed.) Hasan S.A. (ed.) Dwivedi S. (ed.) Kukreja A.K. (ed.) Sharma A. (ed.) Singh A.K. (ed.) Sharma S. (ed.) and Tewari R. (2001). *In vitro* toxicity of plant essential oils against soil fungi. Proceedings of the national seminar on the frontiers of research and development in medicinal plants Lucknow India 16-18 September 2000. *Journal of Medicinal and aromatic plant sciences* 22-23: 177-181.

Ganguly L.K. (1994). Fungitoxic effect of certain plant extracts against rice blast and brown spot pathogen. *Environ. Ecol.* 12: 731-733.

Garcia R.P. and Garcia M.L. (1988). Laboratory evaluation of plant extracts for the control of *Aspergillus* growth and aflatoxin production. Proceedings of the Japanese association of mycotoxicology supplement no. 1 190-193.

Garg S.C. and Siddiqui N. (1992). Antifungal activity of some essential oil isolates. *Pharmazie* 47: 467-468.

Gogoi P., Baruah P. and Nath S.C. (1997). Antifungal activity of the essential oil of *Litsea cubeba* Pers. *J. Essential Oil Res.* 9: 213-215.

Govindachari T.R., Suresh G. and Masilamani S. (1999 b). Antifungal activity of *Azadirachta indica* leaf hexane extract. *Fitoterapia* 70: 417-420.

Govindachari T.R., Suresh G., Banumathi B., Masilamani S., Gopalakrishnan G. and Kumari G.N.K. (1999 a). Antifungal activity of some BD-seco limonoids from two meliaceous plants. *J. Chemical Ecology* 25. 923-933.

Govindachari T.R., Suresh G., Gopalakrishnan, G., Masilamani, S. and Banumathi B. (2000). Antifungal activity of some tetranorterpenoids. *Fitoterapia* 71: 317-320.

Grohs B.M., Wegen H.W. and Kunz B. (1999). Antifungal activity of red cedar extract and hiba oil on wood inhabiting moulds. *Holz-als-Roh-und-Werkstoff* 57: 277-281.

Hidalgo L. and Fernandez A. (1996). Antifungal activity of plant extracts of different species of Cuba. *Revista-de-Proteccion-vegetal* 11: 19-23.

Hitokota H., Morozumi S., Wauke T., Sakai S. and Kurata H. (1980). Inhibitory effects of spices on growth and toxin production of toxigenic fungi. *Applied Environmental Microbiology* 39: 818-822.

Hulloli S.S., Singh R.P. and Verma J.P. (1998). Management of bacterial blight of cotton induced by *Xanthomonas axonopodis* pv. *malvacearum* with the use of neem based formulations. *Indian Phytopath.* 51: 21-25.

Hussain R., Singh P. and Kaul B.L. (2000). Antifungal activities of some aromatic oils against soil and air borne plant pathogens. *J. Mycol. Pl. Pathol.* 30: 255.

Iida Y., Oh K.B., Saito M., Matsuoka H., Kurata H., Natsume M. and Abe H. (1999). Detection of antifungal activity in *Anemarrhena asphodeloides* by sensitive BCT method and isolation of its active compound. *J. Agric. Fd. Chem.* 47: 584-587.

Inamori Y., Shinohara S., Tsujibo H., Okabe T., Morita Y., Sakagami Y., Kumeda Y. and Ishida N. (1999). Antimicrobial activity and metalloprotease inhibition of hinokitiol-related compounds the constituents of *Thujopsis dilabrata* S. and Z. hondai Mak. *Biological and Pharmaceutical Bulletin* 22: 990-993.

Inouye S., Uchida K., Yamaguchi H., Miyara T., Gomi S. and Amano M. (2001). Volatile aroma constituents of three labiatal herbs growing wild in the karakoram-himalaya district and their antifungal activity by vapor contact. *J. Essential Oil Res.* 13: 68-72.

Jagadeeswari B. and Prakasam V. (2000). Antifungal activity of botanicals on the seed mycoflora infection in coriander. *J. Mycol. Pl. Pathol.* 30: 256-257.

Jain R., Alam S., Arora R. and Jain S.C. (1998). Phytochemistry and bioactivity of *Kirganelia reticulata. J. Medicinal and Aromatic Plant Sci.* 20: 740-741.

Karavaev V.A., Solntsev M.K., Yurina E.V., Polyakova I.B. and Kuznetsov A.M. (2001). Antifungal activity of aqueous extracts of the leaves of cowparsnip and comfrey. *Biology- Bulletin of the Russian Academy of Sciences* 28: 365-370.

Kariba R.M., Siboe G.M. and Dossaji S.F. (2001). *In vitro* antifungal activity of *Schizozygia coffaeoides* Bail. (Apocynaceae) extracts. *J. Ethnopharmacology* 74:41-44.

Khan M. and Zubairy H.N. (2001). Mycoses III : antimycotic activity of camphor eucalyptus oil and thymol. *Hamdard-Medicus* 44: 17.

Khan M.I. and Shah N.H. (1992). Antifungal activity of leaf extract of neem on seed mycoflora of wheat. *Bioved* 3: 209-210.

Khan M.S., Nasir M.A. and Bokhari S.A.A. (1998). *In vitro* evaluation of certain neem based products and systemic fungicides against different pathogens responsible for wilt and anthracnose of guava. *Pak. J. Phytopath.* 10: 72-74.

Kishore N. and Dwivedi R.S. (1991). Fungitoxicity of essential oil of *Tagetes erecta* L. against *Pythium aphanidermatum* Fitz. the damping off pathogen. *Flavour and Fragrance J* 6: 291-294.

Kishore R., Shukla D.S. and Kishore R. (1996). Evaluation of some innovatives vis-a-vis powdery mildew of opium poppy incited by *Erysiphe polygoni. J. Living World* 3: 12-17.

Kobaisy M., Tellez M.R., Webber C.L., Dayan F.E., Schrader K.K. and Wedge D.E. (2001). Phytotoxic and fungitoxic acitivities of the essential oil of kenaf (*Hibiscus cannabinus* L.) leaves and its composition. *J. Agric. Fd. Chem.* 49: 3768-3771.

Kolte S.J. and Shinde P.A. (1973). Influence of plant extracts of certain hosts on the growth and sclerotial formation of *Macrophomina phaseoli in vitro. Phytopath. Notes* 26: 351-352.

Kotasthane A.S. and Lakpale N. (1994). Effect of plant and seed extract on growth and sclerotial production of *Rhizoctonia solani* (*Corticium sasakii*) causing sheath blight of rice. *Adv. Pl. Sci.* 7: 407-410.

Kshemkalyani S.B., Telore R., Madhavi B. and Patel G.S. (1990). The effect of allicine and extracts of garlic on *Aspergillus flavus* and *A. parasiticus*. *Indian J. Mycol. Pl. Pathol.* 20: 247-248.

Kumar T.K., Ranjini C.E. and Sasidharan V.K. (1997). Antibacterial and antifungal activity of secondary metabolites from some medicinal and other common plant species. *J. Life Sci.* 2: 14-19.

Kumbhar P.P. and Dewang P.M. (2001). Eco-friendly pest management using monoterpenoids. I. Antifungal efficacy of thymol derivatives. *J. Scientific and Industrial Res.* 60: 645-648.

Kumbhar P.P., Salunkhe D.H., Borse M.B., Hiwale M.S., Nikam L.B., Bendre R.S., Kulkarni M.V. and Dewang P.M. (2000). Pesticidal potency of some common plant extracts. *Pestology* 24: 51-53.

Kurita N., Miyaji M., Kurane R. and Takahara Y. (1981). Antifungal activity of components of essential oils. *Agric. Biol. Chem.* 45: 945-952.

Lakshmanan P. and Mohan S. (1989). Antifungal properties of some plant extracts against collar rot of *Phaseolus aureus*. *Madras Agric. J.* 76: 266-270.

Lal H.C., Upadhyay J.P. and Ojha K.L. (1998). Management of *Alternaria* leaf blight of pigeonpea through plant extracts and chemicals of plant origin. *J. Applied Biology* 8: 80-85.

Letessier M.P., Svoboda K.P. and Walters D.R. (2001). Antifungal activity of the essential oil of hyssop (*Hyssopus officinalis*). *J. Phytopath.* 149: 11-12.

Li T. and Ma Y.P. (2001). Antifungal effects of wild species of *Allium* in Xinjiang. *Chinese Journal of Biological Control* 17: 170-174.

Liu C.H., Mishra A.K., He B. and Tan R.X. (2001). Composition and antifungal activity of essential oils from *Artemisia princes* and *Cinnamomum camphora*. *International Pest Control* 43: 72-74.

Locke C.J. (2002). Identification and development of biological control agents and natural plant products as biopesticides. www.usna.usda.gov/Research/LockeBotanicals

Mahapatra T.K. and Tewari S.N. (1994). *Ocimum sanctum* L. leaf extract toxicity against collar rot (*Aspergillus niger*) and yellow root (*A. flavus*) diseases of groundnut. *Allelopathy J.* 1: 114-117.

Maximo P., Lourenco A., Feio S.S. and Roseiro J.C. (2002). Flavonoids from *Ulex airensis* and *Ulex europaeus* ssp. Europaeus. *J. Natural Products* 65: 175-178.

Meena S.S. and Mariappan V. (1993). Effect of plant products on seed borne mycoflora of sorghum. *Madras Agric. J.* 80: 383-387.

Mehmood Z., Ahmad S. and Mohammad F. (1997). Antifungal activity of some essential oils and their constituents. *Indian J. Natural Products* 13: 10-13.

Mishra M., Malik S.S., Tewari S.N., Tauro P. (ed.) and Narwal S.S. (1992). Allelopathic effect of certain botanicals against six fungal pathogens of rice. Proceedings. First national symposium. Allelopathy in agroecosystems (agriculture and forestry) February 12-14 1992 held at CCS Haryana Agricultural University Hisar-125004 India. 1992 191-193.

Misra S.B. and Dixit S.N. (1976). Fungicidal spectrum of the leaf extract of *Allium sativum. Indian Phytopath.* 29: 448-449.

Misra S.B. and Dixit S.N. (1977). Fungicidal properties of *Clematis gouriana. Indian Phytopath.* 30: 577-579.

Misra D., Misra M. and Tewari S.N. (1997) Toxic effects of volatiles from *Callistemon lanceolatus* on six fungal pathogens of rice. *Indian Phytopath.* 50: 103-105.

Misra N., Khare U. and Shukla D. (2000). Antifungal activity of essential oil of *Cinnamamum zeylanicum* Nees. (Dalchini) against *Stachybotrys chartarum* (Ehrenberg ex. Link) Hughes. *Journal of Essential Oil Bearing Plants* 3: 97-110

Montes B.R. and Carvajal M. (1998). Control of *Aspergillus flavus* in maize with plant essential oils and their components. *J. Fd. Prot.* 61: 616-619.

Moretti M.D.L., Peana A.T., Franceschini A. and Carta C. (1998). *In vivo* activity of *Salvia officinalis* oil against *Botrytis cinerea. J. Essential Oil Res.* 10: 157-160.

Murdock D.I. and Allen W.E. (1960). Germicidal effect of orange peel oil and D-limonene in water and orange juice. *Food Tech.* :441-445.

Nagaoka T., Goto K., Watanabe A., Sakata Y. and Yoshimara T. (2001). Sesquiter- - penoids in root exudates of *Solanum aethiopicum. Zeitschrift fur Naturforschung Section-C Biosciences* 56: 707-713.

Nene Y. L. and Thapliyal P. N. (1993). Fungicides in plant disease control (3rd ed.). Oxford and I B H Publishing Co., New Delhi. Pp. 691.

Nor-Azah M.A., Mastura M., Khozirah S., Mawardi R., Ali A.M. and Jautan I. (2002). Chemical composition and antimicrobial activities of the essential oils of *Cinnamomum aureofulvum* Gamb. *J. Essential Oil Res.* 14: 135-138.

Pandey A.K., Singh H.N. and Singh S.B. (1993). Fungitoxicity of *Trigonella foenum graceium* L. against some sugarcane pathogens. *Sugarcane* 3: 15-18.

Pandey V.N. and Pant D.C. (1997). *In vitro* antifungal activity of some higher plant products against soil borne phytopathogens. *Madras Agric. J.* 85: 149-153.

Pan S. and Deb G. (1997). *In vitro* bioassay of some plant products against some fungal plant pathogens. *Indian Agric.* 41: 277-285.

Pan S., Mukherjee B., Ganguly A., Mitra S.B. and Bhattacharya A. (1985). Antifungal activity of some naturally occuring flavanoids. *Z. Pflkrankh. Pflschutz.* 92: 392-395.

Parimelazhagan T. and Francis K. (1999). Antifungal activity of *Clerodendrum viscosum* against *Curvularia lunata* in rice seeds. *J. Mycol. Pl. Pathol.* 29: 139-141.

Pauletti P.M., Araujo A.R., Young M.C.M., Giesbrecht A.M. and Bolzani V. da S. (2000). Nor-lignans from the leaves of *Styrax ferrugineus* (Styracaceae) with antibacterial and antifungal activity. *Phytochem.* 55: 597-601.

Picman A.K. and Towers G.H.N. (1984). Antifungal activity of sesquiterpene lactone. *Biochem. Syst Ecol.* 12: 13-18.

Pitarokili D., Tzakou O. and Kalamarakia A. (2002). Activity of *Salvia pomifera* L. ssp. Calycina (Sm.) Hayek against soil borne pathogens. *J. Essential Oil Res.* 14: 72-75.

Popova M., Bankova V., Sparsov S., Tsvetkova I., Naydenski C., Vides-Silva M. and Tsartsarova M. (2001). New bioactive chalcones in propolis from El Salvador. *Zeitschrift fur Naturforschung Section-C Biosciences* 56:593-596.

Quarles W. (2002). Least toxic controls of plant diseases. http://www.bbg.org/gar2/topics/sustainable/handbooks/naturaldisease/leasttoxic

Rajappan K., Mariappan V. and Kareem A.A. (1999). Development and evaluation of EC and dust formulation of plant derivatives against green gram powdery mildew. *Annals of Plant Protection Sciences* 7: 110-112.

Rajappan K., Ramraj B. and Natrajan S. (2000). Field evaluation of fungicides and neem formulations against bacterial blight of frenchbean. *Indian Phytopath.* 53: 303-304.

Rao G.N. and Narayanasamy P. (1990). Effect of plant extracts and oils on rice yellow dwarf infection. *Madras Agric. J.* 77: 197-201.

Rao G.P., Sharma S.R. and Singh P.K. (2000). Fungitoxic and insect repellent efficacy of limonene against sugarcane pests. *J. Essential Oil Bearing Plants* 3:157-163.

Rawal P. and Thakore B.B.L. (2000). Effect of plant extracts on Fusarium rot of sponge gourd. *J. Mycol. Pl. Pathol.* 30: 278.

Reyes-Chilpa R., Gomez-Garibay F., Moreno-Torres G., Jimenez Estrada M. and Quiroz-Vasquez R.I. (1998). Flavonoids and isoflavonoids with antifungal properties from *Platymiscium yucatanum* heartwood. *Holzforschung* 52: 459-462.

Rihakova Z., Plockova M., Filip V. and Smidrkal J. (2001). Antifungal activity of lauric acid derivatives against *Aspergillus niger. European Food Res. Tech.* 213:488-490.

Saller R., Berger T., Reichling J. and Harkenthal (1998). Pharmaceutical and medicinal aspects of Australian tea tree oil. *Phytomedicine* 5. 489-495.

Saniewska A. (1998). The use of garlic for protection of snapdragon from rust. *Ochrona Roslin* 42: 8.

Sarvamangala H. S., Govindaiah and Datta R.K. (1993). Evaluation of plant extracts for the control of fungal diseases of mulberry. *Indian Phytpath.* 46: 398-401.

Sata N., Matsunaga S., Fusetani N., Nishikawa H., Takamura S. and Saito T. (1998). New antifungal and cytotoxic steroidal saponins from the bulbs of an elephant garlic mutant. *Bio-science Biotech. and Biochem.* 62: 1904-1911.

Shahi S., Shukla A.C., Dikshit S., Dikshit A., Dehne H.W., Adam G., Diekmann M., Frahm J., Mauler M.A. and Halteren-p-van. (1997). Modified spore germination inhibition technique for evaluation of candidate fungitoxicant (*Eucalyptus* spp.). Proc 4th International symposium of the European foundation for plant pathology. pp. 257-263 Bonn Germany.

Sharma I. and Nanda G.S. (2000). Effect of plant extracts on teliospore germination of *Neovossia indica. Indian Phytopath.* 53: 323-324.

Sharma V.P. and Jarial R.S. (2000). Efficacy of different fungicides and botanicals against false truffle (*Diehliomyces microsporus*) and yield of *Agaricus species. J. Mycol. Pl. Pathol.* 30: 184-187.

Shekhawat P.S. and Prasada R. (1971). Antifungal properties of some plant extracts. *Indian Phytopath.* 24: 800-802.

Shylaja M., Ranakausar Reddy S.M., Srivastava H.P., Purohit D.K. and Reddy R.S. (1997). Antifungal activity of some leaf extracts on the growth and sclerotial production of *Rhizoctonia solani. Microbial Biotech.* 143-144.

Singh D., Shrivastava B. and Garg S.P. (2000 c). Isolation and antimicrobial studies of curcumin from *Curcuma aromatica. Current Agric.* 24: 101-103.

Singh G., Kapoor I.P.S., Singh O.P., Rao G.P., Prasad Y.K., Leclercq P.A. and Klinkby N. (2000 b). Studies on essential oils part 28: chemical composition antifungal and insecticidal activities of rhizome volatile oil of *Homalomena aromatica* Schott. *Flavour and Fragrance J.* 15: 278-280.

Singh G., Rao G.P., Kapoor I.P.S., Singh O.P., Singh G., Kumar S. (ed.) Kukreja A.K. (ed.) Dwivedi S. and Singh A.K. (2000 a). Chemical constituents and antifungal activity of *Lippia alba* Mill. Leaf essential oil. Proceedings of the national seminar on the research and development in aromatic plants: current trends in biology uses production and marketing of essential oils Lucknow India 30-31 july 1999. *J. Medicinal and Aromatic Plant Sci.* 22: 701-703.

Singh H.B. and Handique A.K. (1997). Antifungal activity of the essential oil of *Hyptis suaveolens* and its efficacy in biocontrol measures in combination with *Trichoderma harzianum. J. Essential Oil Res.* 9: 683-687.

Singh H.N.P., Prasad M.M. and Sinha K.K. (1993). Evaluation of medicinal plant extracts against banana rot. *J. Indian Botanical Society* 72: 163-164.

Singh R.K. and Dwivedi R.S. (1987). Effect of oils on *Sclerotium rolfsii* causing foot rot of barley. *Indian phytopath.* 40: 531-533.

Singh U.P., Sarma B.K., Mishra P.K. and Ray A.B. (2000 d). Antifungal activity of venenatine an indole alkaloid isolated from *Alstonia venenata. Folia-Microbiologia* 45: 173-176.

Skaltsa H., Lazari D., Garcia B., Pedro J.R., Sokovic M. and Constantinidis T. (2000 a). Sesquiterpene Lactones from *Centaurea Achaia* a greek endemic species: antifungal activity. *Zeitschrift-fur-naturforschung.-Section-C-Bioscie-nces* 55: 534-539.

Skaltsa H., Lazari D., Panagouleas C., Georgiadous E., Garcia B. and Sokovic M. (2000 b). Sesquiterpene lactones from *Centaurea thessala* and *Centaurea attica.* Antifungal activity. *Phytochem.* 55:903-908.

Sundarraj T., Kurucheve V. and Jayaraj J. (1996). Screening of higher plants and animal faeces for the fungitoxicity against *Rhizoctonia solani. Indian Phytopath.* 49: 398-403.

Suresh G., Narasimhan N.S., Masilamani S., Partho P.D. and Gopalakrishnan G. (1997). Antifungal fractions and compounds from uncrushed green leaves of *Azadirachta indica. Phytoparasitica* 25: 33-39.

Tang H.Q., Hu J., Yang L. and Tan R.X. (2000). Terpenoids and flavonoids from *Artemisia* species. *Planta- Medica* 66: 391-393.

Tellez M., Estell R., Fredrickson E., Powell J., Wedge D., Schrader K. and Kobaisy M. (2001). Extracts of *Flourensia cernua* (L): volatile constituents and antifungal antialgal and antitermite bioactivities. *J. Chemical Ecol.* 27: 2263-2273.

Thiribhuvanamala G. and Narasimhan V. (1998). Efficacy of plant extracts on seed-borne pathogens of sunflower. *Madras Agric. J.* 85: 227-230.

Thompson D.P. (1989). Fungitoxic activity of essential oil components on food storage fungi. *Mycologia* 81: 151-153.

Tiwari R. (1997). Fungitoxicity of volatile constituents of some higher plants against predominant storage fungi. *Indian Phytopath.* 50: 548-551.

Tiwari R., Dixit V. and Dixit S.N. (1994). Studies on fungitoxic properties of essential oil of *Cinnamomum zeylanicum* Breyn. *Indian Perfumer* 38: 98-104.

Torrenegra R.D., Pedrozo J.A., Tellez A.N., Cabeza G., Granados A. and Mendez D. (2000). Chemistry and antifungal activity of *Pentacalia corymbosa* (Asteraceae-Senecioneae). *Revista-Latinoamericana-de-quimica* 28: 31-34.

Tropnikova I.V., Budantsev A.L., Zenkevich, I.G. and Potekhina T.S. (1999). Composition of essential oils of some species of the genus *Nepeta* L. cultivated in the Leningrad region and their antimicrobial activity. *Rastitel'nye-Resursy* 35: 1-10.

Tsao R. and Zhou T. (2000). Antifungal activity of monoterpenoids against post harvest pathogens *Botrytis cinerea* and *Monilina fructicola. J. Essential Oil Res.* 12: 113-121.

Vaughn S.F. and Spencer G.F. (1994). Antifungal activity of natural compounds against thiabendazole-resistant *Fusarium sambucinum* strains. *J. Agric. Food Chemistry* 42: 200-203.

Verma A., Singh R.B. and Verma A. (1997). Inhibitory property of *Clerodendrum fragrans* against *Fusarium oxysporum* f. sp. *udum. Ann. Pl. Prot. Sci.* 5: 190-193.

Verma D.K., Tripathi V.Z., Rana B.K. and Taneja V. (1998). Antifungal activity of triterpenoids from *Lantana indica* root. *Fitoterapia* 69: 188-189.

Vozenilkova B. (1992). The influence of peppermint (*Mentha piperita* Huds.) on healing the soils infected by club root (*Plasmodiophora brassicae* Wor.). *Sbornik-UVTIZ -Ochrana-Rostlin* 28: 237.

Walters D., Meurer-Grimes B. and Rovira I. (2001). Antifungal activity of three spermidine conjugates. *FEMS-Microbiol.-Letters* 201: 255-258.

Wang H.X. and Ng T.B. (2000). Ginkbilobin a novel antifungal protein from *Ginkgo biloba* seeds with sequence similarity to embryo abundant protein. *Biochem. Biophysical Res.Communications* 279: 407-411.

Wang H.X. and Ng T.B. (2001). Purification of Allivin a novel antifungal protein from bulbs of the round cloved garlic. *Life Sci.* 70: 357-365.

Watanabe A., Toshima H., Nagase H., Nagaoka T. and Yoshihara T. (2001). Structural confirmation of 15-norlubiminol and 15-norepilubiminol isolated from *Solanum aethiopicum* by chemical conversion from lubimin and epilubimin and their antifungal activity. *Biosci. Biotech. Biochem.* 65: 1805-1811.

Yamada T., Hiramoto T., Tobimatsu R., Shiraishi T. and Oku H. (1990). Elicitor-like substances present in barley and wheat seeds. *J. Phytopath.* 128: 89-98.

Ye X.Y. and Ng T.B. (2002). A new antifungal protein and a chitinase with prominent macrophage stimulating activity from seeds of *Phaseolus vulgaris* cv. *Pinto. Biochem. Biophysical Res. Communications* 290:813-819.

Ye X.Y., Ng T.B. and Rao P.F. (2001). A bowman-birk-type trypsin-chymotrypsin inhibitor from broad beans. *Biochem. Biophysical Res. Communications* 289: 91-96.

Yin M.C. and Tsao S.M. (1999). Inhibitory effect of seven *Allium* plants upon three *Aspergillus* species. *International J. Food Microbiology* 49: 49-56.

CHAPTER - 20

FUNGAL INFECTIONS IN AIDS PATIENTS

H. C. Gugnani
Department of Medical Mycology, Vallabhbhai Patel Chest Institute, University of Delhi, Delhi-110 007.

INTRODUCTION

The distinguishing feature of acquired immunodeficiency syndrome (AIDS) is a progressive decrease of cell-mediated immunity due to defective functioning of T-lymphocytes or macrophages. Although HIV infection itself can produce disease manifestations, its major effect is the diminution of immunity, leading to appearance of opportunistic infections and tumours. Patients of AIDS are susceptible to a variety of opportunistic infections. Although these patients are primarily infected by bacterial and viral pathogens, during the past two decades, there has been an increase in the frequency of fungal infections causing considerable morbidity and mortality. Indeed the AIDS epidemic has resulted in recognition of a number of new pathogens (Dupont *et al.*, 1994; Kirchner, 1996; Marques *et al.,* 2000). Improvement in the quality and duration of survival of AIDS patients depends upon the efficacy of treatment of opportunistic infections. It is estimated that 58-81% of AIDS patients contract fungal infections at some time, and 10-20% of AIDS patients die as a direct consequence of such infections (Holmberg and Meyer, 1986). The following account concisely reviews the current status of different opportunistic infection associated with AIDS patients. The information on the status of these infections in India is also incorporated.

CANDIDIASIS

Candidiasis is an acute or chronic, superficial or systemic infection caused by species of *Candida. Candida albicans* is the most common species responsible for infection. Other species such as *C. tropicalis, C. parapsilosis, C. glabrata* and *C. krusei* are less common. Infections caused by *Candida* manifesting as oropharyngeal, mucocutaneous or oesophageal candidiasis are recognized as the most frequent opportunistic fungal infections in patients infected with human immunodeficiency virus (HIV) (Dupont *et al.,* 1994; Kirchner, 1996). Observations over the past decade have shown that oropharyngeal *Candida* infection (thrush) appears when the CD4 + cell count is about 200 mm^{-3}. Oesophageal candidiasis is an AIDS defining illness according to 1993 CDC classification system and occurs in patients with CD4+ counts less than 100 mm^{-3}. It may occur alone or in combination with oral candidiasis. Patients may be asymptomatic but frequently complain of odynophagia, dysphagia, and retrosternal pain. Oesophageal candidiasis can be diagnosed on clinical presentation with endoscopy or radiological intervention. However, blind oesophageal brushing via nasogastric tube is much more effective in the diagnosis of oesophageal infection due to *Candida.* Among the surrogate markers of progression of HIV infection, the number of CD4-T lymphocytes roughly parallels the risk of candidiasis: the lower the count, the higher the risk of clinical lesions (Dupont *et al.,* 1994; Kirchner, 1996). Pereiro *et al.* (1997) studied *C. albicans* adherence to buccal epithelial cells of HIV-positive and HIV-negative patients. The adherence capacity of *C. albicans* increased in relation to the progression of HIV infection and decrease of CD4 + cell counts and correlated with the survival of more pathogenic and resistant fungal cells. A variety of molecular techniques have been used to type strains of *C. albicans.* These include DNA hybridization of genomic digests, karyotyping, restriction fragment length polymorphism, and/or random amplified polymorphic DNA (RAPD) (Pfaller *et al.,* 1995). *Candida dubliniensis,* a recently described species is an emerging fungal pathogen in AIDS and other immunosuppressed patients (Sullivan *et al.,* 1995; Ponton *et al.,* 2000; Redding *et al.,* 2000). It is a frequent cause of recurrent episodes of oral candidiasis in AIDS patients. This species can be misidentified as *C. albicans* by virtue of being phenotypically similar to *C. albicans* (Ponton *et al.,* 2000). *Candida vaginitis* occurs when the CD4 + counts are 400 mm^{-3} or lower (Dupont *et al.,* 1994). Recalcitrant *Candida* vaginitis is a major problem in HIV-infected women (Dupont *et al.,* 1994).

Several workers have reported oropharyngeal candidiasis in AIDS patients in India (Giri, *et al.,* 1995; Ayyagari *et al.,* 1999; Baradkar *et al.,* 1999; Bajaj *et al.,* 2000; Ranganathan *et al.,* 2000). Ranganathan *et al.* (2000) reported pseudomembranous candidiasis and erythematous candidiasis in 33% and 14% of AIDS patients respectively. In a study from Tamil Nadu, Kumarasamy *et al.* (1995) also found oral candidiasis to be the second most predominant opportunistic infection. In a report by National AIDS Control Organization (NACO), the prevalence of oral candidiasis was 57%, being next to tuberculosis (62%) amongst the 3551 AIDS cases reported between 1986 and May 1997 (Misra *et al.,* 1998). In an autopsy study of 49 patients of AIDS in Bombay, Lanjewar *et al.* (1996) reported 5 cases of bowel infection and 3 of oesophageal candidiasis. In a study of 150 HIV-infected subjects (Gugnani *et al.*, 2003a), 126 (84%) of the patients were positive for *Candida*. In 78 (61.9%) of the patients positive for *Candida*, plate cultures indicated heavy oropharyngeal colonization by *Candida* (mainly *C. albicans*, followed by *C. tropicalis, C. krusei* and *C. glabrata*) as evidenced by appearance of numerous uncountable colonies in all the plates inoculated with the samples; none of them had any lesions at the time of sampling. This study also demonstrated *Candida dubliniensis* for the first time in India. For the treatment of oropharyngeal candidiaisis in AIDS patients, fluconazole (150 mg) or itraconazole (400 mg) can be administered orally once daily for a period of two to three weeks. A small proportion of AIDS patients, particularly those with the end stage of the disease become refractory to treatment. In such patients, addition of flucytosine to fluconazole or itraconazole or intravenous amphotericin B may be helpful (Dupont *et al.*, 1994). Fluconazole 150 mg daily has been advocated as post therapy prophylaxis for patients with advanced AIDS to prevent relapse.

CRYPTOCOCCOSIS

Cryptococcosis, a cosmopolitan mycosis caused by the encapsulated, basidiomycetous yeast, *Cryptococcus neoformans* manifests clinically in two basic forms, pulmonary and cerebral (mainly meningeal), which manifests by dissemination. The major portal of entry for the fungus is the respiratory tract. The fungus is inhaled with the dust coming from avian habitats or other saprophytic sources. Infection may lead to the saprophytic presence or transient and asymptomatic colonization of the fungus in the respiratory tract. The cryptococci present in the respiratory tract may become virulent any time and produce disease (Kwon-Chung and Bennett, 1992). Pulmonary cryptococcosis is confined to the lung and is usually self-limiting and asymptomatic. In a very small percentage of cases, infection produces a primary complex like that in tuberculosis (Kwon-Chung and Bennett, 1992; Cox and Perfect, 1998). This primary complex consists of a lesion in the lung parenchyma, lymphangitis towards the regional lymph nodes and their involvement (focal lesions). There are no pathognomonic features of pulmonary cryptococcosis. Meningeal cryptococcosis manifests as a subacute meningitis or meningoencephalitis with fever, malaise and headache (Kwon-Chung and Bennett, 1992; Cox and Perfect, 1998).

The epidemiology of cryptococcosis has changed dramatically with the advent of epidemic of AIDS. Cryptococcosis occurs in 2-30% of patients infected with HIV (Cox and Perfect, 1998; Casadevall and Perfect, 1998). The incidence of cryptococcosis worldwide has risen dramatically, due to increased number of immunosuppressed patients at risk for the disease, particularly those with AIDS (Cox and Perfect, 1998; Casadevall and Perfect, 1998). In immunocompromised patients, particularly the AIDS patients, the fungus disseminates. Cryptococcal meningitis is one of the most important neurological manifestations seen in AIDS patients. Its incidence among AIDS patients has been estimated to be 6 to 10% in USA, 3-6% in Europe and 10-30% in sub-Saharan Africa, where it is regarded as a prominent AIDS defining opportunistic infection (Dupont *et al.*, 1994; Cox and Perfect, 1998). The data in India are scanty to give an estimate. Risk factors for development of cryptococcal meningitis in AIDS patients include decreased CD4 + cell count. Fever and headache may be the only signs and symptoms detectable in HIV-infected patients presenting with cryptococcal meningitis, because menigism and photophobia are not common in such patients (30%), perhaps reflecting poor inflammatory response in that population. The presence of respiratory symptoms may precede cryptococcal meningitis. A high index of suspicion is desirable for diagnosis of this infection. Extraneural sites of infection include lungs (up to 20%), skin, bone, urinary tract including prostate and others. Extraneural disease in AIDS patients is common; prostatic involvement can be important in establishing a persistent focus of infection (Casadevall and Perfect, 1998). Three genetically recognized distinct varieties of C. *neoformans* have been recognized. These are *C. neoformans* var. *grubii* (serotype A) (Franzot *et al.*, 1999), *C. neoformans* var. *neoformans* (serotypes D and AD), both with the sexual stage *Filobasidiella neoformans* var. *neoformans* and *C. neoformans* var. *gattii* (serotypes B and C) with the sexual stage *F. neoformans* var. *bacillisporus* (Kwon-Chung and Bennett, 1992; Casadevall and Perfect, 1998). Serotype A is the commonest and is worldwide in distribution; serotype D is less frequent. The serotype AD appears to be a recombinant between the serotypes A and D; it is very rare and is restricted to Europe (Kwon-Chung and Bennett, 1992; Casadevall and Perfect, 1998; Franzot *et al.*, 1999; Cogliati *et al.*, 1999). A great majority of cases of cryptococcosis world-wide are caused by *C. neoformans* var. *grubii* (serotype A) (Franzot *et al.*, 1999), which occurs ubiquitously in nature in association with soil, avian habitats, and decaying wood of several species of trees, viz. pink shower tree (*Cassia grandis*), November shower tree (*Senna multijugga*) and fig trees (*Ficus microcarpa*) (Kwon-Chung and Bennett, 1992; Mitchell and Perfect, 1995; Lazera *et al.*, 1996; Franzot *et al.*, 1999; Cogliati *et al.*, 1999). Randhawa *et al.* (2000) reported recovery of *C. neoformans* var. *neoformans* from the trunk hollow of Tamarind tree (*Tamarindus indica*) in India. The distribution of *C. neoformans* var. *gattii* is geographically restricted to tropical and subtropical regions, where it occurs in association with decaying wood of a number of *Euclayptus* species, viz. *E. camaldulensis* (river red gum), *E. terreticornis* (forest red gum), *E. rudis* (West Australian flooded gum), *E. grandis* (flooded gum) and E. *gomphocephala* (tuart), Almond tree (*Termanalia cattapa*), Pink shower tree (*Cassia grandis*) (Ellis and Pfeiffer, 1990; Casadevall and Perfect, 1998; Franzot *et al.*, 1999; Lazera *et al.*, 1998; Fortes *et al.*, 2001). *Cryptococcus neoformans* var. *grubii* and C. *neoformans* var. *neoformans* predominantly affect immunocompromised hosts, while variety *gattii* mainly affects immunocompetent hosts (Dupont *et al.*, 1994; Speed and Dunt, 1995). With the advent of AIDS, the percentage of cryptococcosis cases caused by variety *gattii* has diminished even in the tropical and subtropical regions because almost all AIDS patients are infected with var. *grubii*. Cryptococcosis due to variety *gattii* is sporadically recorded. Infection caused by var. *gattii* is associated with intracerebral and pulmonary cryptococcal mass lesions and more severe clinical symptoms with higher cryptococcal antigen titers (Speed and Dunt, 1995). Some workers have postulated that exposure to *Eucalyptus*, a host tree is required to initiate infection in humans.

In India, earlier reports of large series of cases by Aikat *et al.* (1967) and Shome *et al.* (1973) indicated a rising incidence of the disease. Subsequently several series of cases were reported from different parts of India. Some of the reports from India show association of cryptococcosis with pulmonary tuberculosis (Shome *et al.*, 1973; Nagrajan *et al.*, 2000). Cryptococcosis is being increasingly recognized as an opportunistic infection in immunocompromised patients, particularly those with AIDS in India (Khan *et al.*, 1986; Khanna *et al.*, 1996; Chakrabarti *et al.*, 2000; Deivanayagam *et al.*, 2001). The first such series of cases were recognized by Khan *et al.* (1986). In a study of 60 cases of cryptococcosis of central nervous system (CNS) from Bangalore (Khanna *et al.*, 1996), 36 (60%) had HIV infection as a predisposing factor. A recent review of cases of cryptococcosis from Chandigarh (Chakrabarti *et al.*, 2000) also showed that HIV infection was the commonest predisposing factor (57.4%) followed by haematological malignancies (6.3%) and renal transplant (4.2%). Very recently several cases of cerebral and disseminated cryptococcosis have been recorded in AIDS patients in Delhi (Gugnani *et al.*, unpublished data). Devianayangam *et al.* (2001) found cryptococcosis with central nervous system manifestations in 196 (12.25 %) patients out of 1600 confirmed cases of HIV-infection with tuberculosis as co-infection. Banerjee *et al.* (2000) have reviewed the cases of cryptococcosis reported in India up to 2000. Occurrence of *C. neoformans* in a natural habitat in India was established by reports of its isolation from old pigeon excreta from Delhi and several other parts of India (Sethi *et al.*, 1966; Gugnani *et al.*, 1967; Gugnani *et al.*, 1976). A few cases of cryptococcal menigitis due to var. *gattii* have been described from India including one case in a patient with AIDS (Padhye *et al.*, 1993; Abraham *et al.*, 1997; Peerapur *et al.*, 2000). There are numerous species of *Eucalyptus* in India; several of them, viz. *E. tereticornis, E. grandis, E. globulus* and *E. camaldulensis* are grown on plantation scale in India. Most of the attempts to recover *C. neoformans* from these species have been unsuccessful, except for two reports of its isolation from flowers of *E. camaldulensis* and the bark of *E. terreticornis* respectively (Chakrabarty *et al.*, 1997; Gugnani *et al.*, 2003b).

Though culture is the gold standard for diagnosis of cryptococcosis, isolation and identification of the organism can take up to at least a week. Immunological diagnosis of cryptococcal meningitis by latex agglutination test (LAT) for cryptococcal polysaccharide antigen is highly specific and has sensitivity of more than 90%. The test is useful in assessing the treatment given to the patient. A positive serum cryptoccocal antigen titer of 1 : 8 or higher is generally regarded as presumptive evidence of cryptococcal infection (Cox and Perfect, 1998). Thus, the detection of cryptococcal antigen in the serum can be used as a screening test for the presence of cryptococcal infection in a febrile-HIV positive patient. The finding of a positive cryptococcal antigen titer in this setting warrants antifungal therapy, even if infection is not subsequently documented (Cox and Perfect, 1998).

Little is known about the molecular epidemiology of *Cryptococcus neoformans* from tropical regions. In a study of 164 clinical isolates of *C. neoformans* from 120 Ugandan AIDS patients with cryptococcal meningitis, Pfaller *et al.* (1998) recognized 43 distinguishable *C. neoformans* DNA types. Thirty patients (25%) in this study were infected with their own unique strain of *C. neoformans*, while 75% of the patients shared their strain with at least one other patient. In a study of genetic relatedness of 12 clinical and 5 environmental isolates of *C. neoformans* from Tamil Nadu, India (Kidd *et al.*, 2001), three major molecular types, VN1, VNII and VNIII were identified; majority of the isolates belonged to the type VNI.

All patients of cryptococcal meningitis should be promptly treated. Positive results have been obtained from a large series of HIV-positive patients, receiving amphotericin B (0.7 mg/IV) for 2 weeks followed by oral fluconazole 400 mg daily for an additional 8 weeks (Casadevall and Perfect, 1998). The addition of 5-flucytosine 100mg/kg/day in conjunction with the amphotericin B phase of treatment does not dramatically improve immediate outcome, but may decrease the risk of relapse and is generally well tolerated. Patients with mild disease and those who present cryptococcosis without meningeal involvement may be treated with fluconazole 400-800 mg or itraconazole 400 mg daily. Combination therapy of fluconazole with 5-flucytosine has also been successfully tried in many cases (Casadevall and Perfect, 1998).

PNEUMOCYSTOSIS

It is an opportunistic pulmonary infection caused by *Pneumocystis carinii*, characteristically occurring in patients with acquired immuno-deficiency syndrome or patients treated with cytotoxic drugs or cortico-steroids for long time. The disease presents as pneumonia with incre-asing dyspnoea, non-productive cough, inability to breathe deeply, chest tightness and night sweats. Prior to 1980`s, *Pneumocystis carinii* pneum-onia (PCP) was largely a sporadic disease. The first large outbreak PCP was reported from St. Jude's Children Research Hospital in USA when seventeen cases were documented during 1968-69. The outbreak of PCP among homosexuals in Los Angeles (USA) in 1981 led to recognition of AIDS as a clinical entity (Cushion, 1998). Seroepidemiological surveys have shown that most adults in USA and Europe have anti-*P. carinii* antibodies and healthy children seroconvert at 4 years of age or else (Cushion, 1998). In another study, 70% of healthy children in Gambia had acquired antibody to *P. carinii* (Wakefield *et al.,* 1990). Malin *et al.* (1995) studied human immunodeficiency virus (HIV) infected patients with pneumonia in Zimbabwe using efficient bronchoalveolar lavage (BAL) methods and found that 33% of these patients had *P. carinii* pneumonia. PCP is associated with less than 200 CD4 + cells mm^{-3} (Kirchner, 1996). The natural habitat of *P. carinii* is not known. Transmission between animals by airborne route has been adequately shown but evidence to support this mode of transmission among humans is lacking. PCP signals associated with *Pneumocystis* in rats without compromised immune system suggest a carrier state or transient colonization. This indicates the possibility of spread of the disease by transmission from human reservoir (Hughes, 1998). The low rate of occurrence *P. carinii* pneumonia in HIV patients in developing countries may be due to the non-recognition of the infection in the face of more serious diseases like tuberculosis.

Data on the prevalence of *Pneumocystis carinii pneumonia* (PCP) in India are scanty. Serological surveys for detection of this infection have not been conducted. The first case of PCP reported from India was a patient of acute non-lymphoblastic leukemia in whom the diagnosis was confirmed on post-mortem transthoracic fine needle aspiration (Dhabbar *et al.,* 1992); another case occurred in a renal transplant recipient on immunosuppressive therapy in whom the diagnosis was made ante-mortem by an open lung biopsy and confirmed at autopsy (Sherke *et al.*, 1992). The third case occurred in a patient of Hodgkin's disease while on steroid therapy (Verma *et al.*, 1994). Date *et al.* (1995) reported several cases of PCP in renal transplant patients in Vellore. In a recent publication from Vellore (Mathew and Mathai, 2000), 24 (11.8%) of 204 samples of bronchial aspirate, induced sputum and suction catheter tip were found to be positive for *P. carinii.* Besides, PCP was diagnosed in 5 of 15 AIDS patients investigated. More recently Rani *et al.* (2000) described six cases of PCP confirmed by demonstration of *P. carinii* cysts in sputum smears stained with Gomori Methenamine Silver. Cup or crescent shaped cysts with capsular thickening are characteristic of *P. carinii* (Kidd *et al.,* 2001). Immunofluorescent (monoclonal) staining of aspirates a sensitive and convenient though expensive method of diagnosis. A HIV-positive person with symptoms of dry cough and a radiological pattern of interstitial pneumonia should be investigated for PCP. The organism can also be identified using a simple stain like toluidine 'O' (Cushion, 1998). The need to equip the laboratories in India for diagnosis of PCP can hardly be overemphasized in view of the ever-increasing number and variety of immunosuppressed patients.

PENICILLIOSIS MARNEFFEI

Penicilliosis marneffei, caused by a dimorphic fungus, *Penicillium marnefffei* is endemic in Vietnam, Thailand, Myanmar (Burma), Vietnam, Laos, Hong Kong, Taiwan, Southern China, Indonesia and Philippines, and northeast India (Duong, 1996). Prior to the AIDS epidemic in

Thailand and elsewhere in Southeast Asia, only 22 cases had been reported in the medical literature (Duong, 1996). A tremendous increase in the number of cases of this disease has occurred in Thailand, exclusively in patients infected with human immunodeficiency virus (HIV). In the Chiang Mai province of Northern Thailand, this potentially fatal infection is regarded as the third most frequent opportunistic infection (15-20%) in HIV-infected patients and is considered as one of the indicator diseases for the diagnosis of AIDS after extra-pulmonary tuberculosis and cryptococcosis (Duong, 1996; Srisanthana and Supparatpinyo, 1998). Imported cases of the disease have been reported among HIV-infected persons in United States, United Kingdom, The Netherlands, Italy, France, Germany, Switzerland, Sweden, Australia, and Japan (Duong, 1996). Clinical symptoms *P. marneffei* infection include fever, weight loss, anemia and characteristic skin lesions. The skin lesions are papules with central umblication and necrosis resembling the skin lesions in molluscum contagiosum. About 30-40 % of patients do not have skin lesions. Chest X-ray manifestations in AIDS patients suffering from disseminated *P. marneffei* infection include alveolar processes, reticulonodular lesions, interstitial infiltrates and pleural effusion (Duong, 1996; Srisanthana and Supparatpinyo, 1998; Viviani and Tortarano, 1998). Microscopic examination of wright stained smears from skin lesion for presence of characteristic fission yeast is a simple method for presumptive diagnosis of pencilliosis in the endemic areas. The diagnosis is confirmed by demonstrating characteristic yeast cells in stained tissue sections and by culturing the fungus. Several serological tests have been developed for diagnosis of the disease (Marques *et al.*, 2000).

Penicillium marneffei has been isolated from the internal organs of four species of bamboo rats, *Rhizomys sinensis*, *R. pruinosus*, *R. sumatrensis* and *Cannomys badius* (Ajello *et al.*, 1995; Chariyalertsak *et al.*, 1996; Deng *et al.*, 1986). A majority of the infected rats are carriers of the fungus and do not show any lesions on their internal organs. Deng *et al.* (1986) recovered *P. marneffei* from soil samples collected from three burrows of bamboo rats. It is not known whether rodents other than the bamboo rats are carriers of the fungus. Two genotypes of *P. marneffei* have been identified by RFLP patterns of whole cell DNA (Vanitanakom *et al.*, 1996). Type I predominates amongst clinical isolates (72-7%) and *R. sumatrensis*, whereas type II has been found in a soil samples, *C. badius* and in a minority of human isolates.

Information on the occurrence of penicilliosis marneffei in India has been lacking. Four autochthonous cases of this disease were described in 1999 from HIV-infected patients in Manipur (Singh *et al.*, 1999). The patients showed symptoms of disseminated infection i.e. fever, weight loss, anemia, hepatospelnomegaly, and characteristic papular skin lesions resembling molluscum contagiosum. Later in 2002, Ranjana *et al.* (2002) reported 46 additional cases of penicilliosis marneffei from Manipur; these cases were diagnosed over a period of 8 months (April

1998-October 1999). It is possible that cases of penicilliosis marneffei occur in other climatically similar areas but are not diagnosed. Bamboo thickets and forests are in abundance in low hills and valleys in Northeast areas of India, namely Mizoram, parts of Meghalaya, Arunachal Pradesh, Nagaland and Manipur. Bamboo rats of the genera, *Rhizomys* and *Cannomys*, and several other species of rats, namely *Rattus rattus brunneuseculus*, *Nesokia* sp., *Mus musculus* and *Golunda* sp are common in there areas (Elerman, 1961; Chauhan and Saxena, 1985). Very recently, Gugnani *et al* (unpublished data) have recovered P. marneffei from internal organs of several bamboo rats of the species, *Cannomys badius badius* in Manipur. With the spread of AIDS in India, penicilliosis marneffei is likely to increase in importance.

Penicilliosis marneffei is a fatal disease in the absence of treatment. The current recommended therapy regimen is to give amphotericin B, 0.6 mg/kg/day iv for 2 weeks, followed by itraconazole, 400 mg/day orally in two divided doses for ten weeks (Srisanthana and Supparatpinyo, 1998).

ASPERGILLOSIS

Invasive aspergillosis previously known to be rare in AIDS patients is being increasingly recorded in such patients from North America, Europe, Africa and Asia (Marques *et al.,* 2000). Approximately 4% of AIDS patients develop aspergillosis. The vast majority of the cases occur in the late stage of the HIV disease, CD4 counts are usually less than 50 CD 4+ cells mm^{-3} (Marques *et al.,* 2000). The risk factors for occurrence of invasive aspergillosis in AIDS patients include neutropenia, therapy with steroids, cytotoxic drugs, and broad-spectrum antibiotics. The predomi-nant species involved in AIDS patients is *Aspergillus fumigatus* (83%). Other species, viz. *A. flavus, A. niger,* and *A. terreus* are occasional agents (Dupont *et al.,* 1994; Marques *et al.,* 2000). Lung is the commonest site of lesions with frequent cerebral involvement. The commonest symptoms include fever, cough, dyspnoea, chest pain, and haemoptysis. The chest X-ray shows localized or diffuse infiltrates, nodules or cavitory lesions with both thick and thin walls present in the upper lobes (Dupont *et al.,* 1994). Most cases occur in patients in an advanced stage of AIDS, when the CD4 + cell count is less than 50 cells mm^{-3}. The mortality rate in acute invasive aspergillosis is over 80% (Marques *et al.,* 2000). The most reliable diagnostic method for acute invasive aspergillosis is demonstration of hyaline, septate hyphae with repeated dichotomous branching, and positive culture. Low concentr-ation of galactomannan, a cell wall component of *Aspergillus* in body fluids, appears to be an indicator of invasive disease. Galactomannan can be detected in serum, and urine of infected patients by Latex agglutination (LA) and enzyme-linked immunoabosrbant assay (ELISA) tests (Marques *et al.,* 2000). The treatment of choice is amphotericin B (AMB) (Dupont *et al.,* 1994; Marques *et al.,* 2000). Itraconazole is suitable for the treatment of invasive aspergillosis in patients with contraindi-cations to receive AMB or as compliment of this drug. The dose recommended for treatment is 400-600 mg/ kg/day (Dupont *et al.,* 1994; Marques *et al.,* 2000).

Acute invasive aspergilloiss of the paranasal sinuses is also seen in AIDS patients with *A. fumigatus* and *A. flavus* being the most frequent etiological agents. The symptoms include fever, headache, nasal discharge, and facial pain. The infection can spread into the orbit and brain, causing thrombosis and infarction (Marques *et al.,* 2000). *Aspergillus* induced tracheobronchitis is seen more often in AIDS patients than in other immunosuppressed patients. The symptoms include dyspnoea, wheezing, cough and fever. Some patients die because of occlusion of the trachea or of a bronchus. Bronchoscopic examination may reveal ulcerative lesions or necrotic pseudomembranes, which prompts the diagnosis.

ENDEMIC MYCOSES IN AIDS PATIENTS

The endemic mycoses, viz. histoplasmosis (caused by *Histoplasma capsulatum* var. *capsulatum*), African histoplasmosis (caused by *H. capsulatum* var. *duboisii*), blastomycosis (caused by *Blastomyces dermatitidis*), paracoccidioidomycosis (caused by *Paracoccidioides brasiliensis*) coccidioidomycosis (*Coccidioides immitis*) and sporotrichosis (caused by *Sporothrix shenckii*) are restricted to geographic areas, where the etiological agents of these occur in soil and other environmental sources. The extent of prevalence of these mycoses varies with geographic locale. The prevalence of these mycoses is slightly greater in HIV-infected patients than that in immunocompetent hosts. However, these mycoses tend to produce disseminated infections in HIV-infected persons (Tan *et al.,* 1993; Marques *et al.,* 2000; Negroni, 2000; Gugnani, 2000). AIDS associated histoplasmosis and coccidioidomycosis are relatively more common than AIDS associated paracoccidioidomycosis and blastomycosis (Marques *et al.,* 2000; Negroni, 2000). Approximately 3-5% of AIDS patients develop disseminated lesions due to infection with *H. capsulatum* var. *capsulatum.* The clinical manifestations are prolonged fever, weight loss, asthania, anorexia, diarrhoea or vomiting, hepatosplenomegaly, multiple adenopathy, skin lesions and pancyto-penia (Negroni, 2000). Mucocutaneous involvement is very frequent. Occasionally symptoms are respiratory in nature (Negroni, 2000). Only a few cases of AIDS associated sporortrichosis are known (Gori *et al.,* 1997; Aarestrup *et al.,* 2001). Information on occurrence of AIDS associated endemic mycoses in India is very scanty. Approximately 60 cases of histoplasmosis including a few AIDS associated cases have been recorded from India (Duraiswamy *et al.,* 1997; Gugnani, 2000; Kakghatgi *et al.,* 2000; Moolcahnd *et al.,* 2000). Regarding the occurrence of *H. capsulatum* var. *capsulatum* in natural habitats in India, there is only one report of isolation of *H. capsulatum* from a sample of soil admixed with bat guano collected from an abandoned room of 350-year-old palatial building, the roosting site of an insectivorous house bat, *Scotophilus heathi* in Serampore near Calcutta, West Bengal (Sanyal and

Thammaya, 1975). Only a few cases of blastomycosis and coccidioidomycosis have been reported in India, there is no record of AIDS associated case of either of these two mycoses (Gugnani, 2000; Verghese *et al.,* 2002). No case of paracoccidioidomycosis is yet known from India (Gugnani, 2000). Sporotrichosis has been described from several parts of India (Gugnani, 2000; Barua *et al.,* 1975; Ghosh *et al.,* 1999) but no case of this infection associated with AIDS has yet been recognized.

AIDS patients of endemic mycoses are at risk for progressive disseminated disease and should be treated. Amphotericin B (AMB) or its other formulations are more effective as initial treatment for disseminated histoplasmosis than itraconazole. The recommended dosage of AMB is 0.7 or 0.8 mg kg^{-1} day^{-1} to complete a total dose of 35-40 mg kg^{-1}. Itraconazole (8 mg kg^{-1} day^{-1}) is effective in mild illness caused by disseminated histoplasmosis, blastomycosis, paracoccidioidomycosis, and coccidioidomyosis (Finquelievich *et al.,* 2000). However, it is a fungistatic explaining its comparatively reduced efficacy. Ketoconazole is effective for treatment of histoplasmosis but appears to be less effective than itraconazole. Fluconazole at 800 mg daily induces good response in 74% of AIDS patients of histoplasmosis (Finquelievich *et al.,* 2000). Fluconazole, in high doses, is effective against blastomycosis, and plays a unique role in the treatment of coocidioidal meningitis because it penetrates the cerebrospinal fluid well. Among the new triazoles, voriconazole and posacoanazole (SCH 65592) have broad-spectrum antifungal activity against the endemic mycoses (Finquelievich *et al.,* 2000). The echinocandins are a new class of antifungal agents that act by inhibition of β(1,3)-glucan synthase, preventing incorporation of glucans into the fungal cell wall (Pfaller *et al.,* 1998). These agents are active against *Pneumocystis carnii, Candida, Aspergillus* but not against other fungi (Pfaller *et al.,* 1998).

INFECTIONS DUE TO OTHER YEASTS, YEAST-LIKE FUNGI AND SOME OTHER MOULDS

Seborrheic dermatitis due to *Malassezia furfur* is regarded as an early marker of AIDS patients and its prevalence may reach 80% (Marques *et al.,* 2000). On the other hand ptyriasis (tinea) versicolor due to *M. furfur* is not more frequent in HIV-infected persons than in non HIV-negative hosts (Marques *et al.,* 2000). No investigations have been carried out to determine the prevalence of *M. furfur* infection in India. A number of unusual infections due to other yeasts and moulds have been described in AIDS patients. These include fungaemia due to *Saccharomyces cerevisiae, Cryptococcus albidus, C. laurentii* and *Rhodotorula mucilaginosa* (Sethi and Mandell, 1988; Kordossi *et al.,* 1998; Johnson *et al.,* 1998; Kiraz *et al.,* 2000), meningitis due to *C. laurentii* and *Rhodotorula mucilaginosa* (Gyaurgieva *et al.,* 1996; Kordossi *et al.,* 1998) skin and subcutaneous infection due to *R. mucilaginosa* (Da Silva *et al.,* 2001). Some rare fungal infections in AIDS patients include rhino-orbital mucormycosis (Hejney *et al.,* 2001), disseminated infection due to *Paecilomyces lilacinus* (Lovell *et al.*, 2002), invasive fungal sinusitis and meningitis due to *Arthrographis kalrae* (Chin-Hong *et al.*, 2001). Gori *et al.,* (1998) described a case of extensive cutaneous infection due to a novel species of *Emmonsia*, viz. *E. pastueriana* in a HIV-infected patient.

DERMATOMYCOSES IN HIV/AIDS PATIENTS

The incidence of dermatophytic infections is not considered to have increased in HIV-infected patients but the lesions are widespread and more severe. Onychomycosis is common in AIDS patients. Tinea unguium due to *Trichophyton rubrum* may occur in up to 50% of cases. (Westerman, 1999). A study in Argentina (Dompmarin *et al.,* 1990) showed onychomycosis to be common with 29.7% of the patients presenting the deep proximal (DP) type of nail invasion as the sole manifestation. In 10.9% of the cases, this type of onychomycosis was associated with distal sub-ungueal (DS) form. DP onychomycosis is regarded as a marker, and sometimes, it is the earliest sign of HIV-infection (Dompmarin *et al.,* 1990).

CONCLUSION

The epidemic of human immunodeficiency virus infection is far from under control despite the advances in diagnosis and treatment of this infection. By the end of 2000, the estimate of people living with HIV was 36 million according to UNAIDS and WHO. More than 95 per cent of all HIV-infected persons currently live in developing countries. AIDS has spread very fast in India posing a major public health concern. In India, according to National AIDS Control Organization (NACO), 3.5 million persons are already infected with HIV/AIDS and 10,000 full-blown cases of AIDS have already been registered. AIDS associated fungal infections are being increasingly recognized. Candidiasis and cryptococcosis occur world wide in AIDS patients. Penicilliosis marneffei is an important life-threatening infection in HIV-infected persons in several south Asian countries. Some endemic mycoses like histoplasmosis occur in disseminated form in AIDS patients. In India, oropharyngeal candidiasis (OPC), cryptococcosis and *Pneumocystis carinii* pneumonia are major AIDS defining fungal infections.

Patients with advanced AIDS are at increased risk for fungal infection. There is great need for developing strategies for prophylaxis or treatment against the commonest opportunistic fungal infections viz. candidiasis, cryptococcosis and *Pneumocystis carnii* pneumonia (PCP) in these patients, and also against penicilliosis marneffei, which appears to be common in some parts of southeastern countries including northeast India. In developed countries, the use of highly active antiretroviral therapy (HAART) has led to considerable reduction of oropharyngeal candidiasis in AIDS patients (Dupont *et al.,* 1994). HAART consists of the combination of at least three drugs that contain minimum of protease inhibitor and /or one non-nucleoside reverse transcriptase inhibitor.

These drugs markedly reduce the HIV replication and increase the CD4-count, and thus reduce the onset of OPC episodes, PCP and cryptococcosis (Dupont *et al.,* 1994). In the absence of availability of very costly HAART in developing countries including India, cost effective surveillance and prophylactic measures have to be evolved for containing the fungal infections in AIDS patients in India. Primary prophylaxis with itraconazole for cryptococcosis is effective if CD4 + cell counts reach 100 cells mm^{-3}. Fluconazole has a role as primary prophylactic agent against *C. albicans* infection and is reasonably effective in secondary prophylaxis of cryptococcal meningitis. Primary prophylaxis with itraconazole at a daily dose of 200 mg has been achieved for cryptococcosis and histoplasmosis for patients with CD4 + cells counts below 200 cells mm^3. Itraconazole has also been demonstrated to be effective in secondary prophylaxis or treatment of penicilliosis marneffei (Graybill *et al.*, 2000). In another study, trimethoprim-sulfamethoxazole prophylaxis though not effective in preventing PCP has been found to markedly reduce the morbidity due to PCP in African children (Madhi *et al.*, 2002).

REFERENCES

Aarestrup, F.M., Guerra, R.O., Viera, B.J. and Cunha, R.M. (2001). Oral manifestations of sporotrichosis in AIDS patients. *Oral Dis* 7 : 134-136.

Abraham, M., Mathew, M., Ganesh, A., John, T.J. and Mathews, M.S. (1997). Infection caused by *Cryptococcus neoformans* var. *gattii* serotype B in an AIDS patient in India. *J Med Vet Mycol* 35 : 283-284.

Aikat, B.K., Chaterjee, B.D. and Banerjee, P.l. (1967). The rising incidence of cryptococcosis. *Indian J Med Res* 55 : 43-46.

Ajello, L., Padhye, A.A., Sukroongreung, S., Nilakul, C.H. and Tantimavanic, S. (1995). Occurrence of *Penicillium marneffei* infections among wild bamboo rats in Thailand. *Mycopathologia* 131 : 1-8.

Ayyagari, A., Sharma, A.K., Prasad, K.N., Dgole, T.N., Kishore, J. and Chaudhary, G. (1999). Spectrum of opportunistic infections in human immunodeficiency virus (HIV) infected cases in a tertiary hospital. *Ind J Med Microbiol* 17 : 78-80

Bajaj, J.S., Singh, A., Aggarwal, S.K., Chattopadhya, D. and Baveja, U.K. (2000). Synergistic immunosuppression by *Candida* in HIV infection: A cytokine based analysis. *J. Commun. Dis* 32 : 1-9

Banerjee U., Datta K., Majumdar T. and Gupta K. (2000). Cryptococcosis in India, the awakening giant? *Medical Mycology* 38 : 51-67

Baradkar, V.P. and Karyakarte, R.P. (1999). Isolation and characterization of *Candida* species in acquired immunodeficiency syndrome. *Indian J. Med. Microbiol* 17 : 42-44.

Barua, B.D., Saikia, T.C. and Bhuyan, R.N. (1975). Sporotrichosis in Assam. *India J Med Sci* 29 : 251-256

Casadevall, A. and Perfect, J.R. (1998) *Cryptococcus neoformans*. ASM Press, Washington DC.

Chakraborty, A., Jatana, M., Kumar, P., Chatha, L., Kaushal, A. and Padhye, AA (1997). Isolation of *Cryptococcus neoformans* var. *gattii* from *Eucalyptus camaldulensis* in India. *J Clin Microbiol* 35 : 3340-3342.

Chakrabarti, A, Sharma, A., Sood, A., Grover, R. and Sakhuja, V. (2000). Changing scenario of cryptococcosis in a tertiary care hospital in North India. *Indian J Med Res* 112 : 56-60.

Chariyalertsak, S., Vanittanakom, P., Nelson, K.E, Sirisanthana, T. and Vanittanakom, N. (1996). *Rhizomys sumatrensis* and *Cannomys badius*, new natural animal hosts of *Penicillium marneffei*. *J Med Vet Mycol* 34 : 105-110.

Chauhan, N.S. and Saxena, R.N. (1985). The phenomenon of bamboo flowering and associated increase in rodent population in Mizoram. *J Bombay Nat Soc* 83 : 644-647.

Chin-Hong, P.V., Sutton, D.A. and Romer, M. *et al.* (2001). Invasive fungal sinusitis and meningitis due to *Arthrographis kalrae* in a patient with AIDS. *J Clin Microbiol* 39 : 804-807.

Cogliati, M., Allaria, M., Tortorano, A.M., Liberi, G. and Viviani, M.A. (1999). Occurrence of diploidy correlated to PCR-finger printing patterns of *Cryptococcus neoformans* var. *neoformans.* 4th International Conference on *Cryptococcus* and Cryptococcosis, London, September 12-16, Abstract PA21.

Cox, G.M. and Perfect, J.R. (1998). *Cryptococcus neoformans* var. *neoformans* and *gattii* and *Trichosporon* species. *In Topley and Wislons`s* Microbiology and Microbial Infections, Vol. 4 Medical Mycology, Ajello L., Hay R.J., Eds. Arnold, pp 461-484.

Cushion, M.T. (1998). *Pneumocystis carinii.* In: Topley and Wilson's. Microbiology and Microbial Infections Vol 4. Medical Mycology, Ajello L and Hay RJ. Eds., Arnold, London, pp 645-684.

Da Silva, B.C.M., Gambia, W. and Auler, M.E. *et al.* (2001). *Rhodotorula rubra* causing skin lesions in a HIV patient. *J Mycol Med* 11 : 149-152.

Date, A., Krishnaswamy, H., John, G.T., Mathai, E., Jacob, C.K. and Shastry, J. (1995). The emergence of *Pneumocystis carinii* pneumonia in renal transplant patients in south Indian hospital. *Trans Roy Soc Trop Med Hyg* 89 : 285-287

Deng, Z., Yen, M. and Ajello, L. (1986). Human penicilliosis and its relation to the bamboo rat (*Rhizomys pruinosus*) *J Med Vet Mycol* 24 : 383-389.

Deivanayagam, C.N., Rajshekharan, S., Senthilnathan, V., Krishnarajasekhar, O.R., Raja K., Chandersekahran C., Palanisamy S., Smauel-Dinesh A., Jothivel G. and Elango S. (2001). Cinico-radiological spectrum of tuberculosis among HIV sero-positive-A Tambaram study. *Ind J Tub* 48 : 123-127.

Dhabbar, B.N., Keller, R.S., Nandkarni, K.S. and Advani, S.H. (1992). *Pneumocystis carinii* in untreated acute nonlymphoblastic leukaemia. *J Assoc Physicians India* 40 : 641-642.

Dompmarin, D., Dompmarin, A., Deluol, A.M., Grosshans, E. and Couland, J.P. (1990). Onychomycosis in AIDS: Clinical and laboratory findings in 62 patients. *Int J Dermatol* 29 : 337 339.

Dupont, B., Denning, D.W., Marriot, D., Sugar, A., Viviani, M.A. and Srisanthana, T. (1994). Mycoses in AIDS patients. *J Med Vet Mycol* 32 (Suppl. 1) : 65-78.

Duong, T.A. (1996). Infection due to *Penicillium marneffei,* an emerging pathogen: review of 155 reported cases. *Clin Infect Dis* 23 : 125-30.

Dupont, B., Denning, D.W., Marriot, D., Sugar, A., Viviani, M.A. and Sirisanthan, T. (1994). Mycoses in AIDS patients. *J Vet Med Mycol* 32 : 65-77.

Duraiswamy, R., Diwan, R.N., Kalghatgi, A.T. and Mukerjee, B. (1997). Unusual pathogens in HIV positive cases. *Med J Armed Forces India* 54 : 1170-1172.

Elerman, J.R. (1961). The fauna of India including Pakistan, Burma and Ceylon, Mammalia; Manager of Publications, Govt. of India, New Delhi 1-884.

Ellis, D.H and Pfeiffer, T.J. (1990). Natural habitat of *Cryptococcus neoformans* var. *gattii. J Clin Microbiol* 28 : 1642-1644.

Finquelievich, J.L., Odds F.C., Queiroz-Telles F. and Wheat L. (2000). Advances in antifungal treatment. *Medical Mycology* 38 : 317-322.

Franzot, S.P, Salkin, I.F. and Casadevall, A. (1999). *Cryptococcus neoformans* var. *grubii*: separate varietal status for *Cryptococcus neoformans* serotype A isolates. *J Clin Microbiol* 37 : 838-840.

Fortes, S.T., Lazera, M.S., Nichikawa, M.M., Macedo, R.CI. and Wanke, B. (2001). First isolation of *Cryptococcus neoformans* var. *gattii* from a native jungle tree in the Brazilian Amazon rainforest. *Mycoses* 44 : 137-140.

Ghosh, A., Chakrabarty, A., Sharma, V.K., Singh, K. and Singh, A. (1999). Sporotrichosis in Himachal Pradesh (North India). *Trans Roy Soc Trop med Hyg* 93 : 41-45.

Giri, T.K., Pande, I., Mishra, N.M., Kailash, S., Uppal, S.S. and Kumar, A. (1995). Spectrum and laboratory characteristics of HIV infection in northern India. *J Commun Dis* 2 : 731-741.

Gori S., Drouhet E., Gueho E., Huerre M., Lofaro A., Parenti M. and Dupont B. (1998). Cutaneous disseminated mycosis in a patient with AIDS due to a novel dimorphic fungus. *J Mycol Med* 8 : 57-63.

Gori S., Lupetti A., Moscato, G., Parenti, M. and Lofaro, A. (1997). Pulmonary sporotrichosis with hyphae in a human immunodeficiency virus-infected patient: a case report. *Acta Cytol* 41 : 519-521.

Graybill, J.R., Tollemar, J., Torres-Rodriguez, J.M., Walsh, T.J., Roilides, E. and Farmaki, E. (2000). Antifungal compounds: controversies, queries and conclusions. *Medical Mycology* 38 (Suppl. 1) : 323-333.

Gugnani H.C. (2000). Histoplasmosis in Africa: a review. *Ind J Chest Dis All. Sc* 42 : 271-277.

Gugnani, H.C. (2000). Epidemiology of systemic mycoses In Proceedings of The First International Symposium on "Mycoses in The New Millenium: Challenges and Strategies" held at Sri Ramachandra Medical College & Research Institute (Deemed University), Porur, Chennai, Tamil Nadu, 9-11 February 2000, pp 28-48.

Gugnani, HC, Becker, K, Fegeler, W, Basu, S, Chattopadhya, D, Satyanarayana, S, Kalghatgi, T and Murlidhar A. (2003a). Oropharyngeal carriage of *Candida* species in HIV-infected patients in India. *Mycoses* 46 (in press)

Gugnani, H.C., Kuamr, A., Basu, S. and Joshi, A. (2003b). Isolation of *Cryptococous neoformans* var. *grubii* and *Cryptococcous neoformans* var. *gattii* from *Eucalyptus* trees. Submitted to *J Mycol Med.*

Gugnani, H.C., Sandhu, R.S. and Shome, S.K. (1976). Prevalence of *Cryptococcus neoformans* in avian habitats in India. *Mykosen* 19 : 1983-1987.

Gugnani, H.C., Shome, S.K., Shrivastav, J.B., Murthy, D.K. and Gupta, N.P. (1967). *Cryptococcus neoformans* (Sanfellice) Vulleimin from natural habitats in India. *Indian J Med Res* 55 : 1290-1294.

Gyaurgieva, O.H., Bogomolova, T.S. and Gorshkova, G.I. (1996). Meningitis caused by *Rhodotorula rubra* in an HIV-infected patient. *J Med Vet Mycol* 14 : 357-59.

Hejney, C., Kerrison, J.B., Newman, N.J. and Stone, C.M. (2001). Rhino-orbital mucormycosis in a patient with acquired immunodeficiency syndrome. *Am J Ophthalmol* 32 : 111-112

Holmberg K. and Meyer R.D. (1986). Fungal infection in patients with AIDS and AIDS related complex. *Scand J Infect Dis* 18 : 179-192.

Hughes, T.R. (1998). Current issues in the epidemiology, transmission and reactivation of *Pneumocystis carinii. Sem Resp Infect* 13 : 283-288.

Johnson, L.B, Breadley, S.F. and Kaufman, C.A. (1998). Fungaemia due to *Cryptococcus laurentii* and a review of non-neoformans cryptococcemia. *Mycoses* 41 : 277-80.

Kakghatgi, A.T, Satyanarayana, S, Gugnani, H.C. and Malviya, A.K. (2000). Disseminated histoplasmosis in AIDS: A case report. *Indian J Med Microbiol* 18 : 42-43.

Khan, Z.K., Katiyar, R., Shukla, R., Kulshrestha, R. and Mathur, A. (1986). Cryptococcosis associated with HIV negative Indian patients and HIV positive Indian patients. *Zbl Bakt* 283 : 369-374.

Khanna, N., Chandramukhi, A, Desai, A. and Ravi, V. (1996). Cryptococcal infection of the central nervous system: an analysis of predisposing factors, laboratory findings and outcome in patients from South India with special reference to HIV infection. *J Med Microbiol* 45 : 376-379.

Kidd, S.E., Hitchcock, C., Maszewska, K., Balakrishnan, P., Hunt, C., Rajendran, P., Usha, A.R. Thyagarajan, S.P., Sorrell, T.C. and Meyer, W. (2001). Molecular epdemiology of clinical and environmental isolates of *Cryptococcus neoformans* from Tamil Nadu, India. *Austral Mycol* 20 : 105-114.

Kiraz, N., Gulbas, Z. and Akgun, Y. (2000). *Rhodotorula rubra* fungaemia due to use indwelling catheters *Mycoses*; 43: 209-210.

Kirchner, J.T. (1996). Opportunistic fungal infections in patients with HIV disease. *P.G. Med* 99 : 209-216.

Kordossi, T., Avlami, A. and Velegraki, A. (1998). First report of *Cryptococcus laurentii* meningitis and fatal case of *Cryptococcus albidus* cryptococcemia in AIDS patients. *Medical Mycology* 36 : 335-339.

Kumarasamy, N., Solomon, S., Jayakar Paul, S.A., Venilla, R. and Amalraj, R.E. (1995). Spectrum of opportunistic infections in AIDS patients in Tamil Nadu, India. *Int J SDT AIDS* 6 : 447-449.

Kwon-Chung, K.J. and Bennett, J.E. (1992). *Medical Mycology* Lea and Febiger, Philadelphia, pp 397-446.

Lanjewar, D.N., Anand, B.S. and Genta, R. *et al.* (1996). Major differences in the spectrum of gastrointestinal infections associated with AIDS in India versus the West: an autopsy study. *Clin Infect Dis* 23 : 482-485.

Lazera, M.S, Cavalcani, M.A.S., Trilles, L., Nishikawa, M.M. and Wanke, B. (1998) *Cryptococcus neoformans* var. *gatti*–evidence for a natural habitat related to decaying wood in a pottery tree hollow. *Medical Mycology* 36 : 119-122.

Lazera, M.S., Piers, F.D. A. Camillo-Coudra, L., Nishikawa, M.M., Bezzera, C.C., Trilles, L and Wanke, B. (1996). Natural habitat of *Cryptococcus neoformans* var. *neoformans* in decaying wood forming hollows in living trees. *J Med Vet Mycol* 34 : 127-131

Lovell, R.D., Moll, M., Allen, J., Cicci, L.G. (2002). *Paecilomyces lilacinus* infection in a patient with AIDS. *AIDS Read* 12 : 212-213.

Madhi, S.A, Ceilland, C., Ismail, K., O` Reilly, C., Manch, A. and Klugmar, K. (2002). Ineffectiveness of sulfamethoxazole–trimethoprim prophylaxis and the importance of bacterial and viral co-infection in African children with *Pneumocystis carnii* pneumonia. *Clin Infec Dis* 35 : 1120-1126.

Malin, A.S., Gwanzuar, L.K.Z. and Klein, S. *et al.* (1995). Infection with *Pneumocystis carinii* in Zimbabwe. *Lancet* 11 : 1258-1262.

Marques, S.A., Robles, A.M., Tortorano, A.M., Tucelt, M.A., Negroni, R. and Mendes, R.P. (2000). Mycoses associated with AIDS in the Third World. *Medical Mycology* 38 (Suppl.1) : 259-268.

Mathew, M. and Mathai, E. (2000). Emerging importance of *Pneumocystis carinii* among immunosuppressed Indian patients. *Indian J Chest Dis Allied Sci* 42 : 323-324.

Misra, S.N., Sengupta, D. and Satpathy, S.K. (1998). AIDS in India: recent trends in opportunistic infections. *Southeast Asian J Trop Med Public Health* 29 : 373-376.

Mitchell, TG. and Perfect, J.P. (1995). Cryptococcosis in the era of AIDS. 100 years after discovery of *Cryptococcus neoformans*. *Clin Mirobiol Rev* 8 : 515-548.

Moolcahnd, R., Singh, A.P. and Salopal, T.K. (2000). Histoplasmosis in acquired immunodeficiency syndrome. *Med J Armed Forces India* 56 : 176-168.

Nagrajan, S., Gugnani, H.C. and Kowshik, T. (2000). Meningitis due to *Cryptococcus neoformans* var. *neofromans* serotype AD associated with pulmonary tuberculosis. *Mycoses* 43 : 67-69.

Negroni, R. (2000). Clinical spectrum and treatment of classical histoplasmosis *In*: Biology of Deramtophytes and related keratinophilic fungi. Kushwaha R.K.S. and. Guarro J. Eds., *Revista Iberoamericana de Micologia*, Apartado, Bilbao, Spain

Padhye, A.A, Chakrabarti, A., Chander, J. and Kaufman, L. (1993). *Cryptococcus neoformans* var. *gattii* in India. *J Med Vet Mycol* 31 : 165-168.

Peerapur, B.V., Mantur, B.G., Smita, S., Mangalagi, S.S. and Birdar, M.S. (2000). Infection caused by *Cryptococcus neoformans* var. *gattii* serotype B in an AIDS patient in Bijapur. *Indian J Med Microbiol* 18 : 45.

Pereiro, M., Losada, A. and Toribio, J. (1997). Adherence of *Candida albicans* strains isolated from AIDS patients. Comparison with pathogenic yeasts isolated from patients without HIV infection. *Br J Dermatol* 137 : 76-80.

Pfaller, M.A., Marco, F., Messer, S.A. and Jones, R.N. (1998). *In vitro* activity of two echinocandin derivatives, LY303346 and MK-0991 (L-743, 749), against clinical isolates of *Aspergillus, Fusarium, Rhizopus,* and other filamentous fungi. *Diagn Microbiol Infect Dis* 30 : 251-255.

Pfaller, M.A., Rhune-Chung, J. and Barry, H.J. *et al.* (1995). Strain variation and antifungal susceptibility among bloodstream infection of *Candida* species from 21 different medical institutions. *Clin Infect Dis* 21 : 1507-1509.

Pfaller, M., Zhang, J., Messer, S., Tumberland, M., Mbidde, E., Jessup, C. and Ghannoum, M. (1998). Molecular epidemiology and antifungal susceptibility of *Cryptococcus neoformans* isolates from Ugandan AIDS patients. *Diagn Microbiol Infect Dis* 32 : 191-192.

Ponton, J., Ruchel, R., Clemons, KV. and Coleman, D.C. *et al.* (2000). Emerging pathogens. *Medical Mycology* 38 (Suppl.1) : 225-236.

Randhawa, H.S., Mussa, A.Y. and Khan, Z.U. (2000). Decaying wood in tree trunk hollows as a natural substrate for *Cryptococcus neoformans* and other yeast-like fungi of clinical interest. *Mycopathologia* 151: 63-69.

Ranganathan, K., Reddy, B.Y. and Kumaraswamy, N. *et al.* (2000). Oral lesions and conditions associated with human immunodeficiency virus infection in 300 south Indian children. *Oral Dis* 6 : 52-157.

Rani, N.U., Reddy, V.V.V.R., Kumar, A.P., Kumar, V., Babu, G.R. and Rao, D.B. (2000). Clinical profile of *Pneumocystis carinii* pneumonia in HIV infected persons. *Ind J Tub* 47 : 93-96.

Ranjana, K.H., Priyaokumar, K., Singh, T.J., Gupta, C.C., Sharmila, L., Singh, P.N. and Chakrabarti, A. (2002). Disseminated *Penicillium marneffei* infection in HIV-infected patients in Manipur State, India. *J Infect* 45 : 268-271

Redding, S.W., Kirkpatrick, W.R., Dib, O., Fothergill, A.W., Rinaldi, M.G. and Patterson, T.F. (2000). The epidemiology of non-albicans *Candida* in oropharyngeal candidiasis in HIV patients. *Spec Care Dentist* 20 : 178-181.

Sanyal, M. and Thammaya, A. (1975). *Histoplasma capsulatum* in the soil of Gangetic plain in India. *Indian J Med Res* 63 : 1020-1028.

Sethi, K.K., Randhawa, H.S., Abraham, S., Misra, S.K. and Damadoran, V.N. (1966). Occurrence of *Cryptococcus neoformans* in pigeon excreta in Delhi. *Indian J Chest Dis* 8 : 207-201.

Sethi, M. and Mandell, W. (1988). *Saccharomyces* fungemia in a patient with AIDS. *N.Y. State J Med* 88 : 278-79.

Sherke, R.L., Elizabeth, M., Hemalatha, K., Jacob, C.K. and Lalitha, M.K. (1992) *Pneumocystis carinii* pneumonia in a renal transplant recipient in India. *J Assoc Physicians India* 40 : 410-412.

Shome, S.K., Sirkar, D.K. and Gugnani, H.C. (1973). Changing spectrum of cryptococcosis in Delhi. *Indian J Med Res* 61 : 23-29

Singh, P.N., Ranjana, K., Sing, Y.I., Singh, P., Sahrma, S.S., Kulachandra, M., Nabakumar, Y., Chakrabarti A., Padhye A.A., Kaufman L. and Ajello L.

(1999). Indigenous disseminated *Penicillium marneffei* infection in the State of Manipur, India: Report of four autochthonous cases. *J Clin Microbiol* 37 : 2699-2702.

Speed, B. and Dunt, D. (1995). Clinical and host differences with the two varieties of *Cryptococcus neoformans. Clin Infect Dis.* 21 : 28-34.

Srisanthana, T. and Supparatpinyo, K. (1998). Epidemiology and management of penicilliosis in human immunodeficiency virus-infected patients. *Int J Infect Dis.* 3 : 48-53.

Sullivan, D.J., Westernberg, T. J., Haynes, K.A., Bennett, D.E. and Coleman, D.C. (1995). *Candida dubliniensis* sp.nov.: phenotypic and molecular characterization of novel species associated with oral candidiasis in HIV-infected individuals. *Microbiol* 41 : 1507-1521.

Tan, G., Kaufman, L., Paterson, E.M., Mada, L. and de la, M. (1993). Disseminated atypical blastomycosis in two patients with AIDS. *Clin Infect Dis* 16 : 107-111.

Vanitanakom, N., Cooper C.R., Chariyalertsak S., Yongchim S., Nelson K.E., Sirisanthana T. and Vanittanakom N. (1996). Restriction analysis of *Penicillium marneffei. J Clin Microbiol* 34 : 1834-1836.

Verghese, S., Arjundas, D., Krishnakumari, K.C., Padmaja, P., Elizabeth, D. and Padhye, A.A. (2002). Coccidioidomycosis in India: report of a second imported case. *Medical Mycology* 40 : 307-309.

Verma, S., Bansal, M.M., Das, A., Kumari, S. and Sharma, B.K. (1994). *Pneumocystis carinii* pneumonia complicating Hodgkin`s disease: a case report. *Indian J Chest Dis Allied Sci* 36 : 91-95.

Viviani, M.A. and Tortarano, A.M. (1998). *Penicillium marnefffei.* In *Microbiology and Microbial Infections.* Vol. 4. *Medical Mycology,* Ajello L. and Hay RJ. Eds., Arnold, Lond. pp 409-422.

Wakefield, A.E., Stewart, T.J. and Moxon, E.R. *et al.* (1990). Infection with *Pneumocystis carinii* is prevalent in healthy Gambian children. *Trans Roy Trop Med Hyg* 84 : 82-84.

Westerman, K. (1999). HIV disease and the skin. Part II. *Spec Med Deramtol* 2 : 105-109.

CHAPTER - 21

KERATINOPHILIC FUNGI : DIVERSITY AND SENSITIVITY TO SOME MEDICINAL HERBS

Shaista Qureshi , S.C. Agrawal** and M.K. Rai**

*Microbiology Research Lab., Danielson College, Chhindwara-480 001, M.P.

**Department of Applied Microbiology and Biotechnology, Dr H.S. Gaur Vishwa Vidhyalaya, Sagar, M.P.

***Department of Biotechnology, Amravati University, Amravati-444 602, Maharashtra

INTRODUCTION

Keratin is the cornified part of the epidermis of vertebrates, which includes feather, hair, nail, horn, wool, etc. After the death of animal or insects the remains are added to the soil. A lot of substrate rich in keratin may also be added to the soil. Keratin can be divided into soft and hard keratin depending on the cystine content in it (Nickerson, 1947). Soft keratin occurs in skin and contains approximately 2 per cent cystine, whereas hard keratin is present in hair and contains approximately 14 per cent cystine.

The ability to degrade keratin is a common and universal phenomenon among fungi. The mold capable of attacking keratinized tissues are termed as keratinophilic fungi. Keratinophilic fungi include the genera *Arthroderma* and *Nannizzia* or *Gymnoascus* which belong to the order eurotiales of primitive ascomycotina. A number of keratinophilic fungi are able to decompose keratin, which is part of man and animals. English (1967) categorized these fungi in two groups:

(i) **Keratinophilic**: Fungus forms able to digest keratinized substrate completely or partly.

(ii) **Keratinolytic**: Fungus forms able to digest keratinized substrate completely.

Some keratinophilic fungi are called dermatophytes because of their capacity to parasitize keratinic tissues in man and animals. A dermatophyte, according to the definition of Hawksworth *et al.* (1983) is a fungus which parasitizes keratinized tissues in man and animals and causes dermatophytosis. The dermatophytosis, which is also known as cutaneous mycosis are infections of keratinized epidermal tissues of human beings. All the dermatophytes are keratinolytic in nature.

Many of the keratinophilic/ dermatophytes are biphasic/ polymorphic and several species have been erected for clinical reasons without considering the comparative morphological studies *in vitro.* This has created problem in identification and in taxonomy of keratinophilic fungi.

On the basis of the morphology and sexual reproduction dermatophytes are included in *Arthroderma* or in *Nannizzia,* two closely related genera within the family gymnoascaceae, order eurotiales in the ascomycotina. Sabouraud (1910) recognized three genera of dermatophytes: *Microsporum, Trichophyton* and *Epidermophyton,* which are based on clinical rather than mycological observations. A significant contribution has been made on classification of dermatophytes (Vanbreuseghem, 1952; Smith and Marples, 1963 and Takashio, 1973).

Keratinophilic fungi and related dermatophytes are increasingly popular subject now-a-days for study but even today, our knowledge regarding the distribution of keratinophilic fungi and related dermatophytes and their variants is meager and fragmentary.

The main aim of the present chapter is to discuss distribution of keratinophilic fungi, *in vitro* perforation test and their sensitivity to plant extracts and oils.

UBIQUITOUS NATURE

A survey of literature reveals that keratinophilic fungi have been reported from different parts of the world (Vanbreuseghem, 1952; Daniels, 1954; Marples, 1955; Duries and Frey, 1955, 1957; Dunna and Morahan, 1964; Otsenasek and Drorak, 1964; Taylor *et al.,* 1964; Varsavasky, 1964; Turner, 1965; Ajello *et al.,* 1965a; Ajello *et al.,* 1965; Evolceanu and Alteras, 1966; Al-Doory, 1967; Al-Doory and Kalter, 1967; Baxter, 1969; Mohammed *et al.,* 1971; Rogers, 1971; Ajello and alpert, 1972; Ajello and Padhye, 1974; Feurman *et al.,* 1975; Caretta and Pointelli, 1975, 1977; Muhammad and Lalji, 1978; Marcantini *et al.,* 1980, 1983; Guarro *et al.,* 1981; Michel-Brun and Monier, 1981; Abdel-Fateh *et al.,* 1982; Ulfig and Koreez, 1983; and Calvo *et al.,* 1984). There are various reports of occurrence of keratinophilic fungi from India (Dey and Kakoti, 1955; Randhawa *et al.,* 1959; Randhawa and Sandhu, 1963; Padhye *et al.,* 1966, 1967; Jain and Agrawal, 1977; Sur and Ghosh, 1980; Verma *et al.,* 1982; Deshmukh and Agrawal, 1983; Jain *et al.,* 1985; Deshmukh, 1985; Randhawa and Sandhu, 1965; Garg, 1966; Kushwaha, 1980a, 1980b, 1983, Kushwaha and Agrawal, 1976a, 1976b and Verma *et al.,* 1982; Nigam and Kushwaha, 1985; Singh *et al.,* 1990;

Malviya *et al.,* 1991, 1992; Qureshi, 1991; Rajak *et al.*, 1991; Rai and Qureshi, 1994; Mitra *et al.*, 1998; Ramesh and Hilda, 1999; Deshmukh, 2002 a, b; Deshmukh and Agrawal, 2003; Kushwaha and Tripathi, 2004) as is mentioned in Table 1.

Table 1. Geographical distribution of Keratinophilic Fungi

Authority	Year	Country
Danials	1954	Great Britain
Duries and Frey	1955	Australia and New Guinea
Duries and Frey	1957	New South wales
Marples	1955	Newzealand and Polynsian Island
Randhawa *et al.*	1959	New Delhi
Randhawa and Sandhu	1962, 1965	India
Dunna and Morahan	1964	Queensland
Otcenasek and Dvorak	1964	Czechoslovakia
Taylor *et al.*	1964	Egypt, Sudan and Ethiopia
Varsavsky	1964	Argentina
Ajello *et al.*	1965	Belgium
Ajello *et al.*	1965	Italy
Caretta and Piontelli	1975	Italy
Turner	1965	Hong kong
Evolceanu and Alteras	1966	Rumania
Al-Doory	1967	Texas
Al-Doory and Kalter	1967	East Africa
Baxter	1969	England
Mohd. *et al.*	1971	West-Pakistan
Rogers	1971	Colombia
Ajello and Alpert	1972	Easter Island
Ajello and Padhye	1974	Galapagos Island
Feuerman *et al.*	1975	Israel
Caretta and Piontell	1975	Italy
Caretta and piontell	1977	Australia and Chile
Mohd. and Lalji	1978	Kenya
Mercantini *et al.*	1980, 1983	Rome
Guarro *et al.*	1981	Tarrangona
Michel-Brun and Monier	1981	Lyon's region
Abdel-Fattah *et al.*	1982	Egypt
Calvo *et al.*	1984	Spain
Filipello	1986	Turin
Chabasse	1988	France
Ali-Shtayeh and Arda	1989	Jorden
Soomro *et al.*	1990	Pakistan
Repova	1990	Czechoslovakia

Del-frate and caretta	1990	Antarctic
Filipello *et al.*	1991	Papua-New-Guinea
Soon	1991	Malaysia
Volz *et al.*	1991	Ukrain USSR
Moubahser *et al.*	1992	Egypt
Youssef *et al.*	1992	Egypt
Caretta	1992	Italy
Mercantini *et al.*	1993	Antarctic
Volz *et al.*	1993	St. Petersburg Russia
El-said	1993	Oman
El-said	1994	Yemen Arab Republic
Abdullah and Hassan	1995	Iraq
Volz	1995	Israel
El-said and Abdel-Sater	1995	El-Bahrain
Al-Musallam	1995	Kuwait
Agut *et al.*	1996	France
Gene *et al.*	1996	Spain
Currah *et al.*	1996	Canada
Gherbawy	1996	Qena city Egypt
Cabanes *et al.*	1997	Spain
Simpanya and Baxter	1997	Zambia
Mancianti *et al.*	1997	Italy
Parihar	1998	India
Ramesh and Hilda	1999	India
Deshmukh *et al.*	2000	India
Deshmukh and Shukla	2000	India
Srivastava *et al.*	2001	India
Deshmukh and Agrawal	2003	India

Soil samples rich in organic matter are more suitable for keratinophilic fungi. Such soils are present in shady and moist sites, e.g., garden, public lawn, heap of compost, sludge and cropfield or from sites rich in dung and keratin substrates. On the other hand, saline soil, dry river sand, beach sand, samples of barren road side soils proved to be poor sources of keratinophilic fungi (Randhawa and Sandhu, 1965).

The enrichment of soil with keratinous material is most conducive to the growth and occurrence of keratinophilic fungi (Padhye *et al.*,1967, Otcenasek, 1978; Mercantini *et al.*, 1980, 1983). The collection sites of Galapagos Island (Ajello and Padhye, 1974) were unusual. The xeric low lying coastal areas and the soil was essentially made up of disintegrated lava, low in organic matter. Report of *C. indicum* and *C. keratinophilum* in soil samples from Antarctic region having a temp of -4^{0}C are considered worthy of attention because of their adaptability and tolerance to temperature (Caretta and Piontelli, 1977).

The keratin bait technique also facilited the discovery of the perfect states of a good number of dermatophytes leading to a natural classification of these fungi. (Vanbreuseghem, 1952; Ajello, 1961; and Ajello and Cheng shu Lau, 1967). Based on habitat, nature and epidemiology, dermatophytes were classified into three broad groups by Ajello (1960):

(i) **Geophilic** : These fungi are saprobic, occur mainly in soil and are rarely pathogenic.

(ii) **Zoophilic** : These fungi are mainly parasitic to lower animals and transmitted through contact.

(iii) **Arthropophilic**: These fungi are mainly parasitic on humans and cause dermatophytosis.

Rai and Qureshi (1994) found that keratinophilic fungi differ in their substrate preference for colonization.

Soil is a good source of most of fungal infections, both for men and animals. Taking into consideration, keratinophilic fungi were isolated from a great variety of soil, particularly from soils containing keratinous substrates as animal and human remains.

Randhawa *et al.* (1959) isolated *M. gypseum* from soil samples collected from rodent burrows and fowl pen. Keratinophilic fungi were not encountered in the salty marshes of coastal Mediterranean areas (Abdel-Fateh *et al.*, 1982). These fungi have been recovered from dung (Caretta *et al.*, 1976), sewage sludge (Ulfig and Koreez, 1983) and school floor (Mercantini *et al.*, 1983). Padhye *et al.* (1967) reported the occurrence of dermatophytes and keratinophilic fungi in marine soils of Bombay (Pugh and Mathison, 1962; Dabrawa *et al.*, 1964 and Padhye *et al.*, 1967).

Keratinophilic fungi were categorized into various groups on the basis of vertical distribution in grassland and forest soils. In both types of soil the upper profiles contained the most diverse population. However, the number of isolates from the deeper layers decreased as the distance from the upper surface increased. Chmel and Vlacilikova (1975) recovered keratinophilic fungi from a depth of 55 cm. The differences in species number in soil profiles may be associated with adaptation to development at low partial pressure of oxygen and high concentrations of CO_2 in deep sites. The vertical distribution also seems to be associated with the concentration of keratinic substrates and with the composition of the soil atmosphere.

A number of surveys revealed the occurrence of keratinophilic and cellulolytic fungi on birds and in their nests (Pugh, 1964, 65; and Rees, 1968). Colonization of *Chrysosporium* species on birds and on their nests have been reported by several workers (Ajello, 1953; Buhimann and Riath, 1962; Pugh, 1965, 66; Otcenasek *et al.*, 1967; Rees, 1968; and Hubalek *et al.*, 1974).

The occurrence and survival of keratinophilic fungi has been reported to be influenced by various climatic factors, such as, chemical composition of organic matter in soil, its quantity as source of nutrition and energy, soil humidity, pH and temperature, the depth of soil profile, soil texture and structure (Chmel *et al.*, 1972 and Vollekova, 1984).

The majority of dermatophytes live saprophytically in soil and every keratinophilic fungus can be considered as a potential pathogen (Ajello, 1953; Mohammed and Lalji,1978; Ali-Shtayeh and Arda, 1985). They are ecologically restricted to keratinous substratum and inhabits in soil debris.

Dermatophytes, a homogeneous group of keratinophilic fungi cause most of the superficial skin diseases. Sabouraud (1893) assumed that dermatophytes are primarily soil saprophytes as they grow on decaying vegetables in soil.

Keratin rich ground is the natural habitat of keratinophilic fungi/dermatophytes. These fungi are saprophytes with a specialized ability to infect susceptible host. It may be worthwhile to verify whether or not the ground in public parks, frequented by children and pets could be a breeding ground for potentially pathogenic fungi.

A number of domestic animals are known to carry dermatophytes on their skin coats without manifesting any sign of infection. The origin of several cases of human ringworm has been traced to animal harboring the etiologic dermatophyte species. Lindquist (1960) reported ringworm due to *M. canis* in children, developing after visiting dog and cat shows where the exhibited animals were found to be lesion free carriers of this dermatophyte. Gentles *et al.* (1965) mentioned many persons having contracted ringworm from a dog with an asymptomatic infection due to *M. canis.* The constant animalization plays a significant role in the establishment of keratinophilic fungi in soil sites not frequented by animals seem to possess less favorable conditions for the growth and survival of such fungi.

The majority of the keratinophilic fungi though seldom involved in dermatophytoses, have the potential for it as one can see natural evolution from keratin utilizing soil saprophytes to association and finally invasion of cornified substrate in living animals and men. Chmel and Vollekova (1981) also observed that atmosphere humidity, temperature, nature and reaction of soil play an important role in the passage from saprophytism to parasitism amongst the keratinophilic fungi. Thus, there is a tendency in keratinophilic fungi to follow human settlement. Some keratinophilic fungi can cause human mycoses due to their colonization habit on dermal appendages like hair. They may become a source of sanitary danger to human health (Hasija *et al.*, 1988 and Rajak *et al.*, 1989). Human disease caused by keratinophilic fungi is difficult to control epidemiologically and presents a growing problem.

In certain circumstances many geophilic fungi can produce opportunistic fungal infections on men and other animals. This lead to many researches on keratin degradation specially keratin present in hair by those fungi. Cano *et al.* (1991) studied the hair destruction by *Aphanoascus fulvescens, A. verrucosus* and *A. keratinophilus.* Fusconi and Fillipello-Marchisio (1991) investigated the pattern of invasion of *C. tropicum. In vitro* hair perforation by eight species of *Chrysosporium* was studied by Bahuguna and Kushwaha (1989). Nigam and Kushwaha

(1990) also studied the hair deterioration by *Myceliopthora* anamorph of *Arthroderma tuberculatum.*

DIVERSITY OF KERATINOPHILIC FUNGI IN SOIL SAMPLES OF CHHINDWARA: A CASE STUDY

A systematic account of all the keratinophilic fungi isolated from various soil samples of Chhindwara using different keratin baits has been described by Qureshi *et al.* (2004).

Cultural characteristics of selected keratinophilic fungi

The morphology and cultural account of all the keratinophilic fungi isolated from Chhindwara is given. In most cases, the cultures were purified by single spore isolation. In all, 9 genera and 14 species of keratinophilic fungi were isolated.

Chrysosporium tropicum Carmichael

The growth of the fungus was slow on sabouraud's dextrose agar. Colony diameter reached up to 4.5 cm in 10 days of incubation at 28^{0}C, colony white, powdery and dense at the center. It is less than 1mm high at the center with fimbriate margin. Reverse of the colony was pale creamy-yellow. Racquet hyphae were absent. Terminal and lateral conidia were sessile or on short protrusions, solitary, subhyaline, smooth and slightly thick-walled. They were obovoid to clavated in shape. Conidia were 1-celled rarely 2-celled, 3.5-7.5 (10) x 3-4.5 µm in size with wide basal scars. Intercalary conidia were absent.

C. indicum (Randhawa and Sandhu) Garg

The growth rate by *C. indicum* exhibited very slow on sabouraud's dextrose agar. Colony diameter attained 3.5 cm in 10 days of incubation at 28^{0}C. Colony was white, thin, dense and up to 2 mm high at the center. The margin of the colony was defined regular fimbriate. Reverse of the colony was cream coloured. Hyphae were 1.5-5 µm wide. Racquet hyphae were present in the colony. Terminal and lateral conidia were subhyaline, smooth to slightly echinulate and thinwalled. Conidia were obovoid to ellipsoid, narrow, often cymbiform and 1-celled. The size of the conidia was 3.5-7.5 x 1.5-3 µm with fairly wide basal scars. Intercalary conidia were less abundant. Chlamydospores were absent.

Microsporum gypseum (Bodin) Gurart and Grigorakis

Colony of *M. gypseum* exhibited moderate growth on sabouraud dextrose agar. It was white and floccose in initial stage, later it became granular to powdery and cinnamon-brown in colour with central umbo. The border of the colony was irregularly fringed. The reverse of colony was buff to reddish-brown in colour. Two type of aleuriospores were present in the colony. Macroaleuriospores abundant, elliptic and spindle shaped with rounded ends, moderately thick-walled and echinulate, 3 to

9-celled, size 30-50 x 8-10μm. Microaleuriospores 1-celled, clavate and sessile along the hyphae.

Trichophyton terrestre Durie and Frey

The colony was moderately fast growing on sabouraud dextrose agar, granular, initially white in colour and fluffy in texture with a pale-yellow granular center, later, it became velvety and yellowish-brown. The reverse of the colony was yellowish-brown colour. Two types of aleuriospores were present in the colony. Macroaleuriospores were abundant, clavate to cylindrical with rounded ends, smooth, thin-walled and sessile, 2-6-celled, size 8-30 x 4-5 μm. Microaleuriospores were typically long and cylindrical, sessile singly or in clusters on hyphae, size 4-6 x 3-4 μm. Chlamydospores were present in the colony.

T. rubrum (Downy type) (Castellani) Sabouraud

The colony of *T. rubrum* was slow growing on sabouraud dextrose agar. It was white, granular to cottony in texture with central umbo and a flat periphery. The reverse of the colony was wine-red. Macroaleuriospores were absent in colony. Microaleuriospores scanty, 1-celled, ovoid to clavate to pyriform, born laterally along the hyphae, size 3-5 x 2-3 μm.

T. simii (Pinoy) Stockdale, Mackenzie and Austwick

The colony was granular with diffuse margin and white to pale-buff in colour. The reverse of the colony was initially colourless, and later, yellowish-brown. Macroaleuriospores, abundant, cylindrical to fusiform in shape, thin-walled and smooth, 3-10-celled, arranged in clusters, size 35-85 x 6-10 μm. In older culture, the formation of endochlamydospores was noted. They were pyrifrom to clavate in shape.

Histoplasma capsulatum Darling

The colony was slow growing on sabouraud dextrose agar, cottony, velvety in texture and whitish-yellow in colour. Colony diameter attained up to 3 cm in 10 days on sabouraud dextrose agar, oval shaped. The reverse of the colony pale-yellow to yellowish-brown. Macroaleuriospores were 1-celled, large, thick-walled with yellowish pigment. spherical, tuberculate or smooth-walled with some bearing finger-like appendages, size 8-12 μm. Microaleuriospores were small round with smooth to echinulate walls.

Gleocladium catenulatum Gilman and Abbott

The colony exhibited moderate growth on sabouraud dextrose agar, white in initial stage, later, becoming greenish-yellow. The reverse of the colony was yellowish-cream. Conidiophores were branched, 50-125 μm long. The size of primary branches was 15-20 x 3.5-4 μm. The measurement for metulae was 7-9 x 1.5-2.5 μm and phialides were 10 -

20 μm long. Conidia elliptical, smooth and greenish-yellow, 4-7.5 x 3-4 μm in dia.

Myrothecium roridum Tode

The colony was brownish-white in initial stage and later it became greyish-black. Sporodochia shield-shaped, sessile and black, 2-6 mm in diameter. Conidiophores were unbranched or forked and bush-like. The conidiophores were 30-40 μm long and 2 μm wide in size. Conidia cylindrical and truncated at both ends, 8-10 x 1.5-2 μm in dia.

Verticillium negrescens Nees

Colony rapid growing, spreading on sabouraud dextrose agar, white in colour, cottony in texture and egg-shaped 2.5-3 cm on sabouraud dextrose agar. The reverse of colony was cream-yellow in initial stage but it became brown in later stage. Conidiophores were branched phialides were laginiform and hyaline. Conidia were 1-2 celled in structure, hyaline. They were ovoid to allantoid in shape. They occur as balls at the apices of the phialides. The condia were 3-10 x 1.5 -2.5 μm in size, 2-celled, 16-18.18 x 3-3.5 μm in dia..

Malbranchia Sp.

Colony was moderately rapid growing on sabourad dextrose agar, flat, powdery and orange in colour. The reverse of the colony was dark-yellow brown to tan in colour. Racquet hyphae were present. Conidiophores were absent. Arthroconidia were 1-celled, cylindrical, straight or curved, truncate and hyaline. Conidia 2.5-5 x 2-3.5 μm in size. Ascocarps formed in old cultures.

Aphanoascus fulvescens (Cooke) Apinis

Colony was moderately fast growing on sabouraud dextrose agar, attained dia. up to 3.5 cm in 8 days of incubation at 28^0 C. Colony was flat, dense and dry. The colour of the colony was white in initial stage, later, it became creamy. The reverse of the colony was cream. Conidia were solitary with pedicels at the tips of the hyphae or intercalary, pyriform, hyaline, smoothwalled, sparsely rough, 10-12 x 7-8 μm in dia. . Chlamydospores were present in culture.

It was interesting to note that thickly populated areas were dominated by species of *Chrysosporium*. On the contrary, *Trichophyton* exhibited poor colonization in populated areas in comparison to school and college campus soil. The soil samples collected from areas of market, parks and gardens exhibited abundance of *Chrysosporium* followed by *Microsporum* and *Trichophyton*. Soil samples collected from hospital campus and nursing home areas demonstrated the rich population of the genus *Chrysosporium* and *Trichophyton* species, whereas, *Microsporum* species was absent in the region. Rich keratinophilic flora was isolated from soil samples obtained from poultry and animal farm. The present findings agrees well with Jain (1983) who also reported *Trichopyton rubrum* and *T. mentagrophyte,* from poultry and cattle farm. The soil from cultivated field was poor in colonization of *Chrysosporium* species (Nigam and Kushwaha, 1990). Similarly, in the present study also, the soil samples obtained from crop fields revealed very low prevalence of keratinophilic fungi.

PATTERN OF MYCELIAL PENETRATION IN KERATIN (HAIR)

Many researches on human hair deterioration by different keratinophilic fungi were carried out by various workers. Mercer and Verma (1963) and Poulain and Biguet (1974) studied hair invasion by *T. mentagrophytes.* Keddie (1964) and Kunert and Hejtmanek (1976) worked on hair deterioration by *M. gypseum.* Kaaman and Forslind (1985) reported *in vitro* hair infection by *T. mentagrophytes* and *T. rubrum.* Hsu and Volz (1975) found penetration of *T. terrestre* in human hair. Takatori *et al.*, (1983) observed human hair infection by *M. ferrugineum.* Deshmukh and Agrawal (1982) and Boxter and Mann (1969) also studied *in vitro* hair invasion by various keratinophilic fungi. The results of all these researches exhibited in general similar patterns of hair invasion but often with some very important variations, which indicate the applicability of these investigations.

Keratinophilic fungi have the ability to attack keratin (hair) *in vitro* (Davidson and Gregory, 1934; Vanbreuseghem, 1949, 1950, 1952; Daniels, 1953; Barlow and Chattaway, 1955; Ajello and Georg, 1957; Lu, 1962; Friedrich, 1964; Galgoczy and Novak, 1966; Pinetti and Lostia, 1966; English, 1969, 1976; Kunert, 1972; Padhye *et al.*, 1980; Deshmukh and Agrawal, 1982, 1985; Kushwaha, 1983; Shrivastava and Kushwaha, 1987 and Bahuguna and Kushwaha, 1989). Vanbreuseghem (1952) recommended that the ability to attack keratin could be of diagnostic value in determining the dermatophytic nature of a fungus. Various dermatophytes attack keratin (hair) in different manner. Ajello and George (1957) demonstrated the practical applicability of perforating nature of keratinophilic fungi. Padhye *et al.* (1980) also reported hair perforation as a diagnostic criterion in the identification of *Epidermophyton, Microsporum* and *Trichophyton* species. After this report, many other workers studied *in vitro* perforation by different keratinophilic fungi.

Qureshi (2001) carried out hair perforation test on *Chrysosporium tropicum, C. indicum. Micorsporum gypseum, Trichophyton terrestre, T. rubrum* and *Malbranchia* species. The ability of fungi to perforate hair was tested by the method of Ajello and Georg (1957). Among tested keratinophilic fungi, *M. gypseum* was the only fungus, which perforated hair. Interestingly, characteristics wedge shaped perforations were observed. When inoculated with *T. rubrum,* it took minimum time (7 days) for complete degradation of hair, while *C. tropicum* and *C. indicum* took 21 days and 20 days respectively for complete degradation of hair. The maximum time (35 days) for complete degradation was recorded when *T. terrestre* was applied for degradation of hair. It was noted that *Chrysosporium* species attacked on medullar region of tips of hair-shaft, while the other test fungi colonized and started degradation of cuticle and cortex region of hair-shaft. Qureshi (2001) concluded that perforating nature of keratinophilic fungi has a clear practical applicability and it is a useful diagnostic tool for differentiating and characterizing fungal flora.

IN VITRO EFFICACY OF MEDICINAL HERBS AGAINST KERATINOPHILIC FUNGI

The introduction of various antimycotic agents for medical treatment has been a common phenomenon in the recent past. The newly developed antimycotic agents are much discussed, not for their usefulness but for their toxic nature. The allergic reactions of the skin are commonly observed due to the high rate of sensitization power of these antifungal agents. These broad spectrum antimycotic agents are much costlier and exhibits side-effect (s). Therefore, the awared people are much attracted towards plant derived medicine. The latter are comparatively safer, cheaper and easily available for the huge population of our country.

There has been an increasing interest in the study of antifungal potential of plant-based drugs. A survey of literature on antifungal therapy reveals that much contribution has been made nationally and internationally (Fawcett and Spencer, 1970; Bhakuni *et al.,* 1971; Dhar *et al.,* 1973, 1974; Acharya and Chatterje, 1974; Tansey and Appleton, 1975; Ray and Majumdar, 1975, 1976; Jain and Agrawal, 1976; Moore and Atkins, 1977; Sharma and Goutam, 1977; Satyanarayana and Rao, 1977; Mishra and Dixit, 1977, 1980; Tripathi and Dixit, 1981; Tripathi *et al.,* 1978; Dixit *et al.,* 1978; Kher and Chourasia, 1978; Wahab *et al.,* 1981; Singh and Agrawal, 1979, Goutam *et al.*, 1980; Ikram and Haq, 1980, 1984; Jain *et al.,* 1980, Pandey *et al.,* 1980; Daphne *et. al,* 1982; Devi and Kumar, 1983; Singh and Deshmukh, 1984; Mares, 1987; Rai, 1987, 1988, 1991, Rai and Upadhyay 1988; Evron *et al.,* 1988; Singh and Pandey 1989; Tyagi and Tyagi, 1989, Kumar and Chouhan, 1991; Chakraborty *et al.,* 1991, Jain and Agrawal, 1992; Sarvamangala *et al.,* 1993; Al-abed *et al.,* 1993; Rahalison *et al.,* 1993; Alam *et al.,* 1994; Muanza *et al.,* 1994; McCutcheon *et al.,* 1994; Grosvenor *et al.,* 1995, Mwosu and Okafor, 1995; Jain, 1996; Navarro *et al.,* 1996; Meurer *et al.,* 1996, Pattnaik *et al.,* 1996; Al-Saleh *et al.,* 1997; Rana *et al.,* 1997; Mehmood *et al.,* 1997; Ahmad-Ikbal *et al.,* 1998; Thopil *et al.,* 1998; Baba-Moussa *et al.,*1999 and Rai and Acharya, 1999, 2000).

Rai *et al.* (1997) evaluated sensitivity of *Chrysosporium tropicum, Microsporum gypseum* and *Trichophyton terrestre* to *Acorus calamus, Adhatoda vasica, Amomum subulatum, Andrographic paniculata, Boerhaavia diffusa, Cassia occidentalis, Centella asiatica, Cymbopogon citratus, Hemidesmus indicus, Hyptis suaveolens, Malvestrum species, Passiflora edulis, Pergularia daemia, Peristrophe bicalyculata, Shuteria hirsuta, Solanum nigrum, Tecoma stans* and *Verbascum chinense.* They found that most of these plants were effective against keratinophilic fungi. *S. hirsuta* is used by Gond and Bharia tribes of Chhindwara for curing conjunctivitis. When tested *in vitro, S. hirsuta* exhibited high antifungal activity against *C. tropicum, M. gypseum* and *T. terrestre. A. calamus, A. subulatum* and *T. stans* were found to be effective against *C. tropicum* and *T. terrestre,* whereas less sensitivity was recorded by *M. gypseum. A. paniculata, Malvestrum species, V. chinense* and *H. suaveolens* showed remarkable efficacy against *M. gypseum.* Santhali and Pahari tribes of West Bengal use *H. suaveolense* medicinally (Iyer, 1992). However, *in vitro* studies showed that *H. suaveolense* was moderately effective against *M. gypseum* and *T. terrestre,* while it showed very low sensitivity against *C. tropicum.* Another plant, *H. indicus* used by these tribes was found to be less effective against keratinophilic fungi. The tribals of Andhra and Orissa use the extracts of roots and leaves of *C. occidentalis* in treating ringworm (Iyer, 1992).

A survey of literature revealed that an excellent contribution has been made on antimycotic potential of essential oils obtained from higher plants (Cardoso and Santos, 1948; Fournier *et al.*, 1978; Chaumont and Burgeois, 1978; Sharma *et al.*, 1979; Ieven *et al.*, 1979; Giesbrecht, 1980; Gutkind, 1981; Craveiro *et al.*, 1981; Prasad *et al.*, 1986; Ugarte, 1987; Benodia *et al.*, 1988; Aranjo, *et al.*, 1988; Shah *et al.*, 1988; Steinmetz, 1988; Lemos *et al.*, 1990; Garg and Muller 1992, 1993).

CONCLUSION AND FUTURE OUTLOOK

The keratinophilic fungi have potential to cause disease in human beings and animals. Therefore, it is the need of the hour to screen the soils of various places. The public parks and play grounds are the breeding grounds of keratinophilic fungi, and hence soils of such places should be screened on priority. Moreover, the proper reasons responsible for their saprophytic and parasitic nature are to be investigated. The biotechnological potential of these keratinophilic fungi should also be worked out. The keratinophilic fungi and other dermatophytes should be tested against medicinal plants and other new metabolites in order to get proper drugs. The identification of these fungi is based on morphological and cultural studies, which is time-consuming and tedious. The use of molecular techniques, such as, oligonucleotide probes specific for particular sequence for particular organism using hybridization, polymerase chain reaction (PCR), restriction fragment length polymorphism (RFLP), randomely amplified polymorphic DNAs (RAPD), AFLP, PCR fingerprinting which is used for the identification of fungi are solely the DNA based and are very accurate and time saving. The keratinophilic fungi can be easily, rapidly and accurately identified by using molecular techniques. Finally, a detailed data base of the keratinophilic fungi should be prepared which can be used as a ready reckoner for a person, who initiates working in the field.

REFERENCES

Abdel-Fateh, H.M., Moubasher, A.H. Maghazy, S.M. (1982). Keratinolytic fungi in Egyptian soils. *Mycopathologia* 79: 49-53

Abdullah, S. K. and Hassan, D.A. (1995). Isolation of dermatophytes and other keratinophilic fungi from surface sediments of the Shatt-Al-Arab river and creeks at Barrah, Iraq. *Mycoses* 38(3-4): 163-166.

Acharya, T.K. and I.B. Chatterjee (1974). Isolation of chrysophonic acid-5-anthrone: A fungicidal compound from *Cassia tora.* Linn; *Sci. and cult.,* 40(7): 316.

Agut, M.M., Bayo, J., Larrondo and M.A. Calvo (1995). Keratinophilic fungi from soil of Brittany, France. *Mycopathologia,* 129 (2): 81-82.

Ahmad - Iqubal, Mehmood Zafar, Mohammad-Faiz (1998). Screening of some Indian medicinal plants for their antimicrobial properties. *J. of Ethno-harmacology,* 62 (2): 183-193.

Ajello, L. (1953). The dermatophyte *Microsporum gypseum* as a saprophyte and parasite. *Invest. Derm.,* 21: 157-171.

Ajello, L. (1960). Geographic distrubution and prevalence of dermatophytes. *Ann. N.Y. A cad. Med. Sci.,* 89: 30-38.

Ajello, L and Georg, L.K. (1957). *In vitro* hair cultures for differentiating between a typical isolates of *T. mentagrophytes* and *T. rubrum. Mycopathol. Mycol. Appl.* 8: 3-17.

Ajello, L. (1961). The ascigerous state of *M. Cookei. Sabouraudia,* 1: 173-177.

Ajello, L., Varsavsky, E. Delvingt, W. (1965a). Keratinophilic fungi from Belgium soil. *Trans. Br. Mycol. Soc.,* 48: 417-442.

Ajello, L., Varsavsky, E., Sotgiu, G., Mazzoni, A. and Mamtovani, A. (1965b). Survey of soils for human pathogenic fungi from the Emilia-Romagna region of Italy. *Mycopath. et. Mycol. Appl.,* 26: 65-70.

Ajello, L. and cheng, Shu-Lan, Y. (1967). The perfect state of *Trichophyton mentagrophytes. Sabouraudia,* 5(3): 230-234.

Ajello, L. and Alpert, E.M. (1972). Survey of Ester Iceland soils for Keratinophilic fungi. *Mycologia,* 64: 161-166.

Ajello, L. and A.A. Padhye (1974). Keratinophilic fungi of the Galapagos islands. *Mykosen,* 17: 239-243.

Al-abed, A.S., Qasem, J.R. and Abu-blan, H.A. (1993). Antifungal effects of some common wild plant species on certain plant pathogenic fungi. *Dirast series B pure and Applied Sciences* 20 (3): 149-159.

Alam, K., Agua, J., Maven, H., Taie, R., Rao, K.S., Burrows, I. Huber, M.E. and Rali, T. (1994). Preliminary screening of seaweeds, seagrass and Lemongrass oil from Papua-New-Guinea for antimicrobial and antifungal activity. *International J. of Pharmacognosy,* 32(4): 396-399.

Al-Doory, Y. (1967). The occurrence of keratinophilic fungi in Taxas soil. *Mycopath. et. Mycol. Appl.,* 33: 105-112.

Al-Doory and Kalter, S.S. (1967). The isolation of *Histoplasma duboisii* and keratinophilic fungi from soils of East Africa. *Mycopath. et. Mycol. Appl.,* 31: 289-293.

Ali-Shtayeh, M.S.A. and Arda, H.M. (1985). Incidence of dermatophytosis in Jordan with special reference to *tinea capitis. Mycopathologia,* 92: 59-62.

Almusallam, A.A., S.S. Al-Zarban, N.A., Al-Sane and Ahmed, T.M. (1995). A report on the predominant occurrence of a dermatophyte species in cultivated soil from Kuwait. *Mycopathologia,* 130(3): 159-161.

Al-Saleh, F.S., Camal-El-Din, A.Y., Abbas, J.A. and Saeed, N.A. (1997). Phytochemical and biological studies of medicinal plants in Bahrain: The family Chenopodiaceae Part-2. *International Journal of pharmacognosy* 359(1): 38-42.

Ali-Shtayeh, M.S. and H.M., Arda (1989). Isolation of keratinophilic fungi from floor dust in Arab elementary and preparatory schools in the west bank of Jordan. *Mycopathologia,* 106: 5-11.

Aranjo, C.C., Paulo, M.Q., Mala, R.F. and Lima, E.O. (1988). Avaliacaoda atividade antimicrobiana do comporto 5.7.2'. 5'-tetra-acetoxi-4-fenilcum-arine obtida a partir de coutaria. Hexandra (Rubiaceae) *Rev. Microbid.* 19: 177-179.

Baba-Moussa, F., Akpagana, K. and Bouchet, P. (1999). Antifungal activities of seven West African combretaceae used in traditional medicine. *Journal of Ethnopharmacology,* 66(3): 335-338.

Bahuguna, S. and Kushwaha, R.K.S (1989). Hair perforation by keratinophilic fungi. *Mycoses,* 32(7): 340-342.

Barlow, A.J.E. and Chattaway, F.W. (1955). The attack of chemically modified keratin by certain dermatophytes. *J. Invest. Dermatol.* 24: 65-74.

Barry, A.L. and Sabath, L.D. (1974). Special test bactericidal activity and activity of antimicrobics in combination. In manual of *Clinical Microbiology* and J.P. Traunt (Ede.) American society of Microbiology, Washington, D.C. P.P: 431-435.

Baxter, M. (1969). Keratinophilic fungi isolated from humans and from soil in the city of Birmingham, England. *Mycopath et Mycol Appl.* 39: 389-397.

Baxter, M. and Mann, P.R. (1969). Electron microscopic studies of the invasion of human hair *in vitro* by three keratinophilic fungi. *Sabouraudia.* 7: 33-37.

Benoudia, A., Hassar, M. and Benjilali, B. (1988). Les propietes antiseptiques des huiles essentielles *in vitro* testees contre des germes pathogenes hospitalieres. *Fitoterapia.* 19: 115-119.

Bhakuni, D.S., M.L. Dhar, B.N., Dhawan, B., Gupta and R.C. Shrimali (1971). Screening of medicinal plants for biological activity. Patr III. *Indian J. of Experimental Biol.* 9: 91-102.

Bhakuni, D.S., M.L. Dhar, M.M. Dhar, B. Gupta and R.C. Shrimali (1977). Screening of Indian plants for biological activity. Part III. *Indian J. Experimental Biol.,* 15: 208-219.

Buhimann, X. and Riath II (1962). Uberdie Erkennung, and Bedeutung von Dermatomykosen bei. Haustieren. Schweizer. *Arch. Tierheilk,* 104: 537-545.

Cabanes, F.J., Abarca, M.L. and Bragulat, M.R. (1997). Dermatophytes isolated from domestic animals in Barscelona, Spain. *Mycopahologia,* 137 (2): 107-113.

Calvo, A., Vidal, M. and Guarro, J. (1984). Keratinophilic fungi from urban soils of Barcelona, Spain. *Mycopahologia,* 85: 145-147.

Cano, J., J. Guarro and Figueras, M.J. (1991). Study of the invasion of human hair *in vitro* by *AphanoaswS species. Mycoses,* 34: 145-152.

Cardoso, H.T. and santos, M.L. (1948). Estudo sobre a presenca de antibioticos nos vegetais. *Brasil med.*, 62: 67-70.

Caretta, G. and Piontelli, E. (1975). Isolation of keratinophilic fungi from soil in Pavia, Italy. *Sabouraudia,* 13: 33-37

Caretta, G., Frate, G.D., Piontelli, E. and Todaro, F. (1976). Keratinophilic mycoflora of hair and dung of cows in fodder and in the soil on farms. Their distrubution. *Riv. Parasite,* 37: 133-161.

Caretta, G. and Piontelli, E. (1977). *M. magellanicum* and *Cunninghamella antarctica,* new species isolated from Australia and Antartic soil of chile. *Sabouraudia,* 15: 1-10.

Caretta, G., A.M. Mangiarotti and Piontelli, E. (1992). Keratinophilic fungi isolated from soil of Italian parks in the province of Pavia. *European Journal od Epidemiology,* 8(3): 330-339.

Chabasse, D. (1988). Taxonomic study of keratinophilic fungi isolated from soil and some mammals in France. *Mycopahologia* 101: 133-140.

Chakraborty, U., Datta, S.K. and Chakraborty, B.N. (1991). Antifungal activity of some plant extracts on phytopathogenic fungi. *Indian Botanical Contactor,* 8(3): 107-112

Chaumont, J.P. and Burgeois, M. (1978). Propriete's antagonistes de cent extraits de plantes superieures vis a vis de sept champignous phytopatogenes. *Lioydia* 41:437-441

Chmel, L., Hasilikova, Anna, Hrasko, J. and Vlacihkova, A. (1972). The influence of some ecological factors on keratinophilic fungi in the soil. *Sabouraudia,* 10: 26-34.

Cheml, L., and Valcilikova, A. (1975). Distribution of keratinophilic fungi in different depths of soil. *Sabouraudia,* 13. 185-191

Cheml, L. and Vollekova (1981). Development of parasitism in keratinophilic fungi *Lyon. Med ,* 245: 37-40.

Craveiro, A.A., Fernandes, A.G., Andrade, C.N.S., Matos, F.J.A., Alencar, J.W. and Macha do, M.I.C. (1981). Oleos Essenciais do Nordeste. Edicoes UFC, *Fortaleza,* 209.

Currah, R.S., Abbott, S.P., and Sigler, L. (1996). *Arthroderma silveraa* sp. nov. and Chrysosporium vallenarense, keratinophilic fungi from arctic and mountain habitates. *Mycological Research* , 100(2): 195-198.

Dabrowa, N., J.W. Landau, V.D., Newcomer and Plunkett, O.A. (1964). A survey of tide washed coastal area of southern California for fungi Potentially pathogenic to man. *Mycopathol. Mycol. Appl.,* 24: 137-150.

Daniels, G. (1953). The digestion of human hair Keratin by *M. Canis Bodin. J.Gen.Microbiol.,* 8: 289-294.

Daniels, G. (1954). Isolation of Keratinomyces ajelloi from soils in Great Britain. *Nature,* 174: 224-226.

Daphne, Mew, Felipe Balza, G.H., Neil Towers and Levy, J.G. (1982). Antitumour effects of the sespuiterpene lactone parthenin. *Planta medica,* 45: 23-27.

Davidson, A.M. and Gregory, P.H. (1934). *In situ* cultures of dermatophytes. *Can. J. Res.,* 10: 373-393.

Del-frate, G. and Caretta, G. (1990). Fungi isolated from Antarctic material. *Polar Biology.* 11(1): 1-8.

Deshmukh, S.K. (1985) Isolation of dermatophytes and other keratinophilic fungi from soil of Mussoorie (India). *Mykosen.* 28: 98-101.

Deshmukh, S.K. (2002a). Incidence of dermatophytes and other keratinophilic fungi in the glacier bank soils of the Kashmir valley (India). *Mycologist.* 16: 165-167.

Deshmukh, S.K. (2002 b). Incidence of keratinophilic fungi from selected soils of Kerala state (India). *Mycopathologia.* 156: 177-181.

Deshmukh, S.K. and Agrawal, S.C. (1983). Prevalance of dermatophytes and other keratinophlic fungi in soils of Madhya Pradesh (India). *Mykosen,* 26: 574-577.

Deshmukh, S.K. and Agrawal, S.C. (2003). Isolation of dermatophytes and other keratinophilic fungi from soils of Jammu. *Mycoses.* 46: 226-228.

Devi. Y.P. and Nand Kumar, N.V. (1983). Herbicidal properties of *Parthenium hysterophorus* root extracts. *Geobios,* 10:134.

Dey, N.C. and Kakoti, L.M. (1955). *Microsporum gypseum* in India. *J. Indian Med. Assoc.,* 25: 160-164.

Dhar, M.L., M.M. Dhar, B.N. Dharwan, B.N. Mehrotra, R.C. Shrimali and J.S. Tondon (1973). Screening of Indian plants for biological activity. Part IV. *Indian J. Exp. Biol.,* 11: 43-54.

Dhar, M.L. B.N. Dhavan, C.R. Prasad, R.P. Rastogi, K.K. Singh and J.S. Tondon (1974). Screening of Indian plants for biological actibity, Part V. *Indian J. Of Exp._Biol.,* 12: 512-523.

Dixit, S.N., H.S., Shrivbastava and R.D. Tripathi (1978). Lawsone,. The antifungal antibiotic from leaves of *Lawsonia inermis* and some aspects of its mode of action. *Indian Phytopath.* 31:131.

Dunna, R. M. and Morahan, R.J. (1964). The occurrence of keratinophilic fungi in Queensland. *Mycopath. et. Mycol. Appl.* 22: 343-349.

Duries, E.B. and Frey, D. (1955). Isolation of *Microsporum gypseum* and *Keratinomyces ajelloi* from Australian soil. *Nature,*176: 936.

Duries, E.B. and Frey, D. (1957). A new species of *Trichophyton* fron New Southwales *Mycologia,* 49: 401-411

El-said, A.H.N. (1993). Keratinophilic and cyloheximide resistant fungi in soils of Oman. *Cryptogamic Mycolgie* 14(4): 271-277.

El-said, A.H.N. (1994). Keratinophilic fungi in soil of Yemen Arab Republic. *Journal of Basic Microbiology,* 34(5): 311-315.

El-said, A.H.N. and Abdel-Sater, M.A. (1995). Fungi associated with the hair of goat and sheep from El-Bahrain. *Qatar University Science Journal* 15(1): 57-64.

English, M.P. (1976). Destruction of hair by two species of *Chrysosporium. Trans. Brit. Mycol. Soc.,* 66: 357-358.

English, M.P. and Southern, H.N. (1967). *Trichophyton versicolor* infection in population of small wild animals. *Sabouraudia,* 5: 302-309.

Evron, R.M., Guzie, U., Zehani and I. Pelacheck (1988). Activities of compound G2 isolated from alfalfa root against dermatophytes. *Antimicrob. Agents chemother,* 32: 1586-1587.

Evolceanu, R. and Altras, L. (1966). The occurrence of *Trichophyton indicum*-Randhawa and Sandhu (1963) in Rumanian soil some data concerning its immunological properties (Preliminary report). *Mycopath et Mycol. Appl.,* 28: 57-62.

Fawcett, C.H. and Spencer, D.M. (1970). *Ann. Rev. Phytopathol,* 8: 403-418.

Feuerman, E., Alteras, I., Honig, E., and Lehrer, N. (1975). The isolation of keratinophilic fungi from soils in Israel. A preliminary report. *Mycopathologia,* 56: 41-46.

Fillipello, M.V. (1986). Keratinolytic and Keratinophilic fungi in children's sandpits in the city of Turin. *Mycopathologia,* 94: 163-172.

Fillipello, M.V, D. Curetti, C. Cassinelli and C. Bordese (1991). Keratinophilic fungi in the soil of Papua New Guiana. *Mycopathologia,* 115(2):113-120.

Fournier, G., Paris, M.R., Fourniat, M.C. and Quro, A.M. (1978). Actibite. Bacteriostatique dhuiles essentielles de *cannabis satiba L. Annales Pharmaceutiques Francaises,* 36: 603-606.

Friedrich, E. (1964). Perforation Von Frauenhaar durch Dermatophytes. *Mykosen,* 7: 53-63.

Fusconi, A. and Fillipelo, M.V. (1991). Ultrastructural aspects of the demolition of human hair *in Vitro* by *Chrysosporium tropicum* Carmichael. *Mycosen,* 34:153-165.

Galcoxy, J. and Novak, E.K. (1966). Investigation of the hair digestion by dermatophyte. *Mycopath. Mycol. Appl.,* 28: 71-76.

Garg, A.K. (1966). Isolation of dermatophytes and other keratinophilic fungi from soil in India. *Sabouraudia,* 4: 259-264.

Garg, A.P. and Muller, J. (1992). Inhibition of growth of dermatophytes by Indian hair oils. *Mycoses,* 35: 363-369.

Garg, A.P. and Muller, J. (1993). Fungitoxicity of fatty acids against dermatophytes. *Mycoses,* 36: 51-63.

Gene, J., Guillamol, J.M., Ulfig, K. and Guarro, J. (1996). Studies on keratinophilic fungi. x. *Arthrographis alba* Sp. nov. *Canadian Journal of Microbiology* . 42(12): 1185-1189.

Gentles, J.C., Dawson, Christine O. and Cannole, M.D. (1965). Keratinophilic fungi on cats and dogs II. *Sabouraudia,* 4: 171-175.

Gherbawy, Y.A.M.H. (1996). Keratinolytic and keratinophilic fungi of Swamp's soil and air in Qena city and their response to garlic extract and onion oil treatments. *Cryptogamie Mycologie* 17 (3-4): 173-185.

Giesbrecht, A.M. (1980). Atividade antibiotica de produtos naturals tese de Livre Docencia do Departamento do Fisiologia e Farmacologia, do Instituto de Ciencias Biomedicas de universidade de Sao Paulo, Brasil.

Goutam, M.P., Jain, P.C. and Singh, K.V. (1980). Activity of some essential oils against dermatophytes. *Indian Drug,* 17: 269-270.

Grosvenor, P.W., Supriono, A. and Grag, O.O. (1995). Medicinal plants from Riau province, Sumatra, Indonesia, part 2. Antibacterial and antifungal activity. *J. of Ethnopharmacology,* 45(2): 97-111.

Guarro, J., Punsolo, L. and Calvo, M.A. (1981). Keratinophilic fungi from soil of Taragona Catalunya. *Mycopathologia,* 76: 69-71.

Gutkind, G.O., Martino, V., Grana, N., Coussio, J.D. and Torres, R.A. Plants for biological actibities. I. Antibacterial and antifungal actibity. *Fitoterapia,* 52: 213-218.

Hasija, S.K., Rajak, R.C., Singh, L.S. and Pandey, A.K. (1988). Mycoflora of the effluent of gelatin factory at Jabalpur. *Proc. Nat. Acad. Sci. India,* 58 (B): IV, 573-578.

Hawksworth, D.L. , Sutton, B.C. and Ainsworth, G.C. (1983). Ainsworth and Bisby's Dictionary of fungi,. Commonwealth Mycological Institute, Kew 1-445.

Hsu, Y.C. and Volz. P.A. (1975). Penetration of *Trichophyton terrestre* in human hair. *Mycopathologia,* 55: 179-183.

Hubalek, Z. (1974). Fungi associated with free living birds in Czechoslovakia and Yugoslovkia. *Acta Scientarium Naturalium Bmo.,* 8:1-62.

Ieven, M. Berghe, D.A.V., Mertens, F., Vlientinch, A. and Lammens, E. (1979). Screening of higher plants of biological activities. I. Antimicrobial activity. *Planta Med.,* 46: 311-321.

Ikram, M. and Inam-ul-haq (1980). Screening of medicinal plants for antimicrobial activity Part II. *Fitoterapia,* LI(6): 281-284.

Ikram, M. and Inam-ul-haq (1984). Screening of medicinal plants for antimicrobial actibities. Part III, *Fitoterapia,* LV(1): 62-64.

Jain, M., Shukla, P.K. and Srivastava, O.P. (1985). Keratinophilic fungi and dermatophytes in Lucknow soils with their global distribution. *Mykosen,* 28: 148-153.

Jain, P.C. and Agrawal, S.C. (1976). Activity of plant extracts against some Keratinophilic species of *Nannizzia. India Drugs,* 13(12): 25-26.

Jain, P.C. and Agrawal, S.C. (1977). Keratinophilic fungi from the soils of Mount Abu (India). *Geobios,* 4: 136-138.

Jain, P.C., Jain, C.K. and Jain, K. (1980). A note on the activity of odoriferous compound against dermatophytes. *Indian Drugs,* 17(12): 397-398.

Jain, P.C. (1983). Prevalence of keratinophilic fungi in soils of Madhya Pradesh and Uttar Pradesh India. *Geobios New Reports,* 2: 10-13.

Jain, P.C. and Agrawal, S.C. (1992). Sporostatic effect of some oils against fungi causing *otomycosis. Indian. J. Medical Sci,* 46: 1-6.

Jain, P.C. (1996). "Herbal medicines biodiversity and conservation strategies." Ed. Rajak, R.C. and M.K. Rai, *International Book Distributors, Deharadun, India,* pp 292.

Kaaman, T. and Forslind, B. (1985). Ultrastructural studies on experimental hair infections *in vitro* caused by *Trichophyton mentagrophytes* and *Trichophyton rubrum. Acta. Derm. Venereol* (Stockholm) 65: 536-536.

Keddie, F.M. (1964). Further investigations of the electorn microscopic examination of human hair naturally infected by *Microsporum gypseum. Sabouraudia,* 6: 1-4.

Kher, A. and Chourasia, S.C. (1978). Activity of essential oils of three medicinal plants against pathogenic and non-pathogenic fungi. *East Pharm* 21 (24): 183-184.

Kumar, M. Suresh and Chauhan, U.K. (1991). A study of antimicrobial activity of *Zizyphus nummularia* leaf extract. *J. Environ. Biol.* 12(3): 273-277.

Kunert, J. (1972). The digestion of human hair by the dermatophyte *Microsporum gypseum* in submerged culture. *Mykosen,* 15: 59-71.

Kunert, J. and Hejtmanek, M. (1976). Raster-Elektronenmikroskopie der menschlichen Haare, befallen durch den dermatophyten *Microsporum gypseum (Bodin) Guiart and Grigorakis in vitro. Mykosen,* 19: 329-336.

Kushwaha, R.K.S. and Agrawal, S.C. (1976a). A new species of the genus *Acrodontium de Hoog* from India. *Mykosen,* 20: 97-100.

Kushwaha, R.K.S. and Agrawal, S.C. (1976b). Some keratinophilic fungi and related dermatophytes from soil. *Proc. Ind. Natl. Sci. Acad.,* 42:102-110.

Kushwaha, R.K.S. (1980a). A new species of *Gleocladium. Curr. Sci.,* 49:74-75.

Kushwaha, R.K.S. (1980b). *Verticillium sp. nov.* A new kerarinophilic species from Indian soils. *Curr. Sci.,* 49: 968-969

Kushwaha, R.K.S. (1983). Taxonomy of dermatophytes and related fungi. *Proc. Natl. Symp. Adv. Pl. Sci.,* 37-38.

Kushwaha, R.K.S. and Tripathi, N. (2004). Human pathogenic fungi : A review. In : *Animal and Human Pathogenic Fungi.* Ed : R.K.S. Kushwaha. *Scientific Publishers,* 185-250.

Latha, N., A., Hilda and Manjula, V.K. (1992). Occurrence of keratinophilic fungi from playground soils in Arakkonam, Tamil Nadu. *J. Environ. Biol.* 13(2): 107-112.

Lemos, T.L.G., Matos, F.J.A., Craveiro, A.A. and Monte, F.Q. (1990). Atividadeanti-microbiana versus composicao quimica de oleos essenciais. XI Simposio de plantas Medicinais do Brasil. 12 a 14 de setembro. *Joao Pessoa, Paraiba. Res.* 3-19.

Lindqvist, K. (1960). Ringworm hos husdyr i Norge. *Nord. Vet. Med.* 12: 21-28.

Lowey, M.D. Parent and M. Leadonx-corbusies (1985). A new combination of *Tolnaphtate* and *Methylparticine (SPA-5-345, Triol) Mykosen,* 28(9): 452-456.

Lu, Y.C. (1962). A new method for the study of hair digestion by dermatophytes. *Mycopathol. Mycol. Appl.* 17: 225-235.

Malviya, H.K., Rajak, R.C. and Hasija, H.K. (1991). Keratin degradation by *Scopulariopsis brevicaulis. Journal of Indian Botanical Society.* 70:213-216.

Malviya, H.K., Rajak, R.C. and Hasija, H.K. (1992). Synthesis and regulation of extracellular keratinase in three fungi, isolated from the grounds of a gelatin factory, Jabalpur, India. *Mycopathologia.* 120:1-4.

Mancianti, F., Mignone, W. and Papini, R. (1997). Keratinophilic fungi from coats of wild boars in Italy. *Journal of wildlife Diseases.* 33(2): 340-342.

Mares, D. (1987). Antimicrobial activity of protoanemonin, a lactone from *ranunculaceous* plants. *Mycopathologia,* 98: 133-140.

Marples, M.J. (1955). The distribution of keratinophilic fungi in soils from New Zealand and from two Polynesian Islands. *Mycopath. et. Mycol. Appl.,* 25: 361-372.

Mcutcheon, A.R., Ellis, S.M., Hancock, R.E.W. and Towers, G.H.H. (1994). Antifungal screening of medicinal plants of British Columbian native peoples. *J. of Ethnopharmacology* 44(3): 157-169.

McNal, E.G. (196a). Biochemical studies on the metabolism of *Griseofulvin. A.M.A. Arch. Derm.,* 81:657.

Mehmood, Z., Ahmad, S. and Mohammad, F. (1997). Antifungal activity of some essential oils and their major constituents. *Indian J. of Natural Products.* 13(2):10-13.

Mercantin, R., Mersella, R., Caprilli, F. and Dovgiallo, G. (1980). Isolation of dermatophytes and correlated species from the soil of public gardens and parks in Rome. *Sabouraudia,* 18: 123-128.

Mercantin, R., Marsella, R., Lambise, L. and Fulvi, F. (1983). Isolation of keratinophilic fungi from floors in Roman primary schools. *Mycopathologia,* 82: 115-120.

Mercer, E.H. and Verma, V.C. (1963). Hair digested by *Trichophyton mentagrophytes. J. Invest. Dermatol,* 48: 357-360.

Meurer-Grimes, B., Mcbeth, D.L., Hallihan, B. and Dalph, S. (1996). Antimicrobial activity in medicinal plants of the *Scrophulariaceae* and *Acanthaceae. International J. of Pharmacognosy* 34(4): 243-248.

Michel-Brun, J. and Monier, M.F. (1981). Keratinophilic fungi from soil in Lyon's region between 1966-1980. Their evolution. *Lyon Medi.* 245: 87-91.

Mishra, S.B. and Dixit, S.N. (1977). Antifungal spectrum of some plant extracts. *Geobios* 4(1): 29-30.

Mishra, S.B. and Dixit, S.N. (1980). Antifungal principle of *Ranunculus sceleratus. Ecors. Bot.,* 34: 362-367.

Mitra, S.K., Sikdar, A., and Das, P. (1998). Dermatophytes isolated from selected ruminants in India. *Mycopathologia.* 142:13-16.

Mohammed, M., Khan, K.A. and Anwar, A.A. (1971). Isolation and study of keratinophilic fungi from West Pakistan soil. *J. Sci. (Karachi),* 1: 144-158.

Mohammed, S.I. and Lalji, N. (1978). The distribution of geophilic dermatophytes in Kenyan soils. *Mycopathologia,* 63: 95-97.

Moore, G.S. and Atkins, R.D. (1977). The fungicidal and fungistatic effect of an aqueous *garlic* extract on medically important yeast like fungi. *Mycologia,* 69: 341-347.

Moubahser, A.H., M.A., El-naghy, H.M., Abdul-Fatteh and S.M. Maghazy (1992). Keratinolytic fungi in Egyptian soils. I baited with hair and wool. *Zentral blatt fuer Mikrobiologie,* 147(8): 529-535.

Muanza, D.N., Kim, B.W., Euler, K.L. and Williams, L. (1994). Antibacterial and antifungal activities of nine medicinal plants from zaire. *International J. of Pharmacognosy,* 32(4): 337-345.

Navarro, V., Villarreal, M.L., Rojas, G. and Lozoya, X. (1996). Antimicrobial evaluation of some plants used in Mexican traditional medicine for the treatment of infectious diseases. *J. of Ethnopharmacology,* 53(3): 143-147.

Nigam, N. and Kushwaha, R.K.S. (1990). Biodeterioration of human hair. *Jour. Ind. Bot. Soc.,* 69: 469-470.

Nigam, N. and Kushwaha, R.K.S. (1985). Eight new keratinophilic fungal records from India. *Proceedings of National Academy of Sciences India.* 55th Ann. Sess. 26.

Nickerson, W.J. (1947). Biology of pathogenic fungi Waltham , MA.

Nwosu, M.O. and Okafor, J.I. (1995). Preliminary studies of the antifungal activities of some medicinal plants against basidiobolus and some other pathogenic fungi. *Mycoses,* 38(5-6): 191-195.

Orr, G.F. and Kuehn, H.H. (1972). Notes on *Gymnoascaceae* II. Some *Gymnoascaceae* and keratinophilic fungi from Utah. *Mycologia,* 64: 55-72.

Otsenasek, M. and Dvorak, J. (1964). The isolation of *Chrysosporium keratinophilum* (Frey) Carmichael (1962) and similar fungi from Czechoslovakia soil. *Mycopath. et. Mycol. Appl.,* 23: 121-122.

Otcenasek, M., Hudec, K., Hubalek, Z. and Dvork, J. (1967). Keratinophlic fungi from the nests of birds i Czechoslovakia. *Sabouraudia,* 5: 350-354.

Otcenasek, M. (1978). Ecology of the dermatophytes. *Mycopathologia,* 65: 67-72.

Padhye, A.A., Mishra, S.P. and Thirumalachar, M.J. (1966). Occurrence of soil inhabiting fungi from soils in Poona. *Hindustan Antibiot. Bull.,* 9: 90-93.

Padhye, A.A., Pawar, V.H., Sukapure, R.S. and Thirumalachar, M.J. (1967). Keratinophilic fungi from marine soils of Bombay, India part I. *Hindustan Antibiot. Bull.,* 10: 138-141.

Padhye, A.A., C. N., Young and Ajello, L. (1980). Hair perforation as a diagnostic criterion in the identification of *Epidermophyton, Microsporum,* and *Trichophyton Spp.* Proceedings of the fifth international conference on the mycoses, Scientific Publications No. 396, Pan American Health Organization, Sci. Pub. No 396, Washinton D.C., pp 115-120.

Page, R.M. (1950). Observationsi on keratin digestion by *Microsporum gypseum. Mycologia,* 42: 591-602.

Pandey, D.K., Tripathi, R.N., Tripathi R.D. and Dixit, S.N. (1980). *Z pflkrankh pteachutz,* 89: 344.

Parihar, P. (1998). Exploitation of non-pathogenic keratinophilic fungi for their potential to degrade keratin. Ph.D. Thesis, Kanpur University, Kanpur, India.

Pattnaik, S., Subramanyam, V.R. and Kole, C.R. (1996). Antibacterial and antifungal activity of ten essential oils *in vitro. Microbios* 86(349): 237-245.

Pinetti, P. and Lostia, A. (1966). Linfezione dermatofiticn del peto *in vitro. Mycopath.* From Rassengna media Sarda. Cagliar, 128-137.

Poulain, D. and Biguet, J. (1974). Etude en microscopie eletronique delaction keratinolytique de *Trichophyton mentagrophytes* dans le poil de Cobaye infecte experimentalement. *Sabouraudia,* 12: 1-6.

Prasad, G. Kumar, A., Singh, A.K. Bhattachrya, A.K., Singh, K. and Sharma, V.D. (1986). Antimicrobial actibity of essential oils of some *ocimum species* and *clove oil. Fitoterapia,* 57: 429-432.

Pugh, G.J.E. and Mathison, G.E. (1962). Studies on fungi in coastal soils III. An ecological Survey of Keratinophilic fungi. *Trans. British. Mycol. Soc.* 45: 567-572.

Pugh, G.J.F. (1964). Dispersal of *Arthroderma curreyi* by birds and its role in the soil. *Sabouraudia,* 3: 275-278.

Pugh, G.J.F. (1965). Cellulolytic and keratinophilic fungi recorded on birds. *Sabouraudia,* 4: 85-95.

Pugh, G.J.F. (1966). Association between bird's nests, their pH and keratinophilic fungi. *Sabouraudia,* 4: 49-53.

Qureshi, S. (2001). Studies on keratinophilic fungi and related dermatophytes from the soils of Chhindwara, Ph.D. thesis (unpublished), Dr H.S. Gaur University, Sagar, M.P.

Qureshi, S., M.K. Rai, and S.C. Agrawal (2004). Isolation of some keratinophilic fungi from soils of District Chhindwara, M.P., In, Microbiology and Biotechnology for Sustainable Development (Ed. P.C. Jain), CBS Publishers & Distributors, New Delhi pp 358-364.

Rahalison, L., Hamburger, M., Hostettman, K., Monod, M., Frenk, E., Gupta, M.P., Santana, A.L., Correa, A.M. and Gonzalez, A.G. (1993). Screening for antifungal activity of Panamanian plants. *International J. of Pharmacognosy* 31(1):68-76.

Rai, M.K. (1987). A first report of *Curvularea verruculosa*; an opportunistic pathogen from scalp of a college student and *in vitro* sensitivity of this fungus to different plant extracts. *Indian drugs,* 24: 518-520.

Rai, M.K. and Upadhyay, S.K. (1988). Screening of medicinal plants of Chhindwara Distt. against *Trichophyton mentagrophytes. Hindustan Antibiotic Bull.* 30: 33-36.

Rai, M.K. (1988). *In vitro* sensitivity of *Microsporum nanum* to some plants extract. *Indian Drugs,* 25: 521-523.

Rai, M.K. (1991). The therapy of dermatophyte infections: A review *Indian Medicine,* 3-4 (4-1): 5-8.

Rai, M.K. and Acharya, D. (1999). Screening of some asteraceous plants for antimycotic activity. *Compositae Newsletter*, 34: 37-43.

Rai, M.K. and Acharya, D. (2000). Search for fungitoxic potential in essential oil of Asteraceous plants. *Compositae Newsletter*, 35: 18-23.

Rai M.K. and Qureshi, S. (1994). Screening of different keratin baits for isolation of keratinophilic fungi, *Mycoses* 37 (7- 8): 295-298.

Rai, M.K., S. Qureshi, and Agrawal, S.C. (1997). *In vitro* evaluation of inhibitory nature of extracts of 18 plant species of Chhindwara against 3-keratinophilic fungi. *Hindustan* Antibiotics Bulletin 39 (1-4): 56-60,

Rajak, R.C., Gupta, S., Hasija, S.K. and Pandey, A.K. (1889). Soil mycoflora of gelatin factory campus at Jabalpur (M.P.) with special reference to keratinophilic fungi. *Proc. Nat. Acad. India.* 59(B)II: 187-192.

Rajak, R.C., Parwekar, S., Malviya, H.K. and Hasija, S.K. (1991). Keratin degradation by fungi isolated from the grounds of a gelatin factory campus in Jabalpur, India. *Mycopathologia.* 114: 83-87.

Ramesh, V.M. and Hilda, A. (1999). Incidence of keratinophilic fungi in the soil of primary school and public parks of Madras city, India. *Mycopathologia.* 143:139-145.

Rana, B.K., Singh, U.P. and Taneja, V. (1997). Antifungal actibity and kinetics of inhibition by essential oil isolated from leaves of *Aegle marmelos. J. of Ethnopharmacology* 57(1): 29-24.

Randhawa, H.S., Nath, A. and Vishwanathan, R. (1959). Isolation of *Microsporum gypseum* from soil in Delhi. *Sci. Cult. (India)* 25(5): 326-327.

Randhawa, H.S. and Sandhu, R.S. (1963). On the occurrence of two new species of *Trichophyton (Chrysosporum)* in Indian soils. *Mycopath. Mycol. Appl.,* 20: 225-238.

Randhawa, H.S. and Sandhu, R.S. (1965). A survey of soil inhabiting dermatophytes and related Keratinophilic fungi in India. *Sabouraudia,* 4(2): 71-79.

Ray, P.G. and Majumdar, S.K. (1975). New antifungal substance from *Alpinia officinarum. Indian J. Exp. Bio.,* 13(4): 489.

Ray, P.G. and Majumadar, S.K. (1976). Antifungal *flavonoid* from *Alpinia officinarum.* Hence. *Indian J. Exp. Biol.,* 14(6):712-714.

Rees, R.G. (1968). Keratinophilic fungi from Queensland II & III. Isolation from feathers of wild birds *Sabouraudia,* 6: 14-28.

Repova, A. (1990). Soil micromycetes from Czechoslovakia. A list of isolated species with bibliography. IV. *Ceska Mykologie,* 44(3): 170-176.

Rogers, A.L. (1971). Isolation of keratinophilic fungi from soil in the vicinity f Bagota. Columbia. *Mycopath. et. Mycol. Appl.* 44, 261-264.

Rosenthal, S.A. and Wise, R.S. (1960). Studies concerning the development of resistance to *Griseofulvin* by dermatophytes. *A.M.A. Arch. Derm.,* 81:684.

Sabouraud, R. (1893). Not surl hypothise dune existence saprophyte des *Trichophytens. Ann. Dermatol, et syphilol,* 3: 561-566.

Sabouraud, R. (1910). Les Teignes, paris, masson.

Servamangala, H.S., Govindaiah and Datta, R.K. (1993). Evaluation of plant extracts for the control of fungal diseases in mulberry. *Indian Phytopath.* 46(4): 398-401.

Shah, M.A., Bokadia, M.M., Menta, B.K. and Jain, S. (1988). Chemical composition and antimicrobial activity of some seed oil. *Fitoterapia,* 59: 126-129.

Sharama, A.K. and Goutam, M.P. (1977). *Indian Drugs and Pharm. Ind.* 12(6):_31-39.

Sharma, G.P., Jain, N.K. and Garg, B.D. (1979). Antimicrobial activity of essential oil from *Glassocardia bosvalia. Planta Med.,* 36: 185-186.

Shrivastava, J.N. and Kushwaha, R.K.S (1987). Hair perforation by *Microsporum. Bull. Bot. Soc.* 33-34: 35-37.

Simapanya, M.F. and Baxter, M. (1997). Isolation of fungi from soil using the keratin-baiting technique. Zambia. *Mycopathologia,* 136(2): 85-89.

Singh, K.V. and Agrawal, S.C. (1979). Efficacy of some plant extracts against some dermatophytes. *Indian Drugs,* 17: 35-36.

Singh, K.V. and Deshmukh, S.K. (1984). Volatile constituents from members of *Liliaceae* and spore, germination of *Microsporum gypseum* complexes. *Fitoterapia* LV(5): 297-299.

Singh, V.K. and D.K. Pandey (1989). Fungitoxic studies on bark extract of *Lawsonia inermis* against Ringworm Fungi. *Hindustan Anti Bull.,* 31 (1-2): 32-35.

Singh, B.S., O.V., Singh, B.J. Singh and Sinha, R.J. (1990). Incidence of keratinophilic fungi and related dermatophytes in Agra soil. *Advances of Plant Sciences,* 3 (1): 8-15.

Singh, B.S., O.V., Singh, B.J. Singh and Sinha, R.J. (1995). A variety of unusual strains of *M. gypseum* in Agra soils. *Mycotaxon,* 54: 465-470.

Singh, C.J., B.J. Singh and B.S. Singh (1996). Morphological, keratin colonizing and degrading characters of some keratinophilic fungi and related dermatophytes isolated from Keoladeo National Park, Bharatpur. *Mycotaxon* 59; 467-473.

Smith, J.M.B. and Marples, M.J. (1963). *T. mentagrophytes* var. *erinacei Sabouraudia,* 3(1), 1-10.

Soomro, I.H., M. Zardari, S. Mangi and H. Abro (1990). Isolation and identification of dermatophytes and other keratinophilic fungi from the soil of hah Abdul Latif Univ. Khairpur, Sindh, Pakistan. *Scientific Khyber* 3(2): 175-182.

Soon, S.H.T. (1991). Isolation of keratinophilic fungi from soil in Malaysia. *Mycopathologia,* 113(3): 155-158.

Steinmetz, M.D. Moulin-traffort, J. and Regli, P. (1988). Transmission and scanning electronmicroscopy study of the action of sage and rosemary essential oils and *Eucalyptol* on *Candida albicans Mycoses,* 31: 40-51.

Sur, B. and Ghosh, G.R. (1980 a). Keratinophilic fungi from Orissa India. I. Isolation from soils. *Sabouraudia,* 18: 269-272.

Sur, B. and Ghosh, G.R. (1980 b). Keratinophilic fungi from Orissa India. II. Isolation from feathers of wild birds and domestic fowls. *Sabouraudia,* 18: 275-280.

Takashio, M. (1973). Une nouvelle forms sexuee du domplexe *T. mentagrophytes, Arthroderma vanbreuseghemii sp. Nov. Annales de Parasitologie* (Paris) 48(5):713-732.

Takatori, K., Udagawa, S., Kurata H. and Hasegawa, A. (1983). Microscopic observation of human hair infected with *Microsporum ferrugineum, Mycopathologia,* 81: 129-133.

Tansey, M.R. and Applenton, J.A. (1975). Inhibition of fungal growth by *garlic* extract. *Mycologia,* 67: 409-403.

Taylor, W.W., Radcliffe, F. and Paenen, V. (1964). The isolation of pathogenic fungi from the soils of Egypt, The Sudan and Ethiopia. *Sabouraudia,* 2: 235-236.

Thoppil, J.E., Tajo, A. and Minija, J. (1998). Antibacterial and Antifungal activity of four varieties of *Ocimum basilicum. Fitoterapia,* 69(2): 191-192.

Tripathi, R.D., Shrivastava, H.S. and Dixit, S.N. (1978). *Experimentia,* 34, 51.

Tripathi, R.D. and Dixit, S.N. (1981). Fungitoxicity of *lawsone,* isolated from leaves of *Lawsonia inermis* and some related *nepthaquinones* of angiosperm origin. Abstract of the paper presented at *3rd Int. Symp. on Plant Pathol. IARI, New Delhi,* Dec. 14-18.

Turner, P.D. (1965). The occurrence of *Microsporum gypseum* in cultivated and uncultivated soils in Hong Kong. *Trans. Br. Mycol. Soc.,* 48: 549-551.

Tyagi, P.D. and Tyagi, V.K. (1989). *Chatak-Chandani* a danger-cum-profit Weed. *Invention intelligence* 24(8): 366-369.

Ugarte, R., Quilhot, W., Diaz, B., Vera, A. and Fieder, P. (1987). Atividade antimicrobiana *in vitro* del Acido Epiforelicol (*Substancia liquenica*). *Acta. Farm. Bonariense,* 6(2): 65-69.

Ulfig, K. and Koreez, M. (1983). Isolation of keratinophilic fungi from sewage sludge. *Sabouraudia,* 21: 247-250.

Vanbreuseghem, R. (1949). Lacultures des dermatophytes *in vitro* sur des Cheveuxisoles. *Ann. Parasitol* 24: 559-573.

Vanbreuseghem, R. (1952). Technique biologique pour l isolament des dermatophytes du sol (biological technique for the isolation of dermatophytes from the soil). *Ann. Soc. Belgs. Med. Trop.,* 32: 173-178.

Vanoorschot, C.A.N. (1980). A revision of *Chrysosporium* and allied genera. Studies in Mycology No. 20, Centraalbureau Voor Schimme Culture. Baarn, The Netherlands.

Varsavsky, E. (1964). Occurrence of Keratinophilic human pathogenic fungi in soils of Argentina. *Mycopath et Mycol. Appl.* 32: 81 - 90.

Verma, T. N., Sinha, B.K. and Das, U.L. (1982). Isolation of keratinophilic fungi from soil in Bihar, *Mykosen,* 25: 449-452.

Volz, P.A., Wlosinski, M.J. and Wasser, S.P. (1991). Sparse diversity of potential pathogenic soil micro-fungi in the Ukraine (USSR). *Microbios,* 65 (264-265): 187-194.

Volz, P.A., A.M., Hamblin, R.W. Han, C.C., Snabes, G.T., Tziahanas and Wasser, S.P. (1993). The keratinophilic fungi from historic sites of St. Petersburg (Russia). *Ukrayins'kyi Botanichnyi Zhurnal,* 50(5):45-55.

Volz, P.A. (1965). Soil micromycetes of selected areas of Israel. *Israel Journal of Plant Sciences,* 43(3): 281-290.

Vollekova, A. (1984). *Microsporum persicolor* and other keratinophilic fungi in soil and in a rodent's den. *Biologia Czechoslovakia,* 39: 899-904.

Wahab, S., Tondon, R.N., Jacob, Z., Sagar, P. and Shrivastava, O.P. (1981). *In vitro* activity of a phytochemical *alantolactone* from *Inula recemosa Hook. F.* against some pathogenic and opportunistic fungi. *J. Indian Bot. Soc.,* 60L: 278-281.

Williams, D.I., Martin, R.H. and Sarkany, I. (1958). Oral treatment of ringworm with *Griseofulvin, Lancet,* 2: 1212.

Williamson, D. and Iyer, S.R. (1992). Dermatomycoses in India with Special reference to Rajasthan *Advances in Medical Mycology* Aditya Publishing house, New Delhi, 131-143.

Youssef, Y.A., El-din, A.A.K. and Hassanein, S.M. (1992). Occurrence of keratinophilic fungi and related dermatophytes in soils in Cairo, Egypt. *Zentral blatt Fuer Mikrobiologie,* 147 (1-2): 80-85.

CHAPTER - 22

CURRENT TRENDS IN AEROMYCOLOGICAL RESEARCH

S.T. Tilak* and B.N. Pande**

*18, Vidyasagar Society, Near Mahesh Society, Bibvewadi, Pune - 411 037 (M.S)
**Department of Environmental Science, Dr. B.A. Marathawada University, Aurangabad, Aurangabad- 431 004 (M.S).

INTRODUCTION

India has the unique distinction of being one of the earliest countries where aerobiological studies in general and aeromycological studies in particular were initiated (Cunningham, 1973). Environmental mycology or aeromycology constitutes one of the major aspects mainly because of the dominance of fungal spores in the ambient air, which has naturally resulted in the accumulation of voluminous literature on aeromycology. However, in general, aeromycological investigations take into account the identification of source, mode of release, dispersal, deposition and impaction and effects of impaction on the various living systems. The fungal spores and hyphal fragments are commonly recorded in the air and are important for the survival and subsequent continuation of generations. Many of the fungal spores are endowed with unique structures and capacity to survive under unfavorable environmental conditions and this probably accounts for their predominance in the air. Nearly all the spores are essentially dispersive units and their significance as gene dispersal units should not be lost.

The spores of fungi propagules, are a variety of forms. All these forms of spores and conidia range from 3-200 μm, most of these are about 10 μm in diameter. The spores are often liberated in the air in massive concentration and remain in the air for a long time. The major component of airspora is the fungal spores. However, the spores fungi have been found to be responsible for allergy.

Cunningham (1973), a surgeon in Indian Medical Service attached to the Sanitary Commission of the Government of India attempted to relate airborne organisms to the incidence of Cholera and other fevers, using an aeroconiscope in Calcutta jails. This catches illustrated in a series of remarkable colour plates, show many fungal spores and pollen grains. However, no correlation with disease could be established although fungal spores appeared to increase in number in wet weather when amounts of inorganic dust decreased. Nearly after about half a century of Cunningham's pioneering work, plant pathologist once again initiated aeromycological studies in India. Chatterjee (1931), and Mehta (1952), studied the occurrence of conidia of *Helminthosporium oryzae*, responsible for the famous Bengal famine, in the air over paddy fields at Cuttack while Chatterjee used a balloon for trapping rust spores. Mehta used kites, aeroplanes and aeroscopes for catching urediniospores of the three rusts of wheat and barley at 62 localities in the country. Two other earlier reports on airborne fungi were those of Rajan *et al.* (1952), at Kanpur and Kalra and Dumbrey (1957) at Pune. Aspergilli were found to be the most dominant in army stores at Kanpur. *Alternaria, Curvularia, Fusarium, Pullularia,* and *Helminthosporium* being the other common types in the atmosphere of army medical college campus at Pune.

Systemic and intensive studies of aerobiology and aeromycology in India can be said to have started with the work initiated by Sreeramulu (1959) at Visakhapatnam using Hirst's volumetric spore trap. The construction of "Automatic Volumetric Spore Trap" by Gregory and Hirst (1952) heralded a new era in aerobiological research all over the world including India. Incidentally two centers, one at Aurangabad and the other at Mysore, came into existence during the middle of the sixties. The center at Aurangabad has been initiated by Tilak in 1966, after obtaining initial training from Sreeramulu; and at Mysore by Ramalingam (1968), former student of Sreeramulu. Aurangabad has a unique place in the development of aerobiology in India. While Sreeramulu was the first to initiate aerobiological studies in a systematic manner, the credit of expansion of aeromycological studies in India and putting India in world aerobiological map goes to Tilak. However, the state of Maharashtra has the unique distinction of having many active centers of aerobiological research in India. Aerobiological studies were initiated during mid seventies at Bose Institute, Calcutta by Chanda and at Bangalore by Agashe. Chanda and his co-workers carried out extensive studies in the foothills of Eastern Himalayas and also places of high attitude in the eastern part of India. Three other centers came into existence more or less at the same time. These were Allahabad (Nautiyal and Manju Midha), Nagpur (Chitaley and Patil) and Gorakhpur (Mishra, Kamal, Shrivastava). The aeromycological studies initiated at Patel Chest Institute, Delhi in 1957 by Shivpuri and his co- workers expanded and strengthened by Gangal, Singh and co-workers. (Table 1)

Table 1. Spore type and their per cent concentration

Sr. No.	Spore Type	Per cent Contribution
1	*Alternaria*	5.4
2	*Aspergillus*	0.9
3	*Asperisporium*	0.01
4	*Beltrania*	0.24
5	*Beltraniella*	0.04
6	*Bispora*	0.12
7	*Botryodiplodia*	0.09
8	*Cephaliophora*	0.01
9	*Cercospora*	1.1
10	*Cladosporium*	31.21
11	*Cladotrichum*	0.04
12	*Coniothecium*	0.01
13	*Corynespora*	0.2
14	*Curvularia*	4.4
15	*Dendryphiopsis*	--
16	*Decoccum*	0.02
17	*Dectyoarthrinium*	0.02
18	*Diplodia*	0.6
19	*Drechslera*	0.42
20	*Epicoccum*	0.25
21	*Exosporium*	0.02
22	*Fusariella*	0.13
23	*Fusarium*	0.24
24	*Haplosporella*	0.6
25	*Helicomyces*	0.04
26	*Helminthosporium*	3.15
27	*Hendersonia*	0.04
28	*Hendersonula*	0.06
29	*Hirudinaria*	0.01
30	*Heterosporium*	0.13
31	*Lacellinopsis*	0.3
32	*Melanconium*	0.11
33	*Memnoniella*	1.05
34	*Nigrospora*	4.3
35	*Acrosporium*	0.14
36	*Arthrinuim*	0.42
37	*Penicillium*	0.51

Continue.....

Sr. No.	Spore Type	Per cent Contribution
38	*Periconia*	3.15
39	*Phaeotrichoconis*	0.2
40	*Phoma*	0.04
41	*Phomopsis*	0.6
42	*Pithomyces*	1
43	*Prathigada*	0.03
44	*Pseudotorula*	1.2
45	*Pyriculari*	0.33
46	*Sadasivania*	0.8
47	*Sclerotium*	0.6
48	*Sirodesmium*	0.3
49	*Spegazzinia*	0.25
50	*Sphacelia*	1.4
51	*Sphaeropsis*	0.02
52	*Spicaria*	0.2
53	*Stachybotrys*	0.04
54	*Steganosporium*	0.01
55	*Stemphylium*	0.2
56	*Tetracoccosporium*	0.03
57	*Tetraploa*	0.12
58	*Torula*	1.8
59	*Trichoconis*	0.03
60	*Trichoderma*	0.56
61	*Trichothecium*	0.04
62	*Trichocladium*	0.05
63	*Zygodesmus*	0.44
64	*Zygosporium*	0.05

Aeromycological studies are being carried out at Chennai by Vittal, Visakhapatnam (by Subba Reddy and Janakibai). Many centers came up in eighties and these include Gwalior (Jain), Jabalpur (Verma), Shantiniketan (Mandal and Bhattacharya), Manipur (Singh), Gauhati (Sarma), Bodh-Gaya (Mishra), Gulbarga (Rajaasab), Raipur (Tiwari), Srinagar (Munshi), Lucknow (Khandelwal), etc.

AEROMYCOLOGY

Fungal spores form an important constituent of bioaerosol. Fungi are often well adapted to airborne dispersal of their spores having either tall conidiophores that penetrate into or through the laminar boundary layer or specialized liberation mechanisms that forcibly project spores through this layer. In the course of evolution the fungi have probably exploited the wind for their dispersal more thoroughly than any other group of organisms and consequently dominate the airspora (80%-90%)

Gregory's experiment of work

Gregory's (1961), experimental aeromycological work, using wind tunnel clearly indicated that the efficiency of impaction directly decreases with spore and wind velocity. It is assumed that, form of the spore is of great significance in relation to deposition, however, there is no experimental evidence of the deposition of elongated spores as compared with the spherical ones so far as the airspora is concerned. It is a common observation, as stated by Tilak (1991), that great many fungi have a more or less a spherical spore with about 10 μm diameter.

Aeromycological researches brought out some salient features of airborne spores. The predominant forms of airspora have significant relevance to shape, size and morphology and moreover these are produced on large scale e.g., *Aspergillus, Cladosporium, Mucor, Penicillium, Phoma, Thielaviopsis* and smut spores. These small sized spores have got tendency to form groups or clusters that form a single dispersal unit, which essentially is an adaptive feature. Large size spore types viz., *Alternaria, Helminthosporium, Curvularia, Cercospora,* etc., have lower contents as compared to their volume in addition to the presence of empty spaces in spores, that also helps to the spores to become airborne and float for a longer time. Bhati and Gaur (1979) stressed the importance of structural aspects in airborne spores. Among the typical shapes of spores, obclavate or muriform *Alternaria*; cylindrical or polymorphous *Cladosporium*; central cell large *Culvularia, Spondylocladiella*; fusiform or falcate *Fusarium;* barrel shape *Geotrichum*; elliptical or curved *Rhizoctonia*; multiradiate with appendages *Tetraploa* etc., are some of the adaptive features which provide buoyancy to float in the air for longer period.

Trapping of the spores in the immediate vicinity of the source has revealed a number of periodic patterns of the recurrence of fungal spores in the air many of which are related to diurnal rhythm of alternating day and night, others are related to water relation of reproductive structures and some reflect changes in the climatic cycles (Tilak, 1991).

The varied components of aeromicrobiota thus are capable of providing useful clues to the environmental parameters like humidity, rainfall temperature, time of day/night, season and pollution. This aspect of various airborne fungal spores has been discussed in detail (Pande, 2001).

SPLASH DISPERSED SPORES

Earlier, it was believed that splash-dispersed spores do not form components of airspora. Tilak (1987), reviewed this aspect and based on his observations supported the earlier observations and reports. He brought out significance of splash-dispersed spores and airspora. However, it must be remembered that the effective range of splash is limited but if combined with wind velocity the range may extend considerably

Splash dispersal is typical of slime spored fungi. However, when a splash drop with conidia falls on and glides over hydrophobic plant surface of a leaf 'non wettable' (dry) conidia e.g. those of *Penicillium*, tend to be deposited near the moving droplet and wettable (slimy) conidia of *Verticillium* are washed away or deposited where droplet comes to rest (Davies, 1959). Many of the foliicolous conidial pathogens, however, are found to be and therefore they may not be fortuitous

Spores of *Colletotrichum* and *Pestalotia* are produced in saucer shaped acervuli. When a thick film of water gets deposited on acervuli efficient splash dispersal is assured. Meier *et al.* (1933) detected viable spores of *Pestalotia* at 18000-ft. altitude. Dispersal by splash mechanism is well known and is very common in *Fusarium, Colletotrichum, Phytophthora, Botrytis, Hemileia, Giberrella, Cercospora, Phyllosticta, Venturia, Nectria,* etc. It is thus evident that for practical purposes splash dispersed spores not only contribute to aeromicrobiota but associated high wind velocity can also transport these to considerable distances. Aeromycological investigations have helped to clarify and modify the earlier concept that splash dispersed spores would not contribute to airspora (Tilak, 1998). It would be of interest to note that a rainfall-simulated model was used on groundnut (*Arachis hypogea*) plots artificially infected with *Puccinia arachidicola* to understand the dispersal of urediniospores caused by various amounts of rainfall. Several trapping methods were used to compare dry, splash and drip dispersal caused by rain (Savery and Jenean, 1986). Reynolds *et al.* (1989) studied splash dispersal of *Phytophthora cactorum* from infected strawberry fruit by simulated canopy drip.

Thus, it may therefore, be concluded that splash dispersed spores, though not truly airborne, but play significant role as effective plant pathogens and human health hazards (Tilak, 1998).

SPORES OF ZYGOMYCOTINA AND MASTIGOMYCOTINA

The group constitutes comparative few types in the airspora and in general most of the spores are recorded during rainy season followed by high humid conditions and low temperatures.

Five spore types are common components of airspora e.g., *Albugo, Cunninghamella, Mucor, Rhizopus* and *Sclerospora*. Conidia of *Albugo* and *Cunninghamella* are common during July-September representing 0.2% of total airspora (Tilak and Babu, 1987). Spores of *Rhizopus* were recorded in maximum concentration (224/m^3) in August when temperature was about 250°C, relative humidity 95-100% with a rainfall of 22 mm. The spores belonging 'Wetspora' group exhibiting day spora pattern reaching a peak in the forenoon hours. A number of workers have also reported the *Rhizopus* spores from high altitude (2400 m). According to Gregory (1973), rapid changes during the morning hours result in the detachment of spores from various members *Mastigomycotina, Zygomycotina and Deuteromycotina.* Various types have been identified by culture plate exposure method and are found efficient in detection and estimation of *Mastigomycotina* and *Zygomycotina* in the airspora. Circadian periodicity studies (Tilak and Babu, 1987; Tilak and Pande, 1997) on *Cunninghamella* indicate their maximum occurrence during rainy days i.e. 1% to the total airspora and exhibits 'Nocturnal' pattern reading a peak at 4am. The high concentration of these spores is associated with low temperature (22^0 C) and high humidity (91%).

Fairly a large number of *Zygomycetous* fungi have been identified by culture plate exposure method from various parts of India (Sahu, 1998; Pande, 1999). These include species of *Blakeslea, Circinella, Mortierella* etc. (Table 2).

Table 2. Average mean concentration of common spores of mastigomycotina and zygomycotina at Aurangabad

Spore Type	% Concentration
Albugo	0.4
Choaenophora	0.6
Mucor	0.6
Rhizopus	2.3
Sclerospora cospores	0.03
Sclerospora sporangia	3.6
Monthly Average	**% of Mstigomycotina and Zygomycotina**
January	1.34
February	2.09
March	1.3
April	1.07
May	0.21
June	1.55
July	1.46
August	6.95
september	4.07
October	1.34
November	0.98
December	1.9

SPORES OF ASCOMYCOTINA

The occurrence and seasonal variation of the ascospores has been extensively investigated which clearly indicate their predominance in airspora during the months from July to September which eventually corresponds with rainy season. Ascospores are freely discharged after the substrate and ascocarp is wetted by rain or dew. Following the rainfall the ascospores are released in the first three hours.

The effect of rainfall may be immediate or delayed. On the basis of circadian periodicity studies, ascospores could be classified in Type A, B, C and D (Sreeramulu and Ramalingam, 1966). Type A: (active discharge between midnight and dawn) *Amphisphaeria, Apiorhynchostoma, Claviceps fusiformis, Hypoxylon, Diatrype, Didymospaeria, Nodulosphaeria, Parodiella, Pleospora, Othia.* Type B (spores with sharp peak at 8am and then rapid fall). *Chaetomium, Leptosphaeria, Pringsheimia Mycosphaerella, Sordaria, Teichosporella* Type C: (peak concentration at 10am and gradual fall till 4 pm) *Leptosphaeria.* Type D: (spores with day time double maxima all fall at 1 pm in between two peaks) *Pleospora, Hysterium, Sporormia, Tryblidaria.*

It may be noted that the maximum ascospore types were recorded during the night hours while few occurred during daytime. Among other meteorological factors, temperature also plays an important role in the release and subsequent occurrence of ascospores in air. In the months of April and May when the temperature is high, the concentration of ascospores is considerably reduced.

Table 3. Ascospore types, their % contribution and their seasonal peaks at Aurangabad

Ascopore Type	% Contribution	Seasonal peak period
Amphisphaeria type	0.6	July - August
Apiorhyncostoma type	0.02	May - June
Bitrimonospora type	0.03	September - October
Calospora type	0.16	Periodical August
Claviceps spp.	0.15	July, September and October
Chaetominum type	0.15	July, September and October
Diymosphaeria type	0.34	August - October
Hypoxylon type	0.2	December - January
Hysterium type	0.03	June - November
Leptosphaeria type	0.23	June - July - November
Lophiostoma type	0.002	August
Melanospora type	0.06	January - December
Mycosphaerella type	0.04	July - September
Nodnlosphaeria type	0.06	January - March
Othia type	0.12	July - September
Parodiella type	0.51	May - October - January
Pleospora type	0.03	July - August - September
Pringsheimia type	—	July - September
Sordaria type	—	August
Sporormia type	—	January - August - December

The observations carried out on the composition of airspora over 15 years have clearly indicated the close relation between rainfall and release of ascospores, (Tilak, 1974, 1980, 1991). The seasonal variation and diurnal periodicity of the various ascospore types have been investigated. This has brought out several new types to airspora. The sudden rise in concentration in April is mainly due to the observations over irrigated fields. On an average, the concentration of ascospores gradually increases from June till it reaches maximum in August corresponding to rainy season (Table 3).

SPORES OF BASIDIOMYCOTINA

Basidiospores are significant contributors to airspora. Most of the basidiospores are prevalent during rainy season when the fructifications are formed and these are released in the air currents. The members of *Agaricales* and *Aphyllophorales* release copious amount of spore load of Basidiospores but the identification of many, often poses a challenge. uredosporest and teleutospores of rust fungi and chlamydospores of smuts are commonly entrapped during the winter season and being pathogenic or allergenic have attracted the attention of aeromycological workers and that probably accounts for extensive literature available. As a matter of fact, *Puccinia* is one of the best and extensively worked out genus with reference to its relevance as allergen, pathogen, the extensive aerial transport of this pathogen, the meteorological factors governing its distribution, occurrence, seasonal variation and its impact resulting in epiphytotics of crop plants.

Some commonly encountered spores are of *Agaricus, Coprinus, Ganoderma, Melampsora, Puccinia, Ravenalia, Sphacelotheca, Uromyces, Ustilago* etc.

In general the prevalence of rust spores in air in a tropical country like India is during the months from October to January. The dry periods favor the liberation of rust spores and the peak concentration of airspora exists during daytime between 10 am and 2 am. The cool, windy and dry periods are responsible for heavy spore load in the airspora. Cadham (1924), reported for the first time inhalant allergy due to rust spores. Sreeramulu and Vittal while studying periodicity of uredospores above and within sugarcane field recorded peak concentration in the months of December and January. Singh *et al.* (2001) while studying the dispersal of *Uromyces fabae* by rain water in the experimental fields at Imphal, reported that rainfall has a pronounced effect not only on the concentration of uredospores of *Uromyces fabae* in air over crop field but also on the dispersal of uredospores within the canopy and through long term dispersal (Table 4).

Chamydospores of smut are encountered during dry, gusty and sunny periods especially in the afternoon. Maximum dispersal of Chamydospores of smut occurs between 11-00 hrs and 15-00 hrs with a peak at 14-00 hrs; while very few are dispersed at nighttime. The prevalence of rainfall and/or high wind velocity results in loss of spores from the infected ears thus reducing the period of smut dissemination. Gregory (1973), pointed out that blowing away occurs commonly with dry spored fungi including moulds, smuts and rusts, because the spores are often present on an elevated sporophore, as stem or leaf pathogen usually being adequately raised on its host tissue. The higher the wind speed more spores are carried away.

Table 4. Average mean % of the spore types of basidomycetes at Aurangabad

Spore Type	**% To The Total Air Spora**
Basidiospores	6.4
Rust spores	1.3
Smut spores	10.9
Monthly Average	**% Of Basidiomycetes**
January	9.7
February	11.8
March	7.3
April	4.1
May	2.7
June	5.7
July	4.1
August	4.7
September	6.6
October	4.9
November	6.1
December	6.6

SPORES OF DEUTEROMYCOTINA

The spores of the fungi imperfectii have been recorded almost throughout the year in India. Maximum spores are reported during the month of November (67%). While the minimum in April-May (40%) which eventually are hot and dry months (Tilak, 1989a). Rainfall influences the concentration of spores in air, the spore of *Fusarium, Pithomyces, Torula, Periconia, Haplosporella* etc. indicate high prevalence with the beginning of rainfall. The spores of *Nigrospora, Spegazzinia, Tetraploa, Aspergillus, Beltraniella* and *Stemphylium* are abundant during morning hours with a rainfall on the preceding day. When the rain was followed by a dry day, there was an abundance of spores of *Alternaria, Helminthosporium, Culvularia, Cladosporium, Sadasivania, Heterosporium, Epicoccum, Fusariella* and *Bispora.*

The spores of *Deuteromycotina* contribute almost up to the total airspora during November and December of which *Cladosporium* (3%), *Alternaria* (5%), *Periconia* (4%), *Helminthosporium* (3%) etc., are some major contributors.

Some other common types recorded in aeromycological studies are reported to be over 125 from India (Tilak, 1989a). Several other spore types have been identified by culture plate exposure method. The spores of *Deuteromycotina* constitute the predominant type in the airspora and the spores of *Cladosporium* contribute 30 to 60% to the total airspora and have been reported throughout the world.

HYPHAL FRAGMENTS

Hyphal fragment constitute a significant population of airspora, airborne hyphal fragments are primarily conidiophores of asexual forms. Culturing of these hyphal fragments revealed that they mostly belong to *Cladosporium, Alternaria, Penicillium* etc., These usually inhabit leaves phylloplane. Fungi of many plants and are dispersed due to scrubbing action of host plants which takes place during wind currents. Hyphal fragments constitute an important airborne material. Due to their viability they also serve as means of propagation. Tilak and Bhalke (1981), described and classified various hyphal fragments. Many of the hyphal fragments are capable of growth and thus act as "Reproductive propagation". The role of hyphal fragments as pathogens and allergens has attracted the attention in the recent years.

Aeromycological survey by Pady and Kramer (1960), recorded 43% of hyphal fragments. The hyphal fragments could be grouped into five morphological types irrespective of their taxonomic group. Simultaneous occurrence and predominance and colour of spore type with hyphal fragments were helpful in identification. It was also confirmed by the occurrence of conidia and sporangia on conidiophores or sporangiophores. Among various genera that would be identified were, *Cladosporium, Alternaria, Aspergillus Cunninghamella, Culvularia, Drechslera, Penicillium, Rhizopus, Trichoderma* and *Verticillium.*

The seasonal variation of the hyphal fragments clearly revealed its close correlation with meteorological parameters like rainfall, temperature, relative humidity and high wind velocity. The lowest contribution of hyphal fragments was in the months of July (3.2%) to September (2.9%) which correspond to rainy season. It is also possible that the wetting of hyphal fragments resisted the fragmentation and 'take off' mechanism. The concentration of hyphal fragment in airspora, however, gradually increased from November (6.4%) to May (8.8%) (Tilak and Bhalke, 1981). It is thus evident that dry, hot and windy weather favored fragmentation and subsequent take off of hyphal fragments. There is significant contribution of hyphal fragments to airspora was during daytime.

Though in general, the percentage contribution of fungal spores of airspora is minimum during summer months, the hyphal fragments indicated maximum concentration in summer. Reasons of such a reciprocal behavior may be due to different mechanism and factors responsible for "take off."

SPORES AS BIOINDICATORS

The superiority of biological indicators over physico-chemical factors has been advocated by many workers as it is a direct method to indicate the prevailing conditions of the environment. Several spore types are closely associated with certain environments or its parameters. *Aspergillus* is common in effluents of distillery and atmosphere of cane sugar factories in India. The abundance of ascospores in the air is closely related to rainfall and high humidity. The conidia of powdery mildews are abundant during winter season when temperatures are low and humidity is over 60-70%. Common moulds like *Botrytis* and *Penicillium* are effectively liberated under low humid conditions associated with high wind speed. Hyphal fragments are recorded in dry spell associated with high wind velocity. A variety of ascospores dominate airspora after rainfall and particularly at nighttime. Numerous examples are available in literature about the role of fungal spores as bioindicators (Tilak, 1989b), close correlation of the abundance of spore types and meteorological parameters has been illustrated and these indicate prevailing environmental conditions or future meteorological conditions to follow. Thus, spores are helpful as 'markers' or 'indicators' which may be of great interest to meteorologists.

As like fungal spores pollen grains also constitute an important part of airspora which can be effectively used as bioindicators in environmental monitoring. The presence and abundance of pollen is related with number of such factors, pollen production, liberation, season and meteorological parameters. Recently, it is recorded that common pollutants like SO_2, NO_2 and CO often alert the structural moiety and their degree of allergenicity. Study of SEM would provide information on heavy metal pollution in air.

The diversified composition of airspora, thus is capable of providing useful clues to the environmental parameters like humidity, rainfall, time of day/night, season and pollution. Tilak (1989b) has discussed on these points at great length.

FUNGAL SPORES AND ALLERGY

Fungal spores as stated earlier, predominate the other bioparticals in the airspora. The source for such fungal spore is the substrate at the ground level. Many of the airborne fungal spores are a significant potential source of allergen. Aeromycological survey in indoors and outdoors, considerably helped to locate the source, its identification, concentration and seasonal variation. Such information, however, provides basic data for the treatment of sensitive individuals suffering from allergy. Many airborne fungal spores are responsible for allergic diseases (Gravesan, 1979). Aeromycological investigations helped to focus attention on such allergenic moulds and the allergen sources. Beside such surveys involves determination of concentration of aeroallergens of a location with its fluctuations and seasonal variation for specific period of time. Data obtained from such survey helps to obtain spore calendar for subsequent allergen avoidance and management.

Various methods of aeromycological techniques have been described by Tilak (1982,1987 a). The spore calendar established on the basis of Air monitoring helps to predict the outdoor fungal spore level which in turn influence the indoor levels. To obtain a picture of the personal exposure of patients to allergenic bioparticals, Boehm's individual pollen collector has been proved to be useful (Tilak *et al.*, 1982; Khillare, 1989; Pande, 1994, 1999).

Common species of *Aspergillus* and *Penicillium* are potent source of nephro and hepatotoxic metabolites that exclude this mycotoxin into various animal foodstuffs and the process is hastened up under moist condition (Table 5).

Table 5. Some aeroallergenic fungal spores

Aceromonium	*Neurospora*
Acremoniella	*Paecilomyces*
Alternaria	*Penicillium*
Arthrinium	*Phoma*
Aspergillus	*Puccinia*
Aureobasidium	*Rhodotorula*
Botrytis	*Rhizopus*
Candida	*Saccharomyces*
Drecshlera	*Scopulariopsis*
Epicoccum	*Stachybotrys*
Epidermophyton	*Stemphylium*
Fusarium	*Trichocladium*
Ecliocladium	*Trichoderma*
Coraphium	*Trichophyton*
Helminthosporium	*Trichothecium*
Humicola	*Trichothecium*
Microsporon	*Ustilago*
Mucor	*Verticillium*

FUNGAL ALLERGY PREVALENCE

Sensitivity of fungal spores was mentioned by Blackley (1873), who described his own allergic reaction to spores of *Chaetomium* and *Penicillium expansum*. The occurrence of seasonal asthma due to fungal spores is now well recognized. *Cladosporium* and *Alternaria* have been also investigated in detail as allergens, because of their worldwide distribution and occurrence. *Leptosphaeria* ascospores though are of insignificant contribution to airspora, are highly potential allergens. Frankland and Gregory (1953), described the allergenic importance of airborne *Didymella exitialis* in an agricultural communities and the relevance of allergic disorders with aeroallergen ascospores (Table 5).

Exposure to the spore dust in the working agriculture environment poses a serious problem of farmers lung disease. Aeromycological investigation in the byres, grinding sheds, threshing and grinding of forage indicates that the higher exposure to organic dusts was in the handling of grains. During this observations, spores of *Cladosporium* (50%), *Yeast* (10%) and *Actinomycetes* (8%) were most commonly encountered as reported by Kotimma *et al.* (1980).

The fungal spores due to their extreme adapting nature are recorded more or less round the year and thus in real sense there is no "aeroallergenic fungal spore free period in the environment". Nevertheless, these aeroallergenic fungal spores show seasonal variation in their occurrence and concentration. Extensive work on allergenicity of fungal spores has been carried out. Among these reference has to be made to the works of Cadham (1924), Chakravarty (1974), Frankland (1977, 1983, 1984), Brown and Jacson (1978), Babu *et al.* (1979), Gesa *et al.* (1980), Kotimma *et al.* (1980), Mandal and Chanda (1981), Chapman (1982), Babu (1983), Gravensen and Jensen (1984), Allit (1986), Nair *et al.* (1986), D'Silva (1988), Tilak (1998) etc.

Fungal spores are highly adapted for their survival and disposal. The wall protects against desiccation and it is often pigmented making them less vulnerable to radiation damage from ultraviolet light in the atmosphere. After liberation, the spores are either transported into air by mild wind currents or may be forcibly ejected and dispersed over great heights. Before sedimentation and germination, spores can be deposited an allergic mucous and thus exhibits symptoms. In indoor atmosphere they may be sedimented together with house dust particles and again be re-dispersed by indoor activities or germinate in suitable places in moist substrate.

MANIFESTATION

In general, it is established that particles of 7.5 μm in size cause Type I allergy and particles of size 5 μm cause Type III allergy and this may be applicable for patients with normed nose breathing. A patient suffering from nasal hindrance may breath by mouth and in that case particles up to 20 μm may be deposited in the bronchiole. *Aspergillus fumigatus* a common airborne mould and with spores of 3 μm diameter provide a good example whose spores may cause infection or a type I or II allergy. Pepys and Devies (1977), revived the etiological aspects of genus *Aspergillus* and the airborne moulds which might be playing similar role.

Thermophilic *Actinomycetes, Micropolyspora fareni* is often linked with extrinsic allergic alveoloitis and mesophilic moulds from farm buildings, infected grains or infested humidifiers have definite role in evoking Type III reactions.

Rhinitis and asthma (in children) are most common manifestations. Urtica (e.g. After ingestion of blue cheese with *Penicillium rogueforti*), Oedma (provoked by Baker's yeast *Saccharomyces cervisiae*) and dermatitis are well documented (Wadhwani, 1994; Verma and Chile,

1994; Tripathi, 1994; Pande, 1994, 1998; Das, 1998; Chakraborty *et al.*, 1998; Jain *et al.* 1998; Kulshrestha and Chauhan, 2001).

A precise and reliable diagnosis of fungal allergy depends on several factors in contrast to pollen allergy. The information from the patient/ clinical history is very much essential especially in connection with the environmental monitoring of his dwelling or work place.

In addition to normal testing procedures by antigens it is always of great significant to find out the sights of exposure and the source of allergic disorders. Find evidence may be obtained with skin testing leading to delayed reactions or provocation test followed by prolonged supervision.

Earlier literature provides ample evidence on allergic manifestations after fungal contact. Clinical data based on patients environment has clearly revealed the fungal allergen source.

The different diagnostic *in vivo* (skin test, nasal/bronchial provocation) and *in vitro* (RAST, histamine-release) tests and the environmental parameters have to be combined in the allergological classification. Skin testing with commercial antigens should be associated with environmental analysis from the individual patient, which is helpful in identifying the source of the allergen.

A number of views are expressed about allergic specificity of various moulds, some favour species specificity on a generic level; while others consider group specificity. Extensive work was carried out by the Agarwal *et al* (1967) on four species of *Aspergillus*. The data on skin, bronchial sensitivity, conjunctival and P.K. test alongwith statistical analysis of data clearly indicated that there existed species specific antigens in fungi, particularly in the genus *Aspergillus*.

Shivpuri (1982), presented some additional information allergy to fungi. In India first ten important allergenic fungi in order to their frequency and allergenic significance are *Curvularia, Alternaria, Helminthosporium, Phoma, Trichoderma, Neurospora, Mucor, Cladosporium, Aspergillus* and *Rhizopus*. These observations also revealed that fungus allergen frequently produces 1+ skin reaction (sometimes 2+) and many times 1+ reaction may be due to presence of histamine releasing substances in the extract giving false reaction.

Allergic bronchopulmonary Aspergillosis is fairly common in India. Sandhu *et al.* (1972) and Khan *et al.* (1976), carried out extensive work, it is commonly reported intendemic form in industries and places producing large amount of moist bagasses, etc., on which *Aspergillus* grow profusely.

Though fungal allergens pose a significant health problem more than half a century of research on mould allergies, has not led to the understanding of the exposure and sensitization of allergic patients to the 2,50,000 mould species present on earth, allergological test results can be found of about (140) (10.06%) (Bronswijk *et al.*, 1986).

PLANT DISEASE FORECASTING

In India, plant disease forecasting service is at its incipient stage. The estimation and assessment of inoculum in the air forms one of the major basis of devising an efficient disease forecasting system.

In the recent revives on aerobiology of cereal crop diseases (Tilak, 1984, 1989), aerobiology of bajra diseases (Tilak and Babu, 1983), aerial transport of plant pathogen and its relevance (Tilak, 1986b), aerobiology and epidemiology of diseases of groundnut (Tilak, 1988), basic research in plant disease forecasting (Tilak, 1989), critical discussions have been presented regarding the utility of air monitoring and its relevance in disease forecasting.

A significant practical utility of aeromycology, which is assuming greater importance is the relevance of such investigations is devising disease forecasting system or warning system to farmers in preventing the impending onset of epiphytotics. Voluminous literature (Young *et al.*, 1978; Malden *et al.* 1978; Jones *et al.* 1980; Tilak, 1980, 1986, 1987b, 1988, 1989a, 1989b; Adams, 1981; Fry, 1982; Nutter *et al.*, 1983; Danneberger *et al.*, 1984; Nagarajan and Joshi; 1985) is available on this aspect clearly indicate a close correlationship between concentration of pathogenic spore load in air, meteorological parameters, growth stage of the crop and the appearance and spread of disease leading to epiphytotics. Aeromycological investigations have significant role in disease forecasting system has helped in preventing epidemics and subsequent crop loss.

Disease forecasting is made more reliable if the reasons for the particular disease development under certain conditions are known.

Among the various considerations a thorough knowledge of interaction of population of pathogen in air and associated environmental factors is essential and would be of significant utility.

A data on aeromycological investigation over groundnut field (Tilak, 1988), clearly brought out the close correlation between the spore load of pathogen in air, growth stages of crop, meteorological factors and disease incidence and its subsequent spread.

Tikka disease of groundnut could be forecasted 15-20 days in advance, considering the meteorological factors and growth stages of crop (Tilak, 1988). Aeromycological investigations for rust of groundnut has also revealed that warning system can be of great utility. Since the pathogenic spore load appears in the air 18-20 days in advance before the first appearance of disease (Naik and Pande, 1998) (Table 6).

Severity and subsequent spread of Tikka disease of groundnut caused by *Cercospora personata* and *Cercospora arachidicola* depends on prevalence of high humidity (75-85%) and low temperature (26-28%). The prevalence of disease after 50-70 days of growth stage of crop. However, the first appearance of pathogenic spore load appears in air 15-20 days before the first appearance of disease. However, the combined observation of all important factors enable us to device a useful warning system.

Table 6. Epidemiology of Groundnut tikka disease

Pathogen	Apperance of spores in air prior to infection (days)	Tempera-ture °C	R-H %	Age of Crop
Cercospora personata	15-20	25-28	75-85	50-70
Cercospora arachdicola	15-20	24-26	60-80	50-70

Bajra Diseases: extensive aeromycological investigations over bajra crop were carried out with reference to epidemiology of bajra diseases (Tilak and Chakre, 1978; Tilak and Mane, 1981; Tilak and Babu, 1984; Tilak and Ramachander Rao, 1987a, b, 1989; Tilak and Pande, 1989). The entire work has yielded significantly useful disease forecasting system.

In case of rust disease of bajra, moderate temperature and high humid conditions are favorable for disease incidence and spread. As per observation made by us, comprising several year data, it could be concluded that the epiphytotics and flowering stage given the above mentioned environmental conditions are viable.

As regards ergot disease the relative humidity of 70% or above and the temperature between 25 and 30^{0}C promoted the liberation of ascospores into the atmosphere, which directly results in the increase of the primary inoculum and hence the increase in disease incidence.

Sorghum Diseases : Shenoi and Ramalingam (1979), conducted detailed aerobiological and epidemiological studies on various disease of Sorghum downy mildew, 6-15 days after the emergence of Sorghum seedlings. These were trapped until the flowering phase of the crop.

The observations would be very much useful in predicting the epidemics of Sorghum downy mildew (Tilak and Jogdand, 1989). Forecasting epiphytotics is possible now for jowar crop, mainly due to the voluminous accrued during the last two decades or so. A comprehensive and composite account of all those diseases is now available.

Rice Blast Disease : Rice blast pathogen *Pyricularia oryzae* is a menacing problem particularly in Tamil-Nadu and Kerala. The perpetuation of the pathogen survives on grasses like *Digitaria, Brachiaria, Pennisetum* etc. which occur along the borders.

The severity of rice blast disease caused by *Pyricularia oryzae* is determined mainly because of abundance of conidia (inoculum), the influence of environmental factors and resistance or susceptibility of the host (Padmanabhan, 1965). The reproductive propagates develop best under high humid conditions (80-95%) low temperature (23-25^{0}C) and associated rainfall of few mm. For the appearance of spore load in air congenial weather parameters are prerequisite and the spore load prevails in the atmosphere 3-5 days in advance before the first appearance of disease on host plant. These findings are most useful in devising useful disease forecasting system of rice. If the congenial meteorological conditions prevail for a period of a weak or more during susceptible phase of crop growth viz., seedling stage, post transplanting stage, tillering stage or neck emergence stage and associated spore load of pathogen, the epidermis are bound to occur within one week (Padmanabhan, 1966).

NEED FOR FORECASTING

Timely warning of possibility of epidemics by disease forecasting system would be of significant help to the farmers to plan preventive methods foe possible loss. The present position of the problem of plant disease forecasting has been reviewed critically by Bedi (1986). The need for the establishment of disease forecasting service in India is a must for vegetable fruits and food crops. To develop a reliable forecasting extensive field and laboratory investigations are necessary for understanding behavior of pathogen and host populations under varying environmental conditions. Forecasting models should be based on holistic field experiments to account for all situations and should be supplemented by huge amount of data for accurate results. Consideration of meteorological data is a must for interpretation of the effect of environment on each and every stage in a disease cycle. For this, epidemiological research must be strengthened and it is here the mycologist can play a key role.

AEROMYCOLOGICAL SURVEY OVER ANTARCTICA

Aeromycological investigations were carried out during III Indian expedition to Antarctica in 1984. The analysis revealed the presence of species of *Aspergillus, Nigrospora, Cladosporium, Didymosphaeria, Memnoniella, Papularia, Petalotiopsis*, Rust and Smut spores (Tilak, 1986). The occurrence of these spore types was due to drifting winds reaching from the main lands of South America, Australia, New Zealand and South Africa.

AEROMYCOLOGY OF INDOOR ATMOSPHERE

Indoor air is usually exchanged fairly and rapidly by ventilation with outside air. Due to this exchange the microbial content of indoor air tends to change. However, pollen and fungal spore's concentration of indoor atmosphere is usually lower than outside. Indoor air also comprises other microbes derived from indoor sources and thrive well due to favorable weather condition and the organic matter providing suitable substrate.

BIODETERIORATION

Many airborne fungal spores are responsible for biodeterioration of storage materials, equipments, library materials and archives. The biodeterioration not only includes mildewing or rotting of substrate but mechanical damage or fundamental impairment of materials which are real manifestation of the interaction of organism and materials.

Greathouse (1950), Kowalik and Sadorska (1956), Kathapalia (1960), Flyate (1968), and Tilak and Saibaba (1984), reviewed role of fungi, bacteria, Actinomycetes (as biological agents) in biodeterioration of library materials. Fungi on papers and books belong to species of *Alternaria, Monilia, Fusarium, Chaetomium, Myrothecium, Torula, Stachybotrys, Cladosporium, Sporidesmium, Rhizopus, Epicoccum* and *Paecilomyces*. Many of these are active cellulose decomposers many producing pigments and stain paper usually with yellow, brown or black spots. Some, however, form colourless colonies. The action of fungi being slow damaged caused by in the early stages escapes the attention. Fungi are also associated with foxing and browning. Increases humid conditions are congenial for growth of fungi resulting in loosening of paste and glue, weakening of fibers of paper and leather. This can be prevented by well ventilated, dry environment, indoors.

Bose and Yadav (1973) reported that the biodeterioration of library books is due to the cellulase enzymes and its correlation with cellulolytic activity. The entire work on biodeterioration of library material has been reviewed by Tilak *et al.* (1982), which signifies the role of airborne fungi, the process of biodeterioration, the cellulolytic activity of book-deteriorating fungi and the role of environment parameters.

It was of great interest to note that the percentage contribution of cellulose destroying fungi in the air inside central library; Dr. Babasaheb Ambedkar Marathwada University was nearly 50% (Tilak and Saibaba, 1984). The spore load during the month of November was maximum (6585 spores/m^3 of air) followed by December (6580/m^3 of air) and January (5525/m^3 of air). The maximum concentration was due to the favorable environment with 60-90% relative humidity and 18-27^0 C range of temperature. Of these two, relative humidity played a key role of the concentration of spore load inside the library. The ability of fungi to secrete cellulose enzymes was both constitutive and adaptive.

AEROMYCOLOGY OF CAVES

The problem of biodeterioration of wall painting has assumed serious proportion and is matter of great concern. Aeromycological investigations were carried out in the caves of Ajanta, Ellora and Aurangabad to assess biodeteriorating fungi, their source, concentration and seasonal variation and impact after deposition (Tilak, 1986a, 1989a).

Several hypomycetous fungi cause deterioration of frescoes and painted canvases e.g., species of *Penicillium, Cladosporium, Fusarium, Gliomastix and Aleurima*. Excessive humid conditions and poor ventilation hastens up the damage. In the study of fungal biodeterioration of frescoes in Churches in Florence in Italy following the floods in 1966. Gargani (1968) reported several species of *Penicillium* as the most common causal moulds like *P. restrictum, P. glabrum, P. decumbens, P. thomi, P. jarithanielum, P. canescans* and *P. chrysogenum*.

Censoi *et al.* (1980), made microbiological survey of indoor and outdoor airspora and particulate deposition on the walls in Chapel

(Padua: Italy), frescoes and made further analysis of quantitative deposition of walls.

Application of nystain controls growth of many fungi and prevent biodeterioration of paintings. Painted surfaces should be washed down with ammonium carbonate before treating with antibiotics. With water soluble nystain the treatment has been found effective in preservation of both affected and unaffected paintings.

Tilak and Kulkarni (1972), Tilak *et al.* (1972), Tilak (1986a, 1989a), in their indoor exploration of biodeteriorating mycoflora, stated that airborne biodeteriogens played significant role in the deleterious effects on cave paintings. It was interesting to note that airspora in the caves was more than outside. *Cladosporium* and *Aspergillus* dominated the mycoflora of air and were also abundant on the deteriorating paintings.

Variation in concentration of spore load was dependent on weather conditions like humidity, rainfall, temperature, topography and season. Aeromycological studies thus have helped to locate and identify the source of biodeterioration and subsequent treatment to preserve and conserve our national heritage.

AEROMYCOLOGY AND VETERINARY SCIENCE

Many diseases of commercial and domestic animals are transmitted by contact or insect bites and through ingestion of contaminated food. However, several other important diseases are definitely airborne over long distances of about 16 km. The foot and mouth diseases, rinderpeste and ephemeral fever in cattle, infectious larynogotrachis and the fowl pest of poultry are common (Ainsworth and Austwick, 1959; Sellers and Parker, 1959; Smith, 1964; Sellers, 1971).

Mammalian aspergillosis, facial eczema, mycotoxicoses and other fungal diseases are receiving attention (Vaidya, 1995). Pathak and Mittal (1966), isolated *Aspergillus fumigatus* from the cervical mucus of a cow suffering from post-parturient metritis which clearly indicates the role of airborne fungi in the infections of domesticated animals in India. Extensive investigations were carried out in New Zealand to locate the source concentration, seasonal variation of the spores and their relation with facial eczema of sheep (Brook and Mutch, 1964). Tilak (1974) also carried out aerobiological investigations in sheep shed. Tilak and Saibaba (1986) investigated airspora inside poultry shed at Aurangabad, the results indicate the presence of about 32 spore types, the dominant pathogenic forms being *Cladosporium* (16%), *Torula* (14%), and *Aspergillus* (11%). The source of the fungal forms was the rice husk, used as litter which if kept for a longer time considerably increase the spore load in the atmosphere. Wankhade (1983) and Meghraj (1989) at Aurangabad and Khillare, (1989) at Ahmednagar carried out detailed investigations inside the cattle shed. Highly contagious skin infection, ring worm may be due to species of *Microsporum*, *Trichophyton* and *Epidermophyton* which are able to digest keratin occurring in horny layer of skin, hairs and nails. *Trichophyton* causes infection of skin between the toes causing athletes foot or skin or scalp itself. *Aspergillus, Penicillium* and *Rhizopus* grow within tissues of various animals. Ihre and Wong (1988) reported that *Microsporum, Trichophyton, Cladosporium, Penicillium and Rhizopus* species were frequently isolated saprophytes. Khillare (1989) encountered species of *Rhizopus, Chaetomium, Cladosporium, Fusarium, Penicillium, and Pithomyces* as significant contributors to airspora in cattleshed. A correlation could be established between pathogenic spore types and disease incidence in cattles.

AEROMYCOLOGY OF POULTRY SHED

Chute and Gershaman (1961) conducted the pioneering aerobiological investigations in poultry. Shed and Chick hatcheries to monitor airborne microbes by using Andersen sampler. These investigations helped in arriving at a clean sanitary programme replacing subjective visual inspection. Rajan and Sivadas (1972) studied the distribution of fungal spores in poultry shed and concluded that the concentration of fungal spore was 10-100 times more as compared to adjacent out door site.

The source of pathogenic forms alongwith saprophytes is the rice husk (used as litter) which if kept for a longer time helps in building up the spore load. The increase in the spore load is due to moistening of litter due to dropping of birds and the substrate gradually becoming alkaline. Frequent change in litter keeps down the spore concentration and this would also help in reducing variable according to season and meteorological parameters.

Tilak and Saibaba (1986) conducted aeromycological surveys at Aurangabad inside the poultry shed and recorded 36 airborne fungal spore types of which *Cladosporium* (16-02%), *Torula* (14.9%), *Aspergillus* (11.24%) were dominant pathogenic forms.

AEROMYCOLOGY OF STORAGE DISEASES

Inside the foodgrain godown the setting of microbes is may be on the stored bags, grains and walls. These settled bioparticles are disturbed during cleaning, handling, aerosol spraying etc., by wind currents entering through doors or windows. Although various views have been expressed about the deterioration of food grains. In storage the possible role of airborne microbes has remained unattended. Tilak and Chakre (1979), carried out aeromycological investigation in stored food grains for evaluating losses. Among the major forms of grain toxic pollutant *Rhizopus* (26.6%), *Penicillium* (7.5%) and *Aspergillus* (6%) were reported while from the phytotoxic biopollutants *Cladosporium* (9.3%), *Alternaria* (7%), Zahid *et al.* (1997), also made survey of ware house to study the grain toxic pollutants.

Helminthosporium (5.3%) and *Curvularia* (4.1%) were predominant forms. The grain toxic biopollutants were more prevalent during wet season when the ambient air temperature was 28-30^{0}C and relative humidity 80-90%. The phytotoxic pollutants appeared more in the air during dry season when temperature ranged from 34-36^{0} C and relative humidity 40-45%. The observations have clearly brought our the correlation between airborne biopollutants and percent of grain loss during storage.

AEROMYCOLOGY AND MISCELLANEOUS SURVEYS

Shastri (1981), Zahid *et al.* (1997), carried out aeromycological investigations inside the vegetable market to assess and establish the close correlation between the concentration of pathogenic spores in air and meteorological factors. The results indicate the dominant forms like *Torula, Aspergillus, Cladosporium, Curvularia, Helminthosporium* and *Alternaria.* Similar observations were also obtained by Sulia and Khan (1980), in the vegetable market at Bangalore. The storage diseases are built up due to dumping of vegetables and in turn cause damage to fresh arrivals, new pathogens are added due to fresh arrival of diseased material. Humidity, temperature inside the vegetable market associated with saprophytic and parasitic forms, play significant role in initiating rots.

Cinema theater, industries, farm houses, etc. are other indoor environments where aeromycological surveys have been carried out. Merjut (1977) monitored the working environments of mill and veneer factory in central Finland. Pohjola *et al.* (1977), reported considerable spore load of *Penicillium* and *Aspergillus* in dumping room of garbage disposal plant in Turku and the concentration of those types was 6,000 times greater as compared to outdoor air. A census of fungi in the airspora of greenhouse (Tilak *et al.*,1982), revealed the presence of 52 spore types of which *Cladosporium* (4.7%), *Sclerotium* (12%), *Alternaria* (5.4%), *Periconia* (4%), *Torula* (2.5%), *Aspergillus* (2.2%) were dominant.

The composition and abundance of aeromycoflora of the dwelling was estimated in Finland by Kapyla (1985). The objective was to identify mould concentration in the indoor air. Fungi isolated from air were species of *Aureobasidium, Aspergillus, Botrytis, Cladosporium, Phoma, Penicillium, Rhizopus, Sporobolomyces* etc. Besides, a puzzling slow growing pigmented hypomycetes was found which would not be identified and thus named as *Scytadium.* Of the fungi found on window frames and wallpapers *Aureobasidium* is known to damage painted surface and wallpapers. *Trichoderma* grew actively on window frames, mould growth appeared to be a common problem with insulated windows, subsequently decaying the wooden frames and creating a possible allergy risk. Similar observations have been found to be true in India, where faulty constructions, unsuitable materials, poor ventilation and excessive dampness due to leakage especially during rainy season favor moldy growth helping the source of allergen.

SPACE AEROBIOLOGY

Microbial forms could be transported both in vertical and horizontal directions for several kilometers. Long distance transport of viable particles constitute a matter of concern and has great practical significance. The terms 'Trans Atlantic' and 'Trance Pacific' flights of viable microbial forms are worthy to note regarding the vertical transport it was one time thought that sky was the limit controlling to this. The recent reports of viable forms in space and outer space have isolated rigorous research activities in space and outer space.

Earlier bacterial forms were recorded at a height of 10 km which survived under most unfavorable environment condition like freezing temperature, O_2 reduced to 1/10 than at ground level and more frequency of UV rays. These organisms are capable not only of surviving but were able to make use of UV rays for food production.

In the year 2001, January 21, over tropical India, a team of International scientists conducted experiments to search of life forms at a height of 41 km above ground level. The joint venture included scientists from Cardiff University, U.K. Indian Space Research organization (ISSRO), Bangalore, Inter-University Center for Astronomy (IUCCA), Pune. Tata Institute of Fundamental Research, Mumbai. Chandra Vikramsinghe presented the results of the investigations in the meet of International Society for Optical Engineering in San Diego, USA. Experiments were conducted to monitor microbes by using special types of balloons and Crysosamplers under aseptic conditions.

The report on the occurrence of viable microbes at a height of 41 km above ground level has attracted worldwide attention.

The occurrence of such living forms at high altitudes has aroused great interest. Activities in space aerobiology are significantly on the increase samples of such biotic forms were brought to laboratory. Unfortunately these could not be grown in culture conditions. However, effects are on for their artificial cultures in laboratory. Vikramsinghe suggested that these may be from outer space. Contrary to the views of Vikramsinghe, many feel that the microbial forms originated from ground level and were lifted higher and higher in the stratosphere.

The occurrence of bacterial form at 41 km above sea level on floating state has no doubt aroused interest in international community. Evidences, however, are inconclusive and one has to wait for further investigations.

Another very interesting report is from the outer space in the galaxy of stars. Space scientist have discovered particles of life. In the scientific article in 'Nature', which describes the observation of life particles in freezing clouds. Microparticles of ice which were seen floating contained life particles. Observations revealed that floating ice particles come in contact with strong UV radiation. This initiated important reactions in particles of ice and atoms are converted into amino acids and protein which constitute basic substances and is prerequisite for life.

Scientist claim that from such ice particles the galaxy and earth may have originated. Experiments conducted in laboratory under similar conditions like –25^{0}C freezing temperature, strong radiation of UV rays. Irradiation with UV rays resulted in glycein, alemine, serene and protein. Investigations of meteorologist "Marchinson" has indicated presence of 70 types of amino acids. Search for possibility of life and its origin in meteorites are being investigated. Fossil bacteria have been reported from rock samples of Mars.

No doubt recent researches in space aerobiology have opened new vistas to this virgin-field which ultimately would help to solve many riddles about life and its origin and also possibility of inter planetary travel of microbes. Bright future awaits space aerobiologists. Microbes have possible record of inter planetary travel much before man could ever think of it.

REFERENCES

Adams, P.B. (1981). Forecasting onion white rot disease, *Phytopathology.* 71: 1178-1181.

Agarwal, M.K., Shivpuri, D.N. and Mukherji, K.G. (1967). Species specific antigens in fungi. *Asp. Allergy Appl. Immunol.* 1 : 69-80.

Ainsworth, G.C. and Austwick, P.K.G. (1959). "Fungal diseases of animals." Comm. Burean of animal health. C.A.B. Famham, Royal Bucks (UK)148 pp.

Allit, K. (1986). Identity of airborne hyaline one septa escosporees and their relation to inhalent allergy. *Trans. Brit. Mycol. Society.* 87: 147-154.

Babu, M. (1983). Indoor air micoflora and its relevance to human allergy. Part II Ph.D. thesis, Marathwada University, Aurangabad.

Babu, C.R, Singh, A.B. and Shivpuri, D.N. (1979). Allergen factors and symptomology of respiratory allergy patients. *J. Asthma Res.*16: 92-102.

Bedi, P.S. (1986). Plant diseases forecasting. Vistas in Plant Pathology. (Verma and Verma, J.P.Ed.) Malhotra Publishing House, New Delhi, 249-261.

Bhati, H.S. and Gaur, R.D. (1979). Studies on aerobiology-atmospheric fungal spores. *New Phytol.* 82: 519-527.

Blackley, C.H. (1873). Experimental research on the causes and nature of *Catarrhus aestivus* (hayfever and Hay asthma). Bailleres Tindall and Cox, London pp 202.

Bose, B. and Yadav, A.S. (1973). Some cellulolytic microfungi and their cellulase activity in vitro. *J. Ind. Bot. Society.* 52 : 218-224.

Bronswijk Van, J.E.M., Rijekaest, G. and Lustgract, B.V. (1986). Indoor fungi distribution and allergenicity. *Acta. Bot. Neerl* 35 : 39.

Brook, P.J and Mutch, G.V. (1964). Field control of facial eczema of Sheep. *NewZealand J. Agri. Res.*7: 138-145.

Brown, H.M. and Jacson, A.F. (1978). The value of aerobiology to an allergy treatment centre. *Proc. 1st Int. Aer. Conf. Munich.* 278-286.

Cadham, F.T. (1924). Asthma due to grain rusts. *J.A.M.A.* 83 : 27

Censoni, A.L.Z., Betteni, C. and Mandrioli, G.P. (1980). Aerobiological research in enclosed spaces of historical and artistic interest. *Proc. 1st Int. Conf. Aero. Munich* 434-438.

Chakravarty, R. (1974). Fungi in air and human allergy. *J. Sci. Club* 28 : 83-106.

Chakraborty, P., Gupta-Bhattacharya, S. and Chanda, S. (1998). Clinical test and chemical analysis of some common aeroallergens from Calcutta. Environment and Aerobiology (Ed. Ashok Jain) Published by research periodicals and book publishing house Texas, USA. pp 239 -244.

Champman, J.A. (1982). The enhancement of the practice of clinical allergy with daily pollen and spore count. *Immunology and allergy practice* 4 :13-18.

Chatterjee, G. (1931). A note on an apparatus for catching spores from upper air. *Ind. J. Agric. Sci.* 1 : 306-308.

Chute, H. L. and Garashman, M. (1961). A new approach to hatchery sanitation. *Poultry science* 49 : 568-571.

Cunningham, D.D. (1973). Microscopic examination of air Government Printers, Calcutta. pp 58.

Danneberger, T.K, Vargans, J.M. Jr and Jones, A.L. (1984). A model for weather forecasting of anthracnose on annual blue grass. *Phytopathology.* 74: 448-457.

Das, R.R. (1998). Aeroallergens : An Environmental view. Environment and Aerobiology (Ed. Ashok Jain), Research periodicals and Book Publishing House, Texas, USA.

Davies, R.R. (1959). Detachment of conidia by cloud droplets. *Nature* pp.1965.

D'Silva, A.M. (1988). Studies in airborne fungi and their role in respiratory allergy. Ph.D. thesis, Bombay University, Mumbai.

Flyate, D.M. (1968). Preservation of documents and papers. Edt. and translated from Russian by J. Schmorak, H. S. Dept. of Commerce, Springfield.

Frankland, A.M. (1977). *Aspergillus fumigatus* : A pathogen and allergen. *Grana.* 16 : 181–182.

Frankland, A.M. (1983). *Fusarium* : A neglected mould *Grana,* 22 : 167-170.

Frankland, A.M. (1984). Allergy problems from non-seasonal moulds with special reference to *Aspergillus fumigatus* and *Fusarium.* sp. Allergy and *Appl., Immunol.* 17 : 107-108.

Frankland, A.M. and Gregory, P.H. (1953). Allergenic and Agricultural implementation of airborne ascospores concentration from fungus *Didymella exitiatis. Nature,* 245 : 336-337.

Fry, W.F. (1982). Disease forecasting: Epidemiological considerations 105-126. *Principles of Plant Disease Management.* Acad. Press. New York pp. 379.

Gargani, G. (1968). Biodeterioration of materials in waters (A.N. and Elphic, J.J. Eds.) 252-257. Elseveier Amsterdam.

Gesa, W., Bergman, I., Rebohle, E., Gemeinhusdt, H. and Thuermer, A. (1980). Occupational respiratory diseases due to moulds in millers and backers. *Abst. Mycol.* 14: 12.

Gravensen, S. (1979). Fungi as a cause of allergic disease. *Allergy* 34: 135-154.

Gravensen, S. and Jensen W. (1984). Atlas of moulds in Europe causing respiratory allergy. Ask Publishers, Denmark, pp.112.

Gregory, P.H and Hirst, J.M. (1952). Possible role of basidiospores as airborne allergens. *Nature* 170 : 414-415.

Gregory, P.H. (1961). Microbiology of the atmosphere. Leonard Hill (Books) Ltd. London.

Gregory, P.H. (1973). Microbiology of atmosphere. 2nd Edition. Wiley Interscience. New York, 251p.

Greathouse, G. (1950). Microbial degradation of cellulose. *Textile Res. J.* 20: 227-238.

Ihre, P.L. and Wong, A. (1988) Cutaneous fungal flora intwenty horses free of skin or ocular disease. *Ann. J. Vet. Res.* 49 : 770-772

Jain, A.K., Datta, T.R. and Narang, K. (1998). Airborne fungi and Hypersensitivity among allergenic patients at Gwalior. Environment and Ecology (Ed. Ashok Jain) Research periodicals and Book Publishing House, Texas, USA, 257-265.

Jones, A.L., Killevik, S.L., Fischer, P.O. and Stebbins, T.C. (1980). A micro-computer based instrument to predict primary apple scab injuction periods. *Plant Diseases* 64:69-71.

Kalra, S.L. and Dumbrey, D.G. (1957). Aerobiology of Army Medical Campus, Poona, *Armed Forces Med.J. (India),* 13:3-16.

Kapyala, M. (1985). Fungi on insulated windows. *Allergy* 40:558-564.

Kathapalia, Y.P. (1960). Deterioration and conservation of Paper-I, Biological deterioration. *Indian pulp and Paper.* 15:112-125.

Khan, Z.U., Sandhu, R.S., Menon, M.P.S. and Duraj, J.S. (1976). Allergen Bronchopulmonary aspergillosis. A study of 46 cases with special reference to labortory aspects.

Khillare, S.S. (1989). Aerobiological investigations at Ahmednagar, Ph.D thesis. Marathwada University Aurangabad.

Kotimma, M., Kavenlampi, L., Tesho, E.O. and Hughman, K. (1980). Exposure to spore dust in agricultural working environments. *Proc.1st Int. Conf. Aerobiol. Munich* 158-164.

Kowalik, R. and Sadurska, J. (1956). Microorganisms destroying Paper, Leather and Wax seals in the air of archives. *Acta Microbiologia polonica* 5:227-290

Kulshreshta, A. and Chavhan, S.V.S. (2001). Aeromycoflora of some Hospitals of Agra City. *Indian J. Aerobiol.* 14 (No.1 & 2):33-35.

Malden L., Pennypacker, S.P. and Nacnab, A.A. (1978). FAST. A forecast system for *Alternaria solani* on tomato. *Phytopathology,* 68: 1354-1358.

Mandal, S. and Chanda, S. (1981). Aeroallergens of West Bengal in the context of environmental polution and respiratory allergy. *Biol. Mem,* 6(1) : 1-61

Meghraj, B. (1989) Airspora of Aurangabad II. Ph.D. thesis Marathwada University, Aurangabad.

Merjut, K. (1977). Airborne spores in a mill and in Veneer Factory. Int. *Aerobiol. News letter* 6 : 8.

Mehta, K.C. (1952). Further studies on cereal rusts in India. Indian coun. *Agri.* Res. *Sci. Mono.* 18 : 368 pp.

Meier, F.C., Stevenson J.A. and Charls V.K. (1933). Spores in upper air. *Phytopath.* 23 : 23.

Naik, M.N. and Pande, B.N. (1998). Aerobiological survey over sunflower fields at Aurangabad, Aerobiology. Proc. 5th Int. Conference, Bangalore (ed. – Agashe S. N.). Oxford & IBH Publishing Co. Pvt. Ltd. New Delhi. pp 285 – 292.

Nagarajan, S. and Joshi, L.M. (1985). Epidemiology in the Indian subcontinent. In the Cereal Rusts Vol. 2. The diseases, their distribution, epidemiology and control (Roelfs A.P. and Bushell W.R., Eds.) pp. 371–402, Academic Press, Oklands.

Nair, P.K.K., Joshi, A.P. and Gangal, S.V. (1986). Airborne pollen, spores and other plant materials of India. CSIR Center Biochemical Delhi, NBRI Lucknow.

Nutter, F.W., Cole, H. and Schein, R.D. (1983). Disease forecasting system for warm weather. Pythium blight of turf grass. *Plant Dis.* 67 : 1126–1128

Pady, S.M. and Kramer, C.L. (1960). Kansas aeromycology- VI, Hyphal fragments. *Mycol.* 52: 681-687.

Pande, B.N. (1994). Indoor aeroallergens of Hospital ward. Current trends in Life sciences Vol. 20. Adv. in mycology and Aerobiolgy (Dr. S.T. Tilak's Commemoration Volume Eds. Talde), 301-313.

Pande, B.N. (1998). Aeroallergens as health hazards to Agricultural workers. *Zagrozenia Biologic zne w Rolnictwie* 173-180.

Pande, B.N. (1999). Aeromycology in India. Frontiers of Microbial Biotechnology and Plant Pathology (Ed. C. Manoharachary). Scientific Publisher (India), Jodhpur pp. 53-68.

Pande, B.N. (2001). Incidence of airborne microbial components over groundnut fields. Frontiers in Fungal Biotechnology and Plant pathology relations (Ed. Manoharachary C.). Allied Publishers Limited. pp 128-135.

Pathak, R.C and Mittal, K.R. (1966). Isolation of *Aspergillus fumigatus* from cervical mucus of a cow suffering from metritis. *Curr. Sci.* 35:312.

Padmanabhan, S.Y. (1965). Studies on forecasting outbreak of blask disease of rice. *Proc. Ind. Acad. Sci.* 28 : 117-126.

Padmanabhan, S.Y. (1966) Prospects of control of plant diseases by forecasting and direct application of fungicides and antibiotics (Abs.) *Ind. Phytopath.*, 19:131.

Pepyas, J. and Davis, R.J (1977). Allergy in Asthma, (Clerk, T.J. and Eoodky S Eds.) Asthama chapter 7. Chapman and Hall, London.

Pohjola, A., Rantis, L. and Makinin (1977). Spore composition of garbage disposal plant. *Int. Aerobiol. News Letter* 6:10

Rajan, A. and Sivadas, C.G. (1972). A study of the distribution of fungal spores in brooder house, S. Kerala. *Vet. Sci.* 3: 186-187.

Rajan, B.S.V, Nigam, S.A. and Shukla, P.K. (1952). A study of the atmospheric fungal flora of Kanpur. *Proc. Ind. Acad. Sci.* 35: 33-37.

Ramalingam, A. (1968). The construction and use of air sampler for routine aerobiology surveys. *Envir.* Health 10 : 61-67.

Reynolds, K.M., Madden, L.V. Richard, D.L. and Ellis, M.A. (1989). Splash dispersal of *Phytophthora cactorum* from infected strawberry fruits by simulated canopy drop. *Phytopathology* 70 : 425-432.

Sahu, S.K. (1998). Aero-phyllo-mycoflora of some Solanaceous crop plants. *Indian J. Aerobiol.*, 11 (1 and 2) : 27-32.

Sandhu, R.S., Mishra, S.K., Randhawa, R.S. and Prakash, D. (1972). Allergic broncho-pulmonary aspergillosis in India. *Scand. J. Resp. Dis.*, 53:289-301.

Savery, Sad and Jenean, J.L. (1986). Rain induced dispersal in *Puccinia arachidis*, studies by means of rainfall simulator. *Neth. J. Plant. Patho* 92: 163-174.

Sellers, R.F. (1971) Quantitative aspects and spreads of foot and mouth disease. *Vet. Bull* 41 : 431-439.

Sellers, R.F. and Parker, J. (1959). Airborne excretion of FMD virus. *J. Hyg. Comb.* 67:671-677.

Shastri, S.D. (1981). Airspora over some fields. Ph.D. thesis, Marathwada University, Aurangabad.

Shenoi, M.M. and Ramalingam, A. (1979). Epidemiology of Sorghum downy mildew. II. Circadian and seasonal periodicity in conidia and uredospores. *Proc. Ind. Acad. Sci.* 88B: 95-102.

Shivpuri, D.N. (1982). Studies in allergy to fungi in India. *Asp. Allergy. Appl. Immunol.* 15:19-30.

Singh, S.R., Devi, N.S. and Devi, P.B. (2001). Dispersal of *Uromyces fabae* (Pers.) de Bary, spores from infected Pea leaves by rain. *Indian J. Aerobiol.*, 14(1&2): 24-27.

Smith, C.V. (1964). Some evidence for the windborne spread of Fowl Pest. *Meteorol. Mag.* 93:257-263.

Sreeramulu, T. (1959). The diurnal and seasonal periodicity of certain plant pathogens in the air. Trans. Brit. Mycol. Soc. 42 : 177-184

Sreeramulu, T and Ramalingam, A. (1966). A two years study of the airspora of paddy fields near Visakhapatnam. *Indian J.Agric.Sci.* 36:112-132.

Sulia, S.B and Khan, K.R. (1980). Airspora of Bangalore city market its relation to the occurrence of market disease. *Advances in pollen spore research* V-VII: 157-169.

Tilak, S.T. (1974). Aerobiology in Maharashtra. *M.V.M. Patrika*, 9:125-131.

Tilak, S.T. (1980). Aeromycology at Aurangabad-I. Ascospores. *Proc. 1st. Int. Conf. (Aerobiology) Munich*, 145-147.

Tilak, S.T. (1982). Aerobiology. Vaijeyanti Prakashan, Aurangabad. Pp. 211.

Tilak, S.T. (1984). Aerobiology and cereal crop diseases. *Rev. Trop. Pl. Path.*1. (S.P. Raychaudhary Ed.) 329-354.

Tilak, S.T. (1986a). Relevance of air monitoring in biodeterioration of paintings in Ajanta. *Res. J. Pl. Rev.* 3 (2): 45-48.

Tilak, S.T. (1986b). Aerial transport of plant pathogens In: Vistas in Plant Pathology (Eds. Verma A and Verma J.P.) MPH New Delhi, 237-248.

Tilak, S.T. (1987a). Splash dispersal and airspora. *Proc. III Nat. Conf. Kalyan. Atm. Biopoll.* 15-19. Env. Pub. Karad.

Tilak, S.T. (1987b). Expanding horizon of aerobiology. Presidential address. *III. Aerobiol. Conf. Kalyan. Atm. Biopoll.* 1-7.

Tilak, S.T. (1988). Aerobiology and epidemiology of ground nut diseases. Perspectives in Mycology and Plant Pathology. (Eds. V.P. Agnihotri, A.K. Sarbhoy and D. Kumar) MPA, New Delhi, 336-352.

Tilak, S.T. (1989). Basic Research on Plant disease forecasting in "*Basic Research for crop disease Management*". P. 234 -242

Tilak, S.T. (1989a). Airborne pollen and fungal spores. Vaijayanti Prakashan, Aurangabad. p 316.

Tilak, S.T. (1989b). Airborne spores as bioindicators *Prof.* G. P. Agarwal. Comm. Vol. Today and Tomorrow Printers, New Delhi. P. 227 – 235.

Tilak, S.T. (1991). Aeromycology aspects and prospects. Fungi and Biotechnology. Recent advances (Ed. H.C. Dube) pp.137-156.

Tilak, S.T. (1998). Aerobiology, Satyajeet Prakashan, Pune. Pp.504.

Tilak, S.T. and Babu, M. (1983). Aerobiology of *Pyricularia* leaf spot disease of bajra in relation to weather conditions. *Curr. Poll. Res. Ind.* 339 –344.

Tilak, S.T. and Babu M. (1984). Aerobiology and epidemiology of certain diseases of bajra. *Indian.J.Bot.* 7(1):77-83.

Tilak, S.T. and Babu, M. (1987). Aeromycology at Aurangabad – IV : Phycomycetes. *Indian. J. Bot.* 10 : 84-88.

Tilak, S.T. and Bhalke, S.P. (1981). Aeromycology at Aurangabad. Hyphal fragments. Advancing frontiers of mycology and pathology. Today and Tomorrows printers and Publishers, New Delhi, pp. 51-54.

Tilak, S.T. and Chakre, O.J. (1978). Atmospheric concentration of *Claviceps fusiformis* over bajra fields in relation to there environmental factors. *IV. Int. Palyno. Conf. Lucknow.* pp.181-182.

Tilak, S.T. and Chakre, O.J. (1979). Indoor microbial pollution of air and its relevance to storage diseases of food grains. *Symp.on Envi.Pollution and Toxicology. Hissar* 45-50.

Tilak S.T. and Jogdand S.B. (1989). Aerobiology, Epidemiology and Forecasting of some jowar diseases. Recent researches in Ecology and Env. Pollution (Ed. Tilak, S.T.) Today and Tomorrow's Printers and Publishers, New Delhi, pp.191-222.

Tilak, S.T. and Kulkarni, R.L. (1972). Microbial content of air inside and outside the caves at Aurangabad. *Curr.Science* 23:850-851.

Tilak, S.T. And Mane, D.A. (1981). Aerobiological approach to rust disease of Bajra. *Proc. Nat. Conf. Env. Biopoll* : 247-250.

Tilak, S.T. and Pande, B.N. (1989). Aerobiology and epidemiology of bajra diseases. Recent reasearch in Ecology and Environment and Pollution (Ed. S.T. Tilak) Today and Tomorrow Printers and Publishers, New Delhi, 3 : 79-96.

Tilak, S.T. and Pande, B.N. (1997). Biodeteriogens inside a library. *Aerobiology 5th Int. Conf., Bangalore.* pp. 177-189.

Tilak, S.T. and Ramchandra Rao, K.S. (1987a). Significans of air monitoring of Ergot of bajra at two places in Marathwada. *Indian. J. Bot.* 10(2):105-109.

Tilak, S.T. and Ramchandra Rao, K.S. (1987b). Relevance of rust spores load over a bajra crop. *Front. Bot.* (1): 101-112.

Tilak S.T. and Ramchandra Rao K.S. (1989). Aerobiology if sunflower diseases. Recent research in Ecology and Env. Pollution (Ed. S.T. Tilak) Today and Tomorrow's Printers and Publishers, New Delhi, pp.41-66.

Tilak, S.T. and Saibaba, M. (1984). Aerobiological approach to book deterioration in libraries. *J. Pl. Nature* 1 (2) : 1-10.

Tilak, S.T. and Saibaba, M. (1986). Indoor aeromycology. *Acta.Bot.Ind.* 13:203-211.

Tilak, S.T., Sharma, B.R.N., Sengupta, S.R. and Kulkarni, R.L. (1972) Studies in the micrbial deterioration of paintings of Ajanta and Ellora studies. Museumology. Baroda University 8 :20-25.

Tilak, S.T., Babu, M. and Bhasale, S.S. (1982). Inhalent allergens inside the hospital ward. *Proc. 1st Nat. Conf. Env. Bio., Aurangabad.* 33-38.

Tripathi, D.M. (1994). Role of aerobiology in treatment of allergic patients. Advances in Aerobiology (Dr.S.T. Tilak commemoration Vol. Ed. Talde). Today and Tomorrow's Printers and Publishers, New Delhi. pp.291-300.

Vaidya, J.G. (1995). Biology of the fungi, Satyajit Prakashan, Pune 29. pp. 676.

Verma, K.S. and Chile, S. (1994). Progress in the studies of fungal allergy-A. Review. Advances in Mycology and Aerobiology (Dr. S.T. Tilak. Commemoration Vol. Ed. Talde) Today and Tomorrow's Printers and Publishers, New Delhi, pp.289-290.

Wadhwani, K. (1994). Air fungal spore and working environment- some medical aspects. Advances in Mycology and Aerobiology (Dr. S.T. Tilak. Commemoration Vol. Ed. Talde) Today and Tomorrow's Printers and Publishers, New Delhi, pp.271-277.

Wankhade, J. S. (1983). Airspora at Aurangabad. Ph.D. Thesis, Marathwada University Aurangabad.

Young, H.C., Jr. Prescot, J.M. and Sarri, E.E. (1978). Role of disease monitoring in preventing epedemics, *Ann. Rev. Phytopath.* 16:263-285.

Zahid, I.H., Pande, B.N. and Aher, S.K. (1997). Incidence of fungal spores in indoor environment of a Ahmednagar (MS) India. *Aerobiolgy. Vth Int. Conf. Bangalore* (Ed. S.N. Agashe) Oxford and IBH Publishing Co. Ltd. New Delhi, pp.149-154.